Handbook of Energy and Environmental Security

Handbook of Energy and Environmental Security

Edited by

Muhammad Asif
King Fahd University of Petroleum and Minerals,
Dhahran, Saudi Arabia

Academic Press is an imprint of Elsevier
125 London Wall, London EC2Y 5AS, United Kingdom
525 B Street, Suite 1650, San Diego, CA 92101, United States
50 Hampshire Street, 5th Floor, Cambridge, MA 02139, United States
The Boulevard, Langford Lane, Kidlington, Oxford OX5 1GB, United Kingdom

Copyright © 2022 Elsevier Inc. All rights reserved.

No part of this publication may be reproduced or transmitted in any form or by any means, electronic or mechanical, including photocopying, recording, or any information storage and retrieval system, without permission in writing from the publisher. Details on how to seek permission, further information about the Publisher's permissions policies and our arrangements with organizations such as the Copyright Clearance Center and the Copyright Licensing Agency, can be found at our website: www.elsevier.com/permissions.

This book and the individual contributions contained in it are protected under copyright by the Publisher (other than as may be noted herein).

Notices
Knowledge and best practice in this field are constantly changing. As new research and experience broaden our understanding, changes in research methods, professional practices, or medical treatment may become necessary.

Practitioners and researchers must always rely on their own experience and knowledge in evaluating and using any information, methods, compounds, or experiments described herein. In using such information or methods they should be mindful of their own safety and the safety of others, including parties for whom they have a professional responsibility.

To the fullest extent of the law, neither the Publisher nor the authors, contributors, or editors, assume any liability for any injury and/or damage to persons or property as a matter of products liability, negligence or otherwise, or from any use or operation of any methods, products, instructions, or ideas contained in the material herein.

ISBN: 978-0-12-824084-7

For information on all Academic Press publications visit our website at https://www.elsevier.com/books-and-journals

Publisher: Charlotte Cockle
Acquisitions Editor: Lisa Reading
Editorial Project Manager: Sara Valentino
Production Project Manager: Sojan P. Pazhayattil
Cover Designer: Mark Rogers

Typeset by TNQ Technologies

To my son
Qasim

Contents

List of contributors

Chrystyane Abreu
Production Engineering Coordination, Centro Federal de Educação Tecnológica Celso Suckow da Fonseca, Nova Iguaçu, Rio de Janeiro, Brazil

Nurfadzilah Ahmad
School of Electrical Engineering, College of Engineering, Universiti Teknologi MARA Shah Alam, Selangor, Malaysia

Ijaz Ahmad
National Skills University, Islamabad, Pakistan

Rosley Anholon
Manufacturing and Materials Engineering Department, Universidade Estadual de Campinas, Campinas, São Paulo, Brazil

Muhammad Asif
Department of Architectural Engineering, King Fahd University of Petroleum and Minerals, Dhahran, Saudi Arabia

Jasim Azhar
Architecture Department, King Fahd University of Petroleum and Minerals, Dhahran, Saudi Arabia

Tri Ratna Bajracharya
Center for Energy Studies (CES), Institute of Engineering, Tribhuvan University, Pulchowk, Lalitpur, Nepal

Maksud Bekchanov
Research Unit Sustainability and Global Change (FNU), Center for Earth System Research and Sustainability (CEN), University of Hamburg, Hamburg, Germany; Center for Development Research (ZEF), University of Bonn, Bonn, Germany

Andy Chan
Department of Civil Engineering, University of Nottingham Malaysia, Semenyih, Selangor, Malaysia

Sajal Chowdhury
ZEMCH EXD Lab, Faculty of Architecture, Building and Planning, The University of Melbourne, Melbourne, VIC, Australia

Ramon Soares Corrêa
Production Engineering Department, Universidade Federal Fluminense, Niterói, Rio de Janeiro, Brazil

Nofri Yenita Dahlan
Solar Research Institute (SRI), Universiti Teknologi MARA, Shah Alam, Selangor, Malaysia; School of Electrical Engineering, College of Engineering, Universiti Teknologi MARA Shah Alam, Selangor, Malaysia

David Dapice
Harvard Kennedy School, Harvard University, Cambridge, MA, United States

Ajith de Alwis
Department of Chemical and Process Engineering, University of Moratuwa, Moratuwa, Sri Lanka

Sandip S. Deshmukh
Department of Mechanical Engineering, Hyderabad Campus, Birla Institute of Technology & Science, Pilani, Telangana, India

Daphne Gondhalekar
Chair of Urban Water Systems Engineering, Technische Universität München, Munich, Germany

Osvaldo Luiz Gonçalves Quelhas
LATEC Laboratory of Technology, Business Management and Environment, Universidade Federal Fluminense, Niterói, Rio de Janeiro, Brazil

Nazatul Faizah Haron
Universiti Sultan Zainal Abidin, Kuala Terengganu, Terengganu, Malaysia

Mabroor Hassan
Department of Environmental Science, International Islamic University, Sector H-10, Islamabad, Pakistan; Green Environ Sol (Private) Limited, Sector H-10, Islamabad, Pakistan

Mazhar Hayat
National Adaptation Process, Ministry of Climate Change, Islamabad, Pakistan

Kasun Hewage
School of Engineering, University of British Columbia (Okanagan Campus), Kelowna, BC, Canada

Nur Iqtiyani Ilham
School of Electrical Engineering, College of Engineering, Universiti Teknologi MARA, Masai, Johor, Malaysia

Hirushie Karunathilake
Department of Mechanical Engineering, University of Moratuwa, Moratuwa, Sri Lanka

Vikrant P. Katekar
Department of Mechanical Engineering, S. B. Jain Technology Management and Research, Nagpur, Maharashtra, India

Muhammad Khalifa
Institute for Technology and Resources Management in the Tropics and Subtropics (ITT), Cologne University of Applied Sciences, Cologne, Germany

Muhammad Irfan Khan
Department of Environmental Science, International Islamic University, Sector H-10, Islamabad, Pakistan

Krishna J. Khatod
Department of Mechanical Engineering, Hyderabad Campus, Birla Institute of Technology & Science, Pilani, Telangana, India

Vikniswari Vija Kumaran
Universiti Tunku Abdul Rahman, Kampar Campus, Perak, Malaysia

Li Lan
School of Design, Shanghai Jiao Tong University, Shanghai, China

Mohd Talib Latif
Department of Earth Sciences and Environment, Faculty of Science and Technology, Universiti Kebangsaan Malaysia (UKM), Bangi, Selangor, Malaysia

Ha-Chi Le
Monash University, Melbourne, VIC, Australia

Phu V. Le
Natural Capital Management Program, Fulbright School of Public Policy and Management, Fulbright University Vietnam, Ho Chi Minh City, Vietnam

Thai-Ha Le
Natural Capital Management Program, Fulbright School of Public Policy and Management, Fulbright University Vietnam, Ho Chi Minh City, Vietnam; University of Economics Ho Chi Minh City, Ho Chi Minh City, Vietnam

Li Li
School of Environmental and Chemical Engineering, Shanghai University, Baoshan, Shanghai, China

Neng-huei Lin
Department of Atmospheric Sciences, National Central University, Zhongli District, Taoyuan, Taiwan; Center for Environmental Monitoring Technology, National Central University, Zhongli District, Taoyuan, Taiwan

Gustavo Naciff de Andrade
Empresa de Pesquisa Energética, Centro, Rio de Janeiro, Brazil

Canh Phuc Nguyen
University of Economics Ho Chi Minh City, Ho Chi Minh City, Vietnam

Masa Noguchi
ZEMCH EXD Lab, Faculty of Architecture, Building and Planning, The University of Melbourne, Melbourne, VIC, Australia

Maggie Chel Gee Ooi
Institute of Climate Change, Universiti Kebangsaan Malaysia (UKM), Bangi, Selangor, Malaysia

Balgis Osman-Elasha
African Development Bank Group, Immeuble Zahrabed Avenue du Dollar, Tunis, Tunisia

Tharindu Prabatha
Department of Mechanical Engineering, University of Moratuwa, Moratuwa, Sri Lanka; School of Engineering, University of British Columbia (Okanagan Campus), Kelowna, BC, Canada

Liliana N. Proskuryakova
National Research University Higher School of Economics, Moscow, Russia

Richard R. Reibstein
Environmental Law and Policy, Department of Earth and Environment, Boston University, Boston, Massachusetts, United States

Abdul Rahim Ridzuan
Universiti Teknologi Mara, Cawangan Melaka, Melaka, Malaysia

Paulo Roberto de Campos Merschmann
Production Engineering Department, Centro Federal de Educação Tecnológica Celso Suckow da Fonseca, Rio de Janeiro, Rio de Janeiro, Brazil

Rehan Sadiq
School of Engineering, University of British Columbia (Okanagan Campus), Kelowna, BC, Canada

Guller Sahin
Kütahya Health Sciences University, Evliya Çelebi Campus, Kütahya, Turkey

Noraina Mazuin Sapuan
Universiti Malaysia Pahang, Pahang, Malaysia

Nur Surayya Saudi
Universiti Pertahanan Nasional Malaysia, Kuala Lumpur, Malaysia

Nobuhiro Sawamura
Asia Pacific Energy Research Centre, Tokyo, Japan

Mohd Shahidan Shaari
Universiti Malaysia Perlis, Perlis, Malaysia

Shree Raj Shakya
Institute for Advanced Sustainability Studies (IASS), Potsdam, Germany; Center for Energy Studies (CES), Institute of Engineering, Tribhuvan University, Pulchowk, Lalitpur, Nepal

Anzoo Sharma
Center for Rural Technology (CRT/N), Kathmandu, Nepal; Center for Energy Studies (CES), Institute of Engineering, Tribhuvan University, Pulchowk, Lalitpur, Nepal

Mayuri Wijayasundara
Faculty of Science Engineering and Built Environment, Deakin University, Melbourne, VIC, Australia

Wei Yang
ZEMCH EXD Lab, Faculty of Architecture, Building and Planning, The University of Melbourne, Melbourne, VIC, Australia

Siti Hajar Yusoff
Electrical and Computer Engineering (ECE), Kulliyyah of Engineering, International Islamic University Malaysia, Gombak, Selangor, Malaysia

Preface

The role of energy has never been more important. Despite the growing realization about the importance of energy in the socio-economic well-being and advancement of societies, the global energy landscape continues to face numerous challenges, above all, energy security issues. The nature of energy security challenges for countries at different socio-economic and technological strata varies significantly. Energy affordability, however, is becoming a global issue with fuel poverty on a rise even in the developed countries. Long-standing geopolitical issues around the major oil and gas-producing regions continue to send frequent shockwaves across the energy industry. The spectrum of energy security challenges facing developing countries is quite broad. Notwithstanding the situation has improved in recent years, access to refined energy fuels remains to be a serious issue as nearly one billion people in developing countries lack access to electricity and over 2.6 billion rely on crude biomass fuels to meet cooking requirements. Issues like poor grid quality, power outages and breakdowns, and planned load-shedding are almost a regular phenomenon. The global environmental scenario is also facing mounting challenges. As climate change is regarded as the biggest threat facing the mankind, hopes to limit the global warming to 1.5°C are fading.

Energy security and environmental security, integral dimensions of sustainable development in the 21st century, are becoming increasingly interwoven areas given their commonalities in terms of dimensions, challenges, implications, and potential solutions. The energy- and environmental security challenges not only affect the socio-economic well-being of masses but also have implications for societies at large including their economic and political systems. The profound challenge the world faces is therefore to meet the rapidly growing energy requirements without inflicting damage to the environment as is manifested by the United Nations Sustainable Development Goals (SDGs). The COVID-19 pandemic has reinforced the importance of energy and environmental security. The world needs to adopt the build back better strategy to sustainably recover from the impacts of the pandemic.

The *Handbook of Energy and Environmental Security* presents a holistic account of the global energy- and environmental security scenarios. It discusses the topics of *Energy Security* and *Environmental Security* separately as well as integratedly from a wide range of perspectives, i.e., fundamental concepts, faced challenges and their solutions, technological, economic, and policy dynamics, and case studies. In terms of structure, apart from the introductory chapter, the handbook is divided into three sections. The first section—*Energy Security*—consists of eight chapters. The second section—*Environmental Security*—contains nine chapters. The third section—*Energy and Environmental security: An Integrated Approach*—consists of eight chapters.

Acknowledgments

The book is a teamwork and I am grateful to the chapter contributors in helping me accomplish it. I would like to thank the reviewers for their time and efforts in reviewing chapter abstracts and manuscripts. Given the COVID-related challenges, all these efforts deserve even more credit. I would also acknowledge the King Fahd University of Petroleum and Minerals (KFUPM) for the provided support.

CHAPTER

Introduction to energy and environmental security

1

Muhammad Asif

Department of Architectural Engineering, King Fahd University of Petroleum and Minerals, Dhahran, Saudi Arabia

Chapter outline

1. Climate change on the March

Climate change is widely regarded to be the biggest threat facing the mankind. Temperature of the planet is rising due to increasing concentration of greenhouse gases (GHGs) in the atmosphere. The phenomenon, termed as global warming, is leading to climate change. GHGs—such as water vapor, carbon dioxide (CO_2), nitrous oxide, and methane—are part of the global ecosystem. Human activities such as burning of fossil fuels, transportation, power generation, and industrial and agricultural processes are increasing the concentration of these gases in the atmosphere. The 18th century industrial revolution is regarded to have triggered the rapid growth in the release of GHGs (Asif, 2021; Qudratullah and Asif, 2020). Since 1880, atmospheric temperature has increased by 1.23°C (2.21°F). The CO_2 concentration in the atmosphere has increased from the preindustrial age level of 280 parts-per-million (ppm) to 415 ppm. The acceleration in the growth of CO_2 concentration can be gauged from the fact that almost 100 ppm of the total 135 ppm increment has occurred since 1960.

Climate change is leading to a wide range of consequences such as seasonal disorder, a pattern of intense and more frequent weather-related events such as floods, droughts, storms, heatwaves and wildfires, financial loss, and health problems (Asif, 2011, 2021; Asif and Muneer, 2007). Climate change is also adversely affecting the water and food supplies around the world. Warmer temperatures are increasing sea level as a result of melting of glaciers. During the 20th century, global sea level rose by around 20 centimeters. The pace of rise in sea level is accelerating every year—over the last two

Handbook of Energy and Environmental Security. **https://doi.org/10.1016/B978-0-12-824084-7.00004-7**
Copyright © 2022 Elsevier Inc. All rights reserved.

decades, it has been almost double that of the last century (NASA). As a result of warmer temperatures, glaciers are shrinking across the world including in the Himalayas, Alps, Alaska, Rockies, and Africa.

An extremely alarming dimension of climate change is that it is growing in momentum. Most of the temperature rise since the industrial revolution has occurred since 1960s. Extreme weather conditions and climate abnormalities are becoming so frequent that the situation is already being widely dubbed as climate crisis. With the recorded acceleration in accumulation of GHGs and consequent increase in atmospheric temperature, the climate change–driven weather-related disasters are becoming more intense and recurrent. Seven most recent years have been observed to be the warmest since records began, while the years 2016 and 2020 are reportedly tied for being the hottest year on record (NASA). The year 2021 witnessed heatwaves, wildfires, storms, and floods across the world. North America, for example, faced intense heat wave, besides recorded high temperatures and huge wildfires. California's Death Valley recorded temperature of 54.4°C (130°F), which is potentially the highest ever temperature recorded on the planet, and British Columbia witnessed temperature of 49.6°C, obliterating Canada's previous national temperature record by 8°C (Samenow, 2021). While the heatwave killed over 500 people in Canada alone, Europe and Asia were hit by unprecedented flooding. High temperatures, heatwaves, and droughts contributed to record-breaking wildfires. The 2019–2020 wildfire in Australia, burnt around 19 million hectares and resulting into an economic loss of over AU$100 billion became the costliest natural disaster in national history (Read and Dennis, 2020). The year 2021 also witnessed heat waves fueling massive wildfires in Australia, North America, and Europe. The year also saw record-freezing conditions across North America and Europe. Massive snow storm and hurricane deprived millions of electricity and other utilities for over a week in the states of Texas and Louisiana. With these highlights from the year 2021, extreme weather events are now considered to be a new normal as experts predict more intense natural calamities including wildfires, storms, floods, and droughts.

2. Energy and environmental security

2.1 Energy security

Energy is a precious commodity that goes through a wide range of flows and transformations pivotal for the existence of human life on the planet. Increasingly extensive and efficient utilization of energy has played a critical role in the evolution of societies, especially since the industrial revolution. Energy has become a prerequisite for almost all aspects of life, i.e., agriculture, industry, mobility, education, health, and trade and commerce. The provision of adequate and affordable energy services is crucial to sustain a modern lifestyle, ensure the development, and eradicate poverty. The per capita energy consumption is an index used to measure the socioeconomic prosperity in any society—the United Nation's Human Development Index has a strong relationship with energy prosperity (Asif, 2021).

In the increasingly energy-dependent age, energy security has become one of the foundational blocks of national and international developmental policies and frameworks around the world. The concept of energy security has been evolving over the time. It can be traced back to the First World War (WWI) era in the context of access to refined and efficient fuels. Energy security played a decisive role on many fronts both during the WWI and WWII. The concept of energy security resurfaced in the political and policy debate in the wake of the 1973 oil embargo (Asif, 2011). During the 1970s and 1980s, the idea of energy security primarily dealt with a stable supply of cheap oil. Attention was also being paid toward better management of energy enterprises and new energy technologies. With the debate around sustainable development in the 1990s, "affordability" became an important aspect of the concept of energy security. Subsequently, the growing concerns over global warming and climate change also started to influence the definitions of energy security. Over the years, the concept of energy security has also varied spatially. It has been transformed in meaning and scope to match the emerging challenges and local conditions. The modern concept of energy security has become an umbrella term encapsulating multiple factors about energy prosperity in a society. Definitions of energy security encompass indicators such as reliability, adequacy, consistency, affordability, sustainability, and environmental acceptability of energy supplies. With the incorporation of economic and environmental dimensions, energy security is synonymous to energy sustainability. The International Energy Agency defines energy security as "the uninterrupted availability of energy sources at an affordable price" (IEA, 2019). The United Nations Development Program regards energy security as "continuous availability of energy in varied forms, in sufficient quantities, and at reasonable prices." The European Union's (EU's) perspective on energy security is viewed as "the uninterrupted physical availability of energy products on the market, at a price which is affordable for all consumers (private and industrial), while respecting environmental concerns and looking toward sustainable development." The concept of energy security has also been discussed under the four "A's" approach. The four "A's" dimensions of energy security include availability, affordability, accessibility, and acceptability (Cherp and Jewell, 2014). Energy security can be improved through indigenous, adequate, and diverse supplies. Energy insecurity undermines the socioeconomic well-being of societies and can be defined as "the loss of welfare that may occur as a result of a change in the price or availability of energy" (Winzer, 2012).

2.2 Environmental security

Environmental security is a relatively new and evolving concept in the backdrop of climate change. Environmental security can be considered as the state of protection of vital interests of the individual, society, and natural environment from threats resulting from anthropogenic and natural impacts on the environment (Landholm, 1998). Fundamentally, environmental security implies striking a balance between the dynamic human–environment interactions. The environment is one of the most transnational issues, and its security is an important dimension of sustainable development. Environmental security is also regarded to be closely linked to national security, owing to the dynamics and interconnections among humans and natural resources. Because of its broader scope and dimensions, environmental security is a normative concept. It is an all affecting phenomenon—it poses threats not only to humans but also to natural resources and the ecosystem. Ensuring environmental security, therefore, means guarding against environmental degradation to preserve or protect human, material, and natural resources at scales ranging from global to local (Zurlini and Muller, 2008). Global warming is regarded as the biggest threat to the environmental security, as its implications including climate change are affecting nations around the world on multiple fronts.

2.3 The energy–environment dilemma

Energy security and environmental security are becoming increasingly interwoven areas especially in the wake of the global drive for sustainable development. There are commonalities both in terms of the challenges faced and the potential solutions. Fossil fuels, for example, besides their crucial contribution in terms of supplying almost 80% of the energy needs of the world, are also a cause of concern when it comes to energy security due to their highly localized nature, depleting reserves, and fluctuating prices. Fossil fuels are also regarded to be a threat to environmental security due to their associated GHG emissions. Similarly, renewable energy for its comparative advantages such as diverse, replenishing, abundant, and widely available resources, declining price trends, and environmental friendliness are regarded to be helpful both toward improving energy and environmental security. Compared with energy security, environmental security is a significantly more complex phenomenon. Environmental security is highly dynamic in nature primarily due to three factors: vibrant human–environment partnership, versatile impacts of environment, and a vast number of parameters constituting and influencing the environment.

There is a growing realization about the importance of energy and environmental security, as is reflected from the fact that while only one of the eight United Nations Millennium Development Goals focused on the environmental sustainability and none on energy sustainability/security, seven of the 17 Sustainability Development Goals (SDGs) have focused on energy and environment as shown in Table 1.1 and Fig. 1.1.

Despite all the realization about the importance of energy in the socioeconomic uplift and advancement of societies, the global energy landscape continues to face numerous challenges including acute energy security issues facing large proportions of the population around the world. Poor and inadequate access to secure and affordable energy is hindering the progress of developing countries. Electricity, for example, is vital for providing basic social services such as education and health, water supply and purification, sanitation, and refrigeration of essential medicines. Electricity can also help support a wide range of income-generating opportunities. Although during the last 25 years, over 1.5 billion people living in developing countries have been provided access to electricity, more than 800 million people still don't have access to it. On top of grid access, reliability of grid is

Table 1.1 Overview of the key SDGs targeting energy and environmental sustainability.

Sustainable development goal	Description
SDG 6: Clean Water and Sanitation	Ensure availability and sustainable management of water and sanitation for all
SDG 7: Affordable and Clean Energy	Ensure access to affordable, reliable, sustainable, and modern energy for all
SDG 11: Sustainable Cities and Communities	Make cities and human settlements inclusive, safe, resilient, and sustainable
SDG 12: Responsible Consumption and Production	Ensure sustainable consumption and production patterns
SDG 13: Climate Action	Take urgent action to combat climate change and its impacts
SDG 14: Life below Water	Conserve and sustainably use the oceans, seas, and marine resources for sustainable development
SDG 15: Life on Land	Protect, restore, and promote sustainable use of terrestrial ecosystems, sustainably manage forests, combat desertification, and halt and reverse land degradation, and halt biodiversity loss

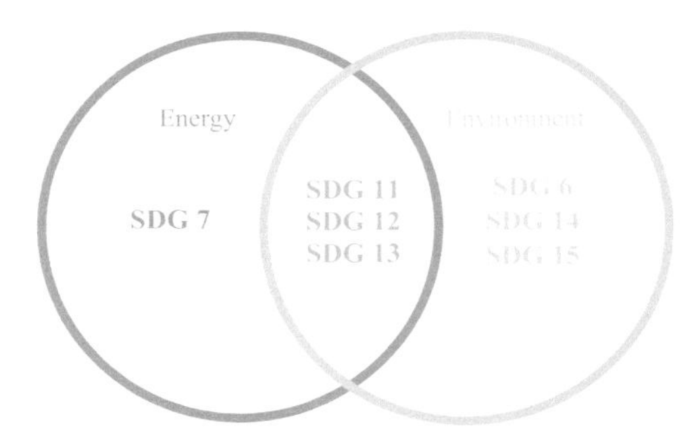

FIGURE 1.1

Distribution of SDGs in terms of the energy and environmental focus.

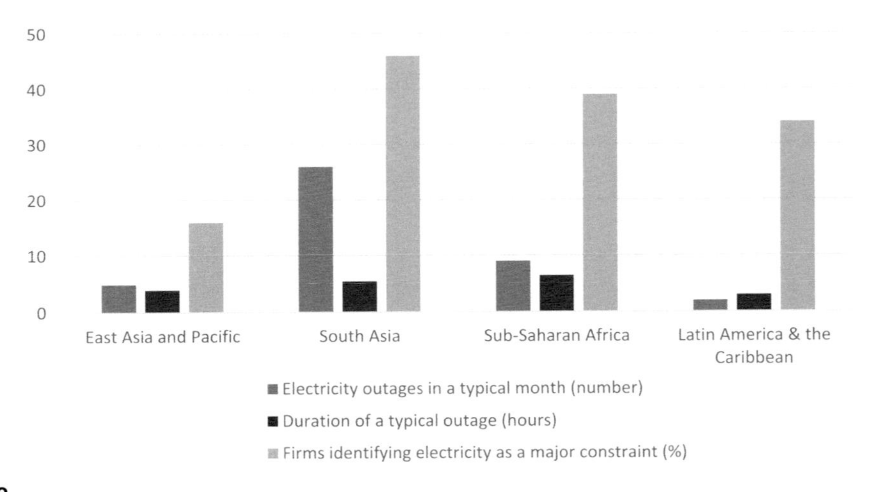

FIGURE 1.2

Overview of the reliability of electricity supply in the four discussed regions.

another major issue. Developing regions face serious gird reliability issues as shown in Fig. 1.2 (Asif, 2021). Furthermore, more than 2.8 billion people rely on traditional biomass, including wood, agricultural residues, and dung, for cooking and heating. Statistics also suggest that more than 95% of people without electricity live in developing regions, and four out of five live in rural areas of South Asia and sub-Saharan Africa. Table 1.2 highlights the countries with major reliance on biomass fuels (Asif, 2021).

The precise nature of energy security challenges for countries at different socioeconomic and technological strata can hugely vary. While a considerable segment of the population in the developed countries continues to experience fuel poverty, electricity deprivation remains a challenge for many in Africa and Asia. Although, globally there has been a significant improvement in recent years in terms of grid penetration and quality of grid, affordability of energy remains to be a critical issue for many in developing countries. Issues such as poor grid quality, power outages, breakdowns, planned load

Table 1.2 Top 20 countries with a share of biomass and combustible waste in energy supplies.

Country	Share in (%)
Ethiopia	92.9
DR Congo	92.2
Tanzania	85.0
Nigeria	81.5
Haiti	81.0
Nepal	80.6
Togo	79.9
Mozambique	79.8
Eritrea	78.2
Zambia	76.9
Ivory Coast	73.6
Niger	73.2
Kenya	72.2
Cambodia	66.9
Myanmar	65.3
Cameroon	65.0
Sudan	62.9
Guatemala	62.8
Zimbabwe	61.8
Republic of Congo	59.2

shedding, and brown shedding are regular phenomena even in major cities in these countries. Lack of access to reliable and refined energy resources can also be gauged from the fact that around 2.6 million lack access to refined cooking fuels. The situation is even more alarming when energy security is looked at from the 4'As perspective: availability, adequacy, affordability, and acceptability (Asif, 2021).

Developing countries are typically reliant on fossil fuels, which are largely imported, to meet their energy requirements, especially for power generation. Fossil fuels are typically extremely localized in terms of their existence and well over 80% of the countries in the world rely on imports, especially in the case of oil and gas. Consumption of fossil fuels entails environmental emissions besides affecting energy security in terms of import dependency and higher energy prices for consumers. The import bill can also be a massive burden on the developing and poor economies. In South Asia, for example, over 90% of the oil and gas supplies are imported with similar consequences. Developing countries experience further challenges, such as aging and inefficient energy infrastructure, resulting in additional monetary and environmental burdens. With the fast growth in energy demand, in the coming decades, they face the task of large capital investments in the energy sector. It is important to channel the investment toward sustainable energy projects focusing on renewables, distributed generation, and energy efficiency to better address the energy and environmental challenges. Renewable and energy efficiency projects, however due to their very nature, are capital-intensive, making them a difficult choice for weak economies.

3. Shared future

The planet is a global village as Marshal McLuhan described it to highlight the idea of global coexistence with influences from international communication, culture, travel, and trade and commerce. The COVID-19 pandemic has most recently cemented the planet's status of a global village as the virus originating from one city virtually paralyzed the whole world. The integrated global energy and environmental scenario is another befitting example to manifest the concept of a global village. Climate change is arguably the most pressing factor advocating for the shared future of the planet in terms of the faced energy and environmental challenges and their potential solutions. On the eve of COP 26, the United Nations Secretary-General Antonio Guterres described climate change and environmental damage being caused by human activities as "We are digging our own graves." Eminent scientists and scholars are also expressing their concern about global warming. In a survey, 50 Nobel Laureates described climate change as the biggest threat facing mankind ahead of issues like disease, nuclear war, and terrorism. Stephen Hawking repeatedly raised alarm bells as "Climate change is one of the great dangers we face, and it's one we can prevent if we act now. We are close to the tipping point where global warming becomes irreversible." He also described energy as one of the key factors that will "lead to the end of life on Earth." According to Noam Chomsky, "we have entered a new geological era, the Anthropocene, in which the Earth's climate is being radically modified by human action, creating a very different planet, one that may not be able to sustain organized human life in anything like a form we would want to tolerate." It is projected that unabated climate change could lead to, millions of species at risk of extinction, costing the global economy at least 5% of GDP each year, and in worst case scenario, the cost could be more than 20% of GDP (Asif, 2021). Climate change can also lead to geopolitical challenges as a recent US intelligence report says that climate change will lead to tension and confrontations between nations (Corera, 2021).

Global warming is a threat to the whole planet; its intensity, however, is not uniformly distributed. Climate change is affecting the whole world, but the poorest countries are suffering the earliest and the most. The majority of the global population suffering from climate change is from developing countries with limited resources to mitigate the challenges and to rebuild their lives after extreme environmental events. It is the low lying and small island countries, also termed as small island developing states (SIDS), which are being hit harder. The United Nations Secretary-General Antonio Guterres also acknowledges that "Climate change is happening now and to all of us. No country or community is immune. And as is always the case, the poor and vulnerable are the first to suffer and the worst-hit" (UN, 2020). For SIDS, global warming poses an enormous set of challenges for their livelihood, safety, and security. Since most of the infrastructure in these countries is on the coast, the damage from consequent erosion and flooding is likely to be hugely burdensome for their typically modest economies. Owing to their smaller land area compared to other countries, they cannot afford to lose land due to surging sea levels. For some, for example, Maldives, it threatens their very existence. Pointing out in this direction, at the COP 26, Mr. Guterres warned that for vulnerable nations and small island states, failure at the climate talks is not an option. "Failure is a death sentence." He further declared that everyone will lose unless humanity makes "peace with the planet" (Seabrook, 2021).

4. Global response

The most significant global response to the issue of climate change is in the form of energy transition. The unfolding energy transition is fundamentally a sustainability-driven energy pathway with the focus on decarbonization of the energy sector by shifting away from fossil fuels. This energy transition, therefore, can also be termed as "sustainable energy transition" or "low-carbon energy transition." The Intergovernmental Panel on Climate Change concludes that in order to meet the climate change targets, set as part of the Paris Agreement, major changes are needed across the four big global systems: energy, land use, cities, and industry. The energy sector is where the greatest challenges and opportunities exist. Following the Paris Agreement, many of the major economies and economic blocks including the United States, the EU, and the United Kingdom have committed to net-zero carbon emissions by 2050. Some other nations like China and the Kingdom of Saudi Arabia have committed to the same target by 2060. Each country or economic block is developing its own plans for incrementally achieving the goals individually, but the common feature is that they will all require transformation of energy sector (IEA, 2021). The EU, for example, has decided to reduce emissions by 55% from the 1990 level by 2030 to go net-zero by 2050. The United States of America has announced to initially cut emissions by 40%–43% by 2030. Some of the notable initiatives in this respect include having 30GW of new offshore wind projects, and to cut the cost of solar energy further by 60% over the next decade in order to achieve target of 100% renewable electricity by 2035. China targets emissions to peak by 2030 to reach carbon neutrality by 2060. Similarly, the United Kingdom has plans to cut emissions by 68% by 2030 to reach the target by 2050. A landmark decision the United Kingdom has made in shifting away from fossil fuels is to close down all coal power plants by 2024, which means within a decade, the country brings down its reliance on coal for power generation from around one third to zero.

Renewable energy has to be at the heart of the ongoing energy transition. Over the last couple of decades, renewable energy has made fast inroads into the global energy landscape. Renewable technologies, especially solar photovoltaic and wind turbines, are making a great progress in terms of technological and economic maturity. For several years now in a row, renewable energy is adding more power generation capacity compared to the combined addition by fossil fuels and nuclear power. In the year 2020, for example, renewables contributed to more than 80% of all new power generation capacity added worldwide. The global installed capacity of renewables increased from 2581 GW in 2019 to 2838 GW in 2020, exceeding expansion in the previous year by almost 50% (REN21, 2021).

In recent years, renewable policy development has also seen a tremendous growth. A large number of countries have already committed to use renewable and sustainable technologies, and the number is growing with each passing year. Benefits of these policies are increasingly being realized around the world as over 10,000 cities and local government bodies have enacted such frameworks. Fig. 1.3 shows the renewable energy policy trend in terms of countries with active policy frameworks. A country is considered to have a policy (and is counted a single time) when it has at least one national or state/provincial level policy in place. Power policies include feed-in tariffs (FITs)/feed-in premiums, tendering, net metering, and renewable portfolio standards. Heating and cooling policies include solar heat obligations, technology-neutral renewable heat obligations, and renewable heat FITs. Transport policies include biodiesel obligations/mandates, ethanol obligations/mandates, and nonblend mandates. By the end of the year 2020, 165 countries in the world have active policies around power sector, heating and cooling, and transportation (REN21, 2021).

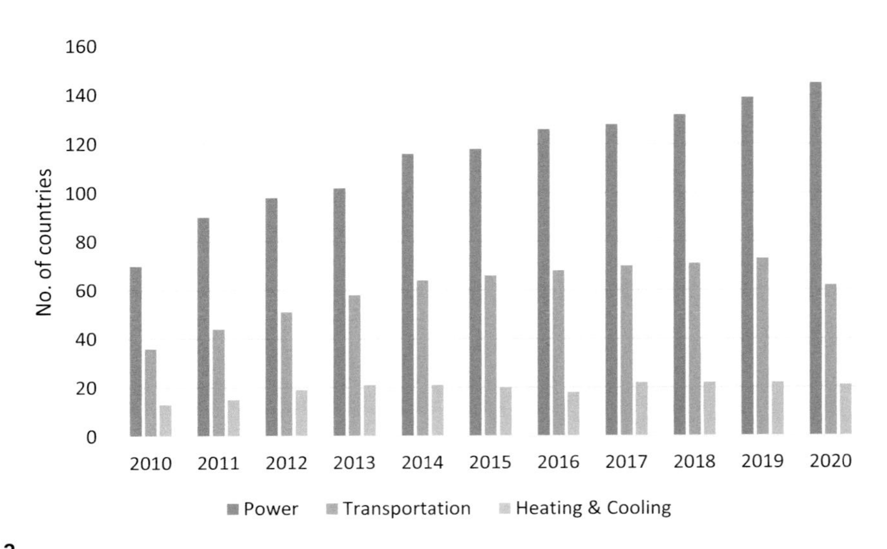

FIGURE 1.3

Number of countries with active renewable energy policies.

While there are positives to take from the global response to the energy and environmental challenges in general and climate change in particular, there are some apprehensions too, above all, words are not duly matched by actions. Rich and industrialized nations, for example, are not living up with their promise of providing $100 billion per year to poor countries in the fight against climate change. At the COP 26, more than 100 world leaders have promised to end and reverse deforestation by 2030. Experts welcomed the move, but warned a previous deal in 2014 had "failed to slow deforestation at all" and commitments needed to be delivered on (Rannard and Gillett, 2021). Reports also suggest that some nations are lobbying to play down the threats of climate change in crucial scientific reports. Also, some wealthy nations are questioning paying more to poorer states to move to greener technologies (Rowlatt and Gerken, 2021).

The solution to climate change should entail climate justice. Focus on climate change should be embedded in development policies at all levels. One of the significant heartbreaks of global warming is that developed and industrialized countries are acutely responsible for the phenomenon, but the heavier price is to be paid by the poor and developing nations. For example, the average value of the per capita energy consumption—an index to measure the contribution toward global warming—in industrialized and developed countries is almost six times greater than that in developing countries. Rich and industrialized countries need to share the responsibility for their emissions by helping the vulnerable and resource-deficient nations in the fight against climate change.

The COVID-19 crisis has underpinned the importance of access to sustainable energy services. From the deep economic recession the pandemic has plunged the world into, it requires a huge rebuilding effort. Given the track record and positive impact of sustainable energy solutions, especially renewable technologies and energy efficiency, as observed over the last couple of decades, have become ever more important. Energy and environmental security through sustainable energy systems should be embedded in the global economic recovery plans. The United Nations has called on governments to seize the opportunity to "build back better" by creating more sustainable, resilient, and inclusive societies in the planning for a postpandemic recovery. The United Nations Secretary-General António Guterres calls upon the nations: "We need to turn the recovery into a real opportunity to do things right for the future," while the UN Climate Chief states more specifically as: "With this restart, a window of hope and opportunity opens, an opportunity for nations to green their recovery packages and shape the 21st-century economy in ways that are clean, green, healthy, safe and more resilient." According to the UN Secretary-General, "The recovery must also respect the rights of future generations, enhancing climate action aiming at carbon neutrality by 2050 and protecting biodiversity. We will need to build back better." The United Nation Development Program Administrator emphasizes that "As we work through response and recovery from the shocks of the pandemic, the SDGs need to be designed into the DNA of global recovery" (UN, 2020). The pandemic has provided governments a unique opportunity to reset their economies, develop new business streams, and create jobs with a focus on sustainable energy and environmental future (IEA, 2020).

5. Conclusions

Energy and environmental security is a critical aspect of sustainable development. The global energy and environmental scenarios face major challenges. The energy sector faces issues like rapidly growing energy demand, depleting fossil fuel reserves, volatile energy prices, lack of universal access to energy, and above all climate change. Climate change, widely regarded to be a consequence of anthropogenic emissions since the industrial revolution, is arguably the most important threat facing mankind. Ironically, the poor and developing countries, despite having a marginal contribution toward the GHG emissions, are mainly at the suffering end when it comes to the implications of climate change. The situation demands collective, concerted, and cohesive efforts at the global level to mitigate the implications of climate change. To meet the Paris Agreement's target of limiting the global warming to 1.5°C, major changes are needed across the four big global systems: energy, land use, cities, and industry. In response to the broader energy security challenges in general and climate change in particular, the global energy scenario is experiencing a decarbonization transition, targeting to be carbon neutral by the middle of the century. Other key dimensions of this transition include decentralization, digitalization, and decreasing use of energy. Renewable energy is at the heart of this sustainable energy transition. There is, however, a realization that despite the several promising initiatives, the progress made at national and international levels is falling short of what is needed to fight climate change. The developed and industrialized nations require to help the poor and developing countries in the fight against climate change through measures such as capacity building, technology and knowledge transfer, and financial assistance.

References

Asif, M., 2011. Energy Crisis in Pakistan: Origins, Challenges and Sustainable Solutions. Oxford University Press, ISBN 978-0-19-547876-1.

Asif, M., 2021. Energy and Environmental Outlook for South Asia. CRC Press, USA, ISBN 978-0-367-67343-7.

Asif, M., 2021. Energy and Environmental Security in Developing Countries. Springer, ISBN 978-3-030-63653-1.

Asif, M., Muneer, T., 2007. Energy supply, its demand and security issues for developed and emerging economies. Renewable & Sustainable Energy Reviews 11 (7).
Cherp, A., Jewell, J., 2014. The concept of energy security: beyond the four As. Energy Policy 75, 415–421.
Corera, G., October 21, 2021. Climate change will bring global tension, US intelligence report says. BBC.
IEA, 2020. Sustainable Recovery. International Energy Agency. https://www.iea.org/reports/sustainable-recovery.
IEA, May 2021. Net zero by 2050: a roadmap for the global energy sector, Flagship report. International Energy Agency.
IEA, 2019. Energy Security. International Energy Agency. https://www.iea.org/areas-of-work/ensuring-energy-security.
Landholm, M., 1998. Defining Environmental Security: Implications for the US Army, AEPI-IFP-1298. Army Environmental Policy Institute. https://apps.dtic.mil/dtic/tr/fulltext/u2/a593191.pdf.
NASA, Climate change: how do we know, Facts, National Aeronautics and Space Administration, Evidence, Facts – Climate Change: Vital Signs of the Planet (nasa.gov).
Qudratullah, H., Asif, M., 2020. Dynamics of Energy, Environment and Economy: A Sustainability Perspective. Springer, ISBN 978-3-030-43578-3.
Rannard, G., Gillett, F., November 2, 2021. COP26: World leaders promise to end deforestation by 2030. BBC.
Read, P., Dennis, R., January 17, 2020. With costs approaching $100 billion, the fires are Australia's costliest natural disaster. Conservation.
REN21, 2021. Renewables 2020 Global Status Report. Renewable Energy Network.
Rowlatt, J., Gerken, T., October 21, 2021. COP26: Documents leak reveals nations lobbying to change key climate report. BBC.
Samenow, J., June 30, 2021. 'Hard to comprehend': Experts react to record 121 degrees in Canada. The Washington Post.
Seabrook, V., November 1, 2021. COP26: 'Failure is a death sentence' warning as climate change summit opens in Glasgow. Sky News.
UN, 2020. Climate Justice. United Nations. https://www.un.org/sustainabledevelopment/blog/2019/05/climate-justice/.
UN, 2020. Climate Change and COVID-19: UN urges nations to 'recover better'. Department of Global Communications. https://www.un.org/en/un-coronavirus-communications-team/un-urges-countries-%E2%80%98build-back-better%E2%80%99.
Winzer, C., 2012. Conceptualizing energy security. Energy Policy 46, 36–48.
Zurlini, G., Muller, F., 2008. Environmental Security, Encyclopedia of Ecology. Elsevier, ISBN 9780444637680.

CHAPTER

2 Dynamics of energy security and its implications

Tri Ratna Bajracharya[1], **Shree Raj Shakya**[1,2] **and Anzoo Sharma**[1,3]

[1]*Center for Energy Studies (CES), Institute of Engineering, Tribhuvan University, Pulchowk, Lalitpur, Nepal;* [2]*Institute for Advanced Sustainability Studies (IASS), Potsdam, Germany;* [3]*Center for Rural Technology (CRT/N), Kathmandu, Nepal*

Chapter outline

1. Energy security

Energy security is an integral parameter to promote sustainable development. It is emerging as a cornerstone of energy policies and frameworks around the world, mainly in industrialized and economically emerging countries. Energy security is an evolving concept, and its scope is expanding and has not reached a universal definition. It has become an umbrella term capturing multiple factors about the energy prosperity of a nation. A widely accepted definition of energy security is based on the notion of an uninterrupted energy supply. But conventionally, the idea of energy security is concentrated on securing access to supplies of oil; however, the idea has subsequently broadened to capture the influence on energy security of other energy sources, price volatility, supply chain mechanisms, political stability of oil nations, sustainability, and other factors. With the emerging challenges, the meaning and scope of energy security has been transformed and the modern definition incorporates a degree of reliability, adequacy, consistency, affordability, sustainability, and environmental acceptability of energy supplies to match the emerging challenges.

As there are wide varieties of energy security definitions available in the literature, there are also different approaches and methodologies toward both qualitative and quantitative assessment of energy

Handbook of Energy and Environmental Security. https://doi.org/10.1016/B978-0-12-824084-7.00019-9
Copyright © 2022 Elsevier Inc. All rights reserved.

security. Mansson et al. (2014) reviewed and suggested that energy security is a multidisciplinary concept rather than interdisciplinary concept with a variety in methodologies according to the researcher's background in multiple disciplines like macroeconomics using methodologies like partial equilibrium models, cost–benefit analysis, etc.; microeconomics using market behavior studies; industrial organization focusing on market risk exposure, financial theory, financial portfolios, real options theory, etc.; engineering researchers focusing on the reliability of power systems, operations research using multicriteria analysis, analytical hierarchy processes, etc.; political science focusing on international relations theory, complex system analysis applying simulation, dynamic system modeling methods, general system analysis using methodologies like energy system scenario analysis, complex indicators and indexes, etc. Cherp and Jewell (2011) distinguished the two most fundamental methodological choices in energy security assessments that are the choice between facts and perceptions in deciding what constitutes a significant energy security concern and concludes that the methodological choice should be systematic, rational, and transparent.

The energy security studies often design multidimensional indicators/metrics for the assessment of energy security and energy systems performance. The International Energy Agency (IEA) developed a different set of metrics to assess the risk of system disruptions, imbalances between supply and demand, regulatory failures, and diversification among a subset of OECD countries (IEA, 2007). The Energy Research Center of the Netherlands has also developed a comprehensive "Supply and Demand Index," a comprehensive approach forwarded by the Energy Research Center of the Netherlands that examines the diversification among energy resources, suppliers, and importers, the long-term political environment in origins of supply, and the depletion rates of resource (Scheepers et al., 2006). Intharak (2007) used the 4A's framework for energy security assessment for the first time with dimensions availability, affordability, acceptability, and accessibility and five indicators: diversification of energy supply sources, net energy import dependency, noncarbon-based fuel portfolio, net oil import dependency, and Middle East oil import dependency under the three elements of physical energy security, economic energy security, and environmental sustainability. Vivoda (2009) attempted to build a "novel methodological" approach to energy security and proposed 11 broad dimensions and 44 attributes that can assess the performance of a nation on energy issues. The analyst used it to assess the energy security in the Asia-Pacific region. The approach of Kruyt et al. (2009) consists of 24 simple and complex indicators. The Association of Southeast Asian Nations (ASEAN) has adopted 4A framework for energy security, namely, availability, acceptability, affordability, and applicability to study the trend of each indicator for 10 member countries (Tongsopit et al., 2016). Ang et al. (2015) conducted a comprehensive literature review and found the dimensions of energy security ranging from 1 to 20 and number of indicators ranging from 1 to 320. Multidimensional criteria system by Chuang and Ma (2013) comprises dependency, vulnerability, affordability, and acceptability and six specified indicators that evaluated the efficacy of Taiwan's energy policies on its energy security. The effect of key policies on 19 key security indicators based on quality function deployment and system dynamics was simulated by Shin (2013). The trend of China's energy security over 30 years of reform is studied by Yao and Chang (2014) employing five metrics. Martchamadol and Kumar (2012) combined social, economical, and environmental aspects and evaluated the energy security of the past and future. Wu et al. (2012) worked with 14 indicators to estimate the correlation between climate protection and China's energy security. Ren and Sovacool (2014) used Fuzzy Decision-Making Trial and Evaluation Laboratory (DEMATEL) methodology and concluded that energy security best consists of the four dimensions: availability, affordability, acceptability, and accessibility to determine the most salient and meaningful dimensions and energy security strategies and apply the results to China, where they conclude that among the four energy security "A's" of availability, affordability, acceptability, and accessibility, availability and affordability are more influential to the energy security metrics than acceptability and accessibility. The energy security of China was examined in a study by Yao and Chang (2014) using 4A's framework of availability of energy sources, applicability of technology, societal acceptability, and the affordability of energy resources, and 25 indicators over 30 years of its economic reform showing that China needs to develop renewable energy resources on a large scale and control emissions. Sovacool (2013) constructed an energy security index to measure national performance on energy security over time under the dimensions of availability, affordability, efficiency, sustainability, and governance using 20 metrics and measured international performance across 18 countries from 1990 to 2010. Wang and Zhou (2017) evaluated energy security of 162 countries under a framework consisting of three energy security dimensions namely the Security of Energy Supply-Delivery dimension (SESD) with energy availability and infrastructure as two major factors, the Safety of Energy Utilization dimension (SEUD) with equity of energy service and environmental sustainability of energy consumption as two major factors, and the Stability of the Political-Economic Environment dimension (SPED) mainly involving a country's political and economic strength. The concept of "sustainable energy security" was proposed by Narula and Reddy (2016) suitable to assess energy security of developing countries where the energy system has been divided into "supply," "conversion and distribution," and "demand" subsystems, each of which are further subdivided into four dimensions of availability, affordability, efficiency, and acceptability.

An efficient and effective policy can be developed through a clear conceptualization of energy security. However, the energy security literature is scattered and is sometimes conflicting on the concept. As the scope of energy is vast and it is essential to perform any kind of task, big or small, various authors from various fields conduct research targeting their area and assess the impacts of the various risks on the security of energy essential to their particular area of interest. The IEA was formed in the 1970s with the objective to "coordinate a robust response to disruptions to oil supplies defines energy security as the uninterrupted availability of energy sources at an affordable price." Though researchers have not reached a consensus on the definitions of energy security provided, they all demonstrate the significance of energy security, its multidimensional impacts, and hence justify why it occupies the policy priority. Since energy is essential for economic development and environmental need, energy security highlights the need for timely investments in energy supply. Also because of the Sustainable Development Goal 7 of the United Nations, "to ensure access to affordable, reliable, sustainable, and modern energy for all" and associated targets, energy security is kept at the core.

1.1 Components of energy security

Ayoo (2020) has identified four different components of energy security as availability, affordability, energy and economic efficiency, and environmental stewardship (Table 2.1).

1.2 Indicators of energy security

Ayoo (2020) has proposed several indicators that can examine the energy security risks of various countries and the impacts of policies on them. Based on the explanation of assessing the energy security, the comparative table is prepared (Table 2.2).

Table 2.1 Components of energy security.

Criteria	Underlying values	Explanation of criterion	Indicators
Availability	Independence, diversification, and reliability	Diversification of the fuels used to provide energy services as well as the suppliers Quick recovery from short-term disruption like dependency on foreign suppliers	Dependency on oil import and natural gas import
Affordability	Equity	Affordable energy and minimum price volatility	Retail price of energy like electricity gasoline/petrol
Energy and economic efficiency	Innovation, resource custodianship, and minimization of waste	Improved efficiency	Energy intensity, per capita electricity use, and on-road fuel intensity of passenger vehicles
Environmental stewardship	Sustainability	Protection of natural environment for future generation	SO_2 and CO_2 emissions

Source: Ayoo (2020).

Table 2.2 Comparative table of indicators and energy security risks.

Indicators	Risks
Energy reserve	The approximate quantity of energy sources (e.g., coal, gas, or oil) known to exist with reasonable available energy resources and the development of infrastructure.
Energy production and consumption	For example, the discovery of vast oil and gas resources in the North Sea had a significant impact on Norway's economy and energy security and made Norway a rich oil-exporting country.
Energy trade balances	Difference between energy imports and exports
Energy prices	Influenced by the supply and demand situations in energy markets Controlled by policies and regulations such as taxes and subsidies
Energy diversity	Diverse energy systems are more secure and flexible to supply shocks For example, the Russian–Ukrainian gas crisis
Share of renewable energy	Abundance portfolio of renewable energy sources implies higher energy security

2. Energy security issues in developed and developing countries

The energy security issue is one of the major agendas of governments around the world. But the energy security can mean different for developed countries, whereas for developing and least developing countries, it would be different. Generally, in developed countries, energy security improvement policies aim at securing low-cost and reliable supplies of petroleum products to generate electricity and fuel transportation. As most OECD countries are heavily reliant on imported fossil fuel, the key concern is to diversify the supplier to lower the risk of shortage and efficient use.

For many developing and least developing countries, along with managing dependency, energy security has an additional function of catering to the basic human needs at the household level and increasing living standards. There lies a huge disparity between the energy consumption (per capital level) and the quality of energy supplied between these countries and the developed countries. Therefore, the strategic policy must focus to expand the renewable energy supply along with demand management and energy efficiency measures.

Overall, for a balanced economy at global, national, and internal development, the need for energy security is paramount. Energy insecurity or vulnerabilities and shortages affect countries in two ways: they handicap productive activities and it discourages investors by threatening production and increasing costs.

2.1 World energy trilemma index

The World Energy Council in partnership with its global consultancy Oliver Wyman, along with Marsh & McLennan Advantage has been preparing "The World Energy Trilemma Index (ETI)" annually since 2010 (World Energy Council, 2020). The World Energy Council defines energy sustainability as being based on three core dimensions: energy security, energy equity, and environmental sustainability of energy systems. A "trilemma" is formed to create a balance among these three important goals, and

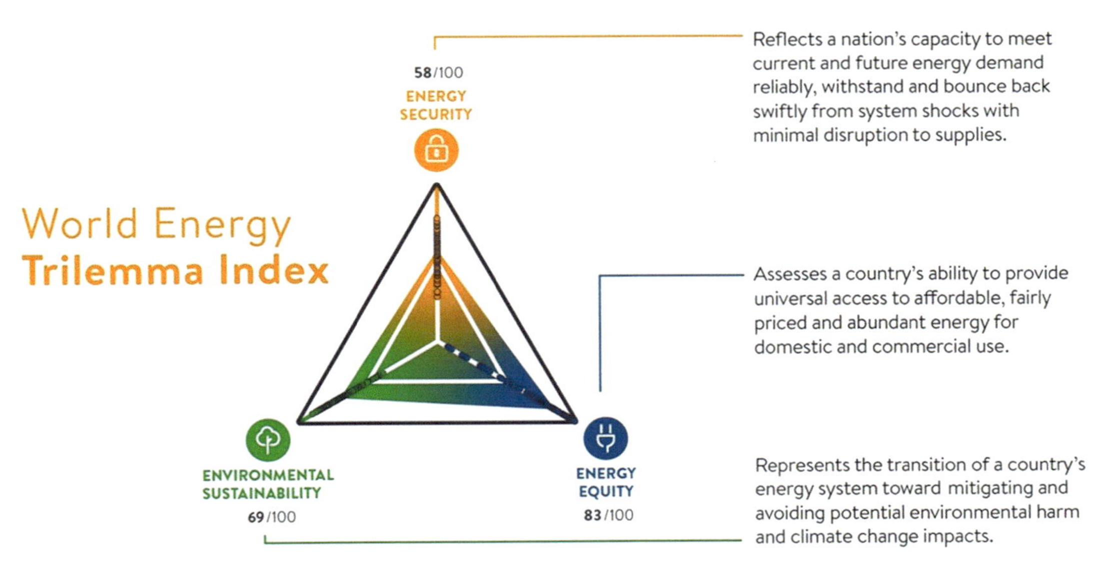

FIGURE 2.1

World energy trilemma index.

Source: World Energy Council.

balanced systems facilitate prosperity and competitiveness of energy systems among individual countries in the world. Fig. 2.1 World Energy Trilemma Index 2020 is published by the World Energy Council in 2021.

The trilemma index dimensions adopted from World Energy Council (World Energy Council, 2020) are presented in tabular form in Table 2.3.

The top 10 overall performers are presented in Table 2.4.

The energy security dimension measures a country's ability to meet the current and future energy demand against supply disruptions. Undoubtedly, as the world relies on fossil fuel for 81% of its primary energy supply, hydrocarbon-rich countries secure the top 10 ranks. The countries that have focused on diversifying and decarbonizing their energy systems such as Canada, Finland, and Romania top the list of best performers. Hydrocarbon resources abundance can be a boon and sometimes it also emerges as a "resource curse"; some hydrocarbon-rich countries' performances are declining as they are reluctant to diversify their energy systems. Diversification of a country's energy mix enhances energy security performance and advances toward a stronger emphasis on increasing the resilience of the system.

According to World ETI, many developing countries are observed to have made commendable improvements in both energy access and energy affordability; the top energy equity improvers have achieved almost an eight-time increase as compared to their baseline 2000 energy equity scores (World Energy Council, 2020). The top 10 energy equity improvers are from sub-Saharan Africa and Southeast Asia, yet the degree of improvement varies significantly. Sub-Saharan African countries: Kenya, Tanzania, Ethiopia, and Niger, and Southeast Asian countries: Cambodia and Bangladesh are

Table 2.3 Trilemma index dimensions.

Trilemma index	Measures	Covers
Energy security	Ability to meet current and future energy demand	Effectiveness of management of domestic/external energy sources
	Withstand and respond to system shocks	Reliability and resilience of energy infrastructure
Energy equity	Access to affordable, reliable, and abundant energy for domestic and commercial use	Basic access to electricity and clean cooking fuels and technologies
		Access to prosperity-enabling levels of energy and affordability
Environmental sustainability	Mitigation/adaptation ability against environmental degradation and climate change impacts	Generation, transmission, and distribution efficiency
		Decarbonization and air quality

Source: World Energy Council.

Table 2.4 2020 top ten overall performers.

Rank	Country	Grade	Score
1	Switzerland	AAAa	84.3
2	Sweden	ABAa	84.2
3	Denmark	AAAa	84.0
4	Austria	AAAa	82.1
	Finland	ABAa	82.1
5	France	AAAa	81.7
	United Kingdom	AAAa	81.7
6	Canada	AABa	81.5
7	Germany	AAAa	80.9
8	Norway	BAAa	80.5
9	United States	AABa	79.8
10	New Zealand	AAAa	79.5

Sources: World Energy Council (2020).

persistent since 2000 in improving their access to electricity and switching from biomass to clean fuel and, therefore, are featured among the top 10 energy equity improving countries. The top 10 overall improvers are presented in Table 2.5 (World Energy Council, 2020). The impact of SDG7 agreed in 2015 has certainly helped some of these countries accelerate their scores.

Globally, the energy transition is sought to bring unprecedented transformation to the energy sector as countries seek to decarbonize. Nevertheless, the energy policies and regulations lag with the incremental step required to bring about the changes. Hence to maintain the relevance to the energy transition, the ETI is required to evolve continually with new or modified indicators.

Table 2.5 2020 top ten overall improvers.

Rank	Country	Grade	Score	Improvements since 2000
1	Cambodia	DDDd	50.8	77%
2	Myanmar	BDCd	54.3	50%
3	Kenya	BDBc	54.3	41%
4	Bangladesh	DDDd	47.8	38%
5	Honduras	CCBc	60.5	36%
6	Ghana	CDBc	55.3	36%
7	Nicaragua	CCBd	57.9	34%
8	Ethiopia	DDCd	43.1	33%
9	Tajikistan	DCCd	57.1	30%
10	Mongolia	DBDc	55.5	28%

Sources: (World Energy Council, 2020)

2.2 Developing countries' perspective

Although most of the developed countries achieved electrification for large populations from the 1950s to 1970s, at a global level, the number of people without access to electricity is still substantial. The population of the developing world is mostly connected to unreliable electricity systems, whereas those with no access to electricity are concentrated in Sub-Saharan African and Southern Asia. Energy consumption in developing countries has soared more than fourfold within the past three decades, and the increase can be expected to continue in the future, as energy creates and expands the services and energy services are the bloodline of economic growth, improved living standards.

The unprecedented growth of energy consumption in developing countries has wide-ranging impacts. Still, the economic development process is fueled by carbon-intensive fuel, and by the increase in industrial countries, the demand is accelerating leading to upward pressures on world oil prices. Though petroleum is consumed globally, its natural distribution is not as such and is increasing the indebtedness level of the developing countries. The quest for energy has somehow contributed to instability in the international money and banking system. The rapid increase in local and regional air pollution, accompanied by increased atmospheric greenhouse gas (GHG) concentration is also attributable to the growing fossil fuel consumption in developing countries. It was well identified in the early 1990s that international efforts are required to control GHG emissions for which active participation from the developing countries are obligatory (U.S. Congress, Office of Technology Assessment, 1991).

3. Energy security policies and its implications in energy transitions

As energy is the driving force of economic growth, for any country, energy security plays an important role in its national security. The primary threats to energy security range from the political instability of own country, the neighboring countries as well as several energy-producing countries. Some other threats include the manipulation of energy supplies, the competition over energy sources, disasters on supply infrastructure caused by vested human interest, accidents, natural disasters, terrorism, and heavy dependence on foreign countries for a strategic commodity like petroleum products.

Energy security is important for every country; therefore, every country develops their energy security policy by developed nations and the developing world as well. In energy security policy, the government focuses on issues related to energy production, distribution, and consumption and strategies to cater to the growing energy demand. These actions include by what means the governments prepare to deal with disruption in energy supply and their efforts to influence the economic growth and accompanied by growth in energy demand.

Most of the developed nations have an energy policy to achieve the demand by transiting to clean and freely available energy resources, i.e., renewable energy resources. The developing nations have adopted renewable energy as an alternative or in small-scale applications only. Energy security policy plays an important role in the implications of energy transitions.

3.1 Energy transition index

The Energy Transition Index (ETI) is an evidence-based scoring developed to enable policy-makers and businesses to outline the course of action for a successful energy transition. The energy system is analyzed and benchmarked annually. The data-driven framework can be generated from the ETI to understand the performance and foster the readiness of energy systems across countries for transition. An effective energy transition can be defined as "a timely changeover toward a more inclusive, sustainable, affordable, and secure energy system that provides solutions to global energy-related challenges, while creating value for business and society, without compromising the balance of the energy triangle." A country's willingness and progress in energy transition can be determined by the extent to which a robust enabling environment is created. The energy transition can be achieved with joint efforts such as political commitment, a flexible regulatory structure, a stable business environment, incentives for investments and innovation on clean energy, consumer consciousness, and their acceptance of new technologies. The ETI measures progress along these dimensions in the transition readiness subindex (Fig. 2.2). The energy transition is not restricted to linear shifts in fuel mix or the substitution of production technologies.

System performance provides an assessment of a country's energy system across three key priorities (World Economic Forum, 2021):

i. The ability to support economic development and growth
ii. Universal access to secure and reliable energy supply
iii. Environmental sustainability across the energy value chain

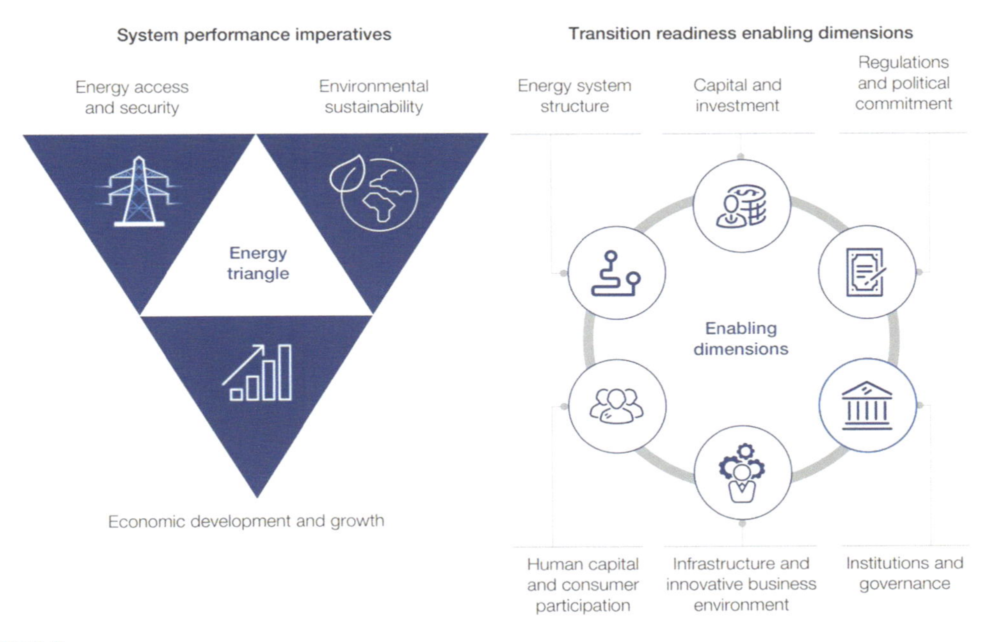

FIGURE 2.2

Energy transition index framework.

Source:(World Economic Forum, 2021)

Comparing the global energy transition condition of 2010 to that of 2020, there has been significant progress with increment in the global investment, the share of electricity from a renewable source, Global EV Share, etc., with a decrease in renewable energy cost, number of the population without electricity as well.

3.1.1 Top 10 ETI ranking countries

Sweden, Norway, and Denmark lead the global ETI rankings. Among the world's 10 largest economies, only the United Kingdom and France have been able to make it through the top 10 (Table 2.6). The highest-placed nonadvanced economy is Latvia (12), which is classed as "Emerging and Developing Europe." The highest-placed developing nation is Costa Rica securing 26th position in the ETI ranking. Interestingly, the top performers in the ETI have managed to stay consistent over the decade. Even though each country designs its energy transition pathway, the commonalities include:

a. Reduce fossil fuel subsidies
b. Diversity of fuel mix and import partners
c. Improving carbon intensity
d. Reduce share of fossil fuels in the energy mix
e. Strong environmental regulatory
Based on the report of Fostering Effective Energy Transition (World Economic Forum, 2021) it can be summarized as follows.

i. It is well understood from (Table 2.7) that Asia houses the fastest emerging and developing countries than other regions. Most of the countries in this region have gained energy access and security. The region is coming in the fastest development of the economy. The energy demand per capita has grown 18% in the last decade and is expected to double by 2050. However, giant Asian economies are coal-dependent as coal continues to dominate the energy mix in some. To reduce the emission by the accelerated deployment of new technologies, there is a need for a robust enabling environment to ensure a return on investments. The mainstreaming renewable energy technology could carry out energy transition pathways, assisting the region to meet future demand in a sustainable climate-friendly environment.

Table 2.6 Top 10 ETI ranking.

Rank	Country	ETI score (2012–21)	SP	TR
1	Sweden	79	84.4	72.7
2	Norway	77	82.7	70.8
3	Denmark	76	74.8	78.2
4	Switzerland	76	79.9	73.0
5	Austria	75	75.2	75.2
6	Finland	73	73.5	73.0
7	United Kingdom	72	75.8	69.2
8	New Zealand	71	76.5	65.6
9	France	71	77.6	64.4
10	Iceland	71	75.0	66.9

Source: World Economic Forum, 2021

Table 2.7 Average Score by Peer Groups (Developing Nations).

Average score by peer groups (developing nations)	ETI score	Improvement (2012–20)	% Of global CO_2 emission	% Of global population	CO_2 per capita in ton
Emerging and developing Asia	54.9	6%	40	47	3.74
Sub-Saharan Africa	50.7	2%	2	8	1.10
Emerging and developing Europe	61.0	5%	3	2	5.20
The Middle East and North Africa	52.8	2%	7	7	3.90
Latin America and the Caribbean	58.6	2.4%	5	8	2.40

Source: World Economic Forum, 2021

ii. Access to electricity and basic energy services remains lowest in the Sub-Saharan African region at 56%. Although the energy access and security in the region are primitive as compared to global advances, its trajectory on the energy transition journey is overwhelming. The region has demonstrated a great potential to catapult by preventing expensive, inefficient, and more polluting energy infrastructure.

iii. The Emerging and Developing Europe region has maintained balanced improvement along three dimensions of the energy transition, and ETI score improved by 5% between 2012 and 2021. The drivers of energy transition in the regions are diversified energy resources, high-quality electricity generation, distribution, improved energy efficiency, and reduced energy intensity. Studies suggest that more than one-third of the energy demand of the region can be catered to by renewable sources, which potentially can lead to energy cost savings, health benefits, and reduced import for primary energy and hence energy dependency.

iv. The Middle East and North Africa have improved the overall trajectory. However, the region is heavily driven by oil revenue and exhibits a major threat to sustainable growth. For the region to improve its prospects, the diversified economy and extended portfolio of the energy system should be taken into consideration. Nevertheless, the member states of the region had come up with ambitious renewables targets for 2030. Hence the region is a fertile land for energy transition investment in the coming that can open meaningful cross-system benefits.

v. The ETI score in Latin America and the Caribbean region remained consistent over a decade. The region stands out in terms of environmental sustainability because of the hydro-based energy system. Further areas for developments can be achieved through improvement in energy affordability. Diversified oil suppliers and diversification in the energy structure can aid in improving energy security.

4. Conclusion

Energy security is an evolving concept, and its scope is expanding and has not reached a universal definition. However, energy security is emerging as a cornerstone of energy policies and frameworks around the world, and a widely accepted definition of energy security is based on the notion of an uninterrupted energy supply. The energy becomes secure when a country has energy reserve, balanced supply and demand, and balanced energy trade. Energy is affordable, and the energy system is diverse with ample amount of renewable share. The energy systems of a country must strike a balance among energy trilemma: energy security, environmental sustainability, and energy equity. The fundamental policies for energy security can vary depending upon the geopolitical status, scale of economy and environmental issues. Most of the developed nations have an energy policy to achieve the demand by transiting to clean and freely available energy resources, i.e., renewable energy resources. The developing nations have adopted renewable energy as an alternative or in small-scale applications only. Energy security policy plays an important role in the implications of energy transitions.

References

Ang, B., Choong, W., Ng, T., 2015. Energy security: definitions, dimensions and indexes. Renewable and Sustainable Energy Reviews 42, 1077–1093.

Ayoo, C., 2020. Towards energy security for the twenty-first century. In: Energy Policy. IntechOpen. https://doi.org/10.5772/intechopen.90872.

Cherp, A., Jewell, J., 2011. Measuring energy security: from universal indicators to contextualized frameworks. In: The Routledge Handbook of Energy Security (Issue January).

Chuang, M., Ma, H., 2013. An assessment of Taiwan's energy policy using multidimensional energy security indicators. Renewable and Sustainable Energy Reviews 17, 301–311.

IEA, 2007. Energy Security and Climate Change Assessing Ineractions. International Energy Agency, Paris.

Intharak, N., 2007. A quest for Energy Security in the 21st century resources and constraints. In: Asia Pacific Energy Research Centre. Institute of energy Economics, Japan. Asia Pacific Energy Research Centre. www.ieej.or.jp/aperc.

Kruyt, B., van Vuuren, D.P., de Vries, H.J.M., Groenenberg, H., 2009. Indicators for energy security. Energy Policy 37 (6), 2166–2181. https://doi.org/10.1016/j.enpol.2009.02.006.

Mansson, A., Johanson, B., Nilsson, L.J., 2014. Assessing energy security: an overview of commonly used methodologies. Energy 17, 1–14.

Martchamadol, J., Kumar, S., 2012. Thailand's energy security indicators. Renewable and Sustainable Energy Reviews 103–122.

Narula, K., Reddy, B.S., 2016. A SES (sustainable energy security) index for developing countries. Energy 94, 326–343.

Scheepers, M., Seebregts, A., Jong, J., Maters, H., 2006. EU Standards for Energy Security of Supply. ECN, Netherlands.

Shin, J.,S.W., 2013. An energy security management model using quality function deployment and system dynamics. Energy Policy 54, 72–86.

Sovacool, B.K., Ren, J., 2014. Quantifying, measuring, and strategizing energy security: determining the most meaningful dimensions and metrics. Energy 76, 838–849.

Sovacool, B.K., Valentine, S., 2013. Sounding the alarm: global energy security in the 21st century. In: Energy Security. Sage Library of International Security, London, pp. 35–78.

Tongsopit, S., Kittner, N., Chang, Y., Aksornkij, A., Wangjiraniran, W., 2016. Energy security in ASEAN: a quantitative approach for sustainable policy. Energy Policy 90, 60−72.

U.S. Congress, Office of Technology Assessment, 1991. Energy in Developing Countries, OTA-E-486. U.S. Government Printing Office, Washington, DC. https://www.princeton.edu/~ota/disk1/1991/9118/9118.PDF.

Vivoda, V., 2009. Diversification of oil import sources and energy security: a key strategy or elusive objective. Energy Policy 4615−4623.

Wang, Q., Zhou, K., 2017. A framework for evaluating global national energy security. Applied Energy 188, 19−31.

World Energy Trilemma Index | World Energy Council. (n.d.). Retrieved June 14, 2021, from https://www.worldenergy.org/transition-toolkit/world-energy-trilemma-index.

World Energy Council, 2020. WORLD ENERGY TRILEMMA INDEX 2020. World Energy Council, p. 11. https://www.worldenergy.org/assets/downloads/World_Energy_Trilemma_Index_2020_-_REPORT.pdf?v=1602261628.

Wu, G., Liu, L., Han, Z., Wei, M., 2012. Climate protection and China's energy security: win-win or tradeoff. Applied Energy 97, 157−163.

Yao, L., Chang, Y., 2014. Energy security in China: a quantitative analysis and policy implications. Energy Policy 67, 595−604.

World Economic Forum (2021), Fostering Effective Energy Transition 2021 edition, Retrieved June 14, 2021, from https://www.weforum.org/reports/fostering-effective-energy-transition-2021.

CHAPTER

Sustainable energy transition in the 21st century

3

Muhammad Asif

Department of Architectural Engineering, King Fahd University of Petroleum and Minerals, Dhahran, Saudi Arabia

Chapter outline

1. Introduction

Energy is the lifeline of modern societies. The industrial revolution of the 19th century has revolutionized the human-energy relationship. Ever since, extensive and efficient utilization of energy has been pivotal in the global development. Energy is becoming an increasingly critical commodity on multiple fronts including technological, socioeconomic, and geopolitical. Energy has attained the status of a prerequisite for all crucial aspects of societies, i.e., mobility, agriculture, industry, health, education, and trade and commerce (Asif, 2021a, b). Energy resources exist in a wide range of physical states, which can be harnessed and capitalized through various technologies. Energy resources can be broadly classified into two categories: renewables and nonrenewables. Renewable energy resources are the ones which are naturally replenished or renewed. Examples of renewable resources include solar energy, wind power, hydropower, and wave and tidal power. Energy resources which are finite and exhaustible are termed as nonrenewable such as coal, oil, and natural gas.

An important dimension of the human use of energy is its contribution toward climate change. Unchecked emissions of GHS are leading to global warming. Climate change, as result of global warming, is regarded as the biggest challenge facing the world. Different types of energy resources, especially fossil fuels, contribute to greenhouse gas (GHG) emissions. Fossil fuels are considered to be

Handbook of Energy and Environmental Security. https://doi.org/10.1016/B978-0-12-824084-7.00023-0

Copyright © 2022 Elsevier Inc. All rights reserved.

the primary reason for the anthropogenic emission of carbon dioxide (CO_2)—the 18th century industrial revolution is considered to have triggered the rapid growth in the release of GHGs. The CO_2 concentration in the atmosphere, for example, has increased from the preindustrial age level of 280 parts per million (ppm) to 415 ppm. The acceleration in the growth of CO_2 concentration can be gauged from the fact that almost 100 ppm of the total 135 ppm increment has occurred since 1960.

The global energy scenario is experiencing a number of other challenges as well as rapid growth in energy demand, depletion of fossil fuel reserves, volatile energy prices, and a lack of universal access to energy. A fast growth in the global energy demand—owing to factors such as surging population, economic and infrastructural development, and urbanization—is adding pressures on the energy supply chain. According to the Energy Information Administration (EIA), between 2018 and 2050, the world energy requirements are projected to increase by 50% (EIA, 2019). Access to refined energy resources remains to be a major challenge for significant proportions of population in the developing countries.

The global energy scenario is experiencing a transition to address the faced energy and environmental challenges. The primary aim of this transition is to shift the global energy system away from fossil fuels. Renewable and low-carbon technologies are at the heart of this energy transformation. This chapter aims to discuss the key dynamics of the energy transition. It has six subsequent sections. Section 3 provides an overview of the four main dimensions of the ongoing energy transition. Sections 3. 4, 5, 6, respectively, discuss decarbonization, decentralization, digitalization, and decreasing use of energy to be followed by the Conclusions section.

2. Dynamics of the energy transition

The use of energy is closely linked to the environment. It is estimated that despite the pledges and efforts by the global community to tackle climate change, CO_2 emissions from energy and industry have increased by 60% since the United Nations Framework Convention on Climate Change (UNFCCC) was signed in 1992 (IEA, 2021). Climate change is already there with its implications such as seasonal disorder, rising sea level, a trend of more frequent and intense weather-driven disasters such as flooding, droughts, heat waves, wildfires, storms, and associated financial losses (Asif, 2021a,b; Qudratullah and Asif, 2020). The situation calls for an urgent and paradigm shift across the entire energy sector. Responding to the challenges on hand and to ensure a supply of energy compatible with the demands of a sustainable future for the planet, the global energy sector is going through a transformation. This energy transition can be defined as "The energy transition is a pathway toward transformation of the global energy sector from fossil-based to zero-carbon by the second half of this century." At the heart of the ongoing energy transition is the need to reduce energy-related CO_2 emissions to limit climate change (IRENA).

In the recent history, there have been two major energy transitions. The first energy transition propelled the industrial revolution as biomass and wood were replaced by coal, a more efficient and effective fuel to drive machines. The second energy transition was a shift from coal to more refined forms of fossil fuels—oil and gas—in the 20th century. The world is now experiencing the third energy transition. This energy transition is much more vibrant, intriguing, and impactful compared with the earlier ones. It is fundamentally a sustainability-driven energy pathway with the focus on decarbonization of the energy sector by shifting away from fossil fuels. This energy transition therefore can also be termed as "sustainable energy transition" or "low-carbon energy transition." Holistically, however, the ongoing energy transition is not just about going low carbon or a shift away from fossil fuels; it is rather much more dynamic, thanks to the enormous changes and developments on the fronts of energy resources and their consumption, technological advancements, socioeconomic and political response, and evolving policy landscape. This energy transition has four key dimensions: decarbonization, decentralization, digitalization, and decreasing use of energy.

3. Decarbonization

Through the Paris Agreement, the world has adopted the first ever universally legally binding global climate deal to avoid dangerous climate change by limiting global warming to well below 2°C. The Intergovernmental Panel on Climate Change (IPCC), however, warns that the world is seriously overshooting this target, heading instead toward a higher temperature rise, asking for major changes in four big global systems: energy, land use, cities, and industry. The energy sector is where the greatest challenges and opportunities exist (Gillam and Asplund, 2021). Different types of energy resources have a huge range of environmental emissions as shown in Fig. 3.1. Following the Paris Agreement, many of the major economies and economic blocks—such as the United States, China, the European Union (EU), and the United Kingdom—have committed to net zero carbon emissions. The US, EU, and United Kingdom are targeting net zero emissions by 2050, while China has targeted it for 2060. Each country or economic block is developing its own plans for incrementally achieving the goals individually, but the common feature is that they will all require transformation of energy sector (Gillam and Asplund, 2021). The EU, for example, has decided to reduce emissions by 55% from the 1990 level by 2030 to go net zero by 2050. The United States of America has announced to initially cut emissions by 40%—43% by 2030. Some of the notable initiatives in this respect include having 30 GW of new offshore wind projects and to cut the cost of solar energy further by 60% over the next decade in

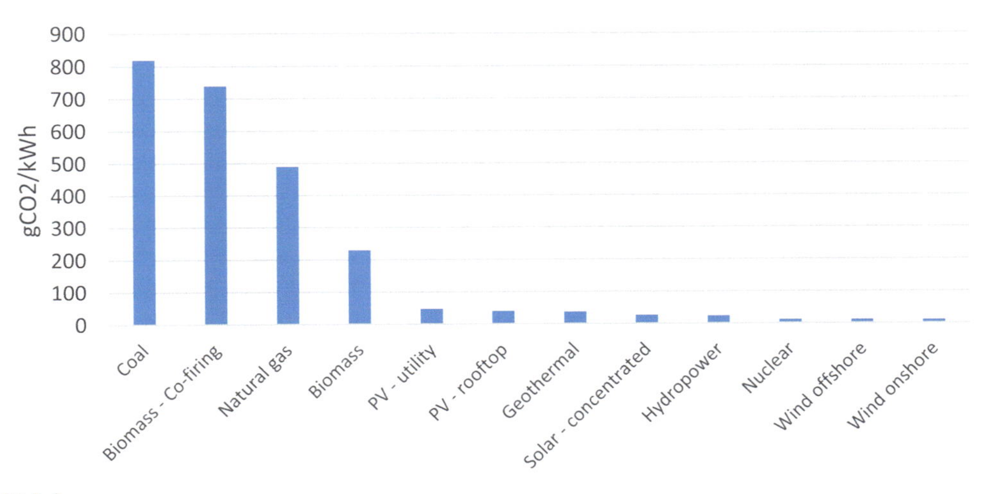

FIGURE 3.1

Comparison of CO_2 emissions from different power generation systems.

order to achieve the target of 100% renewable electricity by 2035 (DOE, 2021). China targets emissions to peak by 2030 to reach carbon neutrality by 2060. Similarly, the United Kingdom has plans to cut emissions by 68% by 2030 to reach the target by 2050. A landmark decision the United Kingdom has made in shifting away from fossil fuels is to close down all coal power plants by 2024, which means within a decade, the country brings down its reliance on coal for power generation from around one third to zero. It is a major step the United Kingdom has taken toward the transition away from fossil fuels and decarbonization of the power sector in order to eliminate contributions to climate change by 2050 (GUK, 2021).

There have been significant decarbonization efforts on the part of other stakeholders as well including energy, banking, and corporate sector. World's leading corporations are becoming increasingly aware of the threats associated with climate change and the business opportunities in taking action. It will help not only reduce their own emissions but also reduce emissions from their business associates. In this respect, a consortium of several international organizations also involving governments, banks, and insurance groups is working on an "energy transition mechanism", a scheme to buy out coal-fired power plants in Asia in order to shut them down within 15 years (From decarbonization).

3.1 Renewable energy

Renewable energy is the backbone of the energy sector's transition toward zero carbon emissions. Over the last couple of decades, renewable technologies, especially solar photovoltaics (PV) and wind turbines, have made a great progress in terms of technological and economic maturity. The global installed capacity of renewables increased from 2581 GW in 2019 to 2838 GW in 2020, exceeding expansion in the previous year by almost 50%. For several years now in a row, renewable energy is adding more power generation capacity compared to the combined addition by fossil fuels and nuclear power. In the year 2020, for example, renewables contributed to more than 80% of all new power generation capacity added worldwide. The growth of the renewable sector is primarily being propelled by solar and wind power, with the two technologies accounting for 91% of the new renewables added during the year (IRENA, 2021). The annual growth and the cumulative installed capacity of solar PV and wind power over the last 10 years is shown in Figs. 2.2 and 2.3, respectively (IRENA, 2021). Renewable energy is already supplying 26% of the global electricity needs. According to IEA, to achieve the net zero emissions by 2050, almost 90% of global electricity generation is to be supplied from renewables. There was over US$ 303 billion invested in renewable energy projects during the year (REN21, 2020). The upward scale of the renewables developments can be gauged from the fact that China has started developing the first 100 GW phase of massive solar and wind power initiative. The initiative is likely to be expanded to several hundreds of GW in capacity as China aims to develop 1200 GW of renewables by 2030 (Scully, 2021). The renewables growth trends are projected to continue, as the annual capacity addition of solar and wind power is set to grow fourfold between 2020 and 2030 (Gillam and Asplund, 2021).

The success of renewables has been propelled by technological advancements, economy of scale, and supportive polices. The wind power industry is massively benefiting from the scientific and engineering advancements both at the manufacturing and the installation end of wind turbines. Besides improvements in aerodynamic designs, advanced and sophisticated materials are helping develop larger, lighter, and stronger wind blades. These developments have enabled wind turbine grow rapidly in size in recent decades as shown in Fig. 3.4.

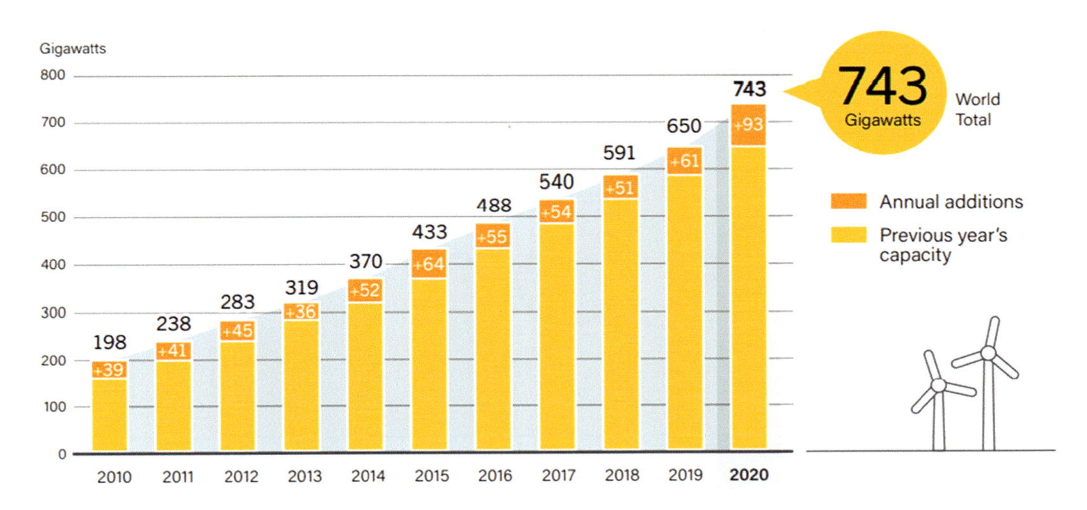

FIGURE 3.2

Growth in solar PV sector between 2010 and 2020.

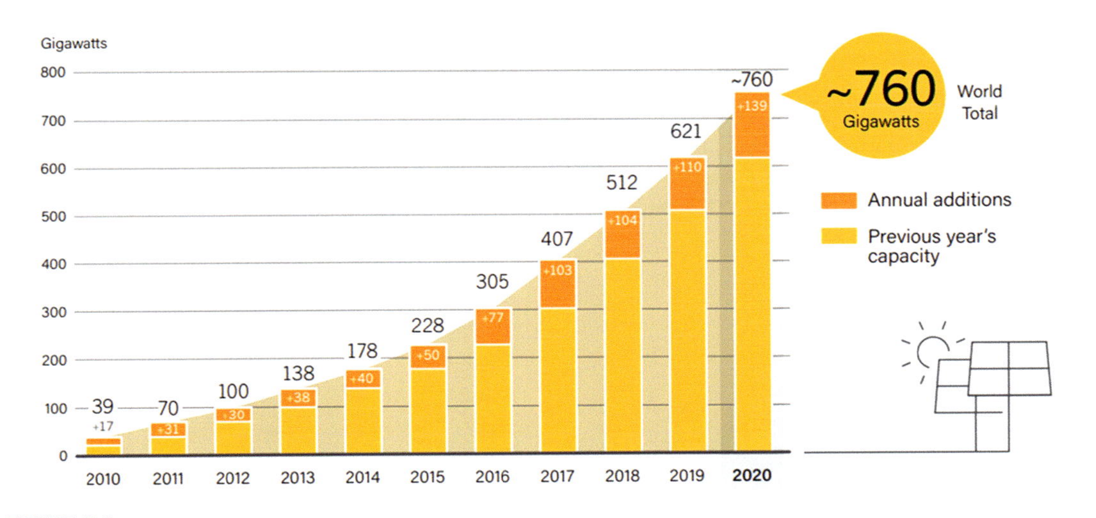

FIGURE 3.3

Growth in wind power between 2010 and 2020.

3.2 Electric vehicles

While some of the decarbonization solutions such as hydrogen, fuel cells, and carbon capture and storage (CCS) are yet to have techno-economic maturity, electric vehicles (EVs) are already making an impact. EVs are environment friendly, require low maintenance due to fewer components, are

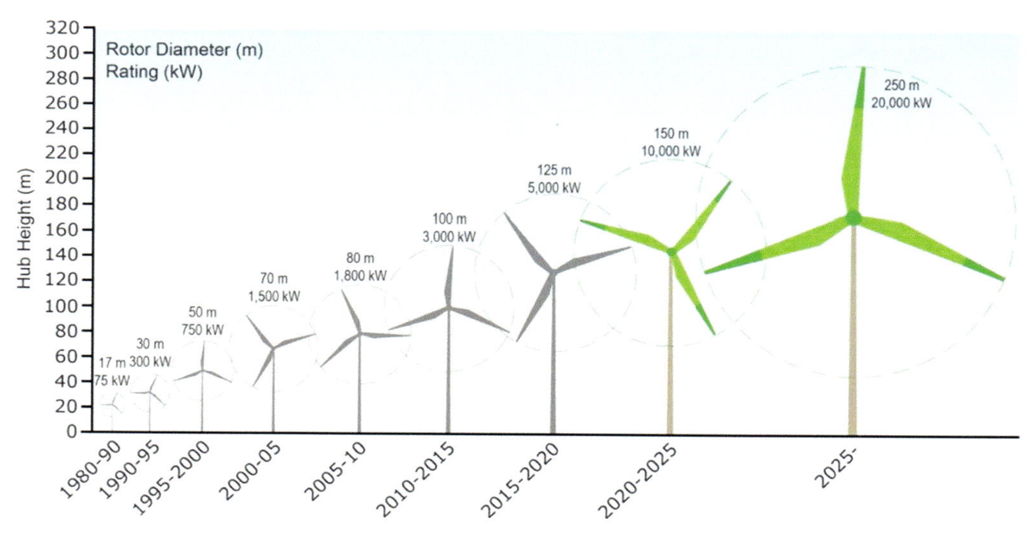

FIGURE 3.4

Capacity growth in wind turbines.

quiet to operate, and offer convenience in urban use. In recent years, EVs have emerged as the new face of transportation. In 2020, the worldwide sale of EVs, for example, increased by 41% despite the COVID-related economic downturn and a drop of 6% in the overall sale of vehicle. During the same year, Europe recorded the registration of new electric cars that increased by 100%, and the number of electric car models available worldwide increased from 260 to 370 (Adler, 2021). The market growth of EVs is being helped by the declining battery prices as shown in Fig. 3.5.

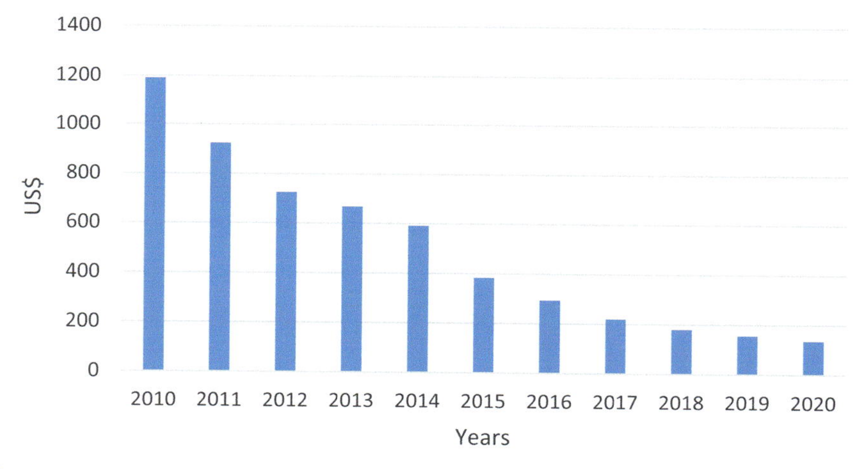

FIGURE 3.5

Declining price trend of lithium batteries.

While electric mobility is also paving its way in the aviation and ship industry, the sale of electric cars is expected to increase from around 3.5 million in 2020 to over 55 million by 2030 (Gillam and Asplund, 2021).

3.3 Energy storage

Renewable energy resources, in general, rely on weather conditions. Intermittency is regarded as one of the biggest drawbacks of renewable energy. Solar radiation, for example, is available only during the daytime. While the daytime availability of solar radiation can be hindered by multiple weather conditions such as rain, snow, fog, and overcast conditions, issues such as dust storms, smog, haze, and smoke from wild fires also affect the intensity of solar radiation. Similarly, the availability of wind is not a constant phenomenon either. Furthermore, even during their spells of availability, solar radiation and wind speed can fluctuate quickly and heavily, accordingly affecting the output from the respective systems. Renewable energy thus needs a backup storage to serve as a reliable source of energy. In terms of energy storage, renewable energy has been helped by the recent developments in the battery technology. The 100 MW lithium battery storage developed by Tesla in Australia in 2017 has been a turning point in the field of large-scale battery storage. The world's largest battery storage system now has a capacity of 300MW/1200 MWh in the United States of America, while the United Kingdom has a 150 MW battery storage system. The world's largest planned battery storage system has a capacity of 1500 MW/6000 MWh to be developed by Vistra in California, USA. The first phase of the lithium-ion battery project with a capacity of 300MW/1,200 MWh started to serve the California unity grid in December 2020. Australia and the United Kingdom are also developing major battery storage projects. Some of these projects include a 1,200 MW project in New South Wales, a 700 MW system by Origin Energy Ltd., a 500 MW system in New South Wales, and a 300 MW facility in Victoria. The United Kingdom has over 1.1 GW of battery storage capacity in operation, while projects of 600 MW of cumulative capacity are under construction. An overview of the leading battery storage projects currently in operation around the world is provided in Table 3.1.

Table 3.1 World's leading battery storage projects.

Project	Capacity (MW)	Technology	Country
Vistra	300	Li-ion	USA
Hornsdale power	150	Li-ion	UK
Stocking pelham	50	Li-ion	Australia
Jardelund	48	Li-ion	Germany
Minamisoma substation	40	Li-ion	Japan
Nishi-Sendai substation	40	Li-ion	Japan
Laurel AES	32	Li-ion	USA
Escondido substation	30	Li-ion	USA
Pomona substation	20	Li-ion	USA

4. Decentralization

Decentralized or distributed generation is the energy generated close to the point of use as shown in Fig. 3.6. Decentralized generation (DG) avoids/minimizes transmission and distribution setup, thus saving on cost and losses. It offers better efficiency, flexibility, and economy as compared to large and centralized generation systems. There are several energy technologies that can be used in DG systems depending on the application and type of project. Based on the type of energy resource, DG technologies can be classified into two categories: renewables-based systems and nonrenewables-based systems. Renewables-based DG systems employ technologies such as solar energy, wind power, hydropower, biomass, and geothermal energy. Some of these technologies can be further classified into different types. Solar technologies, for example, can be categorized into: solar PV, solar thermal power, and solar water heating. Similarly biomass can be used to deliver solid fuels, liquid fuels such as biodiesel and bioethanol, and gaseous fuels. Renewables-based DG systems offer several benefits such as reduced GHG emissions and lower operation and maintenance costs. These systems, however, are typically intermittent and need energy storage to offer reliable solutions. Nonrenewables-based DG technologies are also available in a wide range and may include: internal combustion (IC) engine, combined heat and power, gas turbines, microturbines, Stirling engine and fuel cells. These technologies can use different types of fossil fuels.

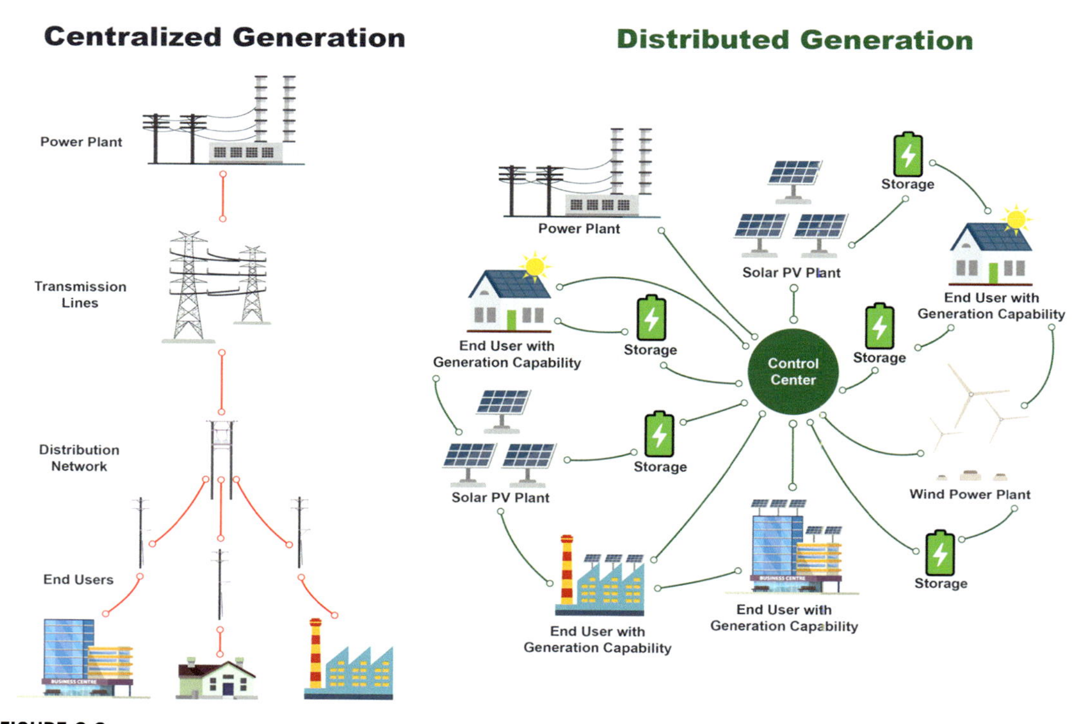

FIGURE 3.6

Overview of central and distributed generation systems.

FIGURE 3.7

Rooftop solar PV.

Renewables like solar and wind power systems are leading the DG landscape. DG is playing an important role in the global electrification efforts and is presenting viable solutions for meeting modern energy needs and enabling the livelihoods of hundreds of millions who still lack access to electricity or clean cooking solutions (Asif, 2019, 2021; Qudratullah and Asif, 2020; Asif and Barua, 2011). Solar PV is one of the most successful DG technology especially at small-scale and off-gird levels. It is estimated that since 2010, over 180 million off-grid solar systems have been installed including 30 million solar home systems. In 2019, the market for off-grid solar systems grew by 13%, with sales totaling 35 million units. Rooftop PV systems, as shown in Fig. 3.7 (NREL), make up 40% of the total PV installations worldwide. Renewable energy also supplied around half of the 19,000 mini-grids installed worldwide by the end of 2019. Efficient biomass systems such as improved cooking stoves and biogas systems are also helping the global efforts toward clean energy access. In 2020, the installed capacity of off-grid DG systems grew by 365 MW to reach 10.6 GW. Solar systems alone added 250 MW to have the total installed capacity of 4.3 GW.

5. Digitalization

Digitalization, also referred as the Fourth Industrial Revolution, is driving the needed fundamental shift in the energy industry, which is also disrupting traditional market players (OWD). Digitalization is a broad term in the context of energy sector. An important dimension of digitalization is the collection and analysis of energy data to optimize energy demand and supply to achieve system efficiency and cost effectiveness. While decarbonization, decentralization, and decreasing use of energy are transforming the energy sector, digitalization—through proliferation of sensors, computing, communication, and predictive and control techniques—is also set to change the way energy services are realized and delivered. This is accomplished through a range of established and emerging technologies, above all artificial intelligence (AI). Digitalization in the perspective of business opportunities created in the energy sector can be regarded as the use of digital technologies to change a business process and enhance efficiency and revenue; it is the process of moving to a digital business.

However it is defined, digitalization is having a profound impact on the global energy scenario. While digitalization is leading to new business models, it is also disrupting existing models of generation, consumption, markets, and businesses and employment, potentially pushing some of the established ones on their way out (EIA, 2019). Digitalization of energy sector employs technologies such as AI, machine learning, big data and data analytics, Internet of Things, cloud computing, blockchain, and robotics and automation. These technologies are at various degrees of techno-economic maturity for their application in the energy sector. Digitalization is revolutionizing the energy sector by improving the productivity, safety, accessibility, and overall sustainability of energy systems. New, smarter ways of modeling, monitoring, analyzing, and forecasting energy production and consumption are helping the sustainable energy transition. With the range of advantages it offers, digitalization is also posing several challenges. Most importantly, the digital transformation heavily relies on large datasets, handling of which is increasingly exposing utilities and energy industry to cyber security risks.

6. Decreasing use of energy

The demand for energy is on a rise across the world, and it is estimated that between 2018 and 2050, global energy requirement will increase by 50%. A one-dimensional approach, of matching the growing energy demands with corresponding capacity addition, is not a sustainable solution, especially when the planet is already overshooting its biocapacity by almost 70%. Any sustainable way forward to satisfy the global energy requirements has to begin with decreasing the use of energy through energy efficiency measures. Energy efficiency is regarded as a better solution to address energy shortages than adding new capacity. A negawatt—a watt of energy *not* used through energy efficiency measures—is considered to be the cheapest watt of energy. To industrial and commercial entities, energy efficiency delivers economic and environmental gains, besides offering competitive edge.

The use of energy can be reduced across all major sectors including buildings, industry, and transport. Buildings account for over one third of the global energy consumption. Energy use in buildings can be reduced through a range of energy efficiency measures. Energy efficient solutions for buildings can be broadly classified as active and passive energy saving measures. The choice of energy efficiency solutions depends on factors such as nature of facility, site condition and local climate, desired levels of comfort and improvement, and financial situation. Through energy efficiency measures, energy demand in existing as well as new buildings can be reduced by 30%–80% (Asif et al., 2017; UNEP, 2009; Asif, 2020). Energy efficiency in the transport sector can be improved through measures such as improving tyre efficiency, incorporating fuel economy standards, and ecodriving (Kojima and Ryan, 2010). Digital technologies can also help save on fuel across the road, air, and sea transportation through optimization of route. The industrial sector also offers a significant potential for energy efficiency especially in the energy-intensive industry. Improvement in energy efficiency enables industrial entities to enhance their productivity and competitiveness, besides contributing toward addressing energy and environmental problems locally, national, and globally. The energy efficiency drive in the industrial sector is also being helped by the digital energy management technologies. It is estimated that with the help of proven and commercially viable technologies, energy use in the manufacturing industry can be reduced by 18%–26% (Fawkes et al., 2016). It is estimated that global economy could increase by $18 trillion by 2035 if energy efficiency is adopted as the "first choice" for new energy supplies, which would also achieve the emission reductions required to limit global warming to 2°C (UNEP). Besides enabling economic growth and improving energy security, energy efficiency can also play a vital role in the fight against climate change as it can deliver more than 40% of the reduction in energy-related GHG emissions over the next 20 years (IEA, 2020).

7. Conclusions

The global energy scenario is experiencing an unprecedented transition, which is going to have a profound impact across the entire energy value chain. The 21st century energy transition is fundamentally a sustainability-driven energy pathway. In the backdrop of the fight against climate change, the main focus of the energy transition is on decarbonization, by shifting away from fossil fuel–based energy systems. Energy transition is being perceived as a pathway toward transformation of the global energy sector from fossil-based to zero-carbon by the second half of this century. In the wake of the Paris Agreement, several major economies and economic blocks—including the United States of America, the United Kingdom, and the EU—have committed to net zero carbon emissions by 2050, while China and the Kingdom of Saudi Arabia have targeted it for 2060. Holistically, however, the ongoing energy transition is not just about going low carbon or a shift away from fossil fuels. It is rather much more vibrant and impactful, thanks to the enormous changes and developments on the fronts of energy resources and their consumption, technological advancements, socioeconomic and political response, and evolving policy landscape. This energy transition has four main and closely linked dimensions: decarbonization, decentralization, digitalization, and decreasing use of energy. Decarbonization of the energy sector is the most important dimension of the energy transition. Reduction in CO_2 and other GHG emissions is fundamental to the fight against climate change. Energy sector can be decarbonized through a range of technologies and solutions including: renewable energy, EVs, hydrogen and fuel cells, CCS, and phasing out of fossil fuels. Replacement of fossil fuels with renewable energy is the most critical part of the decarbonization drive. Renewable energy is already supplying 26% of the global electricity needs. To achieve the net zero emissions by 2050, almost 90% of global electricity generation is to be supplied from renewables. Energy sector can be decarbonized through a range of technologies and solutions including: renewable energy, EVs, hydrogen and fuel cells, CCS, and phasing out of fossil fuels. Renewable energy, having accounted for over 80% of the worldwide newly added power generation capacity in 2020, has already become an important stakeholder in the global energy sector.

References

Adler, K., May 3, 2021. Global Electric Vehicle Sales Grew 41% in 2020, More Growth Coming through Decade. IEA, HIS Markit. https://ihsmarkit.com/research-analysis/global-electric-vehicle-sales-grew-41-in-2020-more-growth-comi.html.

Asif, M., 2019. Energy and Environmental Security, Handbook of Environmental Management. Taylor & Francis.

Asif, M., 2020. Role of energy conservation and management in the 4D sustainable energy transition. Sustainability 12, 10006. https://doi:10.3390/su122310006.

Asif, M., 2021a. Energy and Environmental Security in Developing Countries. Springer, ISBN 978-3-030-63653-1.

Asif, M., 2021b. Energy and Environmental Outlook for South Asia. CRC Press, USA, ISBN 978-0-367-67343-7.
Asif, M., Barua, D., 2011. Salient features of the grameen shakti renewable energy program. Renewable & Sustainable Energy Reviews 15 (9), 5063–5067.
Asif, M., Dehwah, A., Ashraf, F., Khan, H., Shaukat, M., Hassan, M., 2017. Life cycle assessment of a three-bedroom house in Saudi arabia. Environments 4, 52.
DOE, April 19 , 2021. How We're Moving to Net-Zero by 2050. US Department of Energy. How We're Moving to Net-Zero by 2050 | Department of Energy.
EIA, September 24 , 2019. EIA Projects Nearly 50% Increase in World Energy Usage by 2050, Led by Growth in Asia. Today in Energy. EIA projects nearly 50% increase in world energy usage by 2050, led by growth in Asia - Today in Energy - U.S. Energy Information Administration (EIA).
Fawkes, S., Oung, K., Thorpe, D., 2016. Best Practices and Case Studies for Industrial Energy Efficiency Improvement – an Introduction for Policy Makers. UNEP DTU Partnership, Copenhagen.
From Decarbonization Chapter.
Gillam, E., Asplund, R., August 2021. Will Solar Take the Throne. Invesco.
GUK, June 30 , 2021. End to Coal Power Brought Forward to October 2024. Government of UK, Press Release. End to coal power brought forward to October 2024 - GOV.UK. www.gov.uk.
IEA, December 2020. Energy Efficiency 2020. Final Report. International energy Agency, Paris.
IEA, May 2021. Net Zero by 2050: A Roadmap for the Global Energy Sector, Flagship Report. International Energy Agency.
IRENA, Energy Transition, International Renewable Energy Agency, irena.org).
IRENA, April 5, 2021. World Adds Record New Renewable Capacity in 2020. Press Release. International Renewable Energy Agency. World Adds Record New Renewable Energy Capacity in 2020. irena.org.
Kojima, K., Ryan, L., September 2010. Transport Energy Efficiency, Energy Efficiency Series. International Energy Agency.
NREL, Rooftop Solar PV, Photo by Dennis Schroeder, National Renewable Energy Lab.
OWD, Fossil fuel consumption per capita: World, Our World in Data, (Fossil fuel consumption per capita, World (ourworldindata.org).
Qudratullah, H., Asif, M., 2020. Dynamics of Energy, Environment and Economy: A Sustainability Perspective. Springer, ISBN 978-3-030-43578-3.
REN21, 2020. Renewables 2020 global status report. Renewable Energy Network 182–193.
Scully, J., October 12, 2021. China Signals Construction Start of 100GW, First Phase of Desert Renewables Rollout. PV-Tech. China signals construction start of 100GW, first phase of desert renewables rollout - PV Tech. pv-tech.org.
UNEP, Energy efficiency: the game changer, United Nations Environment, https://www.unep.org/explore-topics/energy/what-we-do/energy-efficiency.
UNEP, 2009. Buildings and Climate Change: Summary for Decision Makers. United Nations Environment Programme, Paris, France.

CHAPTER

4

Energy security: role of renewable and low-carbon technologies

Nofri Yenita Dahlan[1,2], Nurfadzilah Ahmad[2], Nur Iqtiyani Ilham[3] and Siti Hajar Yusoff[4]

[1]*Solar Research Institute (SRI), Universiti Teknologi MARA, Shah Alam, Selangor, Malaysia;* [2]*School of Electrical Engineering, College of Engineering, Universiti Teknologi MARA Shah Alam, Selangor, Malaysia;* [3]*School of Electrical Engineering, College of Engineering, Universiti Teknologi MARA, Masai, Johor, Malaysia;* [4]*Electrical and Computer Engineering (ECE), Kulliyyah of Engineering, International Islamic University Malaysia, Gombak, Selangor, Malaysia*

Chapter outline

1. Introduction

Recently, the world's economy has sparked a remarkable rapid growth. The inflation of the population growth together with high interest in transportation has significantly increased the energy demand. This increase in energy means a rise in fossil fuel consumption such as coal, oil, and natural gas that eventually widespread greenhouse gas (GHG) emissions is touted by carbon dioxide (CO_2). A report produced in 2020 stated that electricity and heat production contribute 35.8% of 38.0 $GtCO_2$ total global emissions (Olivier and Peters, 2020). This is closely followed by manufacturing industries and road transport whereby each represents 16.7% and 15.9%. According to the country, China notably

Handbook of Energy and Environmental Security. https://doi.org/10.1016/B978-0-12-824084-7.00015-1
Copyright © 2022 Elsevier Inc. All rights reserved.

emitted the most, accounting for 26.1% of the global CO_2 emissions, followed by the United States at 12.67% (World Resources Institute, 2020).

The Intergovernmental Panel on Climate Change (IPCC) 2019 report has demonstrated that global warming is mainly due to the increase in GHG concentrations (I. P. on C. C. (IPCC), 2019). Subsequently, these lead to the fundamentals of developing new energy using low-carbon technology (Lin, 2011). In recent years, low-carbon technology has been the frequently discussed subject that acts as a key role in the global economic transformation to reduce the worldwide CO_2 level. Low-carbon technology is considered as a technology that helps to reduce GHG emissions, preventing global warming, hence adapting to a low-carbon economy (Lv and Qin, 2016). A low-carbon economy is translated as an economy that is based on low energy consumption with minimal carbon emissions into the atmosphere (He, 2009).

Low-carbon technology is classified into five categories: carbon reduction technology, carbon-free technology, carbon removal technology, carbon management technology, and resource conservation and recycling technology (Xing et al., 2011). Carbon reduction technology refers to the efficient energy use, widely known as energy efficiency (EE) by utilizing energy-saving technology such as a smart meter or energy-efficient light bulb. Meanwhile, carbon-free technology also known as renewable energy (RE) comes from natural resources and continually replenish solar, wind, hydro, or nuclear power to produce electricity. Carbon removal technology is an approach to promote photosynthesis in removing CO_2 such as farming and forestry. Other approaches such as bioenergy with carbon capture and storage and direct air capture are other alternatives to combat climate change. Meanwhile, carbon management technology encourages collaboration between the government and companies to work together in promoting the development of a low-carbon economy. This involves formulating assessment criteria, energy-efficient standards, and carbon emission audit system (Niu, 2011). Resource conservation and recycling technology help to reduce the generated solid waste by converting it into new materials and objects such as garbage power generation technology, recycling waste cement road construction and old building materials (Tseng et al., 2018).

Although there is a wide range of possibilities to generate economic and social development benefits resulting from climate change mitigation, its implementation poses challenges to developing countries. Due to the distinct political, economic, and social settings, the barriers to low-carbon growth in developing countries are quite different from those in developed countries. Apart from the locations, there are similar barriers existing to those in the developed countries such as high upfront investment costs, lack of information, and concerns about discontinuities in production. Furthermore, for access to public–private collaboration and global cooperation, the countries depend on transformative steps such as new policies and market models. The COVID-19 crisis has significantly affected the development of EE globally. However, RE has seen a continuous growth, where many RE plants were developed amidst economic crisis across the globe.

Despite the barriers, countries around the world are progressively focusing on RE resources in their energy transition such as wind, hydropower, solar, and biomass to meet global climate goals and to close energy gaps caused by the phase-out of nuclear and fossil energy output (International Energy Agency (IEA), 2020). The costs of renewable electricity generation continue to fall significantly in 2020 specifically due to the innovation in China (International Energy Agency (IEA), 2020f,g). The innovation includes the development of RE and electric vehicles (EVs) through the application of smart charging, battery recycling, and development of an open power market. China, on the other hand, is not alone in its energy innovations. Most countries, especially in North America, Europe, as well as India, are developing clean energy innovation hubs (IEA, 2020a). The resulting global surplus in RE generation, combined with advancements in EE technology and effective policy, paves the way for low-carbon economy to develop. In this chapter, the development status of RE and EE around the world will be first visited. Next, the role of RE and low-carbon technology in shaping a sustainable power generation mix, improving electrification, and empowering demand side participation will be discussed. Finally, the future roles of RE in enhancing energy security in a decentralized electricity market through peer-to-peer (P2P) energy trading, virtual power plant (VPP), and carbon pricing scheme will be explored.

2. Overview of global low-carbon technology status

In 2019, around 77 countries, 10 regions, and more than a 100 cities have announced their pledge to achieve net-zero carbon emissions by 2050 (REN21, 2020). The European Commission has also proposed a European Green Deal roadmap to establish a carbon-neutral continent by 2050. In 2019, the zero-carbon buildings initiatives were launched by the United Nations Climate Action Summit, aiming to expand the decarbonization roadmaps for buildings. On top of this, more than half of the countries in the world have submitted Nationally Determined Contributions under the Paris Agreement, concerning reducing GHG emissions (world resource institute (WRI), 2019). These joint efforts have remarkably changed the development and penetration status of renewable and low-carbon technologies globally.

2.1 Renewable energy technologies: global trends

In the last decade, the development of RE has been the world's focus due to the depletion and increasing cost of fossil fuels. It was reported by International Energy Agency (IEA) that RE is expected to lead the worldwide electricity sector by 2025 due to economic stimulus focusing on clean energy by global countries (IEA, 2020b). For the last five years, RE has grown three times faster than nuclear and fossil fuels (IEA, 2020c). It was also mentioned in the IEA report; the majority of RE share in global electricity production for the year 2019 came from nonrenewable electricity sources which contributes up to 72.7%. Whereas only 27.3% is from renewable electricity, where 15.9% of this renewable electricity comes from hydropower, 5.9% is generated from wind power, and 2.8% is from solar photovoltaic (PV). Fig. 4.1 illustrates the electricity production per technology with projections for the year 2030 (Ioannis et al., 2020). This projection is based from the European Commission's long-term strategy baseline scenario. It is shown that wind and solar power sector is estimated to generate 51% of electricity followed by nuclear energy, predicted to generate 28% of electricity. By 2030, it is forecasted this RE sector will replace the coal sector and reduce the coal output to 8%.

This statement is also supported by a report published in IEA (Etoro, 2021), where it is estimated that by the year 2025, power generated by RE such as wind and solar PV will surpass the power generated by burning coal by 15.1%. Around this time, it is expected for RE to supply one-third of the world's electricity (IEA, 2020a). This global trend in achieving RE generation 2025 target is also supported by the International Renewable Energy Agency's (IRENA's) Global Renewable Energy Roadmap 2030 policy complied by 87 countries. It is projected that 721 gigawatts of new power generation in wind, solar, and other RE technologies to emerge over the next few years, according to BloombergNEF.

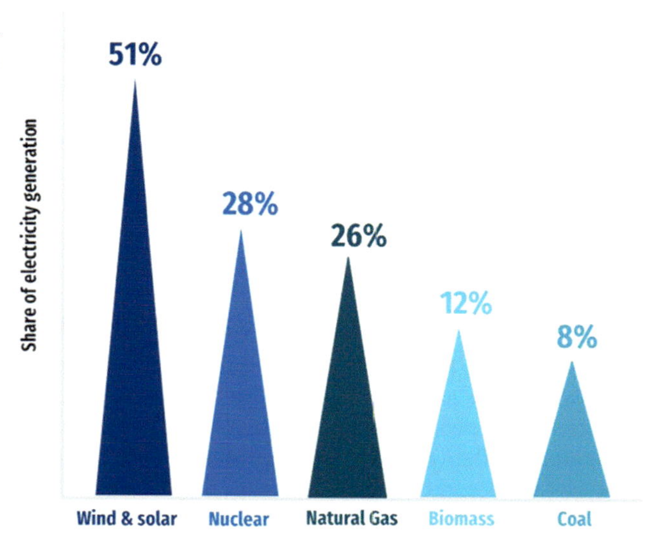

FIGURE 4.1

RE share of global electricity production with projections for the year 2030 (Ioannis et al., 2020).

Fig. 4.2 shows the RE partnership triangle which consists of the government, local community, and private sector. In general, the role of the government is to encourage and promote new development of RE by providing funds, subsidies, and policy support. In 2019, based on the statistic gathered by Statista, China has invested 83.4 billion USD into RE development. Public acceptance is one of the important aspects in determining the success of RE implementation (Kim et al., 2020). Public awareness and trust in the government's decision are the essential elements in determining the RE rate

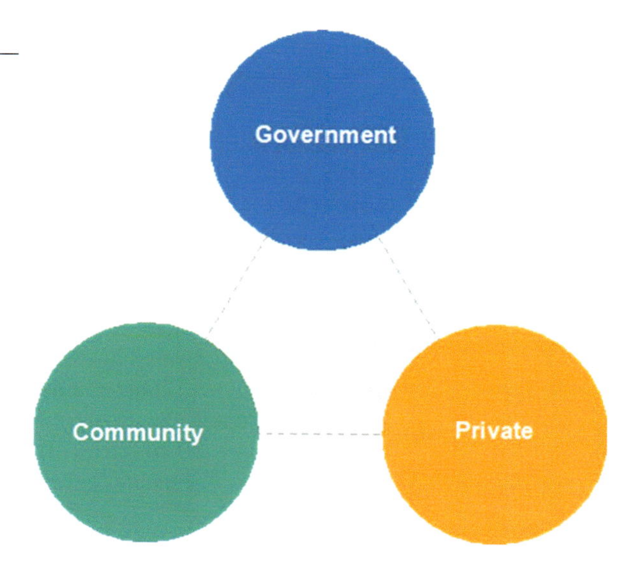

FIGURE 4.2

RE main stakeholder's relationship triangle.

of growth. Besides, private sector's involvement in achieving low-carbon power generation target has significantly accelerated RE growth. A report from Frankfurt School-UNEP Center has disclosed an increasing number of corporations registered to be under the RE100 group. RE100 is a global organization that has set its commitment to source its electricity from 100% RE. It is also disclosed that by end of January 2020, around 220 companies have successfully become part of RE100 (B. UNEP Frankfurt School-UNEP Centre). Corporations from all over the world are making a significant contribution to green energy. This effort is assisted with the introduction to Renewable Energy Certificates (RECs) and Power Purchase Agreement (PPA) (F. S.-U. collaborating Centre, 2020). PPA is a financial agreement between the green energy producer and buyer that allows the buyer to purchase energy from RE. While RECs represents the electricity generated from RE sources.

2.2 Energy efficiency technology global developments

EE plays an important role in clean energy transitions to achieve global climate and sustainability goals. A statement from the United Nations Economic Commission for Europe (UNECE) highlights that EE is the most cost-effective option to meet the growing energy demand in most countries as it contributes to energy security, better environment, quality of life, and economic well-being (UNECE, 2017). Meanwhile, the United Nations' Sustainable Development Goals target 7.3 aims to "double the global rate of improvement in EE by 2030" (UNECE, 2017).

Another supporting statement from UNECE on the global EE technology research, development, and demonstration (RD&D) expenditure has stated a continuous increment in global EE technology RD&D from the year 2017 until 2019. Between 2018 and 2019, a significant growth of 12% from USD 4 billion to USD 4.4 billion has been recorded for EE RD&D expenditure.

It is reported in International Energy Agency (2019a) that national financial incentives of USD 120 billion was invested to support the EE policy governed by 17 countries. These incentives vary from finance loan, grant/subsidy, guarantee, tax relief, direct investment, and equity finance. Apart from these incentives, governments also promote private venture capital funding for startups to develop emerging EE technologies. Majority of the investment was allocated to building technologies which cover the smart building devices, heating and cooling, building energy management system (BEMS), and building envelopes. There was a drop of USD 0.5 billion in investment between 2018 and 2019 in the smart building and other EE sectors observed in the UNECE report (UNECE, 2017). However, other sectors such as industry, BEMS, heating and cooling have managed to maintain their venture capital. The year 2019 has marked new developments of innovative cooling technologies including converting waste heat to power refrigeration and air conditioning, solar storage cooling technologies, and intelligent devices to improve EE for residential air conditioners (International Energy Agency(IEA), 2019).

Another report by IEA on "Renewables 2020-Analysis and forecast to 2025" has disclosed between the year 2010 and 2019 the EE investment has been focused on buildings, transport, and industry. Most of the EE investments are funnelled to buildings which mark 58% of the total global EE investment for the year 2018 followed by 26% for transport, and the other 16% investment is for the energy intensive industry sector (International Energy Agency (IEA), 2020f). The main investors in global EE are the United States of America, the European Union (EU), and China. Based on the report by IEA Sustainable Development Scenario, the year 2018 has seen deployment of EVs, rail and lighting technologies. It is estimated that around USD 250 billion was spent globally in EE across the buildings, transport, and industry for the year 2019 (International Energy Agency(IEA), 2019).

In recent years, digitalization technologies have sparked high potential in improving EE. These digitalization technologies are currently applied in smart BEMS where the system combines the data obtained from various wireless sensors and individual devices. These data are then analyzed using artificial intelligent algorithms to facilitate the consumers' need.

2.3 Impact of COVID-19 pandemic on renewable energy and energy efficiency

The global pandemic outbreak caused by coronavirus 2019 (COVID-19) has caused people to live in limited social freedom and mobility. The closure of commercial buildings, schools, and factories has resulted in a significant drop in energy demand in commercial buildings (Eroğlu, 2020). Due to social distancing and teleworking, electricity usage in residential buildings have grown by 20%–30% (IEA, 2020d). This mobility restriction has disturbed the supply chains and temporarily delayed RE construction projects, equipment supplies, and policy implementation (IEA, 2020a). In the year 2020, the global economic crisis has delayed investments in households' efficienct technologies business and also triggered dramatic changes in markets behaviour which add to uncertainty in EE progress (IEA, 2020d).

Despite the mentioned delays due to COVID-19, RE auctions are breaking new records. In this renewable electricity auctions, governments from various countries call for tenders to install a certain capacity of the RE plant. It is reported in IEA report entitled "Renewables 2020 Analysis and forecast to 2025," there is an increase in the number of global renewable electricity auctions in the first half of 2020. During this term, China has auctioned almost 25 gigawatts renewable electricity followed by India with 12 gigawatts. In addition, 13 countries are awarded almost 50 gigawatts of new renewable capacity starting from 2021 to 2024. It is also recorded in IEA report; in the first half of 2020, a total of 42 gigawatts of solar PV global capacity have been awarded followed by 4 gigawatts onshore wind capacity (International Energy Agency (IEA), 2020f). This surged rate has testified regardless of the pandemic issues; the growth rate of RE and EE is stimulated with continuous financial and policy support from the government. With this increasing incentive from worldwide countries on EE, it is estimated to create an equivalent of 1.8 million full-time jobs between 2021 and 2023 (IEA, 2020d).

3. Roles of renewable and energy efficiency technologies for enhancing energy security

According to IRENA, the global roadmap for the energy transformation is driven by the dual critical factors of limiting climate change and improving energy security for sustainable growth (IRENA, 2019). In addition, the ascending growth of energy demand, fluctuations in fossil fuels price, interruptions in energy import and export (Bekhrad et al., 2020), depletion of fossil fuels reserve, and dependency on a single fuel for electricity generation are the major concerns of countries around the world. Energy security is dependent on the ability to secure fuel for generating electricity. The classical definition of energy security is a stable supply of cheap oil due to embargoes and price manipulations by exporters (William Colglazier and David, 1983; Yergin, 1988). In contrast, the modern energy security is more challenging and not limit to the oil supply issues but also entangled with other energy policy problems such as providing equitable access to modern energy and mitigating climate change (William Colglazier and David, 1983). The contemporary concept of energy security was then introduced by Asia Pacific Energy Research Centre (APERC) through its 4As of energy security which are availability, accessibility, affordability, and acceptability. Furthermore, this section will discuss the roles of RE and EE in shaping sustainable electricity generation mix, increasing electrification, and empowering active consumers (prosumers) through demand side management to enhance global energy security.

3.1 Renewable energy paving world sustainable electricity generation mix

Energy mix diversification is an imperative for energy security and sustainability transitions (Ang et al., 2015; Akrofi, 2021). In electricity sector, power generation mix is a combination of various fuels used to generate electricity in each geographic region. In a regulated electricity industry, the generation mix planning problem is a country's choice and involves determination of types, capacity, and construction time of new candidate generation technologies which should be added to the existing system to meet the growing demand. Meanwhile, in a decentralized electricity industry, the generation mix is resulted from investment decisions of generating companies in a competitive electricity market. The power generation mix varies considerably from one country to another, depending on global markets, national policy, local fuel production and demand. IEA (IEA) reports the evolution of the world electricity generation mix from the year 1971to 2018. Being a cheaper yet abundant resource despite higher CO_2 emission, coal has been the dominant fuel since the earlier years. In fact, coal is still the main source of fuel for world electricity generation especially in the developing countries, accounting for more than 38% of the mix in 2018 (IEA).The share of RE and gas increases each year to accommodate the reduction from oil and nuclear. These specifically happened after the global oil crisis in the 1980s that resulted in increased oil price and Fukushima nuclear plant disaster in 2011 that caused some nuclear plants to close, and many countries deferred or abandoned their nuclear power plan. With the growing demand, it is expected that the implementation of RE projects will be intensified and contribute significantly to the overall generation mix. The IRENA's RE map reveals that the share of renewables in power generation mix would rise from 26% in 2018 to 86% in 2050 and would account for 60% of total energy generation in 2050, particularly from solar PV and wind (IRENA, 2019).

Despite challenges from the COVID-19 crisis faced by the energy industry, the share of renewables in the electricity generation mix rose considerably. This is due to the recent installed capacity of wind and solar PV plants worldwide and priority dispatch received by the plants protecting them from the impacts of lower electricity demand (IEA, 2020a). IEA (IEA, 2020e) reports the renewable technologies' capacity addition from 2013 to 2019 and future projections under the main and accelerated case. The report reveals that solar PV has shown a significant increase every year accompanied by wind and is expected to grow steadily for the next five years. The aggressive growth of renewable technologies would be driven by continuous decline in the cost, strong policy support, and preferential access to many grid systems (IEA, 2020a, e).

Many countries around the world are rapidly pushing forward with their own energy transitions. Low-carbon technologies such as RE and EE have become the choice of these countries in shaping their new energy policy. The EU is among the most vulnerable countries due to its high energy import dependency and scarcity in fossil fuel reserves. Germany and Spain are among the leading countries in utilizing the renewable resources. In Germany, the strongest renewable technology growth is wind power, bioenergy use, and recently solar PV (Hinrichs-Rahlwes, 2013). Meanwhile, in Spain, the 2050

RE transition will be deployed in three phases to replace coal in the first phase, nuclear phase-out in the second phase, and natural gas plants in the third phase (García-Gusano and Iribarren, 2018). The success of Spain in RE efficiency lies in solar investments, wind, and hydro. Unlike Germany, Spain does not have significant savings in oil and natural gas, but has savings in coal (Gökgöz and Güvercin, 2018). Different as in the EU case, the energy security concerns in Southeast Asia are caused by the depletion of indigenous fossil fuels and shortage to meet growing demand. In the worst scenario, some net exporting countries of the fossil fuels are now turning into net importers (IRENA, 2018). Consequently, to ensure the security of fuel supply as well as to meet country climate pledge, Southeast Asian countries have also started their energy transformation by deploying low-carbon technologies, although a slow movement was observed (NBR, 2020). However, the rapid deployment of solar PV in Vietnam and geothermal power in Philippines and Indonesia can be the pilot projects and lesson learned for scaling up the RE transition elsewhere in the region. Large hydro comprises the majority share which is over threequarters of the RE generation. Similar with other regions that rely on large hydro, the Southeast Asia also has a declining share trend of hydro generation due to rapid growth of other renewables technologies such as geothermal and solar PV (IRENA, 2018).

3.2 Renewable energy microgrids for enhancing electrification

IEA reported that 1.1 billion people around the world have lack access to electricity (IEA, 2017). The microgrids system is foreseen to play a critical role in supporting centralized grid system for electrification solutions. Microgrid is a decentralized energy system that supply power from interconnected local distributed energy resources (DERs) over low or medium-voltage distribution networks and usually connected to main utility grid but can function independently (Nosratabadi et al., 2017; Burger, 2016), as depicted in Fig. 4.3. DER includes behind-the-meter generation, RE, energy storage system (ESS), controlled loads, inverters, and EVs. The recent increased activity in the microgrid sector serves not only for electrification but also to remove barriers that inhibit deployment of large-scale renewables and to provide ancillary services to the grid for system reliability (Tjäder and Ackeby, 2018). A small-scale DER could be found in many users known as prosumers that installed PV on the rooftop. A prosumer is an individual who produce and consume energy, enabling themselves to participate in energy trading by selling excess solar energy and buying energy from the grid.

However, overdeployment of renewable DER would be threatening the network as the sources are highly unpredictable and intermittent that could disrupt the supply and demand balance, hence security of the power system. For that reason, various hybrid RE solutions with battery storage and demand response (DR) considering energy management and control strategy have been proposed to improve reliability of microgrid. For instance, a study by Kafetzis et al. (2020) developed an energy management strategy for the control of islanded off-grid hybrid energy systems with ESS, diesel, hydrogen, and RE sources such as solar, wind, hydropower, and biomass for industrial and residential loads. Meanwhile, Mohamed et al. (2021) proposed a renewable microgrids considering solar, wind, and storages in the presence of EVs. Mehdi Hakimi et al. (2020) presented a management strategy of cooling system in buildings in smart grid with high shares of RE. The study found that managing the cooling system would simultaneously increase the reliability of the microgrid. The effect of different prosumers' penetration levels on hybrid electricity dispatch in Australian National Electricity Market is analyzed by Verbič et al. (2019) The simulated case revealed that the increase in renewable penetration requires more energy from gas. With increasing of ESS, the prosumers' self-reliance on the

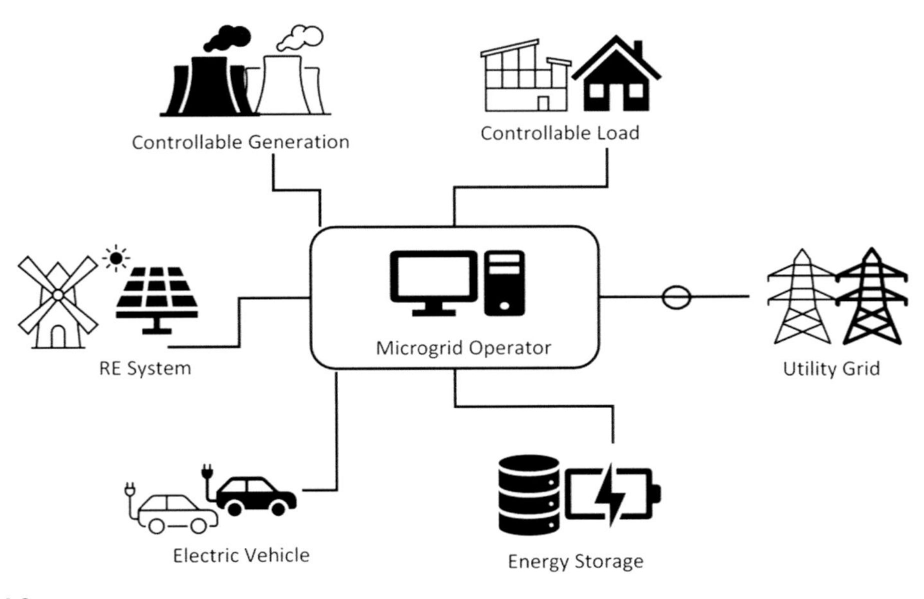

FIGURE 4.3
Visualization of a microgrid system.

grid is reduced. In the absence of storage, the solar PV supplies the load during the day, and the grid supplies the rest. In contrast, with the storage system, RE charges the batteries (mostly by PV during the day and wind during the night) when the grid price is low and discharged in late afternoon to reduce the demand when electricity price is high. In the literature, many authors have also explored the techno-economic performance of various hybrid energy systems as optimal combinations. In general, studies showed that optimized hybrid RE system with battery storage would provide a lower cost of energy while achieving environmental benefits to supply the demand (Kotb et al., 2020; Das et al., 2021; Li et al., 2020).

3.3 Low-carbon technologies through demand side management in buildings

Fig. 4.4 presents different energy savings measures for each level in an energy savings pyramid. The pyramid system suggests that the demand side management strategy in a building should start with energy conservation measures (ECMs) followed by EE and then RE. The ECM can be defined as human's behavior toward the use of energy that results in less energy consumption and usually regards as a low-hanging fruit solution. The awareness is a soft measure but complementary which can potentially lead to more consistent ECM actions. Additionally, an energy audit can be considered as an awareness measure, where it provides technical and detailed feedback on the energy use leading to effective ECM recommendations (Mustapa, 2020). Meanwhile, the ECM involves energy management initiative that can be implemented by introducing good housekeeping measures, regulating the energy use of electrical devices, building automation, control and optimization, DR, zonal heating or

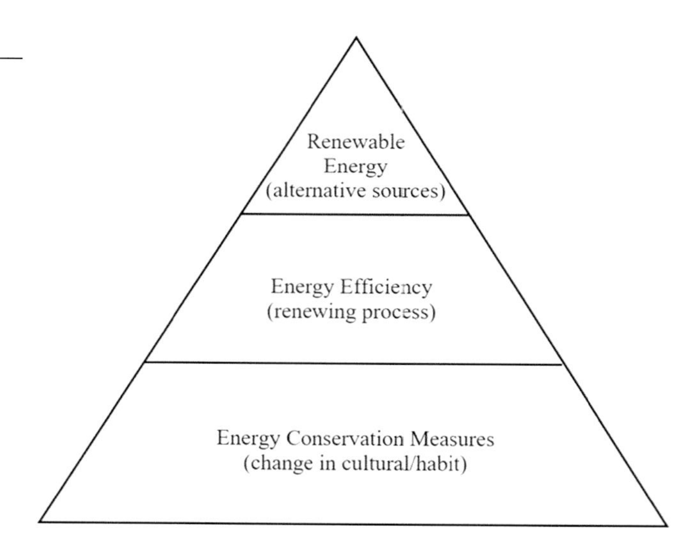

FIGURE 4.4

Energy savings pyramid.

lighting, and fault detection and diagnosis (Mariano-Hernández et al., 2021). A study by Mustapa et al. (2020) evaluates savings from awareness, walkthrough audit, and simple energy management program in a campus building such as switching the lights when not in use and setting air-conditioning temperature to 24°C. The energy savings obtained from the implementation was between 12% and 34%. Meanwhile, Sulaima et al. (2019) investigates simultaneous DR strategies such as valley filling, load clipping, and load shifting for consumers under the Time of Use (ToU) pricing. Findings showed that the commercial and industrial consumers should utilize 20%−50% of load management to acquire optimum benefits of the ToU incentive.

On the other hand, EE is the use of technologies that utilizes less energy to do the same amount of work. The EE requires an inefficient equipment to be replaced with more efficient ones and involves cost to be borne by building owners. Typical EE in buildings include installation of energy-efficient lighting system, lighting controls, variable speed fans and pumps, and efficient heating and air-conditioning system (UNIDO). Building owners may also consider building shell envelope measures such as external walls' insulation, windows' glazing type, air tightness (infiltration), and solar shading (El-Darwish and Gomaa, 2017). Recent emerging business model for EE is Energy Performance Contracting (EPC). EPC is an effective mechanism which can facilitate capital allocation required to retrofit buildings that use the cost savings from reducing energy consumption to repay the capital investment cost. The cost of the ECMs will be borne by Energy Service Companies (ESCOs), and the savings will be shared between the ESCO and building owner on an agreed percentage within a contract period (Aris et al., 2015).

The third level of energy savings measure in buildings is RE. Usually, the building owners employ a small-scale RE such as solar PV panels on their rooftop. The energy generated from the RE is used for own consumption, and the excess can be sold to the grid under incentive schemes such as Net Energy Metering (NEM) or Feed-in Tariff. By optimally managing the production of RE and energy consumption, the prosumer would be able to effectively reduce their energy cost and carbon footprint.

In the NEM scheme, generally the revenue from selling the excess energy will be credited in the bill, hence reducing the energy cost. Additionally, the commercial and industrial consumer can also reduce their maximum demand by consuming the solar energy generated during the daytime. As many countries are now achieving solar grid parity (Martín, 2019), there are emerging business models evolved to drive the expansion of solar rooftop markets. These include solar power purchase agreements (SPPAs) and solar leasing (Tongsopit et al., 2014). The SPPA typically involves customer who is the roof owner, solar developer, and utility company. The contract permits developer to install, own, and operate the solar PV system on the consumer's site and sells the solar electricity to the consumer at a discount, typically 5%–10% lower than the grid electricity tariffs for 20–25 years contract period. Meanwhile, the solar leasing model allows the leasing company (or solar lessor) to enter a leasing contract with the customer (solar lessee) to own, install, and operate a rooftop solar on the customer's roof. The solar lessee will pay for the solar system based on an agreed rate comprising the down payment and monthly instalments. Fig. 4.5 shows the solar PV rooftop installation in seven campuses of Universiti Teknologi MARA (UiTM), Malaysia, with total of 10 MW capacity. The installation is made through SPPA with a private company. It is expected that the solar PV rooftops will generate 66,116 MWh annual energy which provide 45% of UiTM energy consumption and reduce 75,000 tonnes CO_2 annually.

FIGURE 4.5

Solar PV rooftop in seven UiTM campuses, Malaysia, through SPPA

4. Future roles of low-carbon technologies in decentralized energy market

Power networks are enduring with the remarkable energy transition landscape whereby the traditional centralized large-scale generating power plants are now operating in a decentralized manner. Several dominance features are added, enabling bidirectional communication and power flow control. In realizing the actual solutions for low-carbon energy in the future decentralized renewable power landscape, various technological innovations have been embraced. Three enabling technologies toward clean energy and achieving a 1.5°C climate target according to the 2015 Paris Agreement are defined as EVs/electric transportation, ESS, and solar PV (International Energy Agency (IEA), 2017). As these happened, the consumers will become active players in a decentralized energy ecosystem. The deployment of P2P energy trading, VPP, and carbon pricing scheme are currently being the most prominent innovations and actions toward low-carbon solutions to drive the uptake of clean energy technologies (IRENA, 2020). Withal, that requires proper coordination in supplying continuously energy parallel with the economic development and environmental needs along with grid resilience. This is to assure that the power grid will withstand and react promptly to any disruption and changes in the power grid operation system (International Energy Agency (IEA), 2019).

4.1 Peer-topeer energy trading

Due to the high proliferated integration of DER at electricity portfolios, it can potentially create a new approach on electricity being traded at the distribution sides. The P2P energy trading is a business model scheme that emerged to compensate the distribution consumers by allowing them to be prosumers to share the benefits of generating electricity from renewables within the communities bent. Initiated from the P2P economy concept, it can be implemented within a local distribution system (i.e., region, cell, microgrid, and premises) such that the energy trading activities occur among utility providers, prosumers, and consumers (Zhang et al., 2018). P2P energy trading is aiming to transform the grid in becoming more flexible with the usage of RE, provides a balance of energy supply and demand in real time, autonomous in decentralized manner, and empowering the consumers and prosumers. The prosumers in this platform enable them to vigorously manage their DER that includes the interconnection of distributed generation, EVs, ESS, and DR without a mediator, as illustrated in Fig. 4.6 (Liu et al., 2019). Only in some cases, the distribution system operator is responsible for administrating the P2P trading market to ensure the reliable and secure auction operation system happened between prosumers and consumers (Long et al., 2018).

In this energy trading market platform, the bidding process will occur between prosumers and consumers to meet their requirement in terms of electricity price, cost, benefits, energy preference, and demands. In return, the sellers can make profits through higher electricity price sold. At the same time, buyers can save the cost since they are not imposed with other utility charges than the traditional trading mechanism that offered relatively high tariff and low buyback rates through fixed tariffs and ToU tariffs (Park and Yong, 2017). Motivated by this cost-saving factor, various utilities, high technology startup companies, and manufacturers had initiated projects to access further benefits by deploying P2P energy trading. Numbers of trading platform have been developed as the enabling technology for P2P energy trading that applied the centralized and decentralized concepts such as

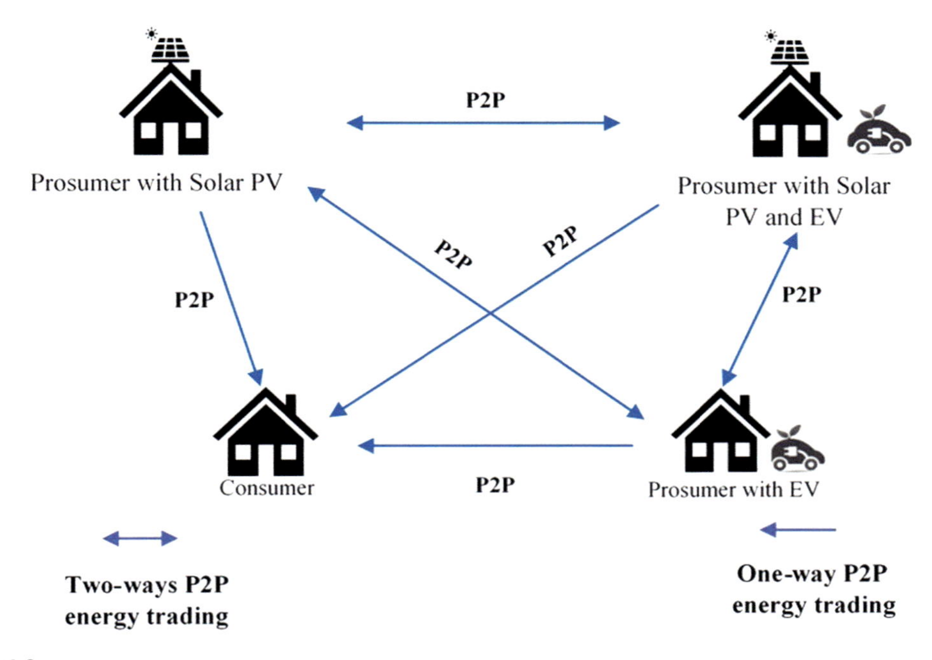

FIGURE 4.6

Structure of P2P energy trading within DER (Liu et al., 2019).

Elecbay and blockchain technology software, respectively (Zhang et al., 2018; Han et al., 2020). Projects such as Brooklyn Microgrid, TransActive grid (the United States), and Centrica plc (the United Kingdom) are examples that deployed blockchain in the trading platform (Mengelkamp et al., 2018).

Through the literature of various P2P projects globally, it was observed that different business models were structure based on market design, trading platform, and information communication technologies (ICTs) infrastructure that provides diverse characteristic for monetizing profits (Zhou et al., 2020). Projects such as Vendebron and sonnenCommunity charged a monthly subscription that amounted to USD 12 and USD 24, respectively, from prosumers and consumers that adopted their trading platform (IRENA, 2020; Park and Yong, 2017). The sonnenCommunity project applied the battery business entity in making profits by binding the agreement for prosumers to buy and use Sonnen's batteries while performing a virtual energy pool to trade the surplus generated renewable electricity instead of feeding it into the grid (sonnenCommunity, 2019). Some projects focused on the ICT infrastructure to support energy sharing as implemented by Smart Watts and PeerEnergyCloud (Germany). Narrowing down to the developing countries, Malaysia and Bangladesh have started in piloting P2P energy trading projects in 2019 and 2015, respectively. Malaysia's pilot project, focusing on highly used commercialized consumers ended in June 2020 and had remarkably gained prosumers profits at about 11% for its excessive solar PV trading that applied blockchain as transaction platform (SEDA, 2019). While in Bangladesh, the SOLshare started its first P2P project in a remote microgrid allowing electricity to be shared within locality to ensure the consistency of power network. This project is envisaged to benefit 2.5 million people in that region by 2023 (SOLshare, 2020).

As P2P energy trading reap exquisite contributions to the power sector, practical implementation is a challenge worldwide. The efficiency and security of virtual layer platform need to be improved regularly since P2P energy trading involved heavy consumption in handling peers and bidirectional energy and communication exchange. Another challenge is applying a conducive regulatory framework and policy to ensure prosumers and consumers trade their RE with relatively cost involved. The EU Directive 2018/2001 enacted in 2018 was the first mandate defined for P2P energy trading using RE under the Clean Energy Package of legislation (European Commission, 2018). While in the United States, only a few projects of P2P energy trading were implemented due to limitation in a regulatory framework that only allowed the trading to be conducted within islanded microgrid (IRENA, 2020). In realizing the P2P concept, the virtual energy trading platform developed should operate efficiently and reliably using distributed ledger technologies. For this purpose, blockchain technology is introduced to assure security, privacy, and prevent information leakage (Shipworth et al., 2019).

Over the decades, world energy consumption has been escalating due to expanding population, rapid revolution in economic growth and industrial development that eventually widespread the emissions of GHG and CO_2. This decentralized renewable market design may lower the environmental impacts on pollution, provide grid resilience, and reduce cost in managing and maintaining electricity infrastructure (IqtiyaniIlham et al., 2017). From the perspective of its business model, the P2P energy trading concept offered a socioeconomic incentive whereby prosumers with surplus energy will sell it among the community in order to generate extra profit while maximizing the energy utilization.

4.2 Virtual power plant

VPP appeared as a new cloud-based model that aggregates a bundle of dispersing connected DERs in the electricity power market operations to improve the grid flexibility, security, and reduce environmental risks. With VPP, better integration of RE is envisaged through supply side flexibility by optimizing power generation sources (i.e., solar PV, wind, combined heat and power and ESS) and demand-side flexibility by aggregating DR (Saboori, 2017). A decentralized energy management system will control all data related to the weather forecast, power supply, DR trends, and wholesale market's electricity price (Maanavi et al., 2019). Fig. 4.7 illustrates an overview of VPP structure that afford several grid services such as forecasting and trading of DER, operating reserve capacity, energy curtailment, and frequency regulation to create a new share of the energy economy. It is a model that creates a new concept that is able to reduce the technical and financial risk contributed by the intermittent nature of RE (Pudjianto et al., 2007).

As many countries in the world have started deploying VPP, the recent global market value for VPP amounted to USD 762 million in 2016, and the number is projected to reach USD 4597 million by 2023 (Fortune Business Insights, 2019). Countries like Australia, the United States, the Netherlands, Germany, Denmark, and the United Kingdom are the most actively involved in VPP activities with an established regulatory framework acknowledging VPP trading. Toward the path in transitioning the world to sustainable energy, in 2018, South Australia (SA) has initiated developing the world's largest VPP by installing 5 kW solar PV rooftop at 50,000 households. Each home was equipped with ESS, smart meters, and a computer system to control the activities between houses and the grid in terms of storage and renewables used (Government of South Australia, 2020). The five-year project positively

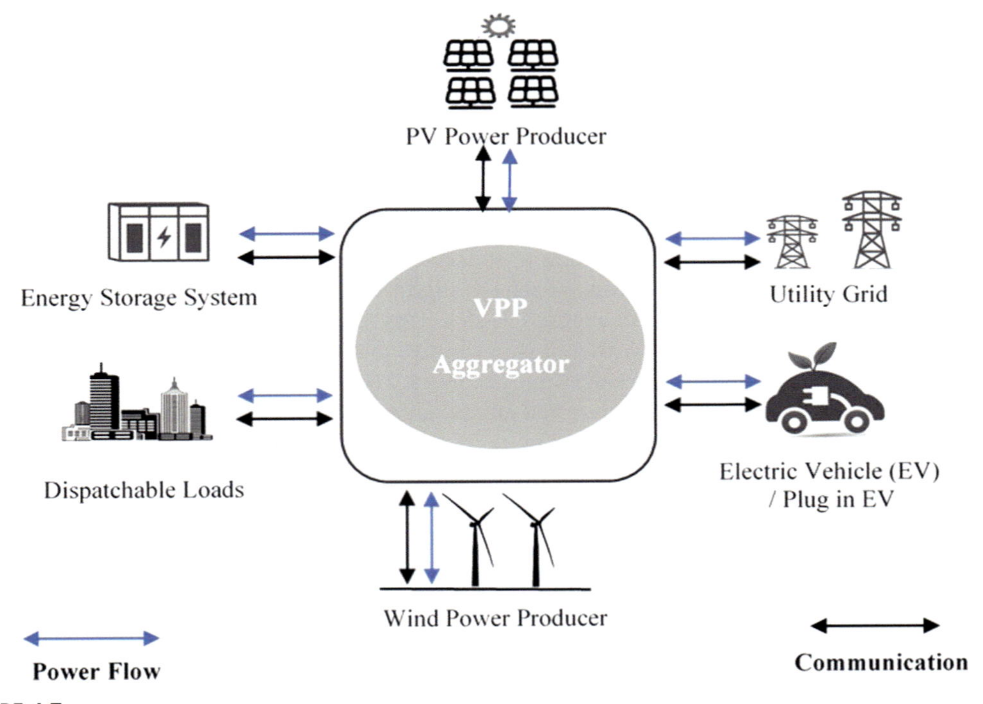

FIGURE 4.7

Overview of VPP structure and its bidirectional flow.

impacted the grid flexibility with RE integration that would add 250 MW during peak capacity to the system. Such stability provided by VPP had support to the significant events during power station trip in Queensland and frequency variation issues happened in 2019. Furthermore, for each 50 MW solar PV capacity brought into the system, the wholesale price is expected to reduce about USD 6/MWh that is equivalent to USD 65 million/year in cost reduction considering engagement from participants fully. Consequently, the VPP consumers can reduce their electricity bills up to 30% (Frontier Economics, 2018).

The VPP model can be designed based on three business cases, namely: (1) forecasting, trading, and energy curtailment with renewables, (2) grid flexibility aggregation from RE, and (3) Demand Response Aggregator. Some projects are also combining all the above-mentioned cases through a large-scale, systematic, and well-coordinated VPP. For instance, the Statkraft's VPP developed in Germany is known as large-scale RE integration into grids that deployed 100% renewable energy system (RES) incorporating about 1300 wind farms, 100 solar power producers, 12 biomass power plant, and eight hydropower plants. The VPP is capable to generate more than 10,000 MW at its peak period comparable with 10 nuclear reactors that provide real-time data management, remotely control of DER, energy forecast, scheduling, and trading based on day-ahead and intraday (Statkraft). Countries in Southeast Asia are also experiencing the rapid growth of economics that motivated them to strengthen their VPP market share anticipation to ensure grid resiliency. Malaysia, for instance, is an exemplary country in that region that has enormous opportunities for VPP deployment considering the high penetration of RE on a large scale and DER as well as its regulatory flexibility to response. Hereof, in 2019, Malaysia's government partnering with South Korea, commenced 30 months copilot project of behind the meter batteries to provide the grid services during peak demand with RES. The VPP pilot project is still operating and is envisaged to contribute to the decisive impact on power sector transformation, especially to consumers and the local network (The New Strait Times, 2019).

Despite various benefits obtained through bundling DER becoming VPP in power system operation and wholesale market, several key challenges to spur the development of VPP can be listed as technology maturity, system integration from various discipline, and regulatory barriers (IqtiyaniIlham et al., 2017; Wang et al., 2019). By changing the electricity generation landscape, it will affect the current network design and control. Withal, it is imperative that the technologies used to enable VPP must be matured and available to assure the supply security. The above-mentioned technologies can be defined as advanced metering infrastructure and advanced forecasting tools that can capture real-time communication data and forecast weather, supply, load, and wholesale prices. Despite all the listed technical key features, government and private sectors' support through financial incentives and flexibility on regulatory framework is essential, enabling aggregators to participate in the wholesale market and contributing to ancillary services. For instance, SA and Grid-Scale Storage Fund supports SA VPP project by giving the financial incentive amounted USD 12 million grants and USD 20 million loans from Renewable Technology Fund (Government of South Australia, 2020). Tesla as the private sector is currently working closely with SA VPP in supplying behind the meter batteries and identifying the participants for this project. Ultimately, without decent incentives and a revised regulatory framework in the energy market, the commercial projects involving VPP will remain a challenge though it has fulfilled the indicators of huge prospects.

4.3 Carbon pricing

The first international carbon market was set up under the United Nation Framework Convention on Climate Change (UNFCCC)—1997 Kyoto Protocol. However, following widespread reports of corruption and abuse of the system, the market collapsed, with the United States refusing to ratify the agreement further. Europe prohibits its member states from buying carbon credit starting in 2012, leaving only a few potential buyers (World Bank Group, 2018). Since then, there has not been a consensus on the best way to implement a carbon market scheme globally. Just during the Conference of Parties (COP) 21—Paris Agreement in 2015, the world unites to fight climate change by limiting global warming to 1.5°C (Intergovernmental Panel on Climate Change (IPCC), 2018). Moreover, unlike the earlier Kyoto Protocol agreement, the 2015 Paris Climate Agreement commits all signatories to impose carbon emission targets. Paving the path for global climate mitigation plan, several ways and policies have been implemented to obtain the same. Initiatives like deploying low-carbon technologies such as EVs, abatement subsidies for fossil fuels, carbon pricing system, and purchasing and manufacturing green energy and EE products are the proactive actions taken by most countries globally to decrease the emissions (Yuyin and Jinxi, 2018).

In the aspect of carbon pricing system, it can be further classified into carbon emission trading system (ETS) and carbon taxes. The ETS is a market-based system that allowed emitters to trade emissions to reduce the countries/businesses' environmental footprint (Narassimhan et al., 2018).

Contrasted from voluntary carbon offsets such as reforestation projects, ETS is a legally binding scheme that caps the total emissions and allows organizations to trade their allocation through cap-and-trade system (Narassimhan et al., 2018). There are several emission trading markets available around the world at both national and regional levels such as European ETS, baseline and credit system (Canada), California cap and trade (the United States), baseline and offset system (Australia), and carbon ETS (China). On the other hand, carbon taxes have been implemented in 25 countries globally, including the EU, Canada, and Japan as one way for climate actions. The carbon tax system defines a tax rate imposed by the respective government to any businesses/organizations that emit the GHG (in per tonne) surpass the allowable rate due to burning fossil fuel activities. As many countries tightened the environmental standards, the total value of global carbon markets grew 20% in 2020, reaching USD 272 billion (Fjellheim, 2021). Currently there are 61 initiatives involving carbon pricing that cover 12 $GtCO_2e$ equal to 22.3% global GHG emissions already in place and scheduled for implementation globally (The World Bank). The number is likely to increase as many countries, cities, and companies worldwide try to meet their ambitious pledge of neutral carbon emissions by 2050 set by UNFCCC.

Every year, in Europe, the government and policymakers will set a cap (carbon credit) to all heavy energy using organizations based on their historical emissions. The allocated cap can be bought and sold on a secondary market to limit the environmental damage. Between 2015 and 2019, about 35% reduction in GHG emissions was recorded through European ETS (European Union, 2020). Apparently, the EU is targeting a 55% net reduction in GHG emission by 2030 in order to achieve climate neutrality by 2050 (European Council, 2019). For the past two decades, the world has started assigning the price in CO_2 emissions. Sweden imposed the highest carbon tax rate globally with USD 126/metric tonne of CO_2 emissions from fossil fuels activities. In turns, the GHG emissions had reduced by 25% since 1995, while at the same time significantly expanding their economy sector by 75% (The Organisation for Economic Co-operation and Development (OECD), 2019).

While in Asia, China has been the world's biggest polluter for more than a decade with three billion tonnes of CO_2 emission on average per year (Ilham et al., 2019). For that reason, the carbon market undoubtedly will be complex. Realizing that fact, the government had started to deploy China ETS (CETS) in handling the climate threat. Nevertheless, the carbon tax is yet to be implemented and still under consideration by the government. The pilot ETS platforms had already started in 2011 covering nine cities and provinces under National Development and Reform Commission. The projects had substantially improved China's green production performance by 10% (Yang et al., 2021). A decade after, in February 2021, the official CETS had rolled out by focusing on eight key emitting sectors starting with the power sector for their first phase implementation. The CETS is targeting the emission reduction at 2200 electric power producers accountable for 30% (3.3 billion tonnes of CO_2) from the total country emission (ICAP, 2018). Paving the carbon neutral path by 2060 will be a challenge for this country, considering the size and complexity of the Chinese national carbon market. For that, the country had built an extensive preparation to improve decarbonization such as constructing more solar park and wind farms, carbon capture technology adoption for heat production and in-house power generation, phasing in the EVs usage by enhancing the batteries technology, and charging infrastructure along with EE innovation in buildings. Apparently, the country had become the world's largest producer of wind and solar energy with 30% of global capacity expansion, and it is leading the deployment of EVs globally reported in 2019 (International Energy Agency (IEA), 2019c, 2020f).

In the past, carbon trading systems have been successful in tackling environmental problems such as reducing acid rain in the United States of America by reducing sulfur dioxide emissions. Compared with carbon taxes and direct regulations, the carbon trading system does not require much interventions from government, thus leaving the businesses/organizations to attain their carbon cap and trade solutions. While for carbon taxes, it requires much involvement from the government/international sectors to ensure the carbon tax is uniform globally. As for now, the European ETS is a valuable model for other countries to emulate. Furthermore, 2021 is remarked with the recent creation of largest carbon trading in China, and the United States react in UNFCCC Paris Agreement, hence remitting the decisive outlook in global carbon pricing to growth. Nevertheless, as long as the cost of emitting GHG is high enough to produce, the carbon pricing mechanism, together with the low-carbon technologies innovations, could be relatively an efficient way to drive global decarbonization for the electricity sector.

5. Conclusion

This chapter discussed the role of RE and low-carbon technology in shaping sustainable energy security. As most countries globally are experiencing an extensive technological breakthrough, fossil fuels still stand as a dominant source for energy propelling. The overdependency on these finite sources has contributed to energy security, sustainable and environmental pollution issues. The continued emissions of CO_2 and GHGs may result in long-term changes worldwide, thus putting people and ecosystems at risk of severe and permanent consequences. However, the world has seen a positive transition of RE growth for the last five years, notably in solar and wind power. RE is expected to lead the worldwide electricity sector by 2025 due to economic stimulus focusing on clean energy to enhance energy security. Despite the unprecedented COVID-19 outbreak, the growth rate of RE and EE projects are stimulated with continuous support from the government, local community, and private sectors.

On a global scale, switching to a low-carbon technology could have substantial benefits for both developed and developing countries in the aspects of environment, society, and economy. The business concept involving DER has inaugurated the electricity wholesale market toward decarbonizing and competitive. Furthermore, the deployment of VPP, P2P energy trading, and carbon pricing are seen as the most prominent innovations and actions to upsurge the energy security and clean energy technologies in global energy transition. Adopting elements such as energy-efficient vehicles, ESS, and DR in a decentralized energy ecosystem could be the prudent path. In molding the future electricity sector, the availability and maturity of technologies and policies imposed in green energy should be appropriately monitored to avoid any delays in progress. The deployment of RE and low-carbon technologies can be an impetus to relook at the electricity industry with the aim of promoting energy security as a long-term efficiency.

References

Akrofi, M.M., 2021. An analysis of energy diversification and transition trends in Africa. International Journal of Energy and Water Resources 5, 1–12.

Ang, B.W., Choong, W.L., Ng, T.S., February 2015. Energy security: definitions, dimensions and indexes. Renewable and Sustainable Energy Reviews 42, 1077–1093.
APERC, "A Quest for Energy Security in the 21st Century: Resources and Constraints.".
Aris, S.M., Dahlan, N.Y., Nawi, M.N.M., Nizam, T.A., Tahir, M.Z., 2015. Quantifying energy savings for retrofit centralized hvac systems at Selangor state secretary complex. Jurnal Teknologi 77 (5), 93–100.
B. UNEP, Frankfurt School-UNEP Centre, "Growth of Corporate Members of RE100.".
Bekhrad, K., Aslani, A., Mazzuca-Sobczuk, T., May 2020. Energy security in Andalusia: the role of renewable energy sources. Case Studies in Chemical and Environmental Engineering 1, 100001.
Burger, A., 2016. Global Microgrid Market to Grow 21%, Exceed $35B by 2020.
Das, B.K., Hassan, R., Tushar, M.S.H.K., Zaman, F., Hasan, M., Das, P., February 2021. Techno-economic and environmental assessment of a hybrid renewable energy system using multi-objective genetic algorithm: a case study for remote Island in Bangladesh. Energy Conversion and Management 230, 113823.
El-Darwish, I., Gomaa, M., December 2017. Retrofitting strategy for building envelopes to achieve energy efficiency. Alexandria Engineering Journal 56 (4), 579–589.
Eroğlu, H., 2020. Effects of Covid-19 outbreak on environment and renewable energy sector. Environment, Development and Sustainability 23 (4), 4782–4790.
Etoro, 2021. Capturing the Renewable Energy Shift.
European Commission, 2018. Directive (EU) 2018/2001 of the European Parliament and of the Council on the Promotion of the Use of Energy from Renewable Sources.
European Council, 2019. European Council Meeting (12 December 2019) – Conclusions, EUCO 29/19.
European Union, 2020. EU Emissions Trading System (EU ETS) [Online]. Available: https://ec.europa.eu/clima/policies/ets_en. (Accessed 16 January 2021).
F. S.-U. collaborating Centre, 2020. Global Trend in Renewable Energy Investment 2020.
Fjellheim, H., 2021. Global Carbon Markets Hit New Highs [Online]. Refinitive Report. Available: https://www.refinitiv.com/perspectives/future-of-investing-trading/global-carbon-markets-hit-new-highs/. (Accessed 16 March 2021).
Fortune Business Insights, 2019. Market Research Report.
Frontier Economics, 2018. South Australia's Virtual Power Plant - Frontier Economics Assessment.
García-Gusano, D., Iribarren, D., October 2018. Prospective energy security scenarios in Spain: the future role of renewable power generation technologies and climate change implications. Renewable Energy 126, 202–209.
Gökgöz, F., Güvercin, M.T., November 2018. Energy security and renewable energy efficiency in EU. Renewable and Sustainable Energy Reviews 96, 226–239.
Government of South Australia, 2020. South Australia's Virtual Power Plant [Online]. Available: https://www.energymining.sa.gov.au/growth_and_low_carbon/virtual_power_plant. (Accessed 14 March 2021).
Han, D., Zhang, C., Ping, J., Yan, Z., 2020. Smart contract architecture for decentralized energy trading and management based on blockchains. Energy 199, 117417.
He, J.K., 2009. Low carbon technology innovation is the key to develop low carbon economy. Greenery 1, 46.
Hinrichs-Rahlwes, R., January 2013. Renewable energy: paving the way towards sustainable energy security. Lessons learnt from Germany. Renewable Energy 49, 10–14.
I. P. on C. C. (IPCC), 2019. Special Report on the Ocean and Cryosphere in a Changing Climate.
ICAP, 2018. Emissions Trading Status Report 2018.
IEA. World Electricity Generation Mix by Fuel, 1971-2018, (Paris).
IEA, 2017. Energy Access Outlook 2017. Paris.
IEA, 2020a. Global Energy Review 2020. Paris.
IEA, 2020b. Renewables 2020 Analysis and Forecast to 2025.
IEA, 2020c. Reductions in Electricity Demand from International Energy Agency (IEA).
IEA, 2020d. Energy Efficiency.

IEA, 2020e. Renewables 2020. Paris.
Ilham, N.I., Hasanuzzaman, M., Mamun, M.A.A., 2019. World Energy Policies. Elsevier Inc.
Intergovernmental Panel on Climate Change (IPCC), 2018. Global Warming of 1.5 °C: An IPCC Special Report on the Impacts of Global Warming of 1.5 °C above Pre-industrial Levels and Related Global Greenhouse Gas Emission Pathways, in the Context of Strengthening the Global Response to the Threat of Climate Chang.
International Energy Agency, 2019a. World Energy Investment 2019.
International Energy Agency (IEA), 2019b. Energy Security [Online]. Available: https://www.iea.org/areas-of-work/energy-security.
International Energy Agency (IEA), 2019c. Global EV Outlook 2019 to Electric Mobility.
International Energy Agency (IEA), 2020f. Renewables 2020-Analysis and Forecast to 2025.
International Energy Agency (IEA), 2020g. Electricity Market Report – December 2020.
Internationl Energy Agency(IEA), 2017. Tracking Clean Energy Progress Tracking Clean Energy Progress.
Internationl Energy Agency(IEA), 2019. Energy Efficiency.
Ioannis, T., Wouter, N., Dalius, T., Pablo, R.C., 2020. Towards Net-Zero Emissions in the EU Energy System by 2050.
IqtiyaniIlham, N., Hasanuzzaman, M., Hosenuzzaman, M., 2017. European smart grid prospects, policies, and challenges. Renewable and Sustainable Energy Reviews 67, 776–790.
IRENA, 2018. Renewable Energy Market Analysis - Southeast Asia.
IRENA, 2019. Global Energy Transformation: A Roadmap to 2050, 2019 edition. Abu Dhabi.
IRENA, 2020. Peer to Peer Electricity Trading.
Kafetzis, A., Ziogou, C., Panopoulos, K.D., Papadopoulou, S., Seferlis, P., Voutetakis, S., December 2020. Energy management strategies based on hybrid automata for islanded microgrids with renewable sources, batteries and hydrogen. Renewable and Sustainable Energy Reviews 134, 110118.
Kim, J., Jeong, D., Choi, D., Park, E., 2020. Exploring public perceptions of renewable energy: evidence from a word network model in social network services. Energy Strategy Reviews 32, 100552.
Kotb, K.M., Elkadeem, M.R., Elmorshedy, M.F., Dán, A., October 2020. Coordinated power management and optimized techno-enviro-economic design of an autonomous hybrid renewable microgrid: a case study in Egypt. Energy Conversion and Management 221, 113185.
Li, J., Liu, P., Li, Z., October 2020. Optimal design and techno-economic analysis of a solar-wind-biomass off-grid hybrid power system for remote rural electrification: a case study of west China. Energy 208, 118387.
Lin, Z.H., 2011. Low carbon technology and its application. Nat Mag 74.
Liu, Y., Wu, L., Li, J., 2019. Peer-to-peer (P2P) electricity trading in distribution systems of the future. Journal of Electronics 32 (4), 2–6.
Long, C., Wu, J., Zhou, Y., Jenkins, N., 2018. Peer-to-peer energy sharing through a two-stage aggregated battery control in a community Microgrid. Applied Energy 226 (June), 261–276.
Lv, J., Qin, S., 2016. On low-carbon technology. Low Carbon Economy 07 (03), 107–115.
Maanavi, M., Najafi, A., Godina, R., Mahmoudian, M., Rodrigues, E.M.G., 2019. Energy management of virtual power plant considering distributed generation sizing and pricing. Applied Sciences 9 (14), 1–19.
Mariano-Hernández, D., Hernández-Callejo, L., Zorita-Lamadrid, A., Duque-Pérez, O., Santos García, F., January 2021. A review of strategies for building energy management system: model predictive control, demand side management, optimization, and fault detect & diagnosis. Journal of Building Engineering 33, 101692. Elsevier Ltd.
Martín, J.R., May 2019. IRENA: Global Solar Months Away from Sweeping Grid Parity. PVTECH.
Mehdi Hakimi, S., Hajizadeh, A., Shafie-khah, M., Catalão, J.P.S., December 2020. Demand response and flexible management to improve microgrids energy efficiency with a high share of renewable resources. Sustainable Energy Technologies and Assessments 42, 100848.
Mengelkamp, E., Gärttner, J., Rock, K., Kessler, S., Orsini, L., Weinhardt, C., 2018. Designing microgrid energy markets: a case study: the Brooklyn Microgrid. Applied Energy 210, 870–880.

Mohamed, M.A., Abdullah, H.M., El-Meligy, M.A., Sharaf, M., Soliman, A.T., Hajjiah, A., July 2021. A novel fuzzy cloud stochastic framework for energy management of renewable microgrids based on maximum deployment of electric vehicles. International Journal of Electrical Power & Energy Systems 129, 106845.
Mustapa, R.F., 2020. Non-Linear Baseline Energy Modelling in Educational Building Using NARX-ANN. Universiti Teknologi MARA.
Mustapa, R.F., Dahlan, N.Y., Yassin, A.I.M., Nordin, A.H.M., May 2020. Quantification of energy savings from an awareness program using NARX-ANN in an educational building. Energy Build 215, 109899.
Narassimhan, E., Gallagher, K.S., Koester, S., Alejo, J.R., 2018. Carbon pricing in practice: a review of existing emissions trading systems. Climate Policy 18 (8), 967–991.
NBR, 2020. Powering Southeast Asia - Meeting the Region's Electricity Needs.
Niu, G.M., 2011. The Core of Developing Low Carbon Economy. Tianjin Daily.
Nosratabadi, S.M., Hooshmand, R.A., Gholipour, E., January 2017. A comprehensive review on microgrid and virtual power plant concepts employed for distributed energy resources scheduling in power systems. Renewable and Sustainable Energy Reviews 67, 341–363. Elsevier Ltd.
Olivier, J.G.J., Peters, J.A.H.W., 2020. Trends in Global CO_2 and Total Greenhouse Gas Emissions 2020 Report Trends in Global CO_2 and Total Greenhouse Gas Emissions: 2020 Report.
Park, C., Yong, T., 2017. Comparative review and discussion on P2P electricity trading. Energy Procedia 128, 3–9.
Pudjianto, D., Ramsay, C., Strbac, G., 2007. Virtual power plant and system integration of distributed energy resources. Renewable Power Generation, IET 1, 10–16.
REN21, 2020. Renewables 2020 Global Status Report.
Saboori, H., December 2017. Virtual Power Plant (VPP), Definition , Concept , Components and Types.
SEDA, 2019. Malaysia's 1st Pilot Run of Peer-To-Peer (P2p) Energy Trading [Online]. Available: http://www.seda.gov.my/2020/11/malaysias-1st-pilot-run-of-peer-to-peer-p2p-energy-trading/. (Accessed 20 December 2020).
Shipworth, D., Burger, C., Weinmann, J., Sioshansi, F., 2019. Peer-to-Peer Trading and Blockchains: Enabling Regional Energy Markets and Platforms for Energy Transactions. Elsevier Inc. no. March 2018.
SOLshare, 2020. Meet SOLshare, Pioneers in Peer-To-Peer Solar Micro-grid Technology [Online]. Available: https://fev.vc/meet-solshare-pioneers-in-peer-to-peer-solar-micro-grid-technology/.
sonnenCommunity, 2019. sonnenCommunity [Online]. Available: https://sonnengroup.com/sonnencommunity/. (Accessed 13 March 2021).
Statkraft, "The Statkraft VPP." [Online]. Available: https://www.statkraft.com/.
Sulaima, M.F., Dahlan, N.Y., Yasin, Z.M., Rosli, M.M., Omar, Z., Hassan, M.Y., August 2019. A review of electricity pricing in peninsular Malaysia: empirical investigation about the appropriateness of Enhanced Time of Use (ETOU)electricity tariff. Renewable and Sustainable Energy Reviews 110, 348–367. Elsevier Ltd.
The New Strait Times, March 2019. TNB and South Korean Partners to Tap Virtual Power Plant Benefits.
The Organisation for Economic Co-operation and Development (OECD), 2019. Taxing Energy Use 2019.
The World Bank, "Carbon Pricing Dashboard." [Online]. Available: https://carbonpricingdashboard.worldbank.org/map_data. [Accessed: 16-Mar-2021].
Tjäder, J., Ackeby, S., 2018. The Role and Interaction of Microgrids and Centralized Grids in Developing Modern Power Systems.
Tongsopit, S., Moungchareon, S., Aksornkij, A., Potisat, T., 2014. Business Models and Financing Options for a Rapid Scale-Up of Rooftop Solar Power Systems in Thailand, Chapter 4 in Financing Renewable Energy Development in East Asia Summit Countries A Primer of Effective Policy Instruments.
Tseng, M.L., Wong, W.P., Soh, K.L., 2018. An overview of the substance of resource, conservation and recycling. Resources, Conservation & Recycling 136, 367–375.
UNECE, 2017. Energy for Sustainable Development in the UNECE Region. Retrieved from. United Nations Economic Commission for Europe.

UNIDO. Sustainable Energy Regulation and Policymaking Training Manual - Module 18: Energy Efficiency in Buildings.

Verbič, G., Mhanna, S., Chapman, A.C., 2019. Energizing demand side participation. In: Pathways to a Smarter Power System. Academic Press, pp. 115–181.

Wang, X., Liu, Z., Zhang, H., Zhao, Y., Shi, J., Ding, H., 2019. A review on virtual power plant concept, application and challenges. In: 2019 IEEE PES Innov. Smart Grid Technol. ISGT 2019, Asia, pp. 4328–4333.

William Colglazier, J.E., David, A.D., 1983. Energy security in the 1980s. Annual Review Energy 8, 415–449.

World Bank Group, 2018. Carbon Markets under the Kyoto Protocol: Lessons Learned for Building an International Carbon Market under the Paris Agreement.

world resource institute (WRI), 2019. Zero Carbon Buildings for All Initiative Launched at UN Climate Action Summit.

World Resources Institute, 2020. World's Top Emitters Interactive Chart I World Resources Institute [Online]. Available: https://www.wri.org/insights/interactive-chart-shows-changes-worlds-top-10-emitters. (Accessed 18 June 2021).

Xing, X., Wang, Y., Wang, J., 2011. The problems and strategies of the low carbon economy development. Energy Procedia 5, 1831–1836.

Yang, L., Li, Y., Liu, H., 2021. Did carbon trade improve green production performance? Evidence from China. Energy Economics 96, 105185.

Yergin, D., 1988. Energy security in the 1990s. Foreign Aff 67 (1), 110–132.

Yuyin, Y., Jinxi, L., 2018. The effect of governmental policies of carbon taxes and energy-saving subsidies on enterprise decisions in a two-echelon supply chain. Journal of Cleaner Production 181, 675–691.

Zhang, C., Wu, J., Zhou, Y., Cheng, M., Long, C., 2018. Peer-to-Peer energy trading in a Microgrid. Applied Energy 220 (March), 1–12.

Zhou, Y., Wu, J., Long, C., Ming, W., 2020. State-of-the-Art analysis and perspectives for peer-to-peer energy trading. Engineering 6 (7), 739–753.

CHAPTER

5 Overcoming the energy security challenges in developing countries

Hirushie Karunathilake[1], Tharindu Prabatha[1,2], Rehan Sadiq[2] and Kasun Hewage[2]
[1]*Department of Mechanical Engineering, University of Moratuwa, Moratuwa, Sri Lanka;* [2]*School of Engineering, University of British Columbia (Okanagan Campus), Kelowna, BC, Canada*

Chapter outline

1. Energy scenario in developing countries

"Ensuring access to affordable, reliable, sustainable, and modern energy for all" is the Goal 07 under the United Nations Sustainable Development Goals (SDGs) (United Nations, 2021). A multitude of factors including population growth, rise in industrial activities, rising expectations on quality of life, and emerging economies have contributed to a significant increase in the global energy demand over the last several decades (Girouard et al., 2011; Sorrell, 2015). Energy has a strong relationship with wealth, economic development, and societal progress, with larger and more affluent societies with complex structures having access to higher energy flows (Sorrell, 2015). Energy prices have a strong relationship with the per capita energy consumption and the gross domestic product (GDP) of a country (Berk and Yetkiner, 2014). Access to cheaper energy generally is a key driving factor in

Handbook of Energy and Environmental Security. https://doi.org/10.1016/B978-0-12-824084-7.00021-7
Copyright © 2022 Elsevier Inc. All rights reserved.

economic growth, ultimately supporting technological progress, better living conditions, and sustainable development (Sorrell, 2015; Ayoo, 2020).

Energy security is a direct policy concern for most countries across the world. In general, energy security is evaluated based on the *four "A's*: *availability, affordability, accessibility*, and *acceptability*" (Cherp and Jewell, 2014; Kruyt et al., 2009). Even though a multitude of definitions have evolved over the years, the concept of energy security has not been strictly defined (Winzer, 2011). This is understandable as the meaning of energy security is very much dependent on people, context, energy systems, and other interrelated policy issues (Cherp and Jewell, 2014). Thus, it is clear that energy security and the pertinent challenges should be analyzed at regional level, considering the macro-environmental factors and stakeholder involvement and priorities.

World Economic Situation and Prospects (WESP), published by the United Nations (UN) Department of Economic and Social Affairs, classifies all countries into three main categories based on the basic economic conditions of countries: namely *developed economies, economies in transition,* and *developing economies* (United Nations, 2020). Other classifications take indicators such as industrial activity and human development index (HDI) for this definition. However, developing countries have a lower per capita gross national income (GNI), lower living standards, and a relatively underdeveloped industrial base compared with "developed economies" in general (Onyiriuba, 2016). These countries have a unique set of challenges related to energy supply and its associated socioeconomic dimensions. As energy is intricately linked with economic and social progress, insufficient access to energy resources, insecure energy supply, extreme dependence on external suppliers, and lack of affordability of energy services can further hinder the growth potential of these countries. While developed countries usually benefit from domestic energy reserves and/or secure energy supplies, high levels of industrial energy use, and the latest innovations in energy technologies, the situation is quite different for the developing world (Ahuja and Tatsutani, 2009). In many parts of the world, universal access to energy is still a challenge; billions of people still do not have access to electricity (The World Bank, 2018). Transitioning to low-carbon clean energy systems while also providing adequate energy access to their populace is a major challenge to developing countries, which often hinders their economic and social development (Ahuja and Tatsutani, 2009).

In developing countries, social well-being is also tied up with a secure and modern energy supply. One example of this is how women in rural areas and "poor" communities in the developing world are relegated to forestry and biomass management out of necessity (Fatona et al., 2013). Women spend a great deal of time on fuel collection and cooking. Due to this, they are deprived of the chance to engage in other income-generating activities, which also curtails a much needed source of income for families affected by poverty (Fatona et al., 2013). With access to a consistently available modern energy supply, traditional energy procurement and related responsibilities that are labor and time intensive can be removed from these women, giving them more opportunities for education, employment, and overall economic independence. Energy security is connected to human security, and ensuring equitable access to energy services is an important goal, which can in turn increase food security, health conditions, educational opportunities, gender equity, and environmental sustainability in vulnerable regions (Kuik et al., 2011). Country-level energy planning efforts and energy policy strategies are thus oriented toward enhancing energy security and energy independence, while also maintaining access to affordable, adequate, and reliable energy supplies (Ayoo, 2020). Yet, these goals can often conflict with each other. Thus, analyzing the complex nexus surrounding energy security concerns and related challenges in developing countries is essential. In doing this, it is useful to first understand the key concepts related to energy supply and demand, global energy markets, and energy crises.

1.1 Current status of supply and demand

While the terms "supply" and "demand" seem simple enough, analyzing them with respect to energy is a complex task. Supply and demand conditions and decisions can be analyzed at various levels, with the involvement of numerous stakeholders. Energy planners and policymakers look at the overall supply and demand from a regional or a national perspective (Bhattacharyya, 2019). Industrial, commercial, and residential end users tend to view energy considering their own individual needs. Energy supply chains extend from supply of energy (via domestic production and imports) and transformation (i.e., refining, processing, and electricity generation) to use (in energy and nonenergy applications Bhattacharyya, 2019). Fig. 5.1 summarizes factors affecting energy supply and demand transitions (Gielen et al., 2019; Girouard et al., 2011; Sorrell, 2015).

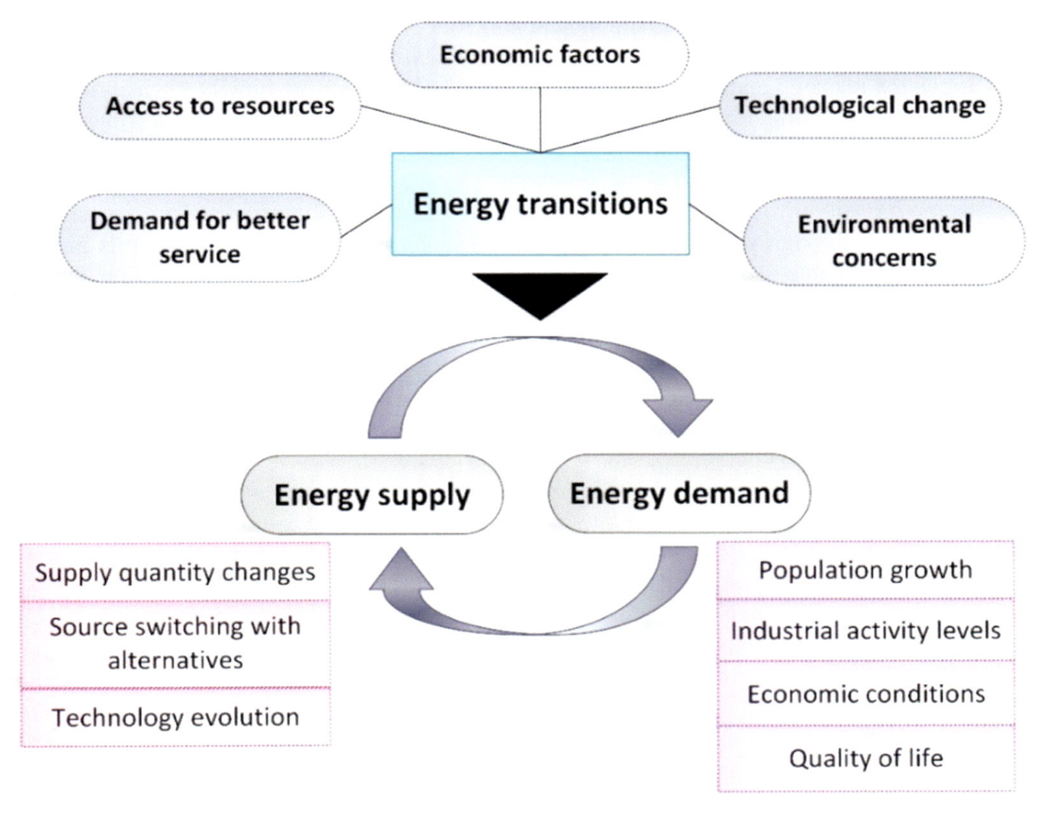

FIGURE 5.1

Energy transitions affecting supply and demand.

At present, the whole world relies greatly on fossil fuel reserves to cater the energy needs. Traditionally, having domestic fossil fuel reserves, be it coal, oil, or gas, has presented an advantageous economic opportunity for a country. Table 5.1 lists the top producers of coal, oil, and gas in the world in 2018, while indicating developing and transitioning economies among them (based on UN WESP classification) (Ritchie, 2017; United Nations, 2020). The countries that contribute over 4% of the total global production in each fuel category are highlighted.

While it appears that there are many "developing" countries among the fossil fuel producers, some interesting dynamics emerge upon further analysis. When economies are classified on the basis of per capita GNI (as per the WESP data), it can be seen that the "poorer" countries falling in low to lower middle-income have limited representation among the top fossil fuel producers (United Nations, 2020). Among the countries that produce over 4% in a fossil fuel category, only India and Indonesia are classified as lower middle-income countries, and no low-income countries are represented. This translates to a realization that the poor and underprivileged populations in these low-income countries are further disadvantaged by the lack of access to a secure domestic energy supply.

The high rate of global fossil fuel consumption means that the proven and economically recoverable fossil fuel reserves will run out at some point in time, even though new reserves are still being identified all over the world and the technological potential of extraction is increasing (Ritchie, 2017). However, the use of fossil fuels may become constrained *before* the world runs out of reserves due to a wholly unrelated reason, i.e., the environment (Ritchie, 2017). With the rapidly intensifying climate concerns, policies are being implemented across the board to mitigate the carbon emissions and other environmental damages associated with conventional energy supplies (Karunathilake et al., 2020). These mitigation measures have led to carbon quotas and taxation being imposed to discourage fossil fuel consumption and encourage the shift to low or zero-emission alternatives. To meet the climate goals currently set through various global conventions such as the Paris Agreement, it is imperative that some of the available fossil fuel reserves remain unused. Fossil fuel firms may undergo a decrease in value if the reserves become "unburnable" due to climate action-initiated market restrictions (Bebbington et al., 2020). Currently, all signs predict that the world will inevitably shift toward a reality in which the demand for fossil fuels is significantly reduced. While this is environmentally beneficial, it is not a good news for developing economies of low and middle income who have traditionally relied on fossil fuel export income or are just beginning to identify and exploit their reserves (Cust et al., 2017).

Overall, the global energy transition patterns indicate that electricity is the favored final energy carrier due to efficiency and the ability to integrate clean and renewable energy (RE) resources (Gielen et al., 2019). It is noticeable that the nonconventional and renewable energy technologies undergo innovation and penetrate more rapidly in developed countries compared with the developing world. The renewable energy potential in developing countries as a whole is quite high, and yet has not been adequately exploited. While the developing world has long relied on traditional noncommercial energies that are not traded in the world energy markets such as fuelwood, animal residue, and crop residue, modern renewable technologies are still underutilized (Bhattacharyya, 2019). In 2017, the share of modern renewables excluding traditional biomass in the final energy consumption was 10.1% for developing countries, compared with 11.6% for North America and 13.5% for Europe (International Energy Agency, 2020b). Developing countries without their own fossil fuel reserves face problems due to extreme dependence on external energy supplies, putting them at the mercy of global geopolitical forces and economic pressures. Especially for these nations, renewables present an

Table 5.1 Top fossil fuel producers in the world.

#	Country		Coal production (TWh)	% of global production	Country		Oil production (TWh)	% of global production	Country		Gas production (TWh)	% of global production
1	China	D	21,272	46.39%	United States		7798	14.93%	United States		8359	21.67%
2	United States		4277	9.33%	Saudi Arabia	D	6697	12.82%	Russia	T	6691	17.35%
3	Indonesia	D	3822	8.34%	Russia	T	6541	12.52%	Iran	D	2383	6.18%
4	Australia		3628	7.91%	Canada		3115	5.96%	Canada		1790	4.64%
5	India	D	3555	7.75%	Iraq	D	2636	5.05%	Qatar	D	1765	4.58%
6	Russia	T	2563	5.59%	Iran	D	2609	4.99%	China	D	1615	4.19%
7	South Africa	D	1665	3.63%	China	D	2196	4.20%	Australia		1301	3.37%
8	Colombia	D	673	1.47%	United Arab Emirates	D	2052	3.93%	Norway		1213	3.14%
9	Kazakhstan	T	592	1.29%	Kuwait	D	1705	3.26%	Saudi Arabia	D	1121	2.91%
10	Poland		550	1.20%	Brazil	D	1628	3.12%	Algeria	D	938	2.43%
11	Germany		439	0.96%	Mexico	D	1188	2.27%	Malaysia	D	773	2.00%
12	Canada		326	0.71%	Nigeria	D	1119	2.14%	Indonesia	D	728	1.89%
13	Mongolia	D	289	0.63%	Kazakhstan	T	1059	2.03%	Turkmenistan	T	615	1.59%
14	Vietnam	D	273	0.59%	Norway		965	1.85%	United Arab Emirates	D	614	1.59%
15	Turkey	D	192	0.42%	Qatar	D	923	1.77%	Egypt	D	586	1.52%
16	Czechia		171	0.37%	Venezuela	D	878	1.68%	Uzbekistan	T	572	1.48%
17	Ukraine	T	163	0.35%	Angola	D	860	1.65%	Nigeria	D	483	1.25%
18	Mexico	D	78	0.17%	Algeria	D	759	1.45%	United Kingdom		405	1.05%
19	Serbia	T	77	0.17%	Libya	D	638	1.22%	Argentina	D	394	1.02%
20	Bulgaria		61	0.13%	United Kingdom		591	1.13%	Oman	D	360	0.93%
21	Greece		50	0.11%	Oman	D	555	1.06%	Mexico	D	352	0.91%
22	Romania		48	0.10%	Colombia	D	529	1.01%	Thailand	D	347	0.90%
23	Thailand	D	44	0.10%	India	D	459	0.88%	Pakistan	D	342	0.89%
24	Brazil	D	27	0.06%	Indonesia	D	459	0.88%	Trinidad and Tobago	D	340	0.88%
25	Zimbabwe	D	27	0.06%	Azerbaijan	T	455	0.87%	Netherlands		323	0.84%
	World		**45,850**		**World**		**52,245**		**World**		**38,575**	

D, developing economies; T, economies in transition.
The production of each fuel type is presented after being converted into primary energy equivalents, using Terawatt-hours as the unit.
Adopted from UN World Economic Situation and Prospects.

escape route from their vulnerable status and allow them to be independent and self-reliant, while also creating new economic sectors and employment opportunities. However, barriers such as the need for high capital investments, low levels of technology-related R&D, lack of expertise, and the potential increase in energy costs to the consumers impede the renewable transition in the developing economies.

While there is a direct link between population growth and overall energy demand increase, the per capita energy demand increase shows signs of slowing down (International Energy Agency, 2020a). The global energy intensities continue to show a declining trend in most countries (International Energy Agency, 2016). This can be attributed to many reasons, including increased efficiency of resource use, outsourcing of energy-intensive activities, and structural changes related to production and consumption activities in the economy (due to the transition from manufacturing to service economies) (International Energy Agency, 2016). Overall energy demand at both global and country-level declines during times of economic and social crises, as seen most recently with the COVID-19 pandemic (International Energy Agency, 2021b; Bilgen, 2014). An interesting trend has been emerging with regard to energy demand and energy footprint[1] in recent times, reflecting the difference of demand patterns between developing and developed economies. The energy footprint of developed countries is higher than the energy demand, and the opposite can be observed in developing countries (Arto et al., 2016). From this, it is clear that the burden of maintaining the high level of development and "lifestyle" of developed countries is spread across the world. New models that can account for this type of phenomena need to be developed in order to support the creation of an equitable global energy policy.

With the advent of emerging economies such as India and China, the global energy markets have undergone rapid structural shifts in supply and demand. While energy is an integral part of economic development, harnessing and utilization of energy poses economic, environmental, and social problems, especially for developing countries. As the concept of energy security does not exist in a vacuum, understanding the position of developing countries in the global energy markets and the problems facing them is critical.

1.2 Energy challenges for developing countries

The global "hunger" for energy is ever increasing. The global population is increasing, and its effect on energy demand has been seen throughout the last century (Vanek and Albright, 2008). However, even when the impact of population growth is not considered, human activities are becoming more energy intense, simply due to urbanization, industrial growth, and increased expectations about the quality of life (Asian Development Bank, 2009). This has been especially true for developing countries, which are only now starting to enjoy the higher quality of life and industrial progress previously seen by developed nations. With this, key energy security concerns such as lack of energy access, insufficient diversification of supply sources, high dependence on traditional fuel, overdependence on imports, increasing gap between supply and demand, and inadequate infrastructure have been identified in developing regions (Asian Development Bank, 2009).

---

[1]Energy footprint: Total energy consumed worldwide in producing the goods and services required by a particular country (In contrast, energy demand is the total energy used by a country.).

The demand for energy from the emerging economies has grown by leaps and bounds in the 21st century. The current growth in global energy demand is mainly attributed to non-OECD[2] countries. Non-OECD Asia is heavily reliant on coal for its energy needs, with a share of 63% in the electricity mix (International Energy Agency, 2020c). In these countries, the industrial sector depends on coal for the recently observed rapid growth, especially in India and China. The emerging economies who have already adopted carbon-intense development pathways show a clear reluctance to adopt nonhydro renewable energy resources, highly preferring cheap and abundant coal instead (Johnsson et al., 2019). High reliance on fossil fuels also comes with a tendency to lock the infrastructure and investment patterns, thus demotivating the penetration of RE (Asian Development Bank, 2009). The market barriers and carbon market risk for fossil fuels (especially coal) can be damaging for both fuel producers and fuel consumers among the developing economies. The established fuel exporters among them, who have neglected to diversify their export portfolios and therefore solely rely on the income from fossil fuel exports, will find the drastically reduced demand quite detrimental to their economy. The nations who have discovered their own reserves more recently may find themselves too late to benefit as expected from those resources (Cust et al., 2017). Strict limitations imposed on fossil fuel use can inhibit the economic growth, especially among the emerging economies who rely on "dirty" coal to fulfill their ever-increasing appetite for energy.

Nonmonetary value of traditional energies often cannot be accurately assessed due to the lack of statistical information, even though this type of energy flows account for a significant fraction of the energy mix in developing countries (Bhattacharyya, 2019). In many developing countries, especially rural populations heavily rely on nontradable traditional energies such as biomass in their energy mix, which result in adverse consequences on income generation, health, and social development (Asian Development Bank, 2009). *Access to affordable, reliable, sustainable, and modern energy for all* as per SDG 07 becomes problematic in this context, and the access to such modern services is lacking (Kuik et al., 2011). Poor and marginalized communities are often overlooked in providing access to adequate modern energy services (Kuik et al., 2011). Developing modern energy infrastructure to uplift the life of these populations is also challenging for developing nations due to the high capital costs involved (Asian Development Bank, 2009). The challenges for increasing modern renewable energy penetration in developing countries are large investments and lack of funding, regulatory barriers, lack of proper incentives, inappropriate tariffs, inadequacy of ancillary services, lack of education and financing options, supply intermittency, uneven playing field, and lack of R&D (Bhattacharyya, 2019). Moreover, split incentives in investments and inadequate risk sharing mechanisms among the players further discourage renewable energy initiatives. Many developing countries are also hampered by the smallness of their markets in the quest for a secure and sustainable energy supply (Asian Development Bank, 2009).

2. The nexus of supply, demand, and energy security

As initially mentioned, energy security is defined based on *availability, affordability, accessibility*, and *acceptability* aspects. For developing countries, the issue of energy security is even more complex and goes beyond the physical quantity and economic price dimensions (Wohlgemuth, 2006). Since access

[2]Organisation for Economic Co-operation and Development is an intergovernmental organization, where the members are generally of developed high-income countries with high HDI.

to modern energies is a clear requirement for economic development and social advancement in a country, developing countries have a vested interest in improving their energy security. For this, policies and plans are being developed at regional, national, and global levels.

One problem often seen in energy security initiatives and policies in various nations is the lack of quantifiable goals (Kruyt et al., 2009). This is perhaps unsurprising given that the definition of energy security is inexact and context dependent by its very nature (Winzer, 2011). Moreover, energy security policies cannot stand alone on their own, but should be linked to other policy issues such as poverty eradication, climate change mitigation, and even foreign affairs (Kruyt et al., 2009; Jakstas, 2020).

Today, energy security has transcended the conventional definitions related to supply and demand and has instead become an interdisciplinary domain. It encompasses dimensions such as climate change, energy efficiency, sustainability, and energy poverty, becoming interconnected with political, social, and national/human security arenas (Jakstas, 2020; Kuik et al., 2011). Thus, it is necessary to identify the parameters governing this complex nexus in order to gain a better understanding and develop attainable energy security goals for developing countries.

2.1 Parameters governing the nexus

Since the basic definitions of energy security revolve around "the continuous availability of energy in various forms at acceptable prices and in adequate quantities," supply and demand factors play a major role in this nexus (Khatib et al., 2000). Higher energy security translates to reduced vulnerability to supply disruptions, whether transient or long-term, which in turn means a country is less dependent on external suppliers. In order for a developing country to not be at the mercy of global geopolitical forces, there should be readily available local resources or capacity to access alternative imported resources at reasonable prices (Khatib et al., 2000). Ultimately, it is obvious that "energy independence" is closely tied with energy security. The global energy and environmental security affect the energy use and supply in developing countries and vice versa (Kuik et al., 2011). For developing countries, achieving this status can be problematic due to several reasons, especially if local energy resources are lacking. Thus, these nations are said to be suffering from *energy vulnerability* (Asian Development Bank, 2009; Shin et al., 2013).

The following parameters listed in Table 5.2 have been identified as key dependent and independent variables of the energy security nexus based on various studies and reports.

Numerous other parameters can be defined in addition to the list presented above. However, it is clear that there are common themes related to national energy security. These parameters have various interrelationships with each other (Fig. 5.2). When it comes to developing countries, it is evident that supply security is of paramount importance. Increased indigenous energy production, supply diversification, maintaining strategic reserves, and secure trade flows all contribute to increased supply security. Demand on the other hand is interconnected with the population growth, quality of life, economic activity, and energy efficiency practices. The supply characteristics determine the economic and environmental impacts of the energy supply in general. However, global level demand variations and impacts on the supply chain can influence energy prices. When the currency and purchasing power of a country degrades (as is the case for many developing countries), the affordability of the energy is detrimentally affected. As previously mentioned, cleaner energy sources are generally associated with higher costs, and thus the acceptability and affordability parameters may be in conflict with each other for developing nations.

Table 5.2 Parameters influencing and being influenced by the energy security nexus.

Category	Parameters	References
Supply and demand	Domestic energy consumption/demand	Shin et al. (2013), Laimon et al. (2019), Bastan and Hamed (2018)
	Demand growth	Shin et al. (2013)
	Indigenous energy production	Asian Development Bank (2009), Koyama and Kutani (2011)
	Energy exports	Staley et al. (2009), Bastan and Hamed (2018)
	Per capita energy intensity	Staley et al. (2009), Bhattacharyya (2019), Bastan and Hamed (2018)
	Energy intensity per GDP unit	Staley et al. (2009), Bhattacharyya (2019), Shin et al. (2013)
	Energy efficiency	Shin et al. (2013), Koyama and Kutani (2011)
	R&D for technology development	Shin et al. (2013), Ahuja and Tatsutani (2009)
Socioeconomic	Population and its growth	Laimon et al. (2019), Bastan and Hame (2018)
	Quality of life and human development indices	Asian Development Bank (2009)
	Industrial activity	Ahuja and Tatsutani (2009), Wohlgemuth, 2006, Bastan and Hamed (2018)
	Economic growth	Asian Development Bank (2009), Khatib et al. (2000)
	Economic competitiveness	United Nations Economic and Social Commission for Asia and the Pacific (2008)
	Rural to urban population ratio	Sovacool (2013)
	Gender equity	United Nations Economic and Social Commission for Asia and the Pacific (2008)
	Purchasing power parity	Asian Development Bank (2009)
	Geopolitical dynamics	Staley et al. (2009), Shin et al. (2013)
	Political stability	Shin et al. (2013)
	Foreign investments in the energy sector	Mangla and Uppal (2014)
Affordability	Energy price	Shin et al. (2013), Laimon et al. (2019)
	Production cost fluctuations	Månsson et al. (2014)
	Relative affordability of energy	Staley et al. (2009)
	Fraction of population suffering from energy poverty	Sovacool (2013)
	National energy expenditure	Mangla and Uppal (2014)
	Consumer welfare	Khatib et al. (2000)
	Carbon taxes and constraints	Asian Development Bank (2009)
Availability	Supply reliability	Khatib et al. (2000), Staley et al. (2009), Koyama and Kutani (2011)
	Uninterrupted supply	International Energy Agency (IEA) (2021)
	Estimated domestic energy resources	Shin et al. (2013), Månsson et al. (2014)
	National strategic energy reserves	

Continued

Table 5.2 Parameters influencing and being influenced by the energy security nexus.—cont'd

Category	Parameters	References
		Khatib et al., (2000), Shin et al. (2013), Koyama and Kutani (2011)
	Redundant facilities	Khatib et al. (2000)
	Technological knowhow/capacity to develop indigenous resources	Khatib et al. (2000)
	Ability to draw on foreign resources	Khatib et al. (2000)
	Diversity of suppliers	Staley et al. (2009), Koyama and Kutani (2011), Månsson et al. (2014)
	Secure trade flows (replacement supply pathways)	Staley et al. (2009)
	Nuclear proliferation	Shin et al. (2013), Staley et al., 2009
	Imports fraction in the supply/self-sufficiency of energy supply	Koyama and Kutani (2011), Asian Development Bank (2009), Bastan and Hamed (2018)
	Diversity of fuels	Staley et al. (2009), Shin et al. (2013), Koyama and Kutani (2011)
Accessibility	Rural electrification	Sovacool (2013)
	Fraction of population with access to modern energy services	Sovacool (2013), Asian Development Bank (2009)
	Energy coverage/electrification	Khatib et al.(2000), Ningi et al. (2020)
	Dependence on traditional sources	Asian Development Bank (2009), Ningi et al. (2020)
	Infrastructure development and robustness	Asian Development Bank (2009), Shin et al. (2013), Månsson et al. (2014)
Acceptability	Carbon intensity of the energy supply	Shin et al. (2013), Koyama and Kutani (2011)
	Noncarbon energy share	Shin et al. (2013)
	Renewable energy use	United Nations Economic and Social Commission for Asia and the Pacific (2008)
	Impact on economic development	Månsson et al. (2014)
	Stakeholder acceptance/social acceptability	Komendantova et al., 2018; Díaz et al. (2017)
	Utilized fraction of renewable energy capacity	Karunathilake et al., (2019a,b)
	Life cycle environmental impacts of energy supply	Karunathilake et al. (2019a,b)
	Health risks related to energy production and consumption	Smith et al. (2013)

Not all of these parameters are measurable or quantifiable due to their qualitative nature and/or associated epistemic and aleatory uncertainties. Moreover, it is challenging to develop models that can encompass all the dimensions and concerns associated with energy security. One problem that has been raised with the four "A's" concept is that it does not consider the risk and resiliency aspects

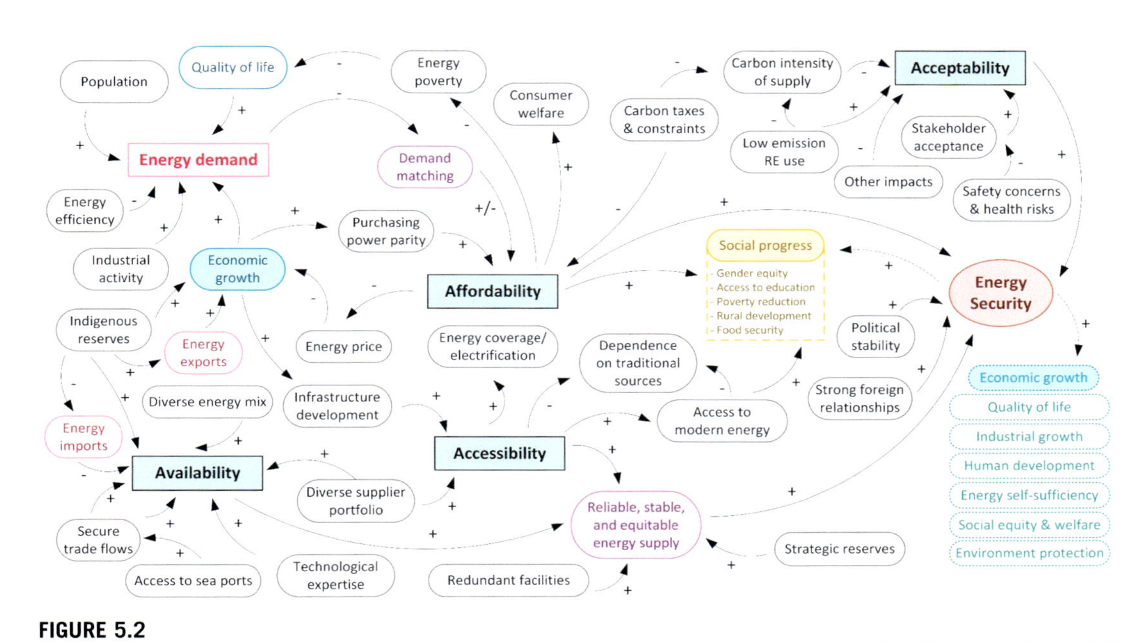

FIGURE 5.2

Cognitive map for interrelationships between parameters.

(Cherp and Jewell, 2014). Similarly, the impact of geopolitical dynamics and foreign policies is ever changing and somewhat difficult to capture through conventional indicators.

2.2 Interrelationships between parameters

Energy supply sources and suppliers are often key determinants of energy security for developing countries. From a four "A's" perspective, depending on energy supplies coming from depleting reserves, politically unstable regions, or unreliable supply routes can affect the "accessibility" and "availability" dimensions negatively (Haar and Haar, 2019). Stable and low-impact energy supplies are often unaffordable. "Acceptability" can vary based on both realized and perceived risks. An energy supply with high emissions and other ecological damages is unacceptable on environmental grounds. However, stakeholder perceptions can also make an energy supply unacceptable. For example, nuclear energy has come to be perceived as "high risk" across the world after past accidents in nuclear reactors (Yim and Vaganov, 2003). Health risks related to energy resource extraction and generation also affect the acceptability dimensions, such as the controversy surrounding the practice of fracking or air pollution associated with coal power plants.

3. Holistic assessment of energy security and sustainable development

Since energy security is closely tied with sustainable development, the dimensions of triple bottom line sustainability should be considered in its assessment (United Nations Economic and Social

Commission for Asia and the Pacific, 2008). While SDG 07 is visibly linked with energy security, it has far reaching consequences for all other SDGs, especially for those such as Goal 8: Decent work and economic growth, Goal 9: Industry, Innovation, and Infrastructure, Goal 11: Sustainable cities and communities, Goal 12: Responsible consumption and production, and Goal 13: Climate action (The United Nations, 2021). Even goals such as no poverty, gender equity, and health and well-being are supported by enhanced energy security, especially in developing countries where unaffordable and unreliable energy supplies coupled with socioeconomic inequities hinder the achievement of such targets (Fatona et al., 2013).

In order to enhance energy security, it is first necessary to have some means of assessing the current status of a country/region and compare it against the rest of the world or self-established benchmarks. This is where an energy security index comes into play.

3.1 Energy security index development

Various energy security indices have been developed in the past, with different geographic coverages, time frames, and assessment scopes (Gasser, 2020). These indices are concerned with one or more of the indicators listed in Table 5.2. Some indices even consider composite indicators that have been defined by combining multiple parameters. An index can focus on a single country, or its coverage can be global. Fig. 5.3 summarizes the key aspects related to energy security index development.

The indicators used in energy security indices need to reflect the expectations of the stakeholders and national development goals. Benchmarks enable self-assessment and comparison between the performance of different countries. Establishing the correlations between different parameters and

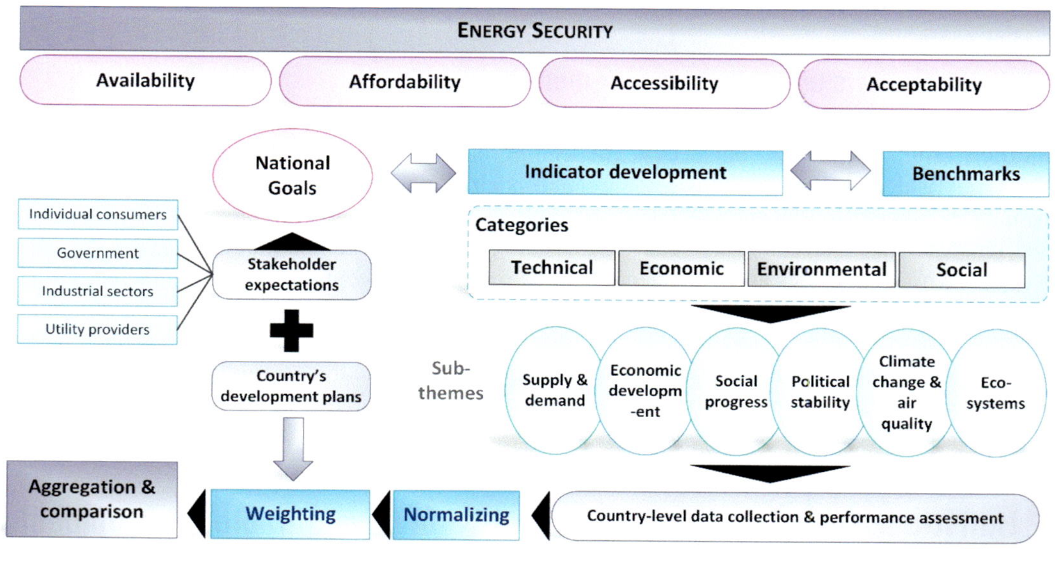

FIGURE 5.3

Energy security index development.

defining their relative importance is quite subjective and differs between various indices (Gasser, 2020). The parameters in the energy security nexus discussed in the previous section are often used in defining the indicators and assessing their performance scores. Normalizing, weighting, and aggregation become necessary when an index is defined based on multiple indicators. Methods for normalizing, weighting, and aggregation purposes listed in Table 5.3 were identified in a previous review on energy security indices (Gasser, 2020).

While index development is similar for both developed and developing countries from a mathematical sense, the reality is far more complicated. Specific indicators need to be selected to reflect the socioeconomic and political realities of developing nations. Data are often incomplete, inaccurate, missing, or simply unavailable. "Fragile" countries or those challenged by social, political, or climate problems are often the most data deprived. In such locations, gathering information in a systemic manner is difficult due to reasons such as limited budgets, natural and man-made disasters, lack of policies and guidelines in place, lagging public services, and low levels of technology and mobility (Hoogeveen and Pape, 2020). Some performance indicators cannot be put into quantifiable forms. In such instances, judgment or preference-based approaches are used instead. In certain cases, the data are uncertain, and systems are highly stochastic. The country situation and context are often changeable and volatile. Most importantly, many of the existing energy security indices focus on a limited scope, and so miss some aspects and dynamics that are important in the context of developing nations (Martchamadol and Kumar, 2013). Correlation development is also challenging, particularly for developing nations due to the effect of various instabilities and compounding parameters. The challenge is to develop holistic frameworks that can capture both the reality of the developing nations and the ever-changing dynamics of the global energy markets.

Holistic energy security indices need to cover the technical, economic, sociopolitical, and environmental dimensions of sustainability. In order to compare between different countries and determine the relative positioning, energy security indicators need to be universally applicable (Cherp and Jewell, 2010). There are various subthemes that come into play in this discussion, spanning over supply and demand, economic and industrial activity, climate change and ecosystems, political environment, and social progress (Martchamadol and Kumar, 2013; Gasser, 2020). Each of these modules have their own nexuses, which requires separate extensive study in order to be accurately defined and quantified.

World energy markets themselves are highly variable. The information available for energy assessments, especially in developing countries, is subject to both aleatory (caused by system variabilities) and epistemic (caused by lack of data) uncertainties (Karunathilake et al., 2019a,b).

Table 5.3 Mathematical methods used for index development.

Normalizing	Weighting	Aggregation
Categorical	Analytic hierarchy process (AHP)	Additive
Distance to a reference	Data envelopment analysis (DEA)	Data envelopment analysis (DEA)
Min-max	Equal weights	Geometric
Percentile rank	Factor analysis	Ordered weighted average
Rank	Import/fuel share	PROMETHEE
Standardized	Principal component analysis (PCA)	Room mean square
	Subjective methods based on preference	Rhombi area

Moreover, the effect of uncertainty in the input parameters propagates through the model, and the combined effect of uncertainties can greatly affect the final outcome. When the volatile situation of developing nations is also considered, a robust energy security index needs to incorporate an uncertainty handling mechanism to the assessment (Gasser, 2020). There are various methods that can be used for such applications, such as fuzzy logic (Bakhtavar et al., 2020). In addition, fuzzy linguistics can be used to handle qualitative or nonquantifiable parameters (Yamanishi et al., 2017). However, the uncertainty and stochasticity aspects are missing most of the time in the currently employed indices and assessment models (Gasser, 2020).

Risk is an important element in the energy security assessment (Martchamadol and Kumar, 2013). Reports published by the Institute for 21st Century Energy of the U.S. Chamber of Commerce indicates that developing countries have higher energy security risk scores compared with developed countries. While risk has declined over the years, the disparity between developed and developing nations still remains (Global Energy Institute, 2020) (Institute for 21st Century Energy, 2013). Resource diversification, supply route management, financial, and infrastructure initiatives can be taken in developing countries to reduce this risk (Månsson et al., 2014).

Easily understandable, communicable, and applicable energy security assessment frameworks and indices are helpful in facilitating energy security enhancement in developing countries (Gasser, 2020). Energy security enhancement decision making and related risk assessment needs a multicriteria decision-making approach, as the goals of energy, economic, environment, and society are directly conflicting with each other (Karunathilake et al., 2020). Policymakers should be able to use such an index for effective communication, and planners should be able to develop long-term strategies on improving the current status based on the information provided through it. All in all, energy security is not a static parameter. Instead, it is a continuously evolving situation. A "good" assessment should not only indicate the comparative standing of a country at a given moment, but it should also shed light on how that standing may be improved with reference to defined goals (Martchamadol and Kumar, 2013). Thus, a system dynamics approach is useful in energy security index development and assessments (Shin et al., 2013).

3.1.1 Classifying countries

Developing countries have varying energy security opportunities and challenges depending on their unique context. The energy security situation can depend on a number of macroenvironmental factors including the size of the country, geographic positioning, political situation, and the current level of development. Some of the definitions used by the United Nations Economic and Social Commission for Asia and the Pacific to classify countries with special challenges in the energy economics discussion are given below (United Nations Economic and Social Commission for Asia and the Pacific, 2008).

Least developed countries: Least developed countries have the common characteristics of low domestic savings insufficient to meet their investment needs, weak governance, political instability, damaged or inadequate infrastructure, limited internal markets and trade opportunities, diverted budgetary resources, and scare foreign investments (United Nations Economic and Social Commission for Asia and the Pacific, 2008). They are also severely affected by adverse environmental impacts, lack of access to natural resources, and undermined human health.

Landlocked developing countries: These countries have no access to the international waters, which hinders their access to foreign markets. Landlocked developing countries have challenges in ensuring secure trade flows, as their energy exports can only reach the country via land borders (Staley et al., 2009). In general, landlocked countries spend more on transport and insurance services compared with other developing as well as developed nations, as supplies have to reach them via transit countries.

Small island developing states: Many of these also belong to the least developed countries. Remoteness, isolation from the rest of the world, small domestic markets, limited natural, financial, and institutional resources, and high dependency on external supplies all contribute to this situation (United Nations Economic and Social Commission for Asia and the Pacific, 2008). Many of these states do not have secure energy supplies of their own and are thus reliant on external sources while also being affected by adverse economic conditions.

Self-sufficient countries: These nations can fulfill their domestic energy demand through indigenous energy production, without relying on external energy imports. Self-sufficiency makes a country less vulnerable to price and supply shocks from variations in the global markets (Koyama and Kutani, 2011).

Importing, exporting, and transit countries: Energy importing countries rely completely or partially on external suppliers to cater their energy needs. These countries are traditionally concerned about the stability and affordability of the energy supply. In contrast, energy exporters, who sell energy supplies to foreign buyers, expect a steady export flow at reasonable prices. Thus, they are concerned about "demand security" (Energy Charter Secretariat, 2015). Thus, these two types of countries have differing definitions of energy security at national level. Transit countries are those located on global transportation routes (Energy Charter Secretariat, 2015). Transit fees can facilitate local economic development for countries that exploit their strategic location successfully.

3.2 Energy security and energy independence: South Asia

It is useful to explore how energy security dynamics play out with real-world examples. Here, South Asian region is taken as a case study to discuss the differences between various countries in terms of energy security. The situations in the eight South Asia (SA) countries can provide a glimpse into how energy dynamics and energy security challenges affect developing countries.

The region consists of eight member states: India, Pakistan, Sri Lanka, Bangladesh, Nepal, Bhutan, Maldives, and Afghanistan. This is one of the most populous regions in the world, with India alone having a population of over 1.36 billion people. Moreover, the population density is very high with over 22% of the world's population occupying just 4% of the global land mass (Sanker et al., 2005). The region also contains three out of the 32 landlocked developing countries in the world. While geographically located close together, these countries differ widely in terms of land area, population, indigenous resources, economic development, social context, and political stability. India, which has the largest population in the region, is among the top five coal producers in the world (Ritchie, 2017). The following Tables 4.4 and 4.5 summarize the country characteristics in this region in terms of population, energy, and economics (The World Bank, 2021). Currently, SA is the world's fastest growing subregion with a collective average annual growth of around 7%, which is higher than economies in East Asia, Southeast Asia, and the Pacific (Song, 2019).

With the economic growth in SA, member countries are steadily moving away from the traditional agricultural economic models and are shifting toward manufacturing-oriented industrial economies. The region has also seen a corresponding increase in the overall energy use, which is further propelled

Table 5.4 South Asia country characteristics.

Country	Population ('000)	GNI -current US$ (Millions)	GDP growth (annual %)	Per capita GDP	Land area (sq.km)	Urban population (%)	Global gender gap index ranking (World Economic Forum, 2019)
Afghanistan	38,041.75	19,597.76	3.9	507.1	652,860.0	26	No data
Bangladesh	163,046.16	316,907.37	8.2	1855.7	130,170.0	37	50
Bhutan	763.09	2301.00	5.5	3316.2	38,140.0	42	131
India	1,366,417.75	2,837,687.53	4.2	2099.6	2,973,190.0	34	112
Maldives	530.95	5083.63	7.0	10,626.5	300.0	40	123
Nepal	28,608.71	30,995.61	7.0	1071.1	143,350.0	20	101
Pakistan	216,565.32	272,611.91	1.0	1284.7	770,880.0	37	151
Sri Lanka	21,803.00	81,590.53	2.3	3853.1	61,864.0	19	102
South Asia	1,835,776.74	3,566,775.35	4.1	1956.6	4,770,754.0	34	-

Table 5.5 Energy situation in South Asia.

Country	Per capita energy use (kg of oil equivalent)	Energy intensity (MJ/$2011 PPP GDP)	Renewables in total energy consumption (%)	Energy imports (% of total use)	Population % with access to electricity	Landlocked	Indigenous fossil fuel reserves (Hou et al., 2019; Sanker et al., 2005)
Afghanistan	No data	2	18.42	3	98.7	Yes	Yes
Bangladesh	229	3	34.75	17	85.2	No	Yes
Bhutan	367	10	86.90	9	100.0	Yes	Negligible
India	637	5	36.02	34	95.2	No	Yes
Maldives	892	4	1.01	59	100.0	No	No
Nepal	434	7	85.26	17	93.9	Yes	No
Pakistan	460	4	46.48	24	71.1	No	Yes
Sri Lanka	516	2	52.88	50	99.6	No	No
South Asia	574	–	38.28	33	91.6	–	–

by improving standards of living and growing urbanization. However, this region is still not at the forefront of the global exports markets (Shakeel and Salam, 2020). As the production sectors evolve and exports increase, a higher demand on the energy sector in the SA countries is inevitable.

Most SA countries except Maldives have relatively lower per capita GDP and energy consumption values. The previous statement stays true even for India, with a GNI of over US$ 2800 billion. Especially the rural communities in these countries lack sufficient access to affordable modern energy. SA is still affected largely by energy poverty, and all of its countries have seen periodic energy shortages leading to disruptions in community activities, industrial processes, and other services (Shakeel and Salam, 2020).

Renewable energy use in SA has been steadily growing in the past decades, and there are many government level initiatives to promote it. While the data provided in Table 5.5 indicate that all SA countries except Maldives have a significant fraction of renewables in the total final energy consumption, it should be noted that most of these countries are highly reliant traditional noncommercial energies such as biomass, and expanding the use of modern commercial energy to increase supply reliability and uplift the living standards of the communities is essential (Sanker et al., 2005). In fact, traditional energies make up 30%–90% of the energy supply in SA countries (Kumar Singh, 2013). This value has fallen to around one-third of the total consumption in India and Pakistan, possibly due to the higher level of industrialization, while it remains over two-thirds in Bangladesh and Sri Lanka (Kumar Singh, 2013).

Hydropower especially is a major energy resource for most South Asian countries (Sanker et al., 2005). Even countries such as Nepal, Bhutan, and Sri Lanka who do not have access to adequate fossil fuel reserves of their own have substantial hydropower resources. Yet, hydroelectricity generation is constrained by seasonal variations (Sanker et al., 2005). Moreover, all South Asian countries are heavily reliant on agriculture, and hydroelectricity generation is further constrained by the competing demand from the agricultural sector (which is generally water-intense) for water especially in the dry season. In addition to major hydro, almost all South Asian countries have seen a significant increase in run-of-the-river small hydro facilities at community level.

The case of Afghanistan is special due to multiple reasons. Even with its own domestic gas reserves, this country is highly reliant of noncommercial energies (Kumar Singh, 2013). Its energy sector growth has been hindered, and the sector remains underdeveloped and vulnerable due to the decades long political instability, conflicts, and weak economic growth (Ahmadzai and McKinna, 2018). Thus, it is clear that while the presence of fossil fuels is a boon to any nation, that alone cannot solve its energy or economic problems. Even though landlocked, Afghanistan is located centrally in a strategic location connecting South and Central Asia. This geographic positioning itself is both an opportunity and a threat to the country. While Afghanistan can potentially gain positive economic outcomes through both energy exports and energy transit fees, the conflicts that have plagued Afghanistan are in fact driven by the need for control over the resources and the location (Ahmadzai and McKinna, 2018). Hence, a strong foreign policy, communal stability, good governance, and regional support and cooperation are needed to ensure a secure energy future for this country.

Nepal has the highest fraction of RE in the final energy consumption among all SA countries, with almost all of its electricity coming from hydropower (International Energy Agency, 2021a). However, its populace is affected by lack of access to modern energy and low supply reliability (Asif, 2020). Being landlocked, all of Nepal's trade routes are through the regional giants, India and China. The fraction of population with access to electricity is also relatively lower compared to similar countries such as Sri Lanka and Bhutan. Due to its high reliance on hydropower, Nepal's energy security is extremely vulnerable to the negative impacts of climate change (Pandey, 2019). If fossil fuel requirement to cater the growing energy demand of the country (as has been the case in the recent past) keeps growing in a rapid phase, Nepal's energy security will be affected by the fuel costs as well as supply constraints due to its landlocked position (Asif, 2020). Inadequate planning, underinvestment in the energy sector, and delays in project implementation have all contributed to energy shortages in the face of the growing demand (Asian Development Bank, 2017). Lack of energy infrastructure and poor diversification of energy supply resources have also been identified as problems for Nepal.

Bhutan is the other landlocked country in the region, again located between India and China. Human well-being and sustainability are core aspects of Bhutan's development policy (Asif, 2020). Bhutan also has bilateral energy trading agreements with India, an arrangement that is rather unique for this region with its complex geopolitical dynamics (Tortajada and Saklani, 2018). Hydropower projects in Bhutan are a reliable resource in managing peak loads in Eastern and Northern India (Tortajada and Saklani, 2018). Bhutan is dependent on India for fossil fuel imports (Asif, 2020). The country has a relatively high per capita GDP, especially considering its small land mass. Overall, even without having access to international waters, Bhutan tends to perform better than its regional peers in terms of both energy security and environmental sustainability and thus can be called a success story among developing countries (Shah et al., 2019). The most important challenge for Bhutan's energy sector is to cater the growing energy demand while also conserving the environment as per their vision. In facing these challenges, Bhutan is attempting to diversify the energy mix and optimize the demand (Asif, 2020).

Out of the two island nations, Sri Lanka has a high population density as well as an increasing demand for energy due to both population growth and rising industrial activity. Currently in a postwar era after a decades long war, the economy is gaining pace with foreign investments. Interestingly, despite almost universal electrification, Sri Lanka is a country with gradually decreasing energy security (Asif, 2020). It is also a climate-vulnerable country, which can further affect both energy demand and the domestic hydroelectricity supply (Shah et al., 2019). However, on the positive side, Sri Lanka has one of the most diverse energy mixes in the region, whereas most of the other SA countries rely on a single source for more than 50% of their electricity (Shah et al., 2019). Sri Lanka has some prospective offshore oil reserves in the Gulf of Mannar, which have not yet been exploited (Asif, 2020). With the current global trends, it may be the case that Sri Lanka is "too late to enter the game" as the carbon constraints and fluctuations in oil prices make the recovery of these reserves uneconomic. The high reliance on imported fossil fuel has become a major economic challenge for the country, leaving the country highly vulnerable to oil price shocks (Murshed et al., 2020). Energy imports have significantly contributed to the country's balance of payments deficit and are draining the foreign reserves, leaving the country in a debt crisis (Moramudali, 2021).

Maldives, made up of 1192 coral islands, has the lowest population in the region. It also falls under the small island developing state (SIDS) definition. Compared with the other SA countries, Maldives enjoys a high level of economic and social progress due to its tourism industry (Asif, 2020). Thus, Maldives understandably has a high per capita GDP and a correspondingly high per capita energy use as seen above. Yet, it is impacted by many of the challenges that plague other SIDS. Inability to obtain favorable credit terms from foreign suppliers affects Maldivian importers. The smaller size of Maldivian companies makes it difficult to use the port and berthing facilities in nearby locations in SA, and the storage capacity on the islands is also limited (Shumais and Mohamed, 2019). Human resource development in the energy sector and reducing the dependence on imported energy are key challenges for the Maldivian energy sector (Revi, 2021).

While densely populated, Bangladesh has one of the lowest per capita energy intensities in the region as well as the world. While the domestic natural gas reserves of Bangladesh are not very substantial, NG is used to cater much of the country's energy needs (Asif, 2020). Since domestic gas production alone is insufficient in catering the demand, imported LNG is used to bridge the deficit (Islam and Khan, 2017). While the country has domestic coal reserves, those are not being exploited much at the moment, although plans for increasing coal power generation are in place (Das et al., 2020). It is clear that the presence of indigenous fossil fuel reserves has contributed to the recently seen economic development of Bangladesh. However, there are difficulties with financing the coal power projects (Nicholas, 2021). There are also concerns about the available domestic gas reserves being soon depleted at the current rate of demand (Asif, 2020). Thus, Bangladesh is attempting to move toward a higher level of renewable energy integration.

Pakistan has the largest population as well as the highest land mass after India in this region. It has sizable indigenous coal reserves, and the production is expected to increase in the future. This country also has a substantial hydropower potential, which is only partially exploited at the moment (Malik et al., 2020). However, Pakistan still relies considerably on imported supplies. Import dependence has also affected Pakistan's trade bills of which oil imports account for a significant fraction (Asif, 2020). Even though the generation capacity has grown significantly in the recent time, some regions in the country have suffered from power blackouts (Bin Abdullah et al., 2020). Energy access gap and energy are very high, with only 71% of the population having access to electricity and the fourth largest unserved population at global level (Asif, 2020). These issues are caused by insufficient investments in the energy sector, substantial demand increase, aging energy infrastructure, and system inefficiencies. Thus, it is seen that even with domestic fossil fuel reserves, social well-being related to energy supply can be lacking in developing countries due to a multitude of other reasons.

India is the giant in the region on all fronts including land area, economy, and population. Due to the vast geographic spread, the conditions and problems vary significantly across the country by state (Sarangi et al., 2019). As one of the major emerging economies and one of the driving forces behind increasing global energy consumption, India's hunger for energy is substantial. Coal is dominant in the Indian energy mix due to domestic availability (Asif, 2020). Yet, this massive coal consumption is also linked with significant environmental damages (Khashimwo, 2020). The growing dependence on oil imports coupled with the sheer scale of the economy and its energy intensity can make India quite vulnerable to the supply shocks and price volatility in the global markets (Kumar et al., 2021). It is predicted that by 2040, China will account for 22% of the global energy consumption while India will account for 11% (Bhowmick and Nalawade, 2020). Even with the rapidly growing economy and increasing quality of life, a significant fraction of Indian population still suffers from lack of equitable access to energy services as well as energy poverty (Gupta et al., 2020).

While ambitious energy policies and plans have been proposed across SA, these goals are challenged by the daunting scale of the required investments as well as the rapid growth of energy demand (Sanker et al., 2005). In many of these countries, poor operational and financial performance of the national energy and electrical governing bodies has been identified as a challenge. Moreover, SA region as a whole is negatively affected by the impacts of climate change.

4. Improving energy security and equity

A secure and equitable energy supply is a major milestone in transitioning an economy from "developing" to "developed" status. National, regional, and global initiatives are needed to enhance the energy security of developing countries. The most critical problem facing the world today that concerns both developed and developing countries is balancing the economic, environmental, and social interests related to energy supply and demand. After discussing the challenges, it is worthwhile exploring the opportunities for improving energy security considering the present-day context.

4.1 Opportunities for an energy secure and sustainable future

Increasing renewable energy integration was a recurring theme in the previous sections of this chapter. It has been noted that many developing nations, especially in the developing Asia-Pacific region have abundant clean energy resources and is also showing commendable progress on adopting energy efficiency and renewable energy (Asian Development Bank, 2009). Most countries in the world, both developing and developed, have pledged to increase the renewable integration in their energy mix in the next few decades, due to both environmental and economic concerns. In essence, all countries should attempt to develop and harness domestically available energy resources to reduce external reliance. Investing in energy storage facilities can also support in increased renewable integration, maintaining a reliable supply without disruptions, and improving system efficiency. Storage is viewed as an important aspect of a secure energy system (Novikau, 2019). In addition, supply/supplier diversification as well as extensive efforts on improving energy efficiency are key steps in achieving a higher level of energy security for developing countries (Novikau, 2019).

Investments and financings as well as innovation and economic competitiveness will assist the energy sectors of developing nations (United Nations Economic and Social Commission for Asia and the Pacific, 2008). R&D is needed for both renewable and nonrenewable energy technologies for increased performance and efficiency. The enormous investments required for infrastructure development and renewable propagation in developing countries cannot be supplied by the governments and traditional sources of funding alone (United Nations Economic and Social Commission for Asia and the Pacific, 2008). Thus, there is a need to novel financial models and partnerships between different entities. Ensuring market resilience is also important in the quest for energy security. Energy investments have risks and issues related to market volatility, funding procurement, and maintaining profitability. Developing countries in particular face problems due to low credit ratings and smaller size of their markets. Partnerships between public and private sectors are a necessity in the context of developing countries for successful energy project deployment (Donastorg et al., 2017). Targeted loans for energy projects (e.g., loans for "green" energy), grants, green bonds, capital subsidies, energy performance contracting, feed-in-tariff schemes, and power purchase agreements are some instruments that can be used to promote projects related to clean and secure energy supplies in developing countries (Donastorg et al., 2017).

As seen in the SA case study, it is likely that no country in this region can be self-sufficient in their energy needs under the current circumstances, especially if they are to maintain their current growth trajectory. Therefore, regional cooperation and cross-border power exchanges mechanisms are likely to be the best strategies in going forward (Sanker et al., 2005; Kumar Singh, 2013). Decentralized energy systems may be the solution to many of the infrastructure-related issues in the SA energy sector. This is true for other developing nations as well. For example, the Asia and Pacific region contains more than 50% of the proven global coal and gas reserves, and yet these resources are not evenly distributed and the region's energy trade is not sufficiently developed (United Nations Economic and Social Commission for Asia and the Pacific, 2008). International energy trade can support the growth of many developing countries if mutually beneficial can be reached. Energy sharing and trade agreements can also resolve the availability and supply reliability issues caused by seasonal variations in resource potential. For example, in SA, summer time sees higher power generation in Nepal and Bhutan, while the demand increases in Bangladesh (Sanker et al., 2005). Energy treaties and agreements between countries and organizations can facilitate energy security enhancement through a cooperative approach. Building transnational secure supply chains and transportation infrastructure is also important (Khatib et al., 2000).

International support is essential for these regions in their quest for reaching an energy secure status while simultaneously achieving climate action goals (Asian Development Bank, 2009). Financing flows from developed countries or international organizations such as grants, loans, investments, and other forms of aid will be highly beneficial (Asian Development Bank, 2009). International loans, especially those coming from global institutions or country-level treaties, are an utmost necessity for developing countries (Donastorg et al., 2017). At present, developing countries do not maintain emergency oil stocks, making them vulnerable to supply disruptions. Thus, easing the flow of fossil fuel supplies is of paramount importance. Energy security of developing can be greatly supported by liberalizing the energy markets among the importers and ensuring an easier and more secure flow of fuel (Khatib et al., 2000).

The governing role of the state is an inescapable reality for developing countries, particularly as energy commodities are almost exclusively state controlled in many such states (Khatib et al., 2000). This is where good governance, fiscal responsibility, consistent market policies, accurate information, and long-term planning becomes necessary. Long-term planning should not only target infrastructure development but also focus on capacity building, demand management, trade agreement and foreign relations building, contingency planning, and stakeholder collaborations (Wang'ombe, 2021). Policy development for energy security and sustainability is often done in the form of isolated endeavors to tackle a single problem at a time. However, the reality is that energy security, poverty, social development, environment, and economy are interconnected. For higher success, policies have to be integrated to ensure that fulfilling the goals in one sector does not adversely impact the other sectors (United Nations Economic and Social Commission for Asia and the Pacific, 2008).

4.2 A global framework for carbon economy

Carbon emissions are an inevitable part of industrial activity and economic growth, at least under the current global energy scenario. As seen over the last one and a half centuries, the countries that exploited fossil fuels and produced most emissions have achieved economic growth and reached a relatively "developed" status. In fact, it could be said that many of these developed nations have already emitted more than their fair share of emissions, while most developing countries are just starting to move toward industrialization, exposing the energy-intense economies. The emerging carbon constraints, if imposed equally on both developed and developing nations, will impede the potential of other "underdeveloped" players who are late to enter the game. Authors of this article would like to name the disadvantages faced by the developing countries due to new emissions reduction policies as the "carbon finance disparity."

As developing countries face many developmental constraints and concerns, their position in the climate change negotiations is problematic (Asian Development Bank, 2009). Developing nations in fact have been asking for technological and financial assistance from developed countries in meeting carbon constraint and climate action targets. Innovative emission transferring models such as internationally transferred mitigation outcomes (ITMOs) have emerged in order to address the issues posed by the carbon finance disparity discussed above (Girouard et al., 2011; Sorrell, 2015). ITMOs enable the countries with higher emissions to purchase ITMOs in order to meet their nationally determined contributions (NDCs) as agreed in the Paris Climate Agreement (Koo, 2017). In general, the developing countries poses more potential to generate higher ITMOs per dollar investment due to lower labor and implementation costs of renewable energy and energy efficiency projects. As many of the South Asian countries are closer to the equator, the yield of the renewable projects such as *solar PV* and *solar thermal* is very high. This also helps reduce the ITMO yield per dollar investment. In summary, promoting ITMOs can motivate the developed countries to invest in the renewable energy projects in the developing world.

The counties who harnessed their fossil fuel reserves and the counties progressed with industrialization using the fossil fuels had the upper hand in economic growth as discussed above. Therefore, it can be seen that the carbon emissions have a time value similar to money. The developed countries continue to benefit from the economic development they achieved with emissions they made due to higher energy use in early stages of the timeline. However, the unused ITMOs brought forward by the developing countries who have not spent their fair share of the emissions over the course of history are being overlooked in the emission mitigation discussion. This leaves the developing countries to battle the climate change, which can be mainly attributed to the counties with higher historical emission footprint, while also being challenged by the poor economic conditions. Therefore, it only makes sense to expand the scope of ITMOs to cover at least part of the historical timeline in order to mitigate the disadvantages faced by the developing world. Considering the historical emissions will bring in the much needed perspective of energy and emission equity to the climate change discussion. Moreover, this will help secure the finances to overcome the higher capital costs associated with the renewable energy and energy efficiency projects. Therefore, a comprehensive emission accounting considering the time value of emissions will help the developing countries to get closer to energy independence, thereby to have a higher energy security, while enabling the developed counties to meet their NDCs in a cost-effective manner.

5. Conclusion

Many qualitative definitions have been provided to define the energy security. The concept of four "*A's*: *availability, affordability, accessibility*, and *acceptability*" provides a general direction about the parameters that determine the energy security. An energy security index can be used for self-assessment of a given country or to compare the performance of countries and regions in a global context. It is important to note that the energy security issues faced by one country or a region may not be the same for another country. Moreover, there is no established energy security index that is globally accepted. Globally relevant performance indicators have to be employed in order for an energy security index to be globally applicable. Even after establishing the performance indicators, the data uncertainties and unavailability can pose challenges in energy security assessment process. Based on literature, this article outlined the procedure that can be followed in developing a globally applicable energy security index. Future research can build on the proposed framework to develop a globally applicable energy security index. The South Asian energy scenario was taken as a case study to reflect how different conditions contribute to the energy security situation in developing countries. The key takeaway from the South Asian analysis was that regional collaboration and long-term planning are very much needed for energy security enhancement in this part of the world.

This study highlights the strong dependence of the energy security of developing countries with both climate change impacts and climate change mitigation policies. It was observed that there is a greater potential in ITMOs, not only in achieving the climate action goals of the countries with a higher environmental food print but also in helping the developing countries to overcome the economic barriers to implement costly renewable energy infrastructure and energy efficiency initiatives. Moreover, the authors would like to emphasize the need of accounting for the time value of emissions when establishing the nationally determined contributions and ITMOs potential.

References

Ahmadzai, S., McKinna, A., 2018. Afghanistan electrical energy and trans-boundary water systems analyses: challenges and opportunities. Energy Reports 4, 435–469.

Ahuja, D., Tatsutani, M., 2009. Sustainable energy for developing Countries. Sapiens 2 (1).

Arto, I., Capellán-Pérez, I., Lago, R., Bueno, G., Bermejo, R., 2016. The energy requirements of a developed world. Energy Sustainable Development 33, 1–13.

Asian Development Bank, 2009. Improving Energy Security and Reducing Carbon Intensity in Asia and the Pacific. Asian Development Bank, Manila.

Asian Development Bank, 2017. Nepal Energy Sector Assessment, Strategy, and Roadmap. Manila, Philippines.

Asif, M., 2020. In: Energy and Environmental Outlook for South Asia, first ed. CRC Press/Taylor & Francis Group, LLC: CRC Press, Boca Raton, FL.

Ayoo, C., 2020. Towards energy security for the twenty-first century. In: Taner, T. (Ed.), Energy Policy, vol. 32. IntechOpen, pp. 137–144.

Bakhtavar, E., Prabatha, T., Karunathilake, H., Sadiq, R., 2020. Assessment of renewable energy-based strategies for net-zero energy communities : a planning model using multi-objective goal programming. Journal of Cleaner Production 272, 122886.

Bastan, M., Hamed, S., 2018. A system dynamics model for policy evaluation of energy dependency. In: Proceedings of the International Conference on Industrial Engineering and Operations Management Paris, pp. 3103–3115.

Bebbington, J., Schneider, T., Stevenson, L., Fox, A., Jan. 2020. Fossil fuel reserves and resources reporting and unburnable carbon: investigating conflicting accounts. Critical Perspectives on Accounting 66, 102083.

Berk, I., Yetkiner, H., 2014. Energy prices and economic growth in the long run : theory and evidence. Renewable and Sustainable Energy Reviews 36, 228–235.

Bhattacharyya, S.C., 2019. Energy Economics. Springer London, London.

Bhowmick, S., Nalawade, H.S., 2020. Sino-Indian Transitions in the Energy Sector. Observer Research Foundation (Online). Available: https://www.orfonline.org/expert-speak/sino-indian-transitions-energy-sector-64894/ (Accessed 25 May 2020).

Bilgen, S., 2014. Structure and environmental impact of global energy consumption. Renewable and Sustainable Energy Reviews 38, 890–902.
Bin Abdullah, F., Iqbal, R., Hyder, S.I., Jawaid, M., Nov. 2020. Energy security indicators for Pakistan: an integrated approach. Renewable and Sustainable Energy Reviews 133, 110122.
Cherp, A., Jewell, J., 2010. Measuring Energy Security: from universal indicators to contextualized frameworks. In: Sovacool, B.K. (Ed.), The Routledge Handbook of Energy Security, first ed. Routledge, London.
Cherp, A., Jewell, J., Dec. 2014. The concept of energy security: beyond the four as. Energy Policy 75, 415–421.
Cust, J., Manley, D., Cecchinato, G., 2017. Unburnable wealth of nations. Finance and Development - International Monetary Fund 54 (1), 46–49.
Das, N.K., Chakrabartty, J., Dey, M., Gupta, A.K.S., Matin, M.A., 2020. Present energy scenario and future energy mix of Bangladesh. Energy Strategy Reviews 32 (November 2018), 100576.
Díaz, P., Adler, C., Patt, A., 2017. Do stakeholders' perspectives on renewable energy infrastructure pose a risk to energy policy implementation? A case of a hydropower plant in Switzerland. Energy Policy 108, 21–28.
Donastorg, A., Renukappa, S., Suresh, S., 2017. Financing renewable energy projects in developing countries: a critical review. IOP Conference Series: Earth and Environmental Science 83 (1), 0–8.
Energy Charter Secretariat, 2015. International Energy Security. Brussels.
Fatona, P., Abiodun, A., Olumide, A., Adeola, A., Abiodun, O., 2013. Viewing energy, poverty and sustainability in developing countries through a gender lens. In: New Developments in Renewable Energy, Hasan Arma, vol. 32(July). IntechOpen, London, pp. 137–144.
Gasser, P., 2020. A review on energy security indices to compare country performances. Energy Policy 139 (December 2019), 111339.
Gielen, D., Boshell, F., Saygin, D., Bazilian, M.D., Wagner, N., Gorini, R., 2019. The role of renewable energy in the global energy transformation. Energy Strategy Reviews 24 (January), 38–50.
Girouard, N., Konialis, E., Tam, C., Taylor, P., 2011. OECD Green Growth Studies: Energy. Paris.
Global Energy Institute, 2020. International Index of Energy Security Risk 2018 Edition: Assessing Risk in a Global Energy Market. Washington DC.
Gupta, S., Gupta, E., Sarangi, G.K., 2020. Household energy poverty index for India: an analysis of inter-state differences. Energy Policy 144, 111592.
Haar, L., Haar, L.N., 2019. A financial option perspective on energy security and strategic storage. Energy Strategy Reviews 25 (May), 65–74.
Hoogeveen, J., Pape, U., 2020. In: Data Collection in Fragile States. Springer International Publishing, Cham.
Hou, Y., Iqbal, W., Shaikh, G.M., Iqbal, N., Solangi, Y.A., Fatima, A., 2019. Measuring energy efficiency and environmental performance: a case of South Asia. Processes 7 (6).
Institute for 21st Century Energy, 2013. International Index of Energy Security Risk. Washington DC.
International Energy Agency (IEA), 2016. Global Energy Intensity Continues to Decline (Online). Available: https://www.eia.gov/todayinenergy/detail.php?id=27032 (Accessed 01 March 2021).
International Energy Agency (IEA), 2021. Energy Security: Reliable, Affordable Access to All Fuels and Energy Sources (Online). Available: https://www.iea.org/topics/energy-security (Accessed 28 April 2021).
International Energy Agency, 2020a. IEA Atlas of Energy (Online). Available: http://energyatlas.iea.org/#!/tellmap/-1118783123 (Accessed 01 March 2021).
International Energy Agency, 2020b. SDG7: Data and Projections (Online). Available: https://www.iea.org/reports/sdg7-data-and-projections/modern-renewables (Accessed 04 March 2021).
International Energy Agency, 2020c. World Energy Balances Overview. Paris.
International Energy Agency, 2021a. Nepal (Online). Available: https://www.iea.org/countries/nepal.
International Energy Agency, 2021b. World Energy Outlook 2020 (Online). Available: https://www.iea.org/reports/world-energy-outlook-2020 (Accessed 03 March 2021).
Islam, S., Khan, M.Z.R., 2017. A review of energy sector of Bangladesh. Energy Procedia 110, 611–618.

Jakstas, T., 2020. What does energy security mean? In: Tvaronavičienė, M., Ślusarczyk, B. (Eds.), Energy Transformation towards Sustainability. Elsevier, Oxford, pp. 99–112.

Johnsson, F., Kjärstad, J., Rootzén, J., Feb. 2019. The threat to climate change mitigation posed by the abundance of fossil fuels. Climate Policy 19 (2), 258–274.

Karunathilake, H., et al., 2020. The nexus of climate change and increasing demand for energy: a policy deliberation from the Canadian context. In: Qudrat-Ullah, H., Asif, M. (Eds.), Dynamics of Energy, Environment and Economy, Lecture Notes in Energy, vol. 77. Springer Nature Switzerland AG, Cham, pp. 263–294.

Karunathilake, H., Hewage, K., Brinkerhoff, J., Sadiq, R., 2019a. Optimal renewable energy supply choices for net-zero ready buildings: a life cycle thinking approach under uncertainty. Energy Build 201, 70–89.

Karunathilake, H., Hewage, K., Mérida, W., Sadiq, R., 2019b. Renewable energy selection for net-zero energy communities: life cycle based decision making under uncertainty. Renewable Energy 130, 558–573.

Karunathilake, H., Bakhtavar, E., Chhipi-Shrestha, G., Mian, H.R., Hewage, K., Sadiq, R., 2020. Decision making for risk management: a multi-criteria perspective. In: Khan, F., Amyotte, P. (Eds.), Methods in Chemical Process Safety, vol. 4. Elsevier, Oxford, pp. 239–287.

Khashimwo, P., 2020. India's quest for energy security: understanding the trends and challenges. Journal of Asian and African Studies, 002190962097244.

Khatib, H., Barnes, A., Chalabi, I., Steeg, H., Yokobori, K., 2000. Energy Security. New York.

Komendantova, N., Ekenberg, L., Marashdeh, L., Al Salaymeh, A., Danielson, M., Linnerooth-Bayer, J., 2018. Are energy security concerns dominating environmental concerns? Evidence from stakeholder participation processes on energy transition in Jordan. Climate 6 (4), 88.

Koo, B., 2017. Preparing hydropower projects for the post-Paris regime: an econometric analysis of the main drivers for registration in the Clean Development Mechanism. Renewable and Sustainable Energy Reviews 73 (February 2016), 868–877.

Koyama, K., Kutani, I., 2011. Developing an Energy Security Index. Jakarta.

Kruyt, B., Van Vuuren, D.P., De Vries, H.J.M., Groenenberg, H., 2009. Indicators for Energy Security. Energy Policy 37, 2166–2181.

Kuik, O.J., Lima, M.B., Gupta, J., 2011. Energy security in a developing world. Wiley Interdisciplinary Reviews: Climate Change 2 (4), 627–634.

Kumar, R., Das, S.K., Nayak, S., Paswan, M., Achintya, May 2021. Sustainable energy security of India based on energy supply. Materials Today Proceedings.

Kumar Singh, B., 2013. South Asia energy security: challenges and opportunities. Energy Policy 63, 458–468.

Laimon, M., Mai, T., Goh, S., Yusaf, T., 2019. Energy sector development: system dynamics analysis. Applied Sciences 10 (1), 134.

Malik, S., Qasim, M., Saeed, H., Chang, Y., Taghizadeh-Hesary, F., 2020. Energy security in Pakistan: perspectives and policy implications from a quantitative analysis. Energy Policy 144, 111552.

Mangla, I.U., Uppal, J.Y., 2014. Macro-economic policies and energy security — implications for a chronic energy deficit country. Pakistan Development Review 53 (3), 255–273.

Månsson, A., Johansson, B., Nilsson, L.J., 2014. Assessing energy security: an overview of commonly used methodologies. Energy 73, 1–14.

Martchamadol, J., Kumar, S., 2013. An aggregated energy security performance indicator. Applied Energy 103, 653–670.

Moramudali, U., 2021. Sri Lanka's Foreign Debt Crisis Could Get Critical in 2021. The Diplomat (Online). Available: https://thediplomat.com/2021/02/sri-lankas-foreign-debt-crisis-could-get-critical-in-2021/ (Accessed 25 May 2021).

Murshed, M., Mahmood, H., Alkhateeb, T.T.Y., Bassim, M., 2020. The impacts of energy consumption, energy prices and energy import-dependency on gross and sectoral value-added in Sri Lanka. Energies 13 (24), 6565.

Nicholas, S., 2021. Why Bangladesh Shouldn't Count on a Fossil Fuel Future. The Daily Star (Online). Available: https://www.thedailystar.net/opinion/news/why-bangladesh-shouldnt-count-fossil-fuel-future-2038469 (Accessed 25 May 2021).

Ningi, T., Taruvinga, A., Zhou, L., 2020. Determinants of energy security for rural households: the case of Melani and Hamburg communities, Eastern Cape, South Africa. African Security Review 29 (4), 299–315.

Novikau, A., 2019. Conceptualizing and achieving energy security: the case of Belarus. Energy Strategy Reviews 26 (August), 100408.

Onyiriuba, L., 2016. Questions in the making of emerging economies and markets. In: Emerging Market Bank Lending and Credit Risk Control. Elsevier, London, pp. 5–24.

Pandey, R., 2019. Energy security and hydropower development in Nepal. Nepal Energy Forum (Online). Available: http://www.nepalenergyforum.com/energy-security-and-hydropower-development-in-nepal/ (Accessed 25 May 2021).

Revi, V., 2021. What Does Energy Security Mean for Maldives? Observer Research Foundation (Online). Available: https://www.orfonline.org/expert-speak/what-does-energy-security-mean-maldives/. Accessed 25 May 2021).

Ritchie, H., 2017. How Long before We Run Out of Fossil Fuels? Our World in Data (Online). Available: https://ourworldindata.org/how-long-before-we-run-out-of-fossil-fuels (Accessed 03 March 2021).

Sanker, T.L., Raza, H.A., Barkat, A., Wijayatunga, P., Acharya, M., Raina, D.N., 2005. Regional Energy Security for South Asia: Regional Report. New Delhi.

Sarangi, G.K., Mishra, A., Chang, Y., Taghizadeh-Hesary, F., Dec. 2019. Indian electricity sector, energy security and sustainability: an empirical assessment. Energy Policy 135, 110964.

Shah, S.A.A., Zhou, P., Walasai, G.D., Mohsin, M., 2019. Energy security and environmental sustainability index of South Asian countries: a composite index approach. Ecological Indicators 106 (66), 105507.

Shakeel, M., Salam, A., 2020. Energy-GDP-exports nexus and energy conservation: evidence from Pakistan and South Asia. Environmental Science and Pollution Research 27 (22), 27807–27818.

Shin, J., Shin, W.-S., Lee, C., 2013. An energy security management model using quality function deployment and system dynamics. Energy Policy 54, 72–86.

Shumais, M., Mohamed, I., 2019. Dimensions of Energy Insecurity on Small Islands: The Case of the Maldives, p. 1049. Tokyo.

Smith, K.R., et al., 2013. Energy and human health. Annual Review of Public Health 34 (1), 159–188.

Song, L.L., 2019. How South Asia Can Continue as World's Fastest Growing Subregion. Asian Development Bank Op-Ed (Online). Available: https://www.adb.org/news/op-ed/how-south-asia-can-continue-world-s-fastest-growing-subregion-lei-lei-song (Accessed 24 May 2021).

Sorrell, S., 2015. Reducing energy demand: a review of issues, challenges and approaches. Renewable and Sustainable Energy Reviews 47, 74–82.

Sovacool, B.K., 2013. Energy Access and Energy Security in Asia and the Pacific, vol. 383.

Staley, B.C., Zyla, K., Goodward, J., Ladislaw, S., 2009. Evaluating the Energy Security Implications of a Carbon-Constrained U.S. Economy. Washington DC.

The United Nations, 2021. The 17 Goals (Online). Available: https://sdgs.un.org/goals (Accessed 11 May 2021).

The World Bank, 2018. Access to Energy Is at the Heart of Development (Online). Available: https://www.worldbank.org/en/news/feature/2018/04/18/access-energy-sustainable-development-goal-7 (Accessed 01 March 2021).

The World Bank, 2021. World Bank Open Data: Free and Open Access to Global Development Data (Online). Available: https://data.worldbank.org/ (Accessed 11 May 2021).

Tortajada, C., Saklani, U., 2018. Hydropower-based collaboration in South Asia: the case of India and Bhutan. Energy Policy 117 (February), 316–325.

United Nations Economic and Social Commission for Asia and the Pacific, 2008. Energy Security and Sustainable Development in Asia and the Pacific. Bangkok.

United Nations, 2020. World Economic Situation and Prospects (WESP). New York.

United Nations, 2021. Goal 07: ensure access to affordable, reliable, sustainable and modern energy for all. Sustainable Devolopment (Online). Available: https://sdgs.un.org/goals/goal7 (Accessed 25 February 2021).

Vanek, F.M., Albright, L.D., 2008. Energy Systems Engineering: Evaluation and Implementation. McGraw-Hill, New York.

Wang'ombe, E., 2021. Energy Sector Considerations in Long-Term Strategies. World Resources Institute (Online). Available: https://www.wri.org/climate/expert-perspective/energy-sector-considerations-long-term-strategies (Accessed 27 May 2021).

Winzer, C., 2011. Conceptualizing Energy Security. Cambridge. CWPE 1151 & EPRG 1123.

Wohlgemuth, N., 2006. Energy security and renewable energy in least developed countries. In: Proceedings of 15th Forum: Energy Day in Croatia: Energy Policy Scenarios to 2050, p. 182.

World Economic Forum, 2019. Global Gender Gap Report 2020. Geneva.

Yamanishi, R., Takahashi, Y., Unesaki, H., 2017. Quantitative analysis of Japan's energy security based on fuzzy logic: impact assessment of Fukushima accident. Journal of Energy 2017, 1−14.

Yim, M.-S., Vaganov, P.A., Jan. 2003. Effects of education on nuclear risk perception and attitude: Theory. Progress in Nuclear Energy 42 (2), 221−235.

CHAPTER

Renewable energy in Latin America and scenarios to the Brazilian energy matrix by 2050

6

Ramon Soares Corrêa[1], Osvaldo Luiz Gonçalves Quelhas[2], Gustavo Naciff de Andrade[3], Paulo Roberto de Campos Merschmann[4], Rosley Anholon[5] and Chrystyane Abreu[6]

[1]*Production Engineering Department, Universidade Federal Fluminense, Niterói, Rio de Janeiro, Brazil;* [2]*LATEC Laboratory of Technology, Business Management and Environment, Universidade Federal Fluminense, Niterói, Rio de Janeiro, Brazil;* [3]*Empresa de Pesquisa Energética, Centro, Rio de Janeiro, Brazil;* [4]*Production Engineering Department, Centro Federal de Educação Tecnológica Celso Suckow da Fonseca, Rio de Janeiro, Rio de Janeiro, Brazil;* [5]*Manufacturing and Materials Engineering Department, Universidade Estadual de Campinas, Campinas, São Paulo, Brazil;* [6]*Production Engineering Coordination, Centro Federal de Educação Tecnológica Celso Suckow da Fonseca, Nova Iguaçu, Rio de Janeiro, Brazil*

Chapter outline

Handbook of Energy and Environmental Security. https://doi.org/10.1016/B978-0-12-824084-7.00005-9
Copyright © 2022 Elsevier Inc. All rights reserved.

1. Introduction: energy and sustainability

The worldwide concern about the consequences of human actions on the environment is notorious. The theme that deals with economic and social development on a sustainable basis began to gain importance and space in the world at the United Nations Conference on the Human Environment in 1972. In Latin America, several countries are showing commitment to the sustainability through the submission of their Nationally Determined Contribution (NDC). The NDC is a United Nations' initiative and is issued by many countries to confirm their commitment to reducing greenhouse gas (GHG) emissions.

In Latin America, many countries have submitted their NDC such as Argentina, Bolivia, Brazil, Chile, Colombia, Ecuador, and Peru (United Nations, 2021). In 2016, on its NDC document (EPE, 2016a), Brazil committed to reducing GHG emissions by 37% until 2025 increasing this rate to 43% in 2030, based on levels of 2005. In December 2020, Brazil confirmed its commitment through the presentation of its updated NDC (United Nations, 2021).

According to Bacha et al. (2010), the importance of the theme sustainability is due, mainly, to the concern created because of the climate change caused by the predatory behavior of humans in their relationship with the environment. Andrade et al. (2016) pointed out that the current discussion regarding climate change considers the CO_2 emissions as one of the main causes of this environmental issue since it contributes to global warming. Both global warming and the use of fossil fuels are some of the most important issues to be faced by humanity in the 21st century, and the deployment of the renewable energies has been presented as a potential solution (Chen et al., 2009). Irfan et al. (2020) state that environmental degradation and climate vulnerability have been led policymakers to realize other possible methods of energy production seeking to minimize fossil fuel—based carbon footprints.

World energy demand is expected to grow significantly. Karatayev et al. (2016) understood that this growth is motivated by two main pressures: the economic development of countries and the population growth. To avoid catastrophic damages to the planet and to the life quality of people, it is inevitable to move toward a low-carbon emission social dynamic and with extensive use of sustainable energy sources (Lee and Zhong, 2014). Baratsas et al. (2021) highlight that governmental agencies, researchers, and organizations consider energy policies and their effects when dealing with the increasing concern in energy scarcity, energy independence, energy sustainability, and pollution related to energy activities.

The energy transition is in the spotlight of the discussions regarding the energy sector and the environment since it seems to be the mandatory pathway to allow substantial reductions of GHG emissions and also to reach energy demand increase in sustainable basis. Companies, society, and government are discussing and acting toward a new paradigm. Vigoya et al. (2020) state that energy transition is the independence of fossil energy sources and the technological development advancing into clean energy sources.

Fuentes et al. (2021) corroborate explaining that this movement is about to transform the energy systems, and it brings new challenges. This independence of fossil sources is generally achieved by changing conventional energy sources with alternative renewable ones and with the adaptation and replacement of the systems already deployed (Vigoya et al., 2020).

EPE (2021) reinforces that the energy transition means to change the composition of the energy matrix, and it implies in a deep changing in terms of technology, social consume pattern, socioeconomical relationships, and reduction and reuse of waste. The Company also explains that the energy transition is characterized by the so-called three "D's" (decarbonization, decentralization, and digitalization) and brings relevant changes in the global energy geopolitics placing challenges and opportunities for different countries.

IRENA (2021a) affirms that energy efficiency and renewable energy measures have the potential to reach 90% of the required carbon reduction. The Agency defines its work on three main pillars: power sector transformation, energy system models and data, and energy planning support.

According to Prasad et al. (2014), energy planning is one of the pillars to create policies for national sustainable development. With planning and public policies focused on the renewable energy sector, several ventures can be materialized promoting significant and consistent economic, social, and environmental gains (Santos and Torres, 2016).

The scenario analysis enables the development of possible future scenarios, making it feasible to deal more effectively and efficiently with the uncertain futures through more robust and long-term strategies (Wortelboer and Bischof, 2012).

The aim of this chapter is to present a brief outlook regarding renewables in Latin America and present a method for future scenario generation combining the tool 2050 Calculator with the Peter Schwartz's method for future scenario building. A case study for the Brazilian energy context is presented. The main question that leads the scenario generation is: is it plausible for the participation of the renewables to surpass the nonrenewables in the Brazilian energy matrix by 2050? Additionally, this work will highlight warning signs to verify along the time which one of the scenarios most closely approximates to the current course of history.

The relevance involving this chapter is also justified by the interest in developing practical and effective dynamics for the construction of future energy scenarios. The method herein described encourages discussions around the energy theme encompassing subjects as energy transition, sustainable development, renewables on energy matrices, and scenario-building models.

2. Future scenarios, 2050 calculator, and renewables in Latin America and Brazil

2.1 Scenarios and the energy sector

The use of scenarios enables the decision-makers to anticipate and respond better to development needs (Moritz, 2004; Lachman, 2011; Chaharsooghi et al., 2015; Nikolaev and Konidari, 2017). For Schmidt-Scheele (2020), scenario planning has become an important and indispensable approach for studying the future pathways to the energy sector.

The wide use of scenarios allows analysts to develop a future vision, particularly when there are uncertainties surrounding a particular problem, helping even, in minimizing costs (Bigerna et al., 2016). Chaharsooghi et al. (2015) emphasize that this technique has been applied in the most diverse contexts, both in issues of a more local nature and in the evaluation of global issues.

In order to corroborate the statement of this last author, several types of research can be found in the literature showing the use of the scenario methods in the most diverse fields of study, such as health (Lehoux et al., 2014), energy (Dias et al., 2016), footwear sector (Blois, 2008), journalism (De Díaz et al., 2013), engineering (Saurin and Ratcliffe, 2011), and environment (Rico et al., 2016).

On undefined systems, where the interrelationships of the variables are less stable and less predictable, the scenario technique proves to have a great value. There are different methods of scenario construction. The works of Carvalho et al. (2011), Moritz (2004), and Oliveira et al. (2018) are some of the research examples that sought to synthesize the published works in terms of methods.

According to Godet (2000), his method and Peter Schwartz's methodology are the approaches frequently adopted in scenario studies. It is possible to observe in the literature the use of these methods in their original form or in an adapted format.

The Schwartz (2006) scenario-building methodology is organized into eight steps, as presented in Fig. 6.1. In this study, this method is used as the theoretical reference to generate the energy scenarios.

The energy field presents great tradition using scenario techniques. The use of the method in the industry began in the 1970s with Pierre Wack, Royal Dutch/Shell planner in London (Schwartz, 2006). Some organizations apply this type of tool such as the European Union and the International Institute for Applied Systems Analysis (Smith et al., 2005).

It is possible to observe in the literature the use of the scenario methods both in studies of nonspecific energy sectors, as can be shown in Söderholm et al. (2011), and in studies for specific forms of energy, such as the energy of oceans (Simioni, 2006), the solar energy (Medina et al., 2015), and the wind energy (Vichi and Mansor, 2009). Pielke Jr. and Ritchie (2021) state that scenarios make important contributions to the evaluation of policy options to be developed. In research in the energy field, scenarios are commonly used to characterize an envelope of expected future conditions and have gained prominence in the fields of study on energy efficiency and climate change (Ghanadan and Koomey, 2005).

Energy scenarios refer to a set of paths that are strategically designed to understand a particularly critical issue (Mcpherson and Karney, 2014). Chen et al. (2009) reinforce that for a great number of uncertainties tied to the business environment and technical development, as in the case of renewable energy, the analysis by scenarios has been widely used.

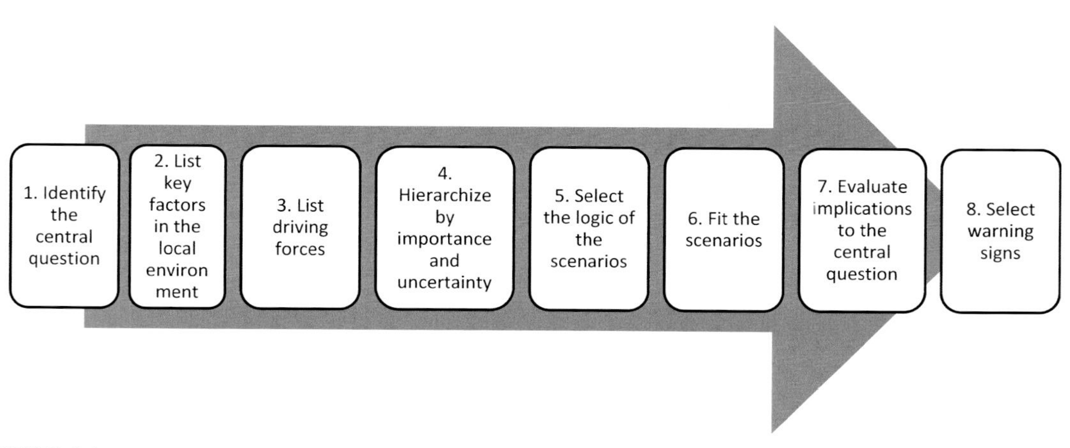

FIGURE 6.1

Peter Schwartz's method for scenarios generation—steps resume.

2.2 The 2050 calculator tool

The 2050 Calculator tool (EPE, 2016b) is an online tool that allows the generation of different energy scenarios until 2050. As mentioned by Strapasson et al. (2020), the success of the initiative started by the United Kingdom has inspired several other nations to prepare their own national calculators. In Latin America, many countries have adapted the 2050 Calculator such as Brazil, Mexico, Colombia, and Ecuador.

The 2050 Calculator has been used in some academic studies in the last years, as can be observed in: Park et al. (2012), studying South Korea; Gambhir et al. (2015), evaluating the Chinese sector; Allen and Chatterton (2013), Lannon et al. (2013), Spataru et al. (2014, 2015), and Newbery (2016), focusing on issues involving the United Kingdom; among others. It is noted that there are still few studies encompassing Latin America. Most of them are related to regions like the United Kingdom and Asia.

The Brazilian Calculator tool was launched in November 2016, during the 22nd Framework Conference of the Parties on Climate Change (COP22) in Marrakesh, Morocco. Through the technical assumptions and the mathematical model available in an MS Excel file on the calculator's own website (http://calculadora2050brasil.epe.gov.br/calculadora.html), this tool can be used to identify a range of technically and physically feasible scenarios for the future, allowing its impacts to be explored. The tool was developed by the British government, and it was adapted for the Brazilian reality by the Brazilian Energy Research Company (EPE), with the support of the Department of Energy and Climate Change, the British Embassy in Brazil, and the Universidade Federal do Rio de Janeiro.

To build the scenarios, the user must define the level of each variable, among the predefined options. Each variable has up to four distinct levels that represent the progressive effort to, on the supply side, evaluate possible increases in the energy supply considering even scenarios with renewable sources of energy and, in the consumption part, explore scenarios of energy efficiency gains and consumption of renewable energies.

Some variables work with levels ranging from 1 to 3, where classification 3 refers to the greater effort, 2 refers to a medium effort, and 1 refers to a minimum effort. Other variables use the letters A, B, C, and D as classification options, indicating different scenarios for the variable, not necessarily one being more difficult than the other.

2.3 Renewable energy in Latin America and Brazil

Countries in Latin America are putting its efforts to move forward on the energy transition. Many international organizations are working in cooperation with Latin American countries and institutions, such as the Latin American Energy Organization (OLADE), in order to strengthen the regional coordination and engagement. Irena (2021b) points out that debates are in progress and many themes are under discussion like long-term policy vision, leveraging hydropower resources, implementation of e-mobility strategies, energy efficiency, best practices related to tendering and risk mitigation mechanisms, improvement of power purchase agreements, increasing of renewable energy project through development partners and international organizations, development of bank financing, and others.

Jakob et al. (2019) address the importance of the "green fiscal reform" that is in progress in some countries, in Latin America. They point out that the initiatives are recent, and because of that, it is not possible to have an accurate conclusion of their effectiveness, however preliminary results are encouraging. They enumerated some critical challenges to the successful implementation of the "green fiscal reform" such as: (i) favorable political conditions, (ii) comprehensive reform plans, (iii) Sequencing of reforms and gradualism, and (iv) an efficient distribution and compensation model regarding subsidies and taxes.

Chile is being presented as a world leader in investment and use of nonconventional renewable energies since it is implementing policies to encourage investments and the consolidation of a low-carbon energy system. The air pollution is one of the most relevant environment issues in Chile, and it is caused mainly due to the widespread use of wood for cooking and heating (Boso et al., 2020).

In Brazil, the energy sector has an important contribution to GHG emissions. The country is committed to advancing gradually to increase the share of renewable energies in the energy matrix. According to EPE (2021), one of the biggest challenges in the country is the transport sector, as it has a strong dependence on fossil fuels. The Company claims that the country is encouraging the use of biofuels such as ethanol and biodiesel, in addition to discussing the increase in the electric vehicle fleet and seeking advances in energy efficiency, especially in the most consuming sectors represented by the industrial and transport sectors.

The use of renewables seems to be an important path to increase the renewables' participation in the energy supply side and to reduce GHG emissions. As per Cremonez et al. (2014), in South America, the liquid biofuels are represented mostly by ethanol and biodiesel with Brazil and Argentina leading the development and consumption. The authors also highlight that some countries are struggling internally to stimulate the use of biofuels as can be seen in Peru and Uruguay.

The energy consumption in Latin America grows at a rate greater than the total energy consumption worldwide since 1980 (Guerrero-Lemus and Shepard, 2017). It is important to consider that in Latin America, it is possible to still have people without access to electricity (Castilho, 2017), which implies that the rate of consumption mentioned could be even greater.

The goals defined by Latin American countries regarding the implementation of renewables and energy efficiency were also prompted by energy security needs. The energy transition is ongoing, and it is being supported by the active development of public policies, tax incentives, the creation of funds, and other initiatives (Castilho, 2017). Castilho (2017) defines that countries that generate more than 1000 GWh of electricity from renewables are considered as advanced ones. In Latin America, the author lists Brazil, Mexico, Chile, Costa Rica, El Salvador, Argentina, Uruguay, Guatemala, and Colombia as advanced countries. He points out that Guatemala, Costa Rica, and El Salvador have a considerable generation from geothermal sources, while in terms of solar and wind generation, Brazil, Chile, and Mexico stand out.

The development and use of renewables in Latin American countries are extremely necessary both to reach the expected energy consumption growth, year by year, on a sustainable basis and also to allow an energy supply system to be more reliable.

The Brazilian energy matrix remains among the most renewable in the world. The calculated percentage of renewable energy in the Brazilian matrix is about 46%, as presented in the national energy balance (EPE, 2020), based on 2019 data. The world average is approximately 14%, while the average of the OECD countries (Organization for Economic Co-operation and Development) is around 10%.

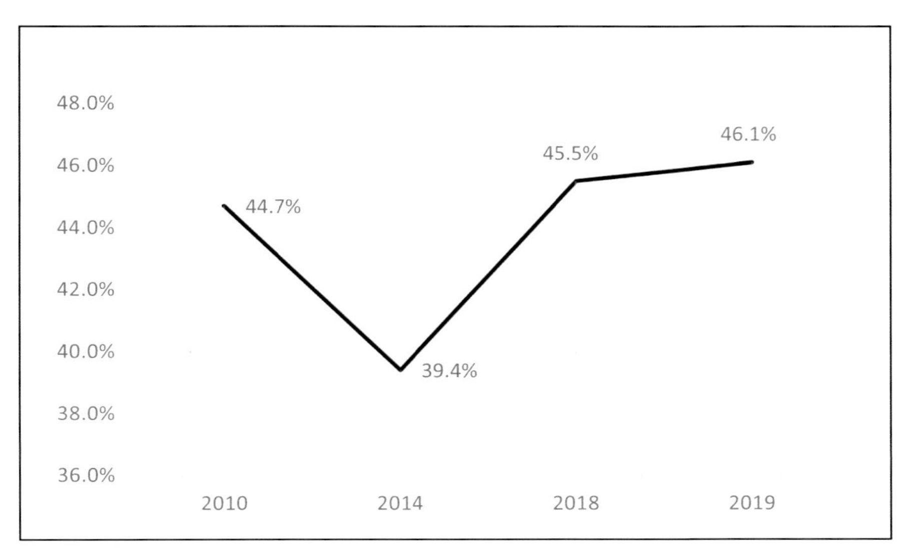

GRAPHIC 6.1

Share of renewables in the Brazilian energy matrix.

The Brazilian's good performance is a result of the abundant and varied sources of renewable energy in the country being hydropower and biomass the most representative ones. Additionally, it is interesting to notice the evolution of the so-called "other renewables" in this nation. According to EPE (2020), from 2018 to 2019, the solar energy supply has an increase of 92%, the wind energy raised 15.5%, and the biogas 31.8%. Although these energies still have small representativeness in the energy matrix, when adding the energy from lixivium, the contribution to the supply side is about 7% of the internal energy offer in Brazil. The data show that investments are being made, and the sector of "other renewables" is in development with a positive perspective in terms of the contribution to increase the share of renewables in the matrix in the coming years.

As per Graphic 6.1, it is possible to see an upward curve from 2014 to 2019. The increase in hydro, wind, biomass, and biofuel generation has an important role in this curve result after 2014. In order to better evaluate where strategic efforts must be applied to increase the share of renewables in the national energy matrix, it is important to understand which are the main sources of energy that compose the matrix, besides seeking to understand who consumes it and in what way this consumption happens.

In terms of energy consumption, according to EPE (2020), the industrial and transport sectors are together representing about 63% of the Brazilian energy demand. Considering this, the concentration of efforts to increase the share of renewables in the energy supply side for these two sectors seems to be fundamental to have a significant increase of renewables in the national energy matrix.

Despite the Brazilian energy matrix figures as one of the most renewable in the world, renewables still have opportunities to be expanded and, in the end, to help the energy matrix to be predominantly renewable in the close future.

3. Study methods

The method used in this study consists basically in combining the methodology of scenario building developed by Schwartz (2006), with the Brazilian 2050 Calculator tool, which means to include the existing 2050 Calculator information in some steps of Schwartz (2006). The subitems below present the path taken to create the energy scenarios.

3.1 First step of Schwartz—definition of the central question

The central question has the function of guiding all the other steps of the scenario construction. In this study, it was defined by the authors. So, the question defined in this work is: is it plausible for the Brazilian energy matrix to become predominantly renewable by 2050?

3.2 Second and third steps of Schwartz—list of the key factors and the driving forces

Steps 2 and 3 of Schwartz (2006) suggest listing the key factors and the driving forces, which, respectively, have influence in the micro- and the macroenvironment of the central question. In order to fulfill this stage, in this study, it was considered as key factors and driving forces the 31 variables already available in the Brazilian 2050 Calculator tool that represents the energy supply and energy demand for the Brazilian context. Thus, it was assumed that the 31 variables are the forces impacting the micro- and macroenvironment around the central question.

3.3 The fourth step of Schwartz—hierarchize the key factors and the driving forces

The objective in this step is to define which of the 31 variables should be considered as guiding ones. Guiding variables (GVs) are the variables that will be fundamentally used to differentiate the future energy scenarios. To perform this task, it is necessary to arrange the variables in the hierarchy.

The hierarchy of the variables was defined using energy experts' opinions. To perform it, a workshop with specialists was developed. It is desirable to use a workshop because it is performed face-to-face (not necessarily physically) with all the participants together on the same day and time. Performing the task this way corroborates with Schwartz's original method which emphasizes the importance of face-to-face meetings to create the strategic conversation environment that is fundamental to the process of scenario generation.

To proceed with the hierarchy definition, it is necessary that the variables of both supply and demand sides are organized in subgroups, grouped by similarity to allow the experts to define which one is the most important in relation to the central question. The subgroups defined in this study using the elements of the Brazilian 2050 Calculator tool are presented below.

For each subgroup, the experts must define, in consensus or by a majority, which of the variables must be considered the most important one in relation to the central question. In other words, which one must be chosen as GV. Also, for each subgroup, the experts can decide that none of the variables should be considered as GV if they understand the ones analyzed have no influence on the central question under discussion. Table 6.1 shows all the variables arranged in subgroups, based on the 2050 Calculator.

Table 6.1 Subgroup's composition.

Demand variables	
Subgroup created	**Key factors and driving forces (from 2050 Calculator)**
1—Passenger transport	Modal choice
	Efficient vehicles
2—Load transport	Modal distribution
	Efficiency
3—Fuels	Preference of ethanol use in flex vehicles
	Share of biodiesel in diesel
4—Residences	Residential sector—energy efficiency
5—Trade and services	Commercial and public sector—energy efficiency
6—Industries	Industry composition
	Energy efficiency
7—Agricultural sector	Energy efficiency
Supply variables	
Subgroup created	**Key factors and driving forces (from 2050 Calculator)**
8—Thermal	Natural gas thermoelectric plants—installed power
	Natural gas thermoelectric plants—CCS
	Coal-fired power plants—installed power
	Coal-fired power plants—CCS
	Thermoelectric to petroleum products
	Use of biomass and biogas
	Use of surplus bagasse
	Priority of biogas use
	Efficiency of biofuel plants
9—Nuclear	Nuclear energy
10—Other renewable sources	Onshore wind energy
	Offshore wind energy
	Ocean energy
	Photovoltaic solar energy
	Heliothermic solar energy
11—Hydraulic	Hydraulic energy
	Importation of binational hydroelectric plants
12—Electrical system	Electrical system safety
13—Gas oil production	Associated oil and gas production
	Production of nonassociated natural gas

Table 6.2 Guiding variable of each subgroup.

Subgroup	Chosen guiding variable (GV)
1—Passenger transport	Passenger transport—efficient vehicles
2—Load transport	Load transport—modal distribution
3—Fuels	Preference of ethanol use in flex vehicles
4—Residences	Residential sector—energy efficiency
5—Trade and services	Commercial and public sector—energy efficiency
6—Industries	Industry composition
7—Agricultural sector	*No GV chosen by the experts*
8—Thermal	Use of biomass and biogas
9—Nuclear	*No GV chosen by the experts*
10—Other renewable sources	Photovoltaic solar energy
11—Hydraulic	Hydraulic energy
12—Electrical system	Electrical system safety
13—Gas oil Production	*No GV chosen by the experts*

Once defined the GVs of each subgroup, the other variables that compose the subgroup will be classified as support variables (SVs). In this study, as defined by the experts, 10 of the 13 subgroups contributed with a GV to differentiate the energy scenarios. Table 6.2 synthesizes the GVs defined by the experts for each subgroup.

The workshop needs to be driven by a person defined as the mediator to guide the experts in the steps of the energy scenario creation. It is interesting to arrange the meeting in two stages: one theoretical and one practical. In the theoretical stage, the context of the research, the objectives, the Schwartz's method for scenario generation, and the 2050 Calculator tool are presented to the participants. The aim of this step is to contextualize and present the method and the tool that will be used in the scenario's generation dynamics.

In this study, the workshop was attended by six experts. The group had professionals from different entities acting in the Brazilian energy sector with experiences in research activities and energy planning. Among them, three have master's degrees and two have doctor's degrees. Three of the six participants work at the Brazilian Energy Research Company (EPE), two at the Brazilian National Bank for Economic and Social Development (BNDES), and one is a Professor at Celso Suckow da Fonseca that is a Federal Center for Technological Education based in Rio de Janeiro (CEFET-RJ). It is stood out that this is the personal opinion of the participants, there is, therefore, no institutional involvement.

According to Schwartz (2006), it is interesting that the group includes people with different roles in order to capture different perspectives. Thus, all the mixed views tend to generate portraits of more complete environments and, consequently, more plausible scenarios.

3.4 Fifth step of Schwartz—defining the logic of the scenarios

In this study, the logic below was defined by the authors to guide the experts on the scenario generation. Three different logics were defined, as described below.

3.4.1 Logics for scenario "Most renewable"

In this scenario, the GVs defined by the experts in the previous step will be configured in the 2050 Calculator tool with the highest effort to the central question. All the other variables (SVs) will be evaluated and defined by the group of experts.

3.4.2 Logics for scenario "Medium renewable"

In this scenario, the defined GVs will be configured in the 2050 Calculator tool with the medium effort to the central question. All the other variables (SVs) will be evaluated and defined by the group of experts.

3.4.3 Logics for scenario "Business-As-Usual"

In this scenario, the defined GVs will be configured in the 2050 Calculator tool with the lowest effort to the central question. All the other variables (SVs) will be evaluated and defined by the group of experts.

3.5 Sixth step of Schwartz—embodying the scenarios

With the GVs and the logics of the scenarios defined, it is time to apply the scenario logics described in item 3.1.4 in the Brazilian 2050 Calculator tool for each one of the scenarios to get the results. After finalizing the GVs configuration and the SVs classification for each scenario, the 2050 Calculator tool generates a URL, and this electronic address can be shared with others or even copied and saved for discussions and future access.

It is important to state that the descriptions for each level of the variables are captured directly from the 2050 Calculator. Behind the Calculator, there is a complex database elaborated with the support of the Universidade Federal do Rio de Janeiro that considers a defined Brazilian macroeconomic environment. It is not an energy model, but a platform for calculating several combinations of supply and demand scenarios, indicating the implications of each one and helping discussions regarding energy planning.

3.6 Seventh step of Schwartz—implications to the central question

At this moment, the scenarios are already formed with all their details and can be consulted using the URLs generated. It means that all the variables of the 2050 Calculator (GVs and SVs) are already classified by the experts and configured in the online tool. The information extracted from the 2050 Calculator tool portrays three different long-term visions for the energy sector. At this moment, it is possible to analyze the energy matrix composition by 2050 for each scenario and verify if any scenario showed a predominance of renewables.

3.7 Eighth step of Schwartz—definition of warning signs

Schwartz (2006) suggests that indicators for monitoring should be defined to serve as warning signs which allows identifying, as soon as possible, which of the scenarios is closest to the current course of the history. Still based on him, with well-chosen indicators, it is possible to be more efficient and effective in strategic decision-making in favor of the central question. Thus, this stage aims to define indicators for monitoring along the time with the function of guiding strategic evaluations, decision-making, and anticipation of actions in relation to the defined central question.

4. Results and discussions

The URLs generated by the 2050 Calculator, using the method described in Section 6., show the details of each one of the three scenarios created. Some of the main information extracted from the scenarios are presented below (Table 6.3).

Graphics 6.2–6.5 below show and compare the behavior of each GV for each scenario based on the data and explanation presented in the 2050 Calculator tool. These graphs help to draw the landscape of each scenario regarding the main variables that impact the share of renewables in the energy matrix by 2050.

In Graphic 6.2, considering the "Most Renewable" scenario, it is possible to highlight the expectation regarding the increase of the electric and hybrid vehicles fleet and also the rising in the use of ethanol in flex vehicles, helping to push the use of renewables from the demand side.

For the scenario "Most Renewable," Graphic 6.3 shows a possible reduction in electricity consumption in the industry sector led by its composition since this scenario expects the industry to be composed of less energy-intensive companies. It is important to highlight that for the GV "Industry Composition," which has level B as the less favorable level to the central question, it was requested by the experts that in scenario "Business-As-Usual (BAU)," the option "A" should be considered with the same level classified for scenario "Medium Renewable." According to the participants, it would be more plausible the description of the option "A" presented in the 2050 Calculator, even in the scenario most unfavorable to the renewability of the energy matrix.

In terms of the supply energy side, as per Graphic 6.4, it is possible to observe an important role of the photovoltaic solar energy in the "Most Renewable" scenario, since it is expected an expressive increase in its power capacity installed.

It is important to highlight that the GV "Hydraulic energy" was the only one that was not configured with the maximum effort level of the 2050 Calculator tool in relation to the central question

Table 6.3 Scenarios' URLs and energy matrix composition summary.

Central question: Is it plausible for the Brazilian energy matrix to become predominantly renewable by 2050?			
Energy matrix composition by 2050	**Scenario "Most Renewable"** Renewable energy: 48% Nonrenewable energy: 52%	**Scenario "Medium Renewable"** Renewable energy: 34% Nonrenewable energy: 66%	**Scenario "BAU"** Renewable energy: 28% Nonrenewable energy: 72%
URL "Most renewable"	http://calculadora2050brasil.epe.gov.br/pathways/011011132121021123223320023303202030303202/energy_security/comparator/01101111111011111111110011101101010101101		
URL "Medium renewable"	http://calculadora2050brasil.epe.gov.br/pathways/01101112212102 11f2122320022202202020201202/energy_security/comparator/01101111111011111111110011101101010101101		
URL "BAU"	http://calculadora2050brasil.epe.gov.br/pathways/01101112121021111121320021101202010101202/energy_security/comparator/01101111111011111111110011101101010101101		

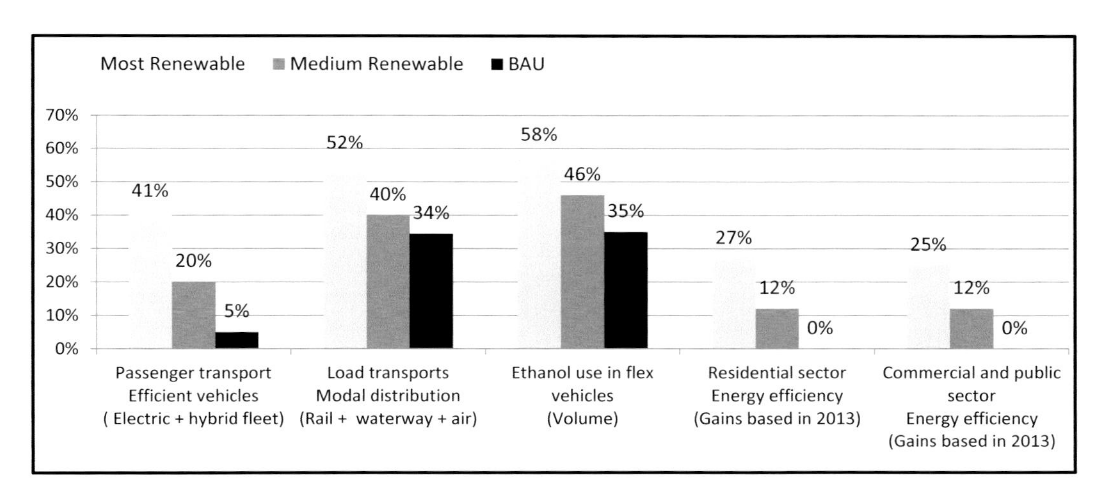

GRAPHIC 6.2

2050 scenarios summary—GVs from demand side.

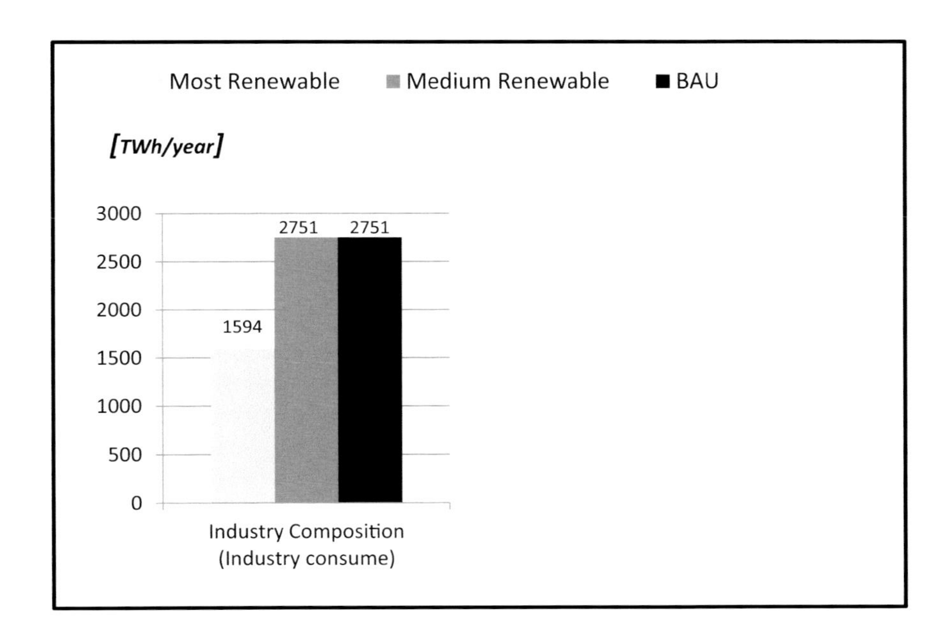

GRAPHIC 6.3

2050 scenarios summary—GV from demand side (cont.).

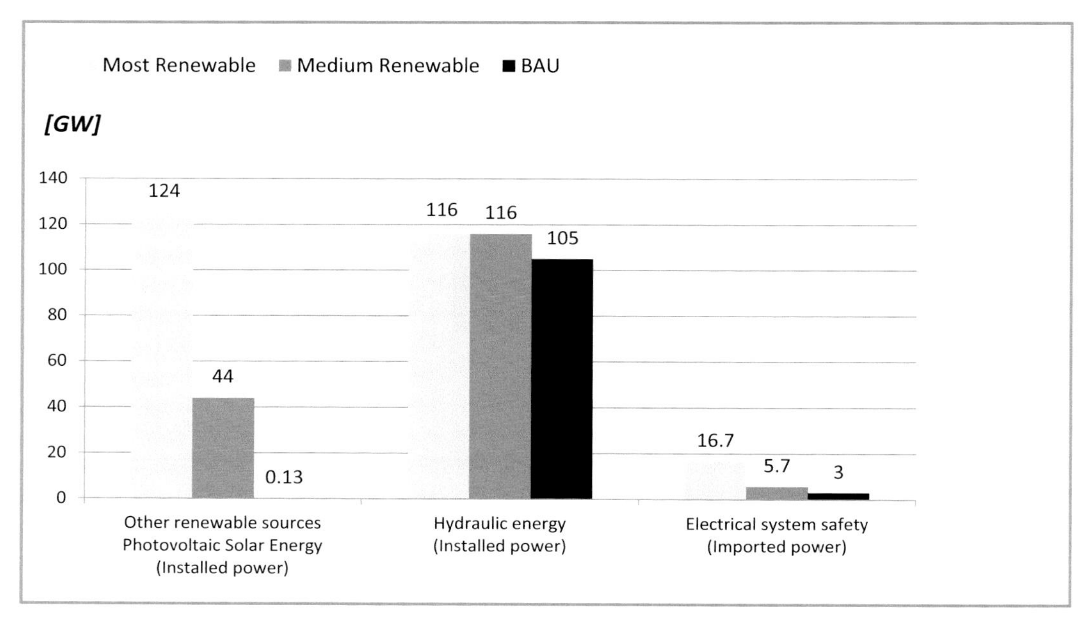

GRAPHIC 6.4

Scenario resume—GVs from supply side.

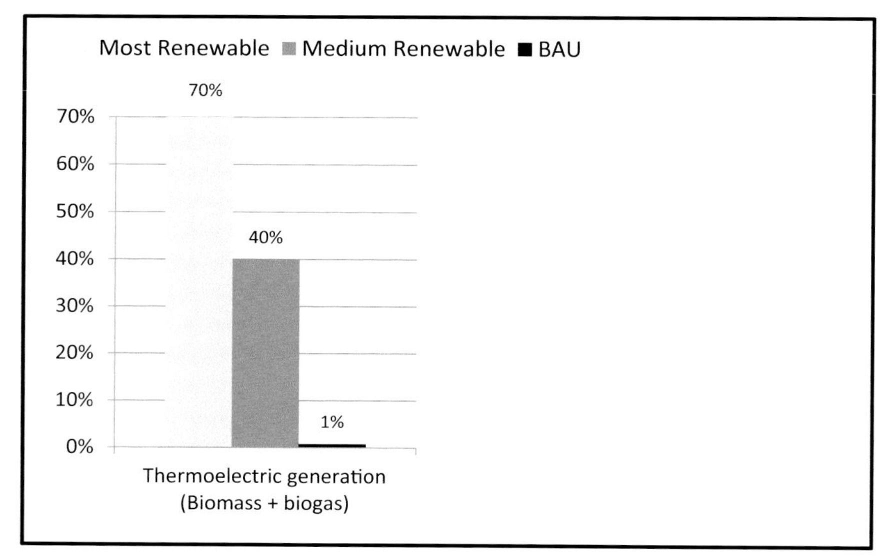

GRAPHIC 6.5

Scenario resume—GVs from supply side (cont.).

for the "Most Renewable" scenario. This happened because, in the experts' understanding, in none of the scenarios this energy source would reach the full inventory potential (172 GW) by 2050, as the tool suggests. The consensus by the experts is that the concerns with the environmental and social impacts of this type of large project should not make it feasible to reach the national full inventory.

Still considering the supply side, as per Graphic 6.5, the "Most Renewable" scenario shows that it is plausible for a remarkable increase regarding the use of biomass and biogas to the thermoelectric generation by 2050, helping to increase the use of renewables on the supply side.

In terms of the SVs, as can be seen by accessing the URLs, under the experts' vision, most of the SVs would follow the same behavior until 2050 regardless the scenario, excepting the SV "Heliothermic solar energy (CSP)." For the scenario "Most Renewable," the participants chose level 2 for this SV, suggesting that by 2050 there would be an expansion of 4.4 GW. For scenarios "Medium Renewable" and "BAU," it was decided to keep effort 1, which means just the development of a pilot project already ongoing.

Graphic 6.6 illustrates the behavior between renewables and nonrenewables in the Brazilian energy matrix for each scenario.

Considering the energy scenarios generated for 2050, it is possible to deduce that the energy matrix could regress in terms of share of renewables since two of three scenarios generated ("Medium Renewable" and "BAU") showed reductions of renewables comparing with the number from 2019. The numbers of these last two scenarios express that efforts may not be done properly to develop renewables to meet the expected energy consumption growth by 2050.

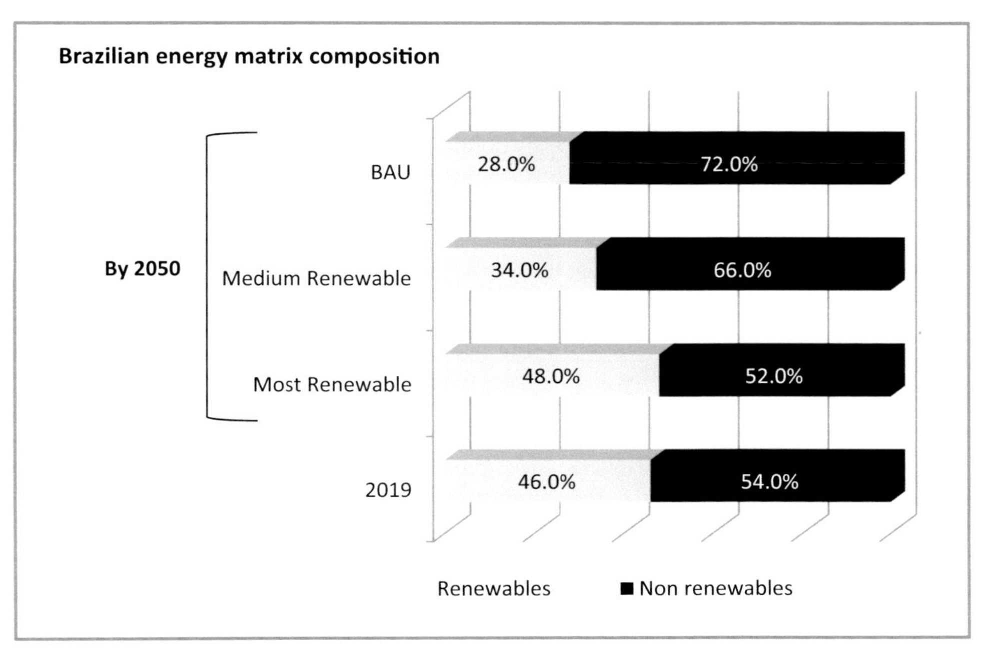

GRAPHIC 6.6

Brazilian energy matrix composition.

Table 6.4 Warning signs.

Item	Subgroup	Variable	Warning signs	Unit
1	–	Brazilian energy matrix	Share of renewables in the Brazilian energy matrix	%
2	Passenger transport	Passenger transport—efficient vehicles	Participation of electric and hybrid vehicles in the national market	%
3	Load transports	Load transport—modal distributions	Percentage of rail, waterway, and air modal against road modal	%
4	Fuels	Preference of ethanol use in flex vehicles	Ethanol use in flex vehicles	%
5	Other renewable sources	Photovoltaic solar energy	Installed power	GW
6	Hydraulic	Hydraulic energy	Installed power	GW

"Most Renewable" scenario was the closest one to an energy matrix composed mainly of renewable sources, with a participation of 48% from renewables and 52% from nonrenewables. In scenarios "Medium Renewable" and "BAU," the share of renewables is reduced to 34% and 28%, respectively.

It is possible to observe that the scenarios have a significant gap in renewables participation among themselves. The difference between scenarios "Most Renewable" and "Medium Renewable" is 14%, while the difference between scenarios "Medium Renewable" and "BAU" is 6%. It suggests that there may be potential gains in terms of matrix renewability the greater the effort employed in the development of the variables identified as guiding (GVs) of the scenarios.

In order to follow up the scenarios along the time, as mentioned, it is interesting to define warning signs to evaluate which one of the energy scenarios most closely approximates the current course of history, helping discussions on this theme. Table 6.4 presents a suggestion for warning signs to be monitored based on the GVs.

5. Conclusions

The negative consequences of human actions on the environment are notorious worldwide, and the energy sector has a significant contribution. The energy from fossil sources implies emission with high-carbon content which is harmful to the environment. The energy transition is mandatory to define a better relationship between humans and the environment.

The energy demand is expected to grow, and Latin America seems to have rates still greater than other parts of the world. To reach the energy demand on a sustainable basis is mandatory the development of renewable sources and investments in infrastructure that enable the transition to the new energy paradigm. Latin America is plentiful in terms of natural resources which brings a positive outlook regarding renewables. It is possible to notice that the renewables' development is ongoing, and it is being supported by active development of public policies, tax incentives, the creation of funds, auctions, and other initiatives. Countries like Brazil, Mexico, Chile, Costa Rica, El Salvador, Argentina, Uruguay, Guatemala, and Colombia emerge as major generators of energy from renewable sources.

The chapter aims to present a brief outlook regarding renewables in Latin America and present a method for future scenario generation combining the 2050 Calculator tool with Peter Schwartz's method for scenario building. A case study considering the Brazilian energy context was developed. The main question that leads the scenario generation was: is it plausible for the participation of the renewables to surpass the nonrenewables in the Brazilian energy matrix?

Combining the methodology of scenario building developed by Schwartz (2006) and the Brazilian 2050 Calculator tool, it was possible to generate three different scenarios for the Brazilian energy matrix, besides to propose warning signs to verify which one of the scenarios most closely approximates to the current course of the history along the time.

Results show that in the scenario "Most Renewable," Brazil reaches a share of 48% of renewables in the energy matrix by 2050. The scenario "Medium Renewable" presents 34% of renewables and the scenario "BAU" shows 28% of renewables' participation. It is possible to observe that "Most Renewable" scenario, although it did not reflect more than 50% of renewables in the energy matrix until 2050, it can be considered a successful and desirable scenario demanding significant efforts by the country on this direction.

It should also be noted that two of the three scenarios presented the participation of the renewables in percentages lower than the one registered in 2019, as seen in Graphic 5.6. This allows inferring that, if the increase in energy consumption is not followed by the development of energy supply from renewable sources, it is quite possible that Brazil will regress in relation to the share of renewable sources in the energy matrix by 2050.

As mentioned, in order to verify which one of the scenarios most closely approximates the current course of history, six warning signs were suggested to be monitored along the time. As per Schwartz (2006), with well-chosen indicators, it is possible to be more efficient and effective in strategic evaluations, decision-making, and anticipation of actions in relation to the defined central question.

It is important to state that the results presented here express the expectations of the consulted experts. Nevertheless, in the case of technological changes and disruptive innovations, there may be a significant impact on the worked variables and, consequently, over the found results regarding the renewability of the matrix.

As a suggestion for future studies, it is quite interesting that other Latin American countries could generate their energy scenarios for 2050 with the method herein described. It could help discussions regarding the path taken to create the scenarios and also could promote the energy planning discussions between the Latin American countries with a focus on the energy transition, on the energy matrix's renewability and on the security for the region's energy system.

The warning signs could be updated periodically in order to get information about the path the warning signs are taken. It could help to understand the energy trends in the country and also to better define which sectors or actions need to be prioritized. Also, new scenarios could be generated with another group of experts to compare expectations regarding renewables and the energy matrix.

It should be noted that, from time to time, the 2050 Calculator may be updated by the institution that manages the tool. In this case, it is also important to generate new scenarios, since the database that supports the tool could bring a new outlook for the energy in the country.

Finally, the proposal of this chapter was to present an easy and practical way to encourage discussions regarding the energy transition. The energy from fossil sources causes high carbon content emissions which is harmful to the environment and to the society. Defining a better relationship between humans and the environment seems to be the best path to build a healthy future.

References

Allen, P., Chatterton, T., 2013. Carbon reduction scenarios for 2050: an explorative analysis of public preferences. Energy Policy 63, 796–808.

Andrade, G., Alves, L., Andrade, F., Mello, J., 2016. Evaluation of power plants technologies using multicriteria methodology Macbeth. IEEE Latin America Transactions 14 (1), 188–198.

Bacha, M., Santos, J., Schaun, A., 2010. Theoretical considerations on the concept of sustainability. In: VII Symposium on Excellence in Management and Technology, p. 14.

Baratsas, S., Niziolek, A., Onel, O., Matthews, L., Floudas, C., Hallermann, D., Sorin, M., Sorescu, S., Pistikopoulos, E., 2021. A framework to predict the price of energy for the end-users with applications to monetary and energy policies. Nature Communications 12, 18.

Bigerna, S., Bollino, C.A., Micheli, S., 2016. Renewable energy scenarios for costs reductions in the European Union. Renewable Energy 96, 80–90.

Blois, H., 2008. Prospective scenarios and systems dynamics: proposal for a footwear industry model. Journal of Business Administration 48 (3), 35–45.

Boso, À., Garrido, J., Álvarez, B., Oltra, C., Hofflinger, Á., Gálvez, G., 2020. Narratives of resistance to technological change: drawing lessons for urban energy transitions in southern Chile. Energy Research & Social Science 65, 101473.

Carvalho, D., Sutther, M., Polo, E., Wright, J., 2011. Construction of scenarios: appreciation of methods most used in strategic management. In: XXXV Meeting of ANPAD, pp. 1–17.

Castilho, L., 2017. Renewable energy and energy efficiency in Latin America: a regulatory vision. Journal of Energy & Natural Resources Law 35 (4), 405–416. https://doi.org/10.1080/02646811.2017.1370175.

Chaharsooghi, S.K., Rezaei, M., Alipour, M., 2015. Iran's energy scenarios on a 20-year vision. International Journal of Environmental Science and Technology 12 (11), 3701–3718.

Chen, T., Yu, O., Hsu, G., Hsu, & F., Sung, W., 2009. Renewable energy technology portfolio planning with scenario analysis: a case study for Taiwan. Energy Policy 37 (8), 2900–2906.

Cremonez, P., Feroldi, M., Feidena, A., Teleken, J., Grisb, D., Dieterb, J., de Rossia, E., Antonellia, J., 2014. Current scenario and prospects of use of liquid biofuels in South America. Renewable and Sustainable Energy Reviews 43, 352–362.

De Díaz, M., Lobo, M., Geraldino, M., 2013. By building new trends and future scenarios for journalism: a prospective vision. Revista Espacios 34 (5).

Dias, M.A.D.P., Vianna, J.N.D.S., Felby, C., 2016. Sustainability in the prospective scenarios methods: a case study of scenarios for biodiesel industry in Brazil, for 2030. Futures 82.

Empresa de Pesquisa Energética (EPE), 2016a. Brazil's Commitment to Combat Climate Change: Production and Use of Energy. http://www.epe.gov.br/sites-en/publicacoes-dados-abertos/publicacoes/PublicacoesArquivos/publicacao-181/NT%20COP%2021%20-English.pdf (Accessed February 2021).

Empresa de Pesquisa Energética (EPE), 2016b. 2050 Calculator Tool. http://calculadora2050brasil.epe.gov.br/calculadora.html# (Accessed February 2021).

Empresa de Pesquisa Energética (EPE), 2020. Synthesis Report on the National Energy Balance 2020. www.epe.gov.br/pt/publicacoes-dados-abertos/publicacoes/balanco-energetico-nacional-2020 (Accessed February 2021).

Empresa de Pesquisa Energética (EPE), 2021. Mudanças climáticas e Transição energética. https://www.epe.gov.br/pt/abcdenergia/clima-e-energia (Accessed September 2021).

Fuentes, S., Villafafila-Robles, R., Rull-Duran, J., Galceran-Arellano, S., 2021. Composed index for the evaluation of energy security in power systems within the frame of energy transitions—the case of Latin America and the Caribbean. Energies 14, 2467.

Gambhir, A., Tse, L., Tong, D., Botas, R., 2015. Reducing China's road transport sector CO_2 emissions to 2050: technologies, costs and decomposition analysis. Applied Energy 157, 905–917.

Ghanadan, R., Koomey, J., 2005. Using energy scenarios to explore alternative energy pathways in California. Energy Policy 33 (9), 1117–1142.

Godet, M., 2000. The art of scenarios and strategic planning: tools and pitfalls. Technological Forecasting and Social Change 65 (1), 3–22.

Guerrero-Lemus, R., Shephard, L., 2017. Current energy context in Africa and Latin America. SpringerLink 38, 39–73.

International Renewable Energy Agency (IRENA), 2021. Energy Transition. https://www.irena.org/energytransition (Accessed September 2021).

International Renewable Energy Agency (IRENA). Energy Transition, 2021. Renewable Energy Deployment in Latin America. https://www.irena.org/events/2018/Aug/Renewable-Energy-Deployment-in-Latin-America (Accessed September 2021).

Irfan, M., Hao, Y., Ikram, M., Wu, H., Akram, R., Rauf, A., 2020. Assessment of the public acceptance and utilization of renewable energy in Pakistan. Sustainable Production and Consumption 27, 312–324.

Jakob, M., Soria, R., Trinidad, C., Edenhofer, O., Bak, C., Bouille, D., Buira, D., Carlino, H., Gutman, V., Hübner, C., Knopf, B., Lucena, A., Santos, L., Scott, A., Steckel, J.C., Tanaka, K., Vogt-Schilb, A., Yamada, K., 2019. Green fiscal reform for a just energy transition in Latin America. Economics: The Open-Access, Open- Assessment E-Journal 13 (2019–17), 1–11.

Karatayev, M., Hall, S., Kalyuzhnova, Y., Clarke, M., 2016. Renewable energy technology uptake in Kazakhstan: policy drivers and barriers in a transitional economy. Renewable and Sustainable Energy Reviews 66, 120–136.

Lachman, D., 2011. Leapfrog to the future: energy scenarios and strategies for Suriname to 2050. Energy Policy 39 (9), 5035–5044.

Lannon, S., Georgakaki, A., MacDonald, S., 2013. Modelling urban scale retrofit, pathways to 2050 low carbon residential building stock. In: 13th Conference of the International Building Performance Simulation Association, vol. 2025, pp. 3441–3448.

Lee, C., Zhong, J., 2014. Top-down strategy for renewable energy investment: conceptual framework and implementation. Renewable Energy 68, 761–773.

Lehoux, P., Gauthier, P., Jones, B., Fishman, J., Hivon, M., Vachon, P., 2014. Examining the ethical and social issues of health technology design through the public appraisal of prospective scenarios: a study protocol describing a multimedia-based deliberative method. Implementation Science 9 (1), 81.

McPherson, M., Karney, B., 2014. Long-term scenario alternatives and their implications: LEAP model application of Panama's electricity sector. Energy Policy 68, 146–157.

Medina, E., Arce, R., Mahía, R., 2015. Barriers to the investment in the Concentrated Solar Power sector in Morocco: a foresight approach using the Cross Impact Analysis for a large number of events. Futures 71, 36–56.

Moritz, G., 2004. Planning for Prospective Scenarios: The Construction of a Case-Based Methodological Framework (Ph.D. thesis). Universidade Federal de Santa Catarina. Florianópolis, Brazil.

Newbery, D., 2016. A Simple Introduction to the Economics of Storage: Shifting Demand and Supply over Time and Space. EPRG Working Paper. University of Cambridge.

Nikolaev, A., Konidari, P., 2017. Development and assessment of renewable energy policy scenarios by 2030 for Bulgaria. Renewable Energy 111, 792–802.

Oliveira, A., de Barros, M., Pereira, F., Gomes, C., Costa, H., 2018. Prospective scenarios: a literature review on the Scopus database. Futures 100, 20–33.

Park, N., Yoo, J., Jo, M., Yun, S., Jeon, E., 2012. Comparative analysis of scenarios for reducing GHG emissions in Korea by 2050 using the low carbon path calculator. Journal of Korean Society for Atmospheric Environment 28 (5), 556–570.

Pielke Jr., Ritchie, J., 2021. Distorting the view of our climate future: the misuse and abuse of climate pathways and scenarios. Energy Research & Social Science 72, 101890.

Prasad, R., Bansal, R.C., Raturi, A., 2014. Multi-faceted energy planning: a review. Renewable and Sustainable Energy Reviews 38, 686–699.

Rico, A., Van Den Brink, P.J., Gylstra, R., Focks, A., Brock, T.C., 2016. Developing ecological scenarios for the prospective aquatic risk assessment of pesticides. Integrated Environmental Assessment and Management 12, 510–521.

Santos, J., Torres, E., 2016. Expansion of Electric Generation in Bahia until 2050 Based on New Renewable Energies. Technical Scientific Congress of Engineering and Agronomy. Paraná, Brazil.

Saurin, R., Ratcliffe, J., 2011. Using an adaptive scenarios approach to establish strategies for tomorrow's workplace. Foresight 13 (4), 46–63.

Schmidt-Scheele, R., 2020. 'Plausible' energy scenarios?! How users of scenarios assess uncertain futures. Energy Strategy Reviews 32, 100571.

Schwartz, P., 2006. The art of the long-term vision: planning the future in a world of uncertainties. In: Best Seller, fourth ed.

Simioni, C., 2006. The Use of Sustainable Renewable Energy in the Brazilian Energy Matrix: Obstacles to the Planning and Extension of Sustainable Policies (Ph. D. thesis). Federal University of Paraná.

Smith, R., Vesga, D., Cadena, A., Boman, U., Larsen, E., Dyner, I., 2005. Energy scenarios for Colombia: process and content. Futures 37 (1), 1–17.

Söderholm, P., Hildingsson, R., Johansson, B., Khan, J., Wilhelmsson, F., 2011. Governing the transition to low-carbon futures: a critical survey of energy scenarios for 2050. Futures 43 (10), 1105–1116.

Spataru, C., Drummond, P., Zafeiratou, E., Barret, M., 2015. Long-term scenarios for reaching climate targets and energy security in UK. Sustainable Cities and Society 17, 95–109.

Spataru, C., Zafeiratou, E., Barret, M., 2014. An analysis of the impact of bioenergy and geosequestration in the UK future energy system. Energy Procedia 62, 733–742.

Strapassona, A., Woodsa, J., Perez-Cirerac, V., Elizondod, A., Cruz-Canoe, D., Pestiauxf, J., Cornetf, M., Chaturvedi, R., 2020. Modelling carbon mitigation pathways by 2050: insights from the global calculator. Energy Strategy Reviews 29, 100494.

United Nations, 2021. Paris Agreement, 2021. Nationally Determined Contributions (NDCs). In: https://unfccc.int/process-and-meetings/the-paris-agreement/nationally-determined-contributions-ndcs/nationally-determined-contributions-ndcs (Accessed February 2021).

Vichi, F., Mansor, M., 2009. Energy, environment and economy. Brazil in the global context. Quim.Nova 32 (3), 757–767.

Vigoya, M., Mendoza, J., Abril, S., 2020. International energy transition: a review of its status on several continents. International Journal of Energy Economics and Policy 10, 216–224.

Wortelboer, F., Bischof, B., 2012. Scenarios as a tool for supporting policy-making for the Wadden Sea. Ocean and Coastal Management 68, 189–200.

CHAPTER

7 Phase change: getting to a sustainable energy future in Vietnam

David Dapice[1], Phu V. Le[2] and Thai-Ha Le[2]

[1]Harvard Kennedy School, Harvard University, Cambridge, MA, United States; [2]Natural Capital Management Program, Fulbright School of Public Policy and Management, Fulbright University Vietnam, Ho Chi Minh City, Vietnam

Chapter outline

1. Background

Vietnam has grown quickly, reduced poverty and provided basic services including electricity to virtually all its people. The country is ruled by the Communist Party which puts a high priority on sustaining rapid growth, but foremost on its own survival. The country is naturally open, having a high ratio of trade to income, many guest workers abroad, millions of foreign visitors, billions of dollars a year in foreign investments, *Viet Kieu* who now return for family visits, and access to Facebook and western news sites that China blocks. These successes, however, have created problems. One is that pollution has been rising and is a public concern. Another is that the cost of expanding energy usage much faster than its neighbors has negative political, health, and economic implications. A third is that

Handbook of Energy and Environmental Security. https://doi.org/10.1016/B978-0-12-824084-7.00024-2
Copyright © 2022 Elsevier Inc. All rights reserved.

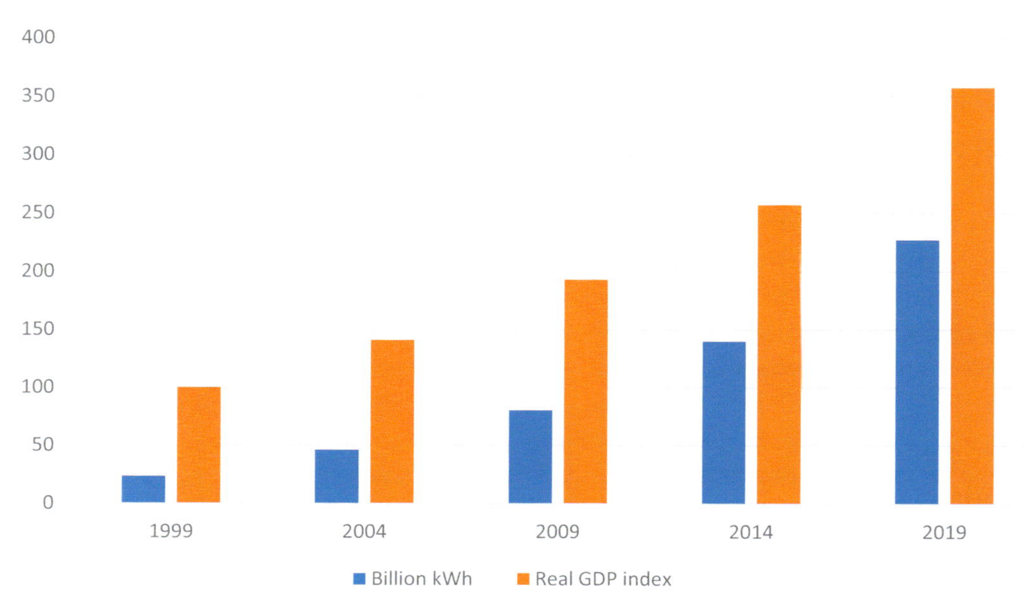

FIGURE 7.1

Growth in electricity production and real GDP.

most new energy use is imported while past supplies were mainly domestic. These factors influence the thinking about future energy policy, particularly with reference to electricity.

Vietnam has increased its electricity output nearly 10 times in the 20 years of 1999–2019, much faster than the rise in real gross domestic product (GDP) of 3.6 times (see Fig. 7.1). A rise of 10% in Vietnam's real GDP produced an 18% rise in electricity, much more than any other major ASEAN (The Association of Southeast Asian Nations) economy or China or India. Even in 2019, Vietnam's GDP rose about 7% and its electricity use by 10%—despite an electricity/GDP ratio that was already higher than China's.[1] China now typically has electricity growth at or below its real GDP growth. This is surprising because China is more industrial and urban than Vietnam, and these factors tend to increase electricity use. Vietnam's conservation policies have not been effective and hampered by low power prices.

Beyond the heavy use and fast growth of electricity, there is the question of how power should be produced. Vietnam started relying mainly on large hydroelectric projects, but the best sites are already developed. Future supplies will largely rely either on fossil fuels or solar and wind generation. Nuclear generation has been ruled out for the time being. Past power development plans (PDPs) relied primarily on coal generation, but more recently, natural gas generation and renewable energy are getting more attention (Baulch et al., 2018). Gas is viewed as cleaner but also more expensive. This is due

[1]The statistics are taken from Asian Development Bank (ADB) Key Indicators Database (https://kidb.adb.org/kidb/) and Vietnam's General Statistics Office (GSO) under "Monthly Statistical Information" and "Main Industrial Products". Some figures are authors' calculations based on raw statistics.

Table 7.1 Electricity shares generated by source in 2014 (actual) and 2030 (projected)[6].

	2014	2030
Hydroelectricity	46.1%	12.4%
Coal	28.6%	53.2%
Gas	21.5%	16.8%
Oil	3.4%	Nil
Renewables (other)	0.3%	10.7%

partly to the high taxation of domestic gas production and low taxation of imported and locally produced coal.[2] Without taxes, the two sources typically have similar financial costs in other countries, but gas has lower pollution costs. In Vietnam, gas from electricity costs about double that in other countries, for reasons that are not clear.[3] New generating plants using either gas or coal will require importing fuel.[4] Vietnam Electricity (EVN), the state utility, has preferred coal over gas due to costs and the easy availability of financing for new coal plants. This preference is reflected in the existing PDP, Vietnam PDP-7 (revised), which is due to be replaced by PDP-8 in 2021.[5]

The composition of electricity generation had historically relied heavily on hydroelectricity, but as demand growth continued and the major dam sites were built out, coal and, to a lesser extent, gas grew in importance. As late as 2014, hydro accounted for nearly half of all power, with 29% for coal and 22% for gas. Small amounts of diesel and even less wind or biogas accounted for the rest. But PDP-7 (revised) suggested a 53% share for coal, 17% for gas, and only 12% for hydro, with other renewables accounting for 11%. These are shown in Table 7.1.

Vietnam had made weak commitments to contain carbon emissions and had come under weak pressure from some donors and local groups to develop its plentiful renewable solar and wind resources. However, as late as 2018, renewable energy was seen as only a minor source of energy, and authorities were surprised to see so much interest by investors when high feed-in tariffs were announced—there were 17,000 MW of new solar proposed by the end of July 2017, primarily in south-central provinces about 200 km from Ho Chi Minh City (HCMC).[7] The lack of transmission capacity meant that even the 5000 MW licensed were about twice as much as could be transmitted and used. This has led to a high degree of curtailment—energy available but not used. A growing concern with air pollution has led to more willingness to encourage renewables to play a larger role in future energy

[2]Domestic coal production in poor northern provinces by the monopoly state enterprise generated substantial employment and was not taxed much. Gas production came largely from foreign firms and was seen as a source of revenue. However, imports of gas (which are minimal now) are not heavily taxed. Neither is imported coal.

[3]Both a Lazard (2018) study on the worldwide Levelized Cost of Energy and a proposal for liquefied natural gas (LNG) fired power in Vietnam show costs of about six cents per kWh, much less than the 10–12 cents per kWh reported by EVN.

[4]Vietnam is producing close to its potential in coal but has extensive offshore gas fields that could be developed if agreement on terms could be reached with foreign companies that would invest the billions of dollars needed to exploit these reserves. It would take five to seven years to produce the gas once contracts were signed.

[5]Refer to https://ieefa.org/wp-content/uploads/2020/09/Vietnams-EVN-Faces-the-Future_September-2020.pdf which presents a solid analysis of PDP-8 and its role as a road map rather than a master plan.

[6]2014 data are from the ADB < Wanniachchi et al., 2016. PDP-7(rev) is for 2030.

[7]See https://english.vietnamnet.vn/fms/business/185331/hundreds-of-solar-projects-registered-to-invest-in-vn.html.

plans, though costs should be lower than the 8–10 cents per kWh that 2018 feed-in tariffs provided.[8] Meanwhile, according to Lazard (2019), in other countries such as India, unsubsidized renewable energy is sold for half those prices.[9]

EVN is a state enterprise and by law is the only entity allowed to transmit power, though third-party generation is permitted if the power is sold to EVN or used by the firm producing it. EVN is not allowed to set prices of electricity itself. This makes some sense since it is a monopoly for most customers (some generate their own power), but the regulated prices have been held well below the cost of generating new power and delivering it. Many foreign direct investment (FDI) projects and state enterprises get very low prices for electricity, and this holds the average price down to 7–8 cents per kWh. Electricity Generating Authority of Thailand (EGAT), an efficient and unsubsidized Thai utility, charges 11–13 cents per kWh, and this is a reasonable indicator of where prices should be set even if low-cost hydroelectricity has been available.[10] Low electricity prices reduce any incentive for conservation and starve EVN of resources it needs to upgrade its grid and generating stations. Because Japan and China make loans easily available for investing in foreign coal generation, EVN often chooses (with government approval) to borrow and invest in new—but not terribly efficient or clean—coal plants. As pollution levels rise, this has caused popular complaints, and several provinces have said they do not want new coal plants. This has caused a rethinking of past energy expansion plans. But what policies would allow Vietnam to continue its brisk growth, attract FDI, and clean up its air and water without raising costs drastically? These are the questions that need to be answered.

2. The demand side

Recent scholarship on Vietnam's energy demand includes an overview in the Vietnam Energy Outlook Report (2019) by the Ministry of Industry and Trade (MOIT) and Danish Energy Agency; a cross-country study on excessive electricity intensity in Vietnam (Hien, 2019) and a study of enterprise energy use substituting for capital (Le, 2019). These studies find that the main sources of energy demand are from industry, transport, and residences. Electricity is a major energy input for industry, commercial firms, and residences, while fossil fuel predominates in transport. Le (2019) found strong evidence that low energy prices are substituting for capital, and that unsubsidized electricity prices would result in energy-saving capital investments—particularly if banks willingly lent for efficiency-boosting investments. The Vietnam Energy Outlook Report found that much greater emphasis on conservation would more than pay off in lower energy use and cost savings. Hien (2019), who conducted the cross-country study, found Vietnam to be the most electricity-intensive economy in the Asia-Pacific region, primarily due to low electricity prices for industry and a lack of effective conservation initiatives in industry and in other sectors.

[8]See https://asialawportal.com/2019/04/06/vietnam-proposed-new-solar-feed-in-tariff/.

[9]The weak contracts with EVN (they buy only what they can transmit) make international lenders leery of supplying credit, so local capital markets provide much of the capital. The terms of 8%–10% annual interest and ten year loans make renewable energy more expensive than 3%–4% loans for 20 years allow elsewhere.

[10]The authors based on the EGAT's financial statements to calculate and observe the cost of electricity. EGAT is a state-owned company but runs as a normal profit-making company. It makes a profit and charges 3.61 baht/kWh or 11.2 cents per kWh as shown in its 2018 annual report. In a commercial utility, price is equal to the marginal cost of producing and delivering new power, not the average cost of power.

It is not credible that such a high and growing energy intensity is needed, or even helpful, for sustaining rapid economic growth. China, India, Laos, and Malaysia all saw GDP growth about as fast or faster than Vietnam in the last decade but with much less reliance on cheap electricity. It is cheap electricity that causes GDP growth to drive rapid energy use growth. But most other fast-growing economies do not need and have not used such a path. Hien found that Vietnam's industry sector used 10 times as much electricity as the service sector but generated less GDP value-added.

A review of energy efficiency and conservation policies in Vietnam by Luong (2015) is also relevant. He argued for better data collection, analysis and management, capacity building, and interministerial coordination. Meanwhile, a more recent study of the decoupling of Chinese electricity consumption and output by Meng et al. (2018) gives a clear roadmap for needed steps starting with conservation measures in the industrial sector, although improving residential and commercial building efficiency is also very important. Overall, as Le (2019) found, enterprises will respond to higher energy prices and invest in efficiency if credit is available, but efficiency standards and information will also speed the response of both industry and building construction and retrofitting. It remains to be seen how big a role potential enhanced conservation plays in the new power development plan (PDP-8) expected in mid-2020.

3. The supply side

If GDP continues to grow at 6%–7% a year, energy use will grow even if efficiency improves. Of course, the pricing of energy, conservation policies and standards, and the structure of economic output will influence energy demand as well. But for any given level of demand, the question is how to supply electricity customers. Current plans for demand to grow at 8%–9% a year, but this could be reduced.

3.1 PDP-7 revised

The MOIT and the Institute of Energy, with input from EVN, develop long-range PDPs for building generating capacity. The latest, PDP-7 (revised in 2017), envisioned most incremental growth coming from coal plants. The projections for PDP-7, revised, were as follows (in billion kilowatt-hours and yearly growth). Growth rates are for midpoints in totals (Table 7.2).

This plan also allowed for 5.7% coming from nuclear energy in 2030, but plans for nuclear plants were later shelved (Projected imports accounted for 2.4% of total production and imports in 2020 and 1.2% in 2030). Since both coal and nuclear generation are used for base-load generation (the generators are on most of the time), it is reasonable to add most of the 5.7% share of nuclear in 2030 to coal, though other sources might be used. That would add up to 34 billion kWh to coal and raise coal-fired electricity's production growth rate to over 10% a year.

EVN has a major influence on the PDPs. Its priorities are to deliver power and avoid blackouts while keeping costs down, since there is government reluctance to raise electricity prices—in real terms, they have declined in several recent years. However, local resistance to new coal plants has made planning more difficult. The government wants to promote growth, but many local governments put more stress on avoiding air pollution. Overall, there is a mix of priorities by different actors, and these are negotiated not only in the plans but also in approvals for particular projects.

Table 7.2 PDP-7 revised projections for electricity and sources.

	2020	2030	Growth
Total production and import (billions kWh)	265–278	572–632	8.5%
Coal share of power output (%)	49.3%	53.2%	
Coal-fired production (midpoint) (billions kWh)	134	320	9.1%
Gas-fired share of power (%)	16.6%	16.8%	
Gas-fired production (midpoint) (billions kWh)	45	101	8.6%
Hydro share of power (%)	25.2%	12.4%	
Hydro power production (billions kWh)	68	75	1.0%
Renewable energy share (%)	6.5%	10.7%	
Renewable power production (billions kWh)	18	64	13.6%

Source: Vietnam Power Development Plan for the period 2011–2020: Highlights of the PDP 7-revised.

A new PDP is being prepared and will be released in mid-2020. It seems likely that a larger amount of natural gas generation, including from LNG imports, and renewable energy will be inserted, as well as a somewhat greater stress on conservation, broadly consistent with the Vietnam Energy Outlook Report (2019) referenced earlier. However, coal will likely continue to be a significant share of the growth.

4. Recent developments

There are several developments since 2017 that should and will be considered by the new plan. One is that energy demand is less robust than expected. According to Vietnam's GSO, power production in 2020 was only 233 billion kWh. That would have 2020 production at 12% below the low demand estimate. If conservation programs were pushed, the new 2030 demand would be much less than that projected by PDP-7. This also creates some uncertainty about medium and long-term demand growth.

A second development is the previously mentioned reluctance of many provinces to approve coal-fired generation. This comes partly from awareness of rising air, water, and ground pollution due to growing industrialization, vehicle use, and electricity generation. There have been some excellent Vietnamese studies of the costs of pollution, particularly on health. Credible estimates of pollution deaths in Vietnam run up to 60,000 a year from total air pollution (World Health Organization, 2018) and 20,000 per year in 2030 just from projected coal burning (Koplitz et al., 2017). Resistance is also due to awareness of the falling costs of wind and solar power. Power is a priority, but pollution is not. If it is possible to have one without the other, then why should EVN insist on building coal plants? This reluctance puts pressure on EVN which does not have sufficient funds to develop alternatives since its prices do not cover total marginal costs of expansion.

A third development is the rapid decline in worldwide costs of wind and solar which are not yet reflected in the feed-in tariffs for renewable energy. The government was surprised at the response of investors to these high feed-in tariffs (8–10 cents per kWh), and by the end of 2019, about 5000 MW of licensed solar alone will join with wind, which is expected to have 800 MW installed (as projected) by the end of 2020. In other countries with access to low international interest rates and longer-term loans, solar utility scale power is being sold without subsidies at 3.2–4.2 cents per kWh and wind at 2.8 to 5.4 cents per kWh. This compares to coal where new plants in Vietnam typically charge 7–8 cents per kWh. Gas-fired power in Vietnam costs 10–12 cents per kWh, for reasons that are not clear—a bidder on a new LNG plant suggested costs of only six cents/kWh. If taxes are ignored, gas costs from 4.4 to 6.8 cents per kWh in other countries according to Lazard's 2019 annual report.

EVN would very much prefer to pay three or four cents for solar electricity than 9.35 cents—the feed-in tariff that had been available to investors. But it already must curtail (not buy) renewable energy supplies that are available in south-central Vietnam due to its limited transmission capacity. Unless or until the grid is strengthened, there will be considerable risk for any developer building solar or wind capacity. This limits the inflow of international capital and raises the cost of renewable energy in Vietnam. There are proposals to allow private investment in the grid, but this requires a change or flexible interpretation of the constitution and mutual agreement on contractual terms.

Even if the transmission issues were resolved—and the cost of transmission should not be more than 1 to 1.5 cents per kWh[11]—there is the problem of intermittent supplies of renewable energy. The grid is set up to supply power from a few large and stable sources, not to mix variable renewable output with historic sources like hydro or fossil fuel generators. This is a soluble problem, but it takes a complicated mix of investments and policies to make it work reliably. Elements of the needed response are a "smarter" grid with more signals, prediction (of demand and variable energy supply), battery storage, and response of gas and hydro generators which can be cycled up and down quickly. A spinning reserve of coal plants (so they can respond more quickly) might also be needed. In addition, negotiating short periods of load shedding would add flexibility to the system. An office or hotel might be able to go for 10 min without air conditioning, for example. Putting these elements together will take time and new management skills. Utilities are heavily criticized when the power goes off, so they are conservative.

A fourth development is the unexpected trade policy of the United States, which under President Trump attached great importance to bilateral trade balances. Vietnam has benefitted from the United States–China trade dispute, and many factories have set up to assemble goods for export which had been in China. The value-added of these goods from Vietnamese labor or other inputs is quite low—usually in the single digits—but trade data do not distinguish this. In any case, the US Treasury has suggested that if the bilateral trade imbalance is not reduced—and it is now growing sharply—Vietnam may face US tariffs (It remains to be seen how President Biden handles this issue). This would upset FDI inflows and slow growth. However, the United States has a large supply of liquified natural gas which is competitively priced and could be used instead of coal to generate electricity. Importing LNG would reduce the bilateral trade surplus Vietnam is running with the US. It would also be easier to place the gas generating plants since they generate much less local pollution as well as less carbon dioxide.

[11]The margin is authors' conservative estimates based on "The Economics of Electricity Transmission" from the Institute for Energy Research (IER) available at https://www.instituteforenergyresearch.org/electricity-transmission/. This source is for the United States which has long distances for many generator–consumer pairs. Distances are usually shorter in Vietnam except for very large hydro plants. In the United States, distances are typically many hundreds of miles rather than the few hundred km typical for renewable energy. Land costs are also likely to be lower in Vietnam where the government can take over rural land cheaply.

4.1 A view from the Vietnam business forum

The Vietnam Business Forum represents private and mainly foreign businesses and investors in Vietnam. In 2016, they produced a "Made in Vietnam Energy Plan" (MVEP) arguing for conservation, greater private involvement in generation and transmission—requiring more favorable regulatory frameworks including prolonged feed-in tariffs—and greater use of gas for power generation. In December 2019, they produced a 43-page update, "MVEP 2.0" in which their proposals and analysis were further developed. The revised study added investment in battery storage which has grown cheaper and more widely used by utilities since 2016. It requested that companies be allowed to generate up to 50 MW of their own renewable energy without a license and to contract directly with renewable independent power producers. They also renew arguments for conservation, getting electricity prices gradually up to their marginal costs, and much greater use of LNG for generation. They note that combined cycle gas plants can adjust within 10 minutes to fluctuating demand or variable supply while coal can take hours (Hydroelectricity can also adjust quickly). If combined with modest amounts of battery storage, they argue that a conservation/renewables mix would allow zero growth of new coal plants. The paper argues that nonhydro renewable energy could provide 30% of supply by 2030. The new paper does not refer to plans to phase in electricity auction markets in the 2020s which would reduce the use of feed-in tariffs and perhaps slow renewable investment.

5. Challenges and opportunities

5.1 Transmission issues

There is a very large potential for wind and solar energy electricity production in Vietnam—the proposed 17,500 MW of solar electricity in 2018 when the total 2030 renewable target from PDP-7 (revised) was just 12,000 MW only hints at how much could be produced. If firm contracts ("take or pay") allowed international capital to flow in and reflect a low risk of default, it is likely that the auction price of electricity will fall below five cents per kilowatt-hour (The cost will also reflect a longer term of loans than those usually extended in Vietnam from local capital markets, though the currency will be in dollars, euros, and yen rather than dong).

The problem is that the current transmission structure was not set up to carry vast amounts of electricity. Most electricity demand was satisfied by power plants close to power consumers, though some large hydro projects were exceptions. It was mainly balancing loads that were carried even by the high voltage lines. It would be expensive to upgrade the grid to the point that it could take very large amounts of renewable energy from the south-central coastal provinces like Binh Thuan and Ninh Thuan where solar and wind conditions are very favorable. Distances of the required upgraded transmission lines run from 180 to 260 km (110–160 miles). Existing high voltage transmission lines would have to be upgraded significantly. The time needed for such an upgrade will depend on a variety of factors such as whether or not new land would be needed, or existing or improved towers could carry extra lines. EVN has argued it would take five years or more to extend the strengthened grid in the south, while others argue it might be done in 2–3 years. A new coal plant, if approved immediately and built without delays, takes 4–5 years to build. A combined cycle gas plant will take three years and solar less than one year.

5.2 Auction markets

There is a plan to shift from negotiated prices or feed-in tariffs to auction markets over the next several years. There are many issues to sort out, and it will probably be in the middle years of the decade before auctions held by EVN for power producers become the predominant mode of supplying electricity. First of all, those producers who signed fixed or variable price supply contracts would have to be used first unless curtailments were possible. Second, new private investment into electricity generation might stall if the historic coal plants owned by generating companies spun off from EVN but still regarded as linked to them were somehow favored (This is less of a problem with hydroelectric plants since they have very low marginal costs and could legitimately bid competitively). Third, the transmission issues would have to be sorted out since there is not now enough transmission capacity to carry existing renewable energy to the major markets in HCMC. These and other details of how the auctions are conducted will eventually be worked out but not right away.

If new renewable investments were able to access global capital markets with very low interest rates and longer repayment periods, the cost of electricity from wind and solar would likely approach that of the levelized cost estimated by Lazard or the observed unsubsidized bids in India and other developing countries with comparable insolation. Since the marginal costs of coal are 3—4 cents per kWh while those of hydro and renewables are close to zero, it is likely that coal would only be able to compete at peak periods. However, this would raise the average cost of coal-fired electricity since its high fixed costs and difficulty of rapidly adjusting output would place coal plants at a disadvantage. In India and the United States, many coal plants are being shut down or getting subsidies to be available in case of shortages or unavailability of power from other sources. Auction market would hasten this process. However, for EVN to sign "take or pay" contracts with any producer—if they won the auction—the transmission grid would have to be much more capable than it is now. Only then would low-cost international capital be made available and the risk of curtailment be reduced.

5.3 Costs of pollution

Under the last PDP (PDP7, revised in 2016) to 2030, coal power will see an unexpected jump. Coal thermoelectricity is now a major source of the power mix, with the total installed capacity reaching 20 GW out of 55 GW system capacity by 2019, burning approximately 54 million tons of coal in 2019. By 2025, these figures are 47 GW and burning 95 million tons of coal. By 2030, coal power will increase to 55 GW and consuming about 129 million tons of coal, with most coal is to be imported. By then, Vietnam will add 37 GW of new coal power. Coal power will contribute almost a half of the total generation. One clear goal of this coal centric development is that Vietnam aims to provide cheap and reliable electricity to almost anywhere in the country.

However, the consequence is that pollutant emissions such as sulfur oxides, nitrous oxides, and dust particles will quadruple. For example, coal power contributes to the national average PM2.5 concentration by approximately 0.9 μg per cubic meter in 2017. If all coal power plants were to be built according to the last PDP7, the contribution will be four times higher. Coal power emissions include CO_2, a greenhouse effect, with a long life cycle in the atmosphere. Particle pollution, primarily incombustible fine dust including PM2.5 and PM10, is considered to be most harmful to human health. Sulfur dioxide is a product of combusting sulfur in coal. SO_2 is the primary cause of acid rain and a secondary source of PM2.5 dust. NO_x is produced by nitrogen in the air reacting with oxygen at high temperature in the combustion chamber. NO_x is also a precursor to the formation of ozone (O_3) in the atmosphere near the ground, which is a secondary PM2.5 air pollutant. Coal power also produces massive amount of ash and slag. Despite plans to use coal ash as cement supplements and building materials, very little has been actually used. Most ash is dumped in the open ground at nearby sites. There has been some major incident such as in Vinh Tan when coal ash was blown by high winds to residential areas, causing widespread local pollution and provoking social unrests. The Red River Delta is the region with the highest concentration of PM2.5 due to a high number of the coal power plants near coal mines and industrial hubs in the north.

The total number of premature deaths from coal-fired power to the country in 2017 is estimated at 4359 cases. However, if all went according to the revised PDP7, the total number will rise to almost 28,000 cases (GreenID, 2020). However, under the most aggressive abatement and efficiency drive, the potential number of deaths will drop to about 4400 cases, almost the same level as of 2017. Not only causing pollution and deaths inside Vietnam, coal power is also the largest contributor of transboundary pollution in the region. If the plan to build coal power in the revised PDP7 went ahead, the projection shows that there would be nearly 9000 premature deaths in China in 2030 due to transboundary PM2.5 pollution.

Of course, these figures assume that future plants would use the same technology as of today. Better regulatory requirements and improved efficiency standards may reduce potential emissions significantly. Vietnam has employed various technologies in coal-fired power plants to control the amounts of emissions. Among 26 coal-fired power plants in operation, about a third using circulating fluidized bed while the other pulverized combustion accounting for the remaining two thirds of all plants. All plants are designed for subcritical and supercritical boilers. Future plants are expected to use supercritical boilers with increased efficiency; however, there may not be much enthusiasm in ultra-supercritical boilers due to cost issues. Currently, most coal power plants in Vietnam claim to have installed the ESP system with over 99% filtration efficiency. SO_2 is treated in wet flue gas desulfurization tower using lime water or seawater. Incorporating energy efficiencies and tightening environmental standard will reduce much of the increased pollution to the 2017 level.

A published study found high levels of PM 2.5 accounted for more than 40,000 excess deaths per year in Vietnam averaged over 2009–14 (Shi et al., 2018). While the study did not project deaths, a linear relationship combined with rapid growth in coal would suggest at least a doubling over the 2020 decade and more than a tripling from 2014. PM2.5 is highly correlated with nitrogen oxides, associated with coal. A recent research has found that PM2.5 will be the leading cause of death by 2030 in Vietnam and regional countries (Koplitz et al., 2017) if plans for coal generations proceed to plan. Note that the number of deaths is not solely from the increased use of coal, as other causes might exacerbate health effects, such as rapid migration to a few densely populated cities leading to higher levels of exposure.

The main damage is health impacts, or the number of premature deaths, in economic values. This involves a concept—the value of a statistical life or VSL. There are two approaches to quantify these costs: one relying on the loss of potential income of an average working person due to an early death (the forgone output), and another by measuring the willingness to pay (WTP) to avoid an early death. For example, if a person earning USD 2500 a year loses 30 years of productive life, assuming wage rises at 3% a year, and a discount rate of 6%, the present value of the forgone output is approximately USD 52,000 (Some argue a lower discount rate should be used in dealing with human lives). The WTP is a subjective welfare concept and often not easily observed or calculated in developing countries.

That leaves a comparative approach. For example, using a VSL of USD 5–10m in the United States as the baseline, the VSL in Vietnam is between USD 241,000 and 304,000. Other authors have used vastly different values, from as low as USD 58,000–98,000 in China in 2000 to as high as USD 1m. Converting these figures to monetary values give the approximate health cost of coal-fired electricity at 2–4 cents/kwh.[12]

The costs of carbon pollution are harder to estimate since the main impact is on global warming. Rising sea levels and weather/temperature issues are major threats to Vietnam, but Vietnam is not, by itself, going to have a major impact on global temperature levels. However, there may be a global compact that essentially negotiates a carbon cost for all fossil fuel users. This tax, if it occurs, would be paid to the national government but would show up as an extra financial charge on each fossil fuel user. A metric ton of thermal coal produces about 2.5 tons of carbon dioxide when burned, so the question becomes, what is a reasonable estimate of the cost of carbon dioxide? No one knows but there are many estimates. One recent and plausible estimate by a well-known academic expert put it at USD 31 per ton of CO_2 (2010 prices) or USD 35 now. If this were the amount negotiated, it would cost coal producers in Vietnam about USD 4.2 billion in 2016 rising to USD 10.5 billion by 2025. This would add 4.6 cents per kWh to the cost of coal-fired electricity. Of course, there may be no such carbon agreement, or if there is one, it might be for a lower initial amount. But applying even half of this estimate to the cost of coal plants would make them uncompetitive. Natural gas would also be hit, but much less. A million BTU of gas generates 53 kg of carbon dioxide and creates 150–160 kWh of power. So, it would take 2900 kWh of gas-fired electricity to generate a ton of carbon dioxide, or 1 to 1.2 cents per kWh with a carbon tax of USD 35 per ton. Renewable energy would pay even less—only for the fossil fuels used in making the panels on windmills, batteries, and transmission lines.

Regardless of which methods to use, the point is that, if carefully and fully calculated, coal is clearly a lot more expensive than other low-emission sources, regardless of which set of parameters used. Taking a precautionary approach, it is entirely justifiable to use a higher damage number to avoid apparent and irreversible damages should emerging cleaner technologies become more affordable in the future, as they have done and are projected to do. Even without considering the cost of carbon to climate change, the health cost already justifies noncoal power. Adding carbon cost on top of pollution makes coal even less affordable. So, with a proper environmental accounting, coal power is already not the least cost option for electricity in Vietnam. In fact, coal plants would become dinosaurs, unable to compete early in their productive lives and become a huge drain on the national budget and EVN.

A final consideration for EVN and any investor is the rise of provincial government opposition to approving the construction of new coal plants. As there is rising awareness of the health and economic costs of air pollution, many provinces are arguing for other sources of electricity including LNG. This can increase delays in approving a new coal plant or even cause the application to fail. The situation is fluid, but it appears that coal is no longer the automatic default and will play a more modest role in PDP-8 when that plan is released in mid-2020.

[12]If we anticipated 100,000 early deaths in 2030 and a VSL of USD 200,000 per life in that year, the value would be USD 20 billion or (for 500 billion kWh), or four cents per kWh in pollution costs, excluding any health treatment costs.

5.4 Pricing of electricity

Vietnam has tended to charge average rather than marginal costs for power. There are legacy low-cost hydroelectric plants, often financed with soft loans, that bring down the average cost of electricity, but this source is essentially exhausted, and new demand must be met with thermal or solar/wind generation which costs more—typically 7–9 cents per kWh. Distribution costs add 3–4/kWh cents more.

After more than two years of keeping the electricity tariff stable, at the end of March 2019, the average retail price of electricity was increased to VND1,864/kWh, equivalent to an increase of 8.4% or about eight cents per kWh. The main reasons for the MOIT to request the Government to adjust the electricity price were due to the sharp rises in the input prices as well as the costs of production and investment in the electricity distribution system.

But in context, this 2019 price increase merely brought the real price of electricity slightly above the 2009 level. Through 2018, inflation had been 75%, and electricity prices had risen only 72% from 2009. In addition, many costs of imported fuel and loan repayments are in foreign currency, and the 33% depreciation (2009–2018) against the dollar reduced the "surplus" of EVN to offset the rising marginal costs of new demand.

Electricity prices in Vietnam are still relatively low compared with those of neighboring countries, even when compared with other countries of the same income group. Table 7.3 compares the electricity tariffs in Vietnam with the countries in ASEAN, up to the time of the recent price increase. It can be seen that the electricity tariff for business (commercial) consumers in Vietnam is quite high, but that for industrial consumers are still very low and that for household prices are relatively low compared with those of other countries in the region. This is true even after taking into account the recent adjustment of electricity tariff.

Given the relatively low price of power and its heavy use and rapid growth, raising the electricity price to its actual marginal cost is an obvious and necessary step for conservation. There is no good reason Vietnam has to be such an intensive user of electricity at so low an income level. As Le (2019) demonstrated in his paper, there are available ways to substitute capital for energy, and doing so would have considerable environmental benefits and, as the Vietnam Energy Outlook Report (2019) finds, also be a leas cost way of meeting demand. Yet increasing the real price of electricity is done with great reluctance and only very gradually.

Table 7.3 Comparison of electricity tariffs in Vietnam with some countries in the region (as of July 2018–before the latest increase in electricity tariff).

	Electricity tariff in the United States cents per kWh					
	Vietnam	Indonesia	Malaysia	Thailand	Singapore	Philippines
Residential	10.59	11.00	10.00	12.41	19.97	18.67
Business—medium	13.44	11.00	13.58	11.00	14.30	12.23
Business—large	12.36	8.36	9.60	11.00	14.02	11.98
Industrial—medium	7.81	8.36	8.29	8.36	13.05	11.69
Industrial—large	7.41	7.47	7.76	8.36	12.72	11.63

Source: Digital Energy Asia (2018).

So why is electricity price increase in Vietnam so hard? With the current institutions in Vietnam, it is not easy. EVN is a state-owned enterprise and has to accomplish too many goals. First, it must provide adequate and stable electricity supply for the economy and must ensure national energy security. In addition, a stable electricity price is often regarded as a tool of macroeconomic stability, meaning that when all commodity prices rise, electricity prices must remain stable. Only when consumer prices are stable, electricity prices are allowed to increase. EVN also has political duties, that is, to provide electricity at all costs to all parts of the country, including mountainous villages with just over a dozen households and with rugged terrain. EVN, like some other big state-owned enterprise (SOEs), is also a pillar of the economy, and thus their nonindustry investment has been arguably related to pursuing this goal. Besides, EVN is also an enterprise. They must also do business and have reasonable profits.

Despite being a monopoly, EVN's scope of autonomy and self-determination is very limited. For important matters of a business, including raising the price of electricity, EVN must consult the governing body and the Government. To a certain extent, EVN is not the "cause" but the "consequence" of the mechanism of Vietnam's electricity market. Because, certainly, no business can address all of the above goals at once.

Specifically, with regard to electricity price increases, since the price had been stable for some time, the recent price adjustment was needed. In the coming time, EVN will no longer be a major producer of electricity, but only focus on power transmission and distribution. Since electricity generation activities will gradually shift to other components or businesses, investment resources must be mobilized. To obtain investment resources, one of the necessary conditions is that the electricity price must be sufficiently high to be profitable. In addition, it is also important to adjust prices to encourage economical use of energy. But that does not mean EVN will not feel pressure to do business effectively.

In fact, with the recent increase in electricity tariff, the public also wondered about the abnormality when EVN was claimed to be "in debt" but still leftover VND 42,000 billion of demand deposits in banks (up nearly VND 10,000 billion compared to the end of 2017) and nearly VND 20,000 billion of short-term financial investments as of June 30, 2018. Although EVN explained it, the question from the public is whether that is due to weak financial management or some other motive. If EVN manages the cash flow better and restructures its deposit portfolio at commercial banks, this huge financial resource will be used more effectively, increasing EVN's revenue by hundreds of billion Vietnamese Dong. At the end of 2017, EVN was once concluded by the Ministry of Finance Inspectorate that it recorded nearly VND 2000 billion in wrong accounting but increased the electricity price to cover the wrongly recorded losses.

The business investment of State capital in enterprises like EVN should thus be supervised and transparent. Only by so doing, the new electricity price increase will be found acceptable and justifiable by the public. It is necessary to require EVN to be transparent about its costs and business processes so that consumers can monitor how they buy and distribute electricity. They may also curtail nonutility aspects of their business. Previously, the input costs for electricity production were rarely mentioned. Typically, there are seven to nine factors calculated into the input, including: depreciation, raw materials, materials, salary (bonus), overhaul, outside services, financial expenses (loan interests and exchange rate differences), customer development costs, and other cash costs. Among the above input costs, there are two types of costs, namely, salary and depreciation costs that are set by the State, while other expenses are entirely decided by EVN. The fact that EVN can decide these expenses by itself, according to many people, can lead to nonobjective factors, which are beneficial to the producer at the cost of the consumers.

Therefore, there should be a more flexible mechanism for adjusting electricity prices. At the same time, it must be more transparent about the price structure, the price level, and the goals that are to be achieved so that stakeholders can monitor EVN effectively. Only when there is effective supervision will there be pressure for objective evaluation tools and a greater acceptance of needed price changes.

In any case, price increases for electricity will be required, and educating the public that there is a cost to clean and reliable energy will be necessary. There has been great political reluctance to raise real electricity prices, and as a result, conservation suffers from a lack of incentives while EVN lacks resources to make critical investments.

So, there is much to do but many positive possibilities. The interest groups of established state enterprises, domestic businesses, international business, and public opinion interact in coalitions while the Party and government juggle and manage these groups. International aid agencies also play a role, working to provide technical assistance and finance to shape things. It is the interaction of this dynamic and complicated interaction that will determine how much conservation, particular types of generation, and transmission and grid management strategies will combine.

6. Concluding observations

Utilities are conservative by nature. They are tasked with, first of all, supplying electricity reliably. EVN knows how to build thermal power plants and operate them along with hydroelectricity. Except for some transmission and distribution issues, they have gotten better at keeping the lights on. Given their inability to charge enough to cover marginal costs of expansion, this is no small thing. On the other hand, EVN has been used to getting its way in putting power plants where it wants to and has not seriously dealt with externalities. As social media and even local governments become aware of the total costs, not just the financial costs, of power, EVN's ability to operate as before is diminishing.

Renewable energy injects considerable uncertainty into their operating model. While a modest amount of battery storage may make solar and wind a steadier supply source with more predictable output profiles, it is still a big change. It also involves tremendous grid investments, and this is money EVN does not have. In addition, the move to a "smart grid" and the prospect of negotiating short periods of interrupted power in advance are new ideas that add additional uncertainties. It is not surprising that there is some resistance and a desire to operate in more traditional ways.

A research question is why LNG fired power with combined cycle gas generators is so expensive in Vietnam (around 12 cents a kWh) relative to coal. If the Lazard cost figures are accurate for the rest of the world—and they were informally confirmed by an LNG generation investor—it would be good to identify the reasons that Vietnam has roughly double the costs of other places. It could be higher loan rates, shorter amortization periods for things like regasification facilities, or other reasons. Until the details of cost estimates are made public, it will be hard to bring down gas-fired electricity costs to those of coal, as Lazard suggests other places have done. Since gas has lower pollution costs, it would normally be preferred if its financial costs were equal to coal. And since both gas and coal will be imported for expansions of output, there is no argument for self-sufficiency. If offshore gas could be developed, domestic gas supplies could be used.

It is also not a normal activity of EVN to push conservation. In some countries, utilities have taken the lead and gotten credits for spending on demand reduction. In Vietnam, conservation policies and funding are not comparable with China's (for example) and, of course, low electricity prices discourage investing in more efficient equipment. Gradually raising electricity prices (some reportedly get off-peak power at just two cents a kWh) would provide an incentive to invest in efficiency. A much more activist and aggressive conservation policy, with credits available for efficiency-raising machinery, would help to lower Vietnam's electricity elasticity to one or less rather than the 1.3 or more it has been. Having demand grow at 6% a year in this decade would raise demand by 80%, while 8.5% annual power growth sees demand grow by 126%. Reducing demand growth by a third would be a huge help, and several studies suggest that the return on investment would pay for itself. This is a fundamental change, though it is not clear if EVN is the best institution to take the lead.

The need for better transmission is clear, but the best way to provide for it is not. Private investors are willing to build new or upgraded transmission lines, but the terms of the contract are naturally in dispute. EVN would like maximum flexibility to determine when it uses the "extra" capacity, and if it switches to auctions, it would want to use the lowest cost power. Investors seek assurances that their new capacity will be heavily used and sometimes wish to tie their own production capacity to "their" transmission lines. These differences can be narrowed but may not be resolved soon. It is the contractual terms, as much as the physical building out of the transmission capacity, that is likely to hold up the heavy use of renewable energy, at least in the south. An alternative would be to build solar or wind closer to HCMC and lessen transmission requirements but increase generation costs.

One aspect of renewable energy that should make it more attractive is that it is falling in cost each year. At a time of low global interest rates, falling solar panel prices and even faster declines in battery storage, the short turnaround time for placing renewable energy into production—normally less than a year—has great advantages if demand growth is uncertain. A coal plant is normally many hundreds if not a few thousand megawatts and has to be started 4–5 years before it is needed. This plus the delays from local opposition and the uncertainties of climate-driven carbon taxation makes it very risky. LNG need not take as long, especially if floating regasification units are used, but is still about three years and tends to require large generating plants. Since it is possible to wait until power demand is clear and the costs of renewables have fallen even more, the proper calculation is the risk-adjusted cost of future renewable energy against the risk-adjusted cost of current fossil fuels. Advances in grid management and integration of renewables with storage should also make EVN more comfortable relying less on large central fossil generation, especially if financial costs are similar.

All of this is a process, and there are many moving parts. The conveners of a December 2019 conference asked the authors of this paper for "success stores" from each country. In Vietnam, the successes are there but incomplete and lead to other failures. For example, the setting of high feed-in tariffs caused a huge interest in investing in renewable energy. PDP-7 (revised) expected just 18 billion kWh in 2020 from renewable energy, but if all 17,000 MW of solar licenses applied for after the feed-in tariffs were announced, and if their output could be transmitted, there would have been more than 25 billion kWh just from solar and more from wind. So, the feed-in tariffs were successful in attracting investors to supply renewable energy—far beyond the amount thought likely, which was less than 1000 MW of solar capacity. But they highlighted the limitations of the transmission grid and the difficulty in keeping it stable without storage.

Likewise, the rise of NGOs and social media were successful in communicating to provincial officials (and even some in Hanoi), the unease with which ordinary people viewed the increasing use of coal. PDP-8 is likely to shift priorities away from coal. But new coal plants will still be needed, and its use will increase, even if less than previously projected. And the important details of LNG costs remain cloudy. If EVN shifts to gas, it may have to raise electricity tariffs or get larger subsidies to avoid financial problems—unless costs can be brought down to those in other countries. The slower growth of demand due to the corona virus may give Vietnam time to push the institutional changes needed to resolve these issues.

7. Lessons for other countries

Vietnam has been successful in extending electricity to essentially all its citizens and in producing enough to satisfy demand while reducing blackouts and shortages. It is working on making these successes cleaner, less energy intensive, and more financially sustainable. The lessons for other countries at perhaps an earlier stage of development might be to do some of the following.

1. Integrate conservation, including full-cost pricing for most users, into the original plan rather than waiting until later. Machinery, buildings, and appliances can have long lifetimes, and it is best to bake in efficiency early.
2. Invest in grid capacity and intelligence along with renewable energy. Typically, solar and wind sites will not be very close to major users, so connecting them as they are built will avoid wasting potential production due to curtailment.
3. Worry about stranded costs. Investing in large fossil fuel plants that last a long time is increasingly dangerous. Falling renewable and storage costs along with likely charges to carbon use will make them financially obsolete long before they wear out.
4. Work out soft infrastructure of contract terms, contingencies, and structures for auction markets so that low-cost capital can finance energy investments. Investors will want certainty, but the scale of renewable energy is so large that high feed-in tariffs are overly generous. Transparency about how auctions are run and fair procedures to resolve differences will build confidence without raising costs. Given that renewable energy has low operating costs, lowering the cost of capital is critical.
5. Be politically adroit when raising electricity prices. This step is often treacherous for a government, so gradual increases and careful justification of them are needed. Offering "lifeline" prices for poor users with modest consumption will help to ease this bitter pill. Fair regulation and publication of prices in other countries may help.

References

Baulch, B., Duong Do, T., Le, T.-H., 2018. Constraints to the uptake of solar home systems in Ho Chi Minh City and some proposals for improvement. Renewable Energy 118, 245—256.

Digital Energy Asia. Competitive Indonesian Electricity Rates in the ASEAN Region. Published on August 25, 2018. Retrieved on February 3, 2020. https://digitalenergyasia.com/competitive-indonesian-electricity-rates-in-the-asean-region/.

Green Innovation Fund (GreenID), 2020. Impact of Coal Power on Air Pollution and Health in Vietnam (Vietnamese: Tác Động Của Nhiệt Điện than Tới Chất Lượng Không Khí Và Sức Khỏe Tại Việt Nam). Hanoi, Vietnam.

Hien, P.D., 2019. Excessive electricity intensity of Vietnam: evidence from a comparative study of Asia-Pacific countries. Energy Policy 130, 409–417.

Koplitz, S.N., Jacob, D.J., Sulprizio, M.P., Myllyvirta, L., Reid, C., 2017. Burden of disease from rising coal-fired power plant emissions in Southeast Asia. Environmental Science and Technology 51 (3), 1467–1476.

Lazard. Lazard's levelized cost of energy analysis - version 12.0. Published in November 2018. Retrieved on February 3, 2020. https://www.lazard.com/media/450784/lazards-levelized-cost-of-energy-version-120-vfinal.pdf.

Lazard. Lazard's levelized cost of energy analysis - version 13.0. Published in November 2019. Retrieved on February 3, 2020. https://www.lazard.com/media/451086/lazards-levelized-cost-of-energy-version-130-vf.pdf.

Le, P.V., 2019. Energy demand and factor substitution in Vietnam: evidence from two recent enterprise surveys. Journal of Economic Structures 8 (1), 1–17.

Luong, N.D., 2015. A critical review on energy efficiency and conservation policies and programs in Vietnam. Renewable and Sustainable Energy Reviews 52, 623–634.

Meng, M., Fu, Y., Wang, X., 2018. Decoupling, decomposition and forecasting analysis of China's fossil energy consumption from industrial output. Journal of Cleaner Production 177, 752–759.

Shi, Y., Matsunaga, T., Yamaguchi, Y., Zhao, A., Li, Z., Gu, X., 2018. Long-term trends and spatial patterns of PM2. 5-induced premature mortality in South and Southeast Asia from 1999 to 2014. Science of the Total Environment 631, 1504–1514.

Vietnam Energy Outlook Report 2019. Published by the Ministry of Industry and Trade and Danish Energy Agency. Retrieved on February 13, 2020. Available at: https://ens.dk/sites/ens.dk/files/Globalcooperation/vietnam_energy_outlook_report_2019.pdf.

World Health Organization. More than 60,000 deaths in Viet Nam each year linked to air pollution. Published on May 2, 2018. Retrieved on February 13, 2020. Available at: https://www.who.int/vietnam/news/detail/02-05-2018-more-than-60-000-deaths-in-viet-nam-each-year-linked-to-air-pollution.

Wanniachchi, A. K., Bui, D. T., Lee, H., Jung, C. S., & Tuan, A. M. (2016). Viet Nam Energy Sector Assessment, Strategy, and Road Map. *Asian Development Bank and Asian Development Bank Institute*. ISBN: 978-92-9257-312-6 (Print); 978-92-9257-313-3 (e-ISBN). Available at: https://www.adb.org/documents/viet-nam-energy-sector-assessment-strategy-and-road-map

Further reading

Asian Development Bank (ADB). Key Indicators Database, 2019. Retrieved on February 3, 2020. https://www.adb.org/publications/key-indicators-asia-and-pacific-2019.

Electricity Generating Authority of Thailand (EGAT). 2018 annual report. Retrieved on February 3, 2020. https://www.egat.co.th/en/images/annual-report/2018/egat-annual-eng-2018.pdf.

Vietnam Business Forum (Power and Energy Working Group). Made in Vietnam Energy Plan (MVEP) 2.0. Published on December 1, 2019. Retrieved on February 20, 2020. https://asiafoundation.org/wp-content/uploads/2020/02/Made-in-Vietnam-Energy-Plan-2.0_EN.pdf.

Vietnam Business Forum (Power and Energy Working Group). Made in Vietnam Energy Plan (MVEP). Published in October 2016. Retrieved on February 20, 2020. https://static1.squarespace.com/static/5b7e51339772aebd21642486/t/5bae623d71c10b08c07f999e/1538155071467/Made-in-Vietnam-Energy-Plan-MVEP-v12.pdf.

Vietnam Power Development Vietnam Power Development Plan for the period 2011 – 2020: Highlights of the PDP 7 revised. Retrieved on February 13, 2020. http://www.gizenergy.org.vn/media/app/media/legal%20documents/GIZ_PDP%207%20rev_Mar%202016_Highlights_IS.pdf.

CHAPTER

Energy security challenges of developing countries: a pragmatic assessment

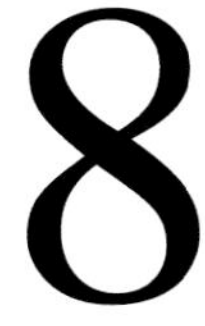

Krishna J. Khatod[1], Vikrant P. Katekar[2] and Sandip S. Deshmukh[1]

[1]*Department of Mechanical Engineering, Hyderabad Campus, Birla Institute of Technology & Science, Pilani, Telangana, India;* [2]*Department of Mechanical Engineering, S. B. Jain Technology Management and Research, Nagpur, Maharashtra, India*

Chapter outline

Handbook of Energy and Environmental Security. https://doi.org/10.1016/B978-0-12-824084-7.00009-6
Copyright © 2022 Elsevier Inc. All rights reserved.

1. Introduction

Recognizing the 21st century's expanding energy and environmental issues has become a critical component of national and international developmental frameworks. With energy serving as the lifeblood of modern civilizations, concerns concerning its security are becoming increasingly prevalent. Energy security is seen as a significant concern for global socioeconomic progress. Energy and environmental situations are inextricably intertwined in the age of fossil fuels. Global warming is one of the world's most serious environmental problems, with potentially disastrous repercussions (Jambhulkar et al., 2015).

Around 1.5 billion people in developing countries—about 16% of the world's population—do not have electricity, and nearly three billion people rely heavily on solid fuels for electricity generation. According to the latest UN estimates, South Asia is home to approximately two billion people as of June 2021, one of the most densely inhabited regions in the world (Worldometer, 2021). Historically, countries in the region have been classified as developing or undeveloped.

1.1 South Asia

SouthAsiaisacollectivetermfortheregion,includingIndia,Pakistan,Bangladesh,Maldives,Nepal,SriLanka, Afghanistan,andBhutan.Thesenationscollectivelyaccountformorethanone-sixthoftheworld's population. It is the birthplace of one of the world's oldest civilizations and, at the same time, one of the most lively and significant hubs of global culture economically, politically, and socially. It is a location that is gaining prominence and will offer chances and vocations throughout the world (Sciences, 2021). It is critical to better research South Asian countries to understand all other developing countries worldwide.

South Asian countries are confronted with a slew of energy and environmental issues. This chapter gives a pragmatic assessment of South Asia's energy security challenges. Specifically, to thoroughly examine the developing countries in South Asia from various perspectives, the current research focuses on Bhutan, India, Pakistan, and Sri Lanka, which are among the most and least densely populated countries in the region. Concerning overall development, the countries highlighted above are in the best position currently and offer a better outlook than other developing countries in the area when analyzed by development measurement statistical indices such as per capita income (per person), life expectancy, gross domestic product (GDP) per capita, freedom index, literacy rate, and others (U. N. in India, 2021).

In the present study, extensive energy initiatives have been explored at regional and national levels. Each country's supply mix, energy resources, cooking fuels, and access to power are discussed in detail. New developments and trends in renewable energy are also discussed. In addition, the environmental outlook for the region has been presented in terms of climate change and other ecological problems. The region's perspective on the Sustainable Development Goals (SDGs) has also been investigated in the subsequent sections. It is challenging to present the data and its analysis of all South Asia; hence, this study focuses on the two largest (India and Pakistan) and the two smallest (Bhutan and Sri Lanka) countries.

1.2 Regional overview

In Fig. 8.1, South Asia is portrayed as a sub-Himalayan region that includes several countries on the southern portion of the Asian continent. About 24.89% of the world's population (Worldometer, 2021) lives in the region, covering only about 3% of the entire land area. As indicated in Fig. 8.1, the region has traditionally comprised eight countries: Afghanistan, Bangladesh, Bhutan, India, Maldives, Nepal, Pakistan, and Sri Lanka. In 1985, these nations established the South Asian Association for Regional Cooperation (SAARC) to foster regional economic cooperation and growth. Afghanistan was accepted to the SAARC on an official basis in 2007 (Katekar et al., 2021). According to the latest available UN Figures, India, with over 1.39 billion people, is the most populated country in South Asia as of June 2021. Pakistan and Bangladesh, with populations of 224 million and 166 million, respectively, are the region's two most populous countries after India. Moreover, there are countries with only a tiny proportion of these populations in the region. For example, the Maldives and Bhutan have about 0.54 million and 0.77 million people (Worldometer, 2021).

South Asia is a vital region globally regarding its vast population and considerable energy and environmental issues. Countries have many energy issues in the broader area, ranging from inadequate access to refined fuels to unreliable grid stability to expensive energy costs and import dependency. According to the International Energy Agency (IEA), South Asia's energy demand is expected to increase at nearly twice the rate of the rest of the globe in the years ahead. However, the region faces resource shortages and a scarcity of well-thought-out policies and legislation (Alagh, 2006). Energy poverty is a severe impediment to socioeconomic progress for a sizable South Asian population. These countries also face significant environmental difficulties, which are discussed in the subsequent sections of this chapter.

1.3 Energy, demand, and resources in South Asian countries

All of nature, including life and nonlife, depends on energy. The growth of economies is determined by many factors, the most critical social and economic development (Bhujade et al., 2017). Due to energy shortages, the South Asia area will have to manage economic growth carefully. A significant portion of the population is located in isolated rural regions with no access to reliable electricity and renewable energy. Mainly, biomass has been used for cooking and residential purposes (Katekar and Deshmukh, 2021a). South Asian countries depend on fossil fuels to fulfill the demand for electrical energy. The significant expense they incur importing fossil fuels is funded by them (Alagh, 2006). This part depicts South Asian countries' energy situation with Bhutan, India, Pakistan, and Sri Lanka.

FIGURE 8.1

Map of South Asian countries (Group, 2021).

The countries of South Asia are very different in terms of commercial resources and commercial energy needs. The fact that a country has an enormous resource base does not always suggest that it has enough energy, as resource-rich countries are just as energy-dependent as other countries, as illustrated by the compound annual growth rate statistics in Table 8.1 (Susantono et al., 2020).

Table 8.1 The compounded annual growth rate in energy demand: South Asia.

Country name	Energy demand in 2010 (GWh)	Energy demand in 2020 (GWh)	Compounded annual growth rate (%)
Bhutan	1749	3430	7
India	938,000	1,845,000	7
Pakistan	95,000	246,000	10
Sri Lanka	10,718	21,040	7

Source: Asian Development Bank (Susantono et al., 2020).

1.3.1 Bhutan

Today, about 30% of Bhutan's energy demand is met by renewable energy sources in which hydropower plants provide the majority of the electricity. Other energy demands are often met with fuelwood (traditional biomass), which contributes to the carbon footprint. Hydropower is the country's highest revenue earner and is green. Despite its mountainous terrain, it has ample solar and wind resources in many areas. According to the Renewable Energy Management Master Plan of 2016, Bhutan could generate enough renewable energy as solar energy with a capacity of 12 GW and wind power with a capacity of 760 MW. However, aside from large hydro plants, the country's current installed renewable energy capacity is only 9 MW (International Renewable Energy Agency, 2019).

Fig. 8.2 shows that thermal energy accounts for 72% of Bhutan's energy demand, with electricity accounting for only 28%. Biomass fuelwood, biogas, and briquettes are the most significant supply of thermal energy, which satisfies 36% of overall energy demand. Electricity production is mainly by hydropower.

1.3.2 India

With a population of 1.39 billion people and one of the world's fastest-growing major economies, India is critical to the future of global energy markets. Access to electricity and other essential amenities such as sanitary cooking has improved due to the Indian Government's work. Additionally, it has implemented various market reforms related to the energy sector, including acquiring a considerable amount of renewable energy in solar power.

Coal continues to be India's principal source of energy and electricity generation, as shown in Fig. 8.3. In addition to being a vital source of electricity, a robust coal sector is also critical for the industry in industries such as steel, cement, and fertilizers (Alagh, 2006).

India ranks third in crude oil output, fourth in crude oil refining, and exports refined goods, making it one of its three most prominent producers, refiners, and net exporters. Over the next few years, India's oil consumption is predicted to exceed that of China, making India a significant oil consumer (World Energy Outlook, 2021). Now, it is an excellent time to invest in a refinery with all of these factors in mind. India's refining capacity must increase with expected demand to maintain its position as a refining center.

Table 8.2 depicts India's renewable energy mix, revealing that wind and solar power are the most prevalent total installed capacity.

Many sources indicate that the Indian Government has confirmed that around 300 million people are still without electricity, and approximately 800 million lack access to safe cooking fuels. The

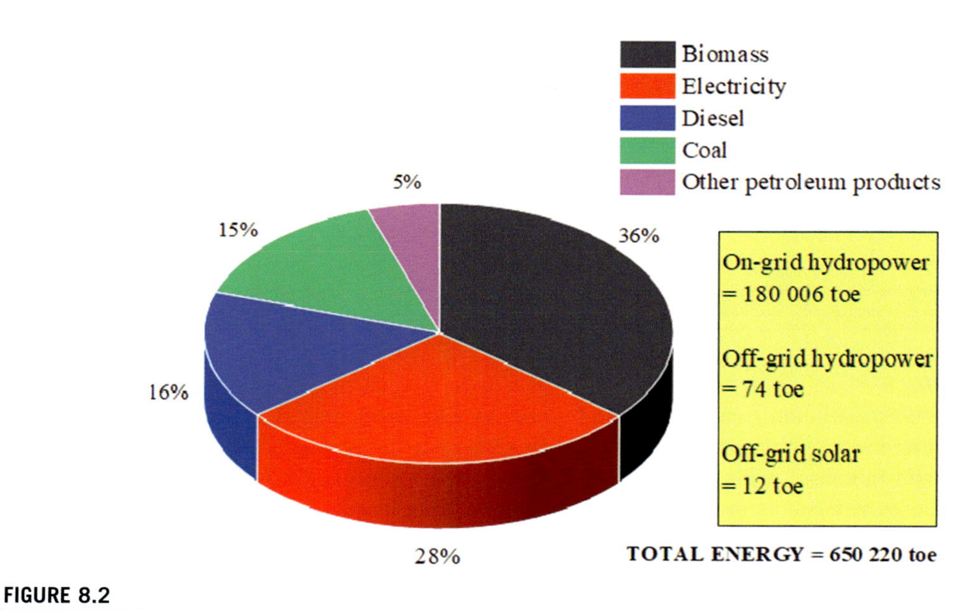

FIGURE 8.2

Fuel mix in the economy of Bhutan (toe).

Source: DRE—MOEA (International Renewable Energy Agency, 2019).

recent announcement of 450 GW of renewable energy capacity, which is expected to be operational by 2030, represents a significant increase in ambition. It is more than five times India's present renewable capacity (82.5 GW) and more than its total installed power capacity from all sources, i.e., 362 GW (Jaiswal, 2019).Table 8.3 presents India's projected electricity capacity by 2030.

While India's renewable energy ambitions are aggressive, it is possible to meet these goals if implemented as illustrated above, including on-demand creation, revenue certainty, risk minimization, and system integration.

1.3.3 Pakistan

Most of Pakistan's primary energy demands are provided by oil and gas. About 64% of the country's electricity is generated from oil and gas. Fossil fuels account for 64%, 27%, and 7%, respectively, in the Pakistan Economic Survey 2018–2019 (Irfan et al., 2020). As seen in Fig. 8.4 below, renewable energy accounts for only 2% of overall energy generation.

Pakistan now has a 33 GW installed total capacity. Fig. 8.5 depicts the past and current electricity demand and supply trends.

Until 2035, natural gas will be the primary source of energy. By 2025, 43.8% of the primary energy mix will be produced. Domestic natural gas production is expected to fall from 38.4 BCM to 13 BCM by 2035. Pakistan will therefore be increasingly reliant on imported gas in the future.

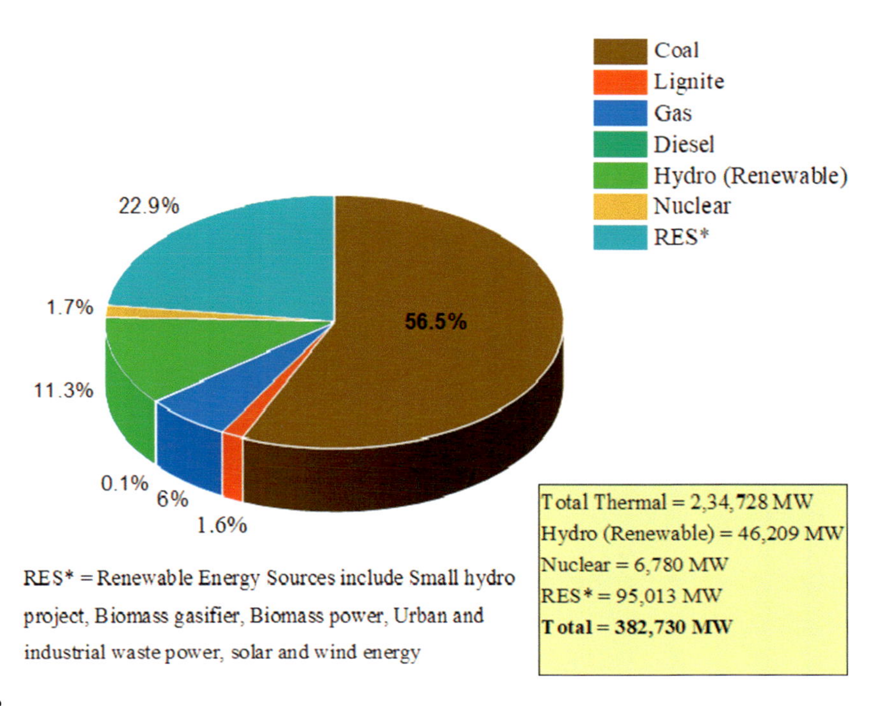

FIGURE 8.3

India's installed power capacity by fuel, April 2021.

Source: G. of India (2021).

Table 8.2 India's total installed capacity of renewable energy in 2020 (India, 2019).

Renewable energy	India's total installed capacity in 2020 (MW)
Wind power	38,789
Small hydropower	4683
Solar power—ground mounted	31,980
Solar power—roof top	6100
Biomass power (biomass, gasification, and bagasse cogeneration)	10,150
Waste to power	168.64

1.3.4 Sri Lanka

It is one of the most populous countries in South Asia with around 21.5 million, which is equivalent to 0.27% of the total world population and an area of 62,710 km^2 as of June 2021 worldometer

Table 8.3 The Indian Government's roadmap is to meet the country's energy demand by 2030 (Manoj Kumar Upadhyay, 2017).

Type of fuel	Capacity (GW)	Percentage (%)
Hydro (large, small, and imports)	73.44	9
Gas	24.35	3
Coal + lignite	266.82	32
Nuclear	16.88	2
Solar	300	36
Wind	140	17
Biomass	10	1
Total Installed capacity	831.5	100
Total nonfossil fuel (hydro, nuclear, solar, wind, and biomass)	540.32	65
Total renewable (wind, solar, and biomass)	450	54

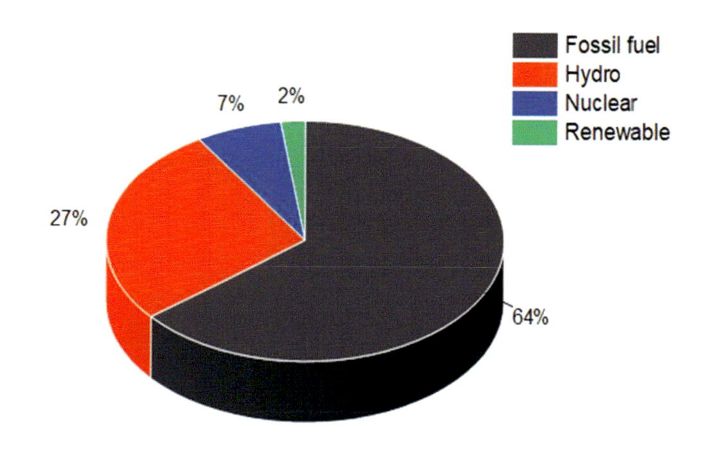

FIGURE 8.4

Share energy production's fuel sources (Irfan et al., 2020).

elaboration of the latest UN data (Worldometer, 2021). With 341 inhabitants per square km, Sri Lanka has a high population density. The oil equivalent capacity of Sri Lanka was utilized at 12.8 million tonnes in 2017. With $7.5 of economic output per kilogram of oil equivalent in 2010, Sri Lanka stands high among nations with comparable per capita economic output (Asian Development Bank, 2019).

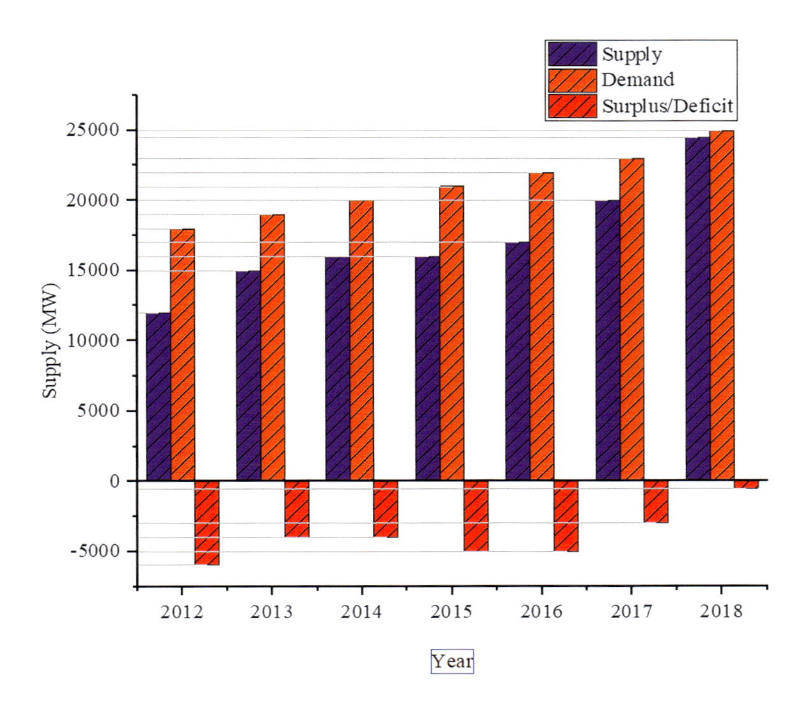

FIGURE 8.5

Pakistan's 2012–2018 energy supply trend (Irfan et al., 2020).

Biomass, hydropower, sun, and wind are renewable indigenous energy sources. Petroleum and coal are imported into the country. The principal energy sources are biomass and oil. Petroleum accounted for around 43% of primary energy supply in 2017, with biomass accounting for 37% and coal accounting for 11% (Asian Development Bank, 2019).

Fig. 8.6 depicts the share of primary energy supply provided by various sources. In 2017, 12,850 kilotons were equivalent to the total direct energy supply (ktoe). In 2017, 0.58 tones/person was predicted to be the per capita primary energy supply (Asian Development Bank, 2019).

Bioenergy is used in both commercial and domestic settings in Sri Lanka. The biomass utilized in 2017 consumed 12 million tonnes. The coal consumption for Sri Lanka is anticipated to increase by 4.9 MTOE between 2010 and 2030. As forecasted by Asian Development Bank, by 2030 the oil will become the dominant energy source, with 42.9% in the primary energy mix (Asian Development Bank, 2019).

Liquefied petroleum gas (LPG), biomass, kerosene, and electricity are essential energy sources within households in Sri Lanka. In the urban area, 95% of homes use LPG, whereas 95% use biomass for cooking in rural areas. The population of the semiurban area employs a combination of all sources (LPG-70% and biomass-85%). Day biomass is now substituted by LPG and kerosene, reducing noxious pollutants in the kitchen air considerably (Asian Development Bank, 2019).

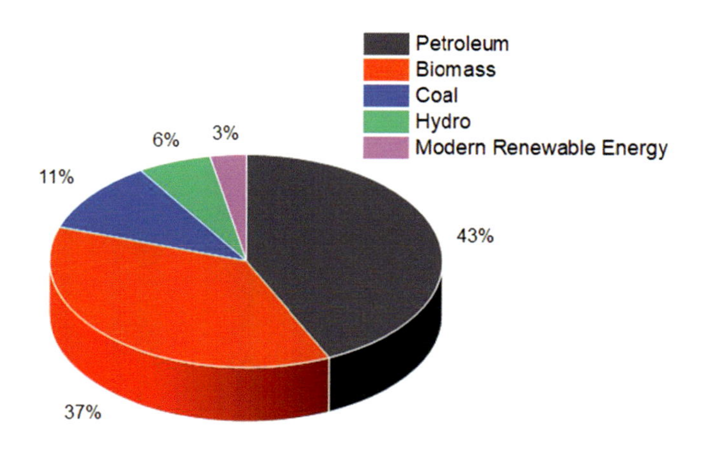

FIGURE 8.6

Share of Primary energy supply by various sources in Sri Lanka (Portale and de Wit, 2014).

1.4 Climate change—South Asia

This is a vast territory, covering the countries of Bhutan, Nepal, Bangladesh, and Sri Lanka and the islands of the Maldives, but especially India's peninsula and the country's famous jewel-like islands of Sri Lanka and the Maldives. Even though this region is experiencing many climate change effects, the most prominent of which is the glacial melt, the area is also suffering from numerous other changes, including forest fires, rising sea levels, mountain and coastal soil erosion, and saline water intrusion, which are standard across all climatic zones. Frequent and intense storms and atypical monsoon patterns have increased the impact of natural disasters. The vast majority of the planet's impoverished people, over 600 million of them, reside in the region, accounting for more than half of the world's total poor population. These people primarily subsist on climate-sensitive businesses such as agriculture, forestry, and traditional fishing (Sudan, 2021).

1.4.1 Climate change in Bhutan

Bhutan's reliance on hydropower to supply its electricity demands exposes the country to the long-term consequences of climate change, creating concerns about energy security. Future and present hydropower projects could be put in danger due to changes in rainfall, snow, and ice melting related to climate change. Hydropower is prone to electricity supply problems because of seasonal weather patterns and weather extremes (International Renewable Energy Agency, 2019).

Rainfall diminishes in Bhutan during the winter months, causing reduced meltwater runoff from snow and ice (Fig. 8.7). Due to Bhutan's run-of-river dams, the electricity volume fluctuated with the season. Electricity imports from India are necessary to supply 4% of total domestic consumption in 2017 due to low river flows in the winter (International Renewable Energy Agency, 2019).

Due to fossil fuel energy consumption and the manufacturing of cement, CO_2 emissions in Bhutan are mostly coming from burning fossil fuels. Carbon dioxide emissions are produced when fossil fuels are burned and manufactured cement. The combustion of solid–liquid generates carbon dioxide, gaseous fuels, and gas flaring. CO_2 emissions (metric tonnes per capita) in the world were 4.4835

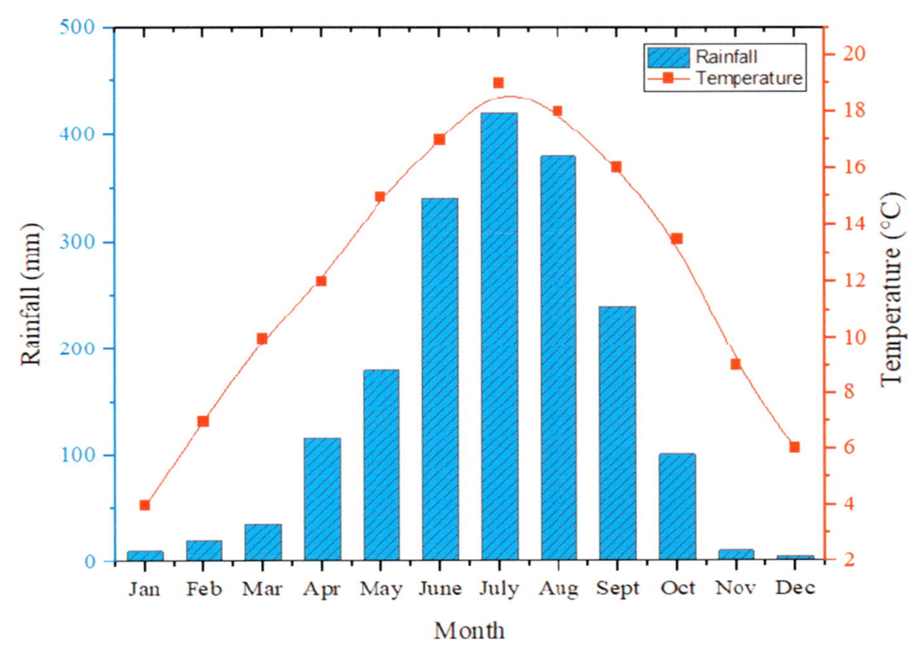

FIGURE 8.7

Annual rainfall and temperature averages for the Kingdom of Bhutan.

Source: SARI/EI and IRADe (2016) (International Renewable Energy Agency, 2019).

metric tons in 2018, according to the World Bank's collection of development indicators gathered from publicly available data. Fig. 8.8 illustrates Bhutan's annual CO_2 emissions from 2000 to 2018 (W. Bank Group, 2019).

1.4.2 Climate change in India

According to the report, India had experienced a tremendous increase in the rate of climate change impacts between 1996 and 2015, when it placed fourth among the countries most hit by climate change. With over one billion, India generates approximately two and a half tonnes of CO_2 equivalents per person, half the global average. It is estimated that the region contributes 7% of world carbon emissions. As the temperature rises on the Tibetan Plateau, glacial activity along the Himalayan mountain range forces glaciers to recede, putting the flow of important rivers such as the Ganges, the Brahmaputra, and the Yamuna in peril; these rivers support the livelihoods of hundreds of thousands of people (Vld and Carlsson, 2010). In 2007, the World Wide Fund for Nature (WWF) reported that the Indus River might eventually run dry due to overirrigation. Heatwaves are on the rise, both in frequency and intensity. Heatwaves are occurring more frequently and intensely in India because of climate change (Alagh, 2006).

We have found that CO_2 emissions declined by 1.4% in the most recent fiscal year using the indicators for consumption of coal, oil, and natural gas (Fig. 8.9). Compared to the same months the

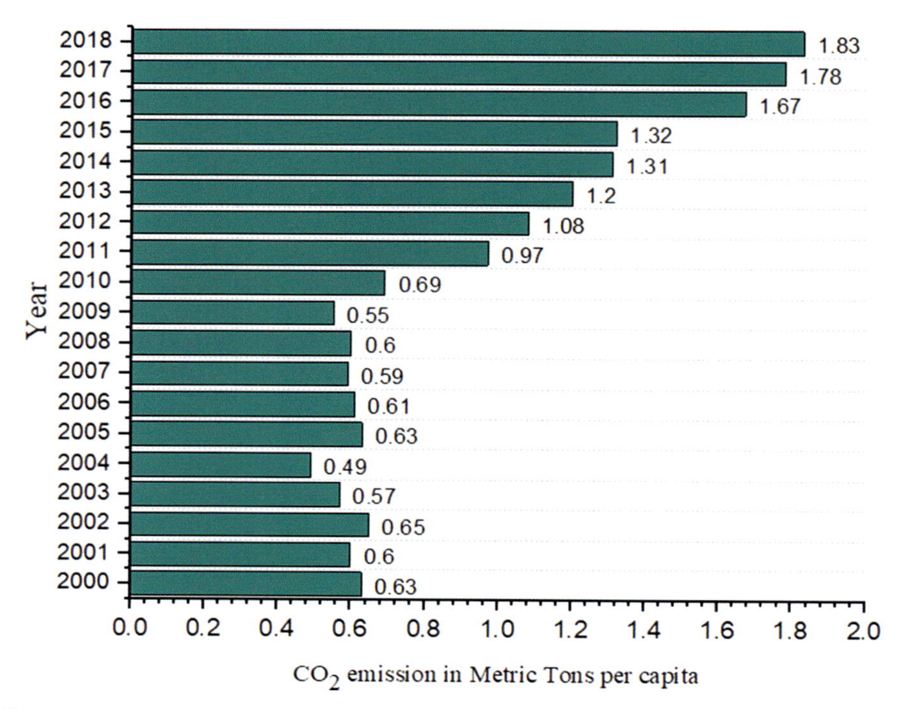

FIGURE 8.8

CO_2 emission of Bhutan in Metric Tons per capita (W. bank Group, 2019).

previous year, in March and April of this year also had 15% and 30% reductions in emissions, respectively. The April estimate is calculated based on daily generation data, which measures the emissions from the electricity sector (Hickman et al., 2020).

This fiscal saw consumption growth, the worst in 22 years, due to the coronavirus pandemic and already reduced growth earlier in the year. Even with the 5.5% increase in natural gas use, usage is predicted to decline by 15%–20% when the shutdown begins. Even while crude oil production in India was down 5.9% from the previous financial year, natural gas production was 5.2%. In contrast to the year prior, crude oil production at the refinery was down 1.1% for the year.

Crude steel output declined 22.7% in March 2020 compared to the previous month, while the financial year 2019–2020 experienced a 2.2% decrease compared with the year earlier, according to the Ministry of Steel statistics (Hickman et al., 2020).

1.4.3 Climate change in Pakistan

Changes in Pakistan's climate and environment are anticipated to have many long-term repercussions. A combination of climate change and continuous instability has led to increasingly variable weather in Pakistan in recent decades, and this pattern is projected to continue in the future. Among other things, Pakistan's most important rivers face increasing water supply issues and heat, drought, and extreme weather conditions due to the melting of the glaciers in the Himalayas. Pakistan was among the five

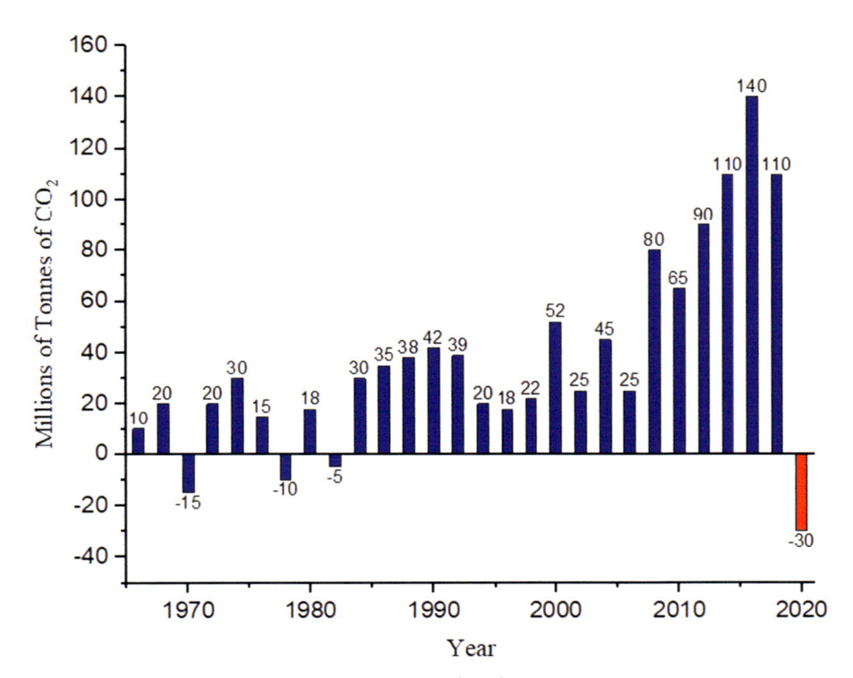

FIGURE 8.9 India's CO_2 emissions have fallen for the first time in four decades.

(Annual CO_2 emission fell 30 million tons of CO_2, i.e., 1.4% in the financial year 2019–2020).

Source: Indian government data analysis and the BP World Energy Statistical Review (Hickman et al., 2020).

most adversely impacted countries in terms of drastic climatic change due to climate change from 1999 to 2018. Pakistan has a low amount of GHG emissions, yet its position concerning climate change is relatively unstable (Ur Rehman et al., 2019).

Fossil CO_2 emissions in Pakistan were 178,013,820 tons in 2016. CO_2 emissions per capita in Pakistan were equivalent to 0.87 tons per person (based on a population of 203,631,353 in 2016). It increased by 0.06 over the number of 0.82 CO_2 tons per person registered in 2015; this represents a change of 6.9% in CO_2 emissions per capita (Pakistan, 2016). Figs. 8.10 and 8.11 present the fossil CO_2 emissions of Pakistan by sector and year, respectively.

1.4.4 Climate change in Sri Lanka

Climate change is a big issue in Sri Lanka, endangering both human and natural systems. In the west, south, and south-west portions of the island, about half of the island's 22 million population already live in coastal areas vulnerable to future sea-level rise due to rising sea levels. Climate change in Sri Lanka's coastal regions, particularly the Northern and Northern Western provinces, is worrisome (Shrestha et al., 2012). Climate change could have disastrous impacts on numerous levels and threaten the biodiversity of Sri Lanka. Among some of the long-term implications of this behavior include problems for agriculture, raising the risk of natural disasters, and helping the spread of infectious illnesses (Fig. 8.12). Most of the natural disasters in Sri Lanka have been caused by climate change,

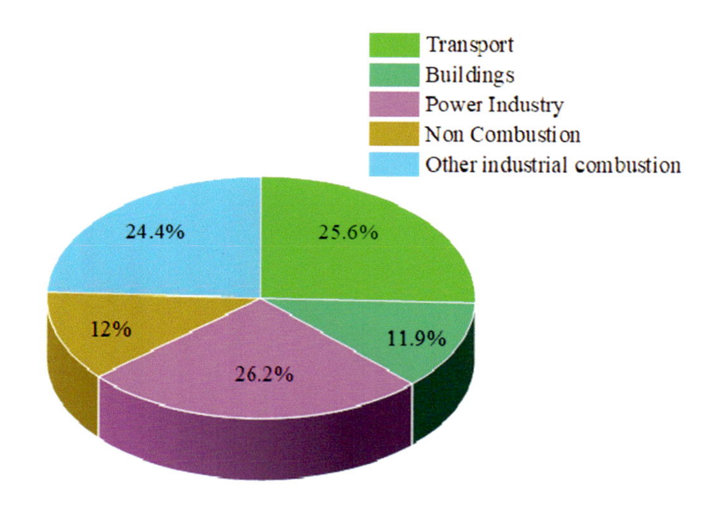

FIGURE 8.10

Fossil CO_2 Emissions of Pakistan by sector.

Source: EDGAR (Pakistan, 2016).

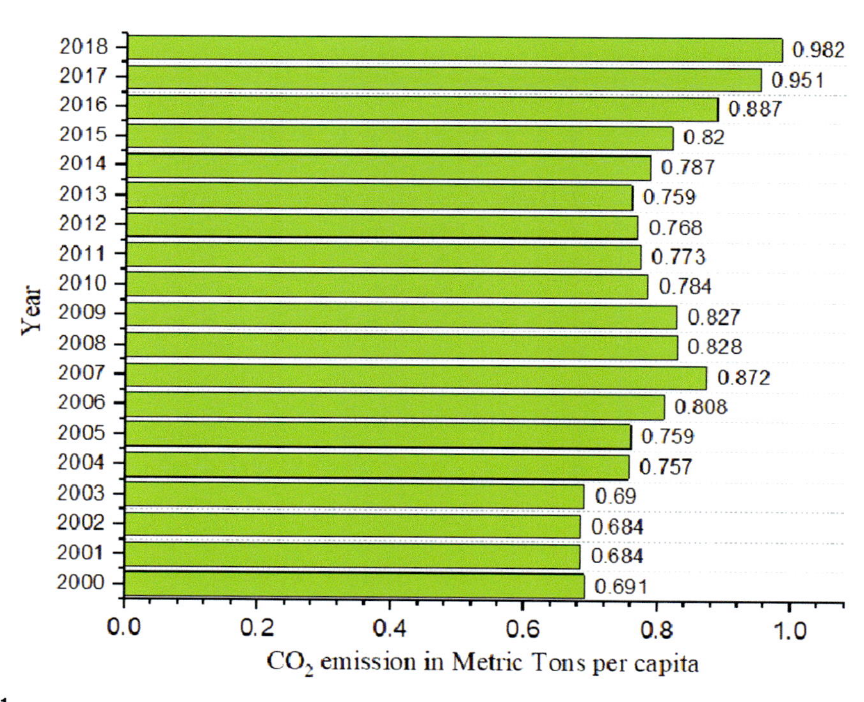

FIGURE 8.11

Pakistan CO_2 emissions in Metric Tons per capita by year (Hutfilter et al., 2019).

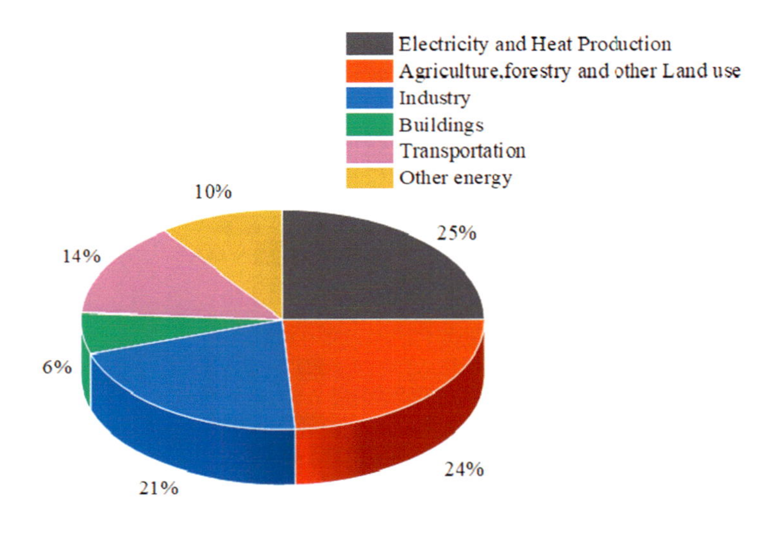

FIGURE 8.12

Global greenhouse gas emissions of Sri Lanka by economic sector (Illangasinghe, 2012).

and it ranked fourth in the most impacted countries by this phenomenon in 2016 (Manpreet Singh, 2017). As a result, we have a more vital ability to respond to climate change by bolstering disaster preparedness and intervening earlier.

Sri Lanka had 1.31 tonnes of CO_2 emissions per capita in 2019. As seen in Table 8.4, CO_2 emissions per capita in Sri Lanka climbed from 0.25 tonnes per capita in 1970 to 1.31 tonnes per capita in 2019, expanding at an average yearly rate of 3.96%.

Table 8.4 Sri Lanka's CO_2 emissions per capita.

Date	Value	Change in %
2019	1.31	8.01
2018	1.21	0.07
2017	1.21	9.56
2016	1.11	6.56
2015	1.04	16.39
2014	0.89	19.43
2013	0.75	−13.93
2012	0.87	8.58
2010	0.68	5.15
2009	0.65	−4.53
2008	0.68	

Source: World Data Atlas—Sri Lanka (2019).

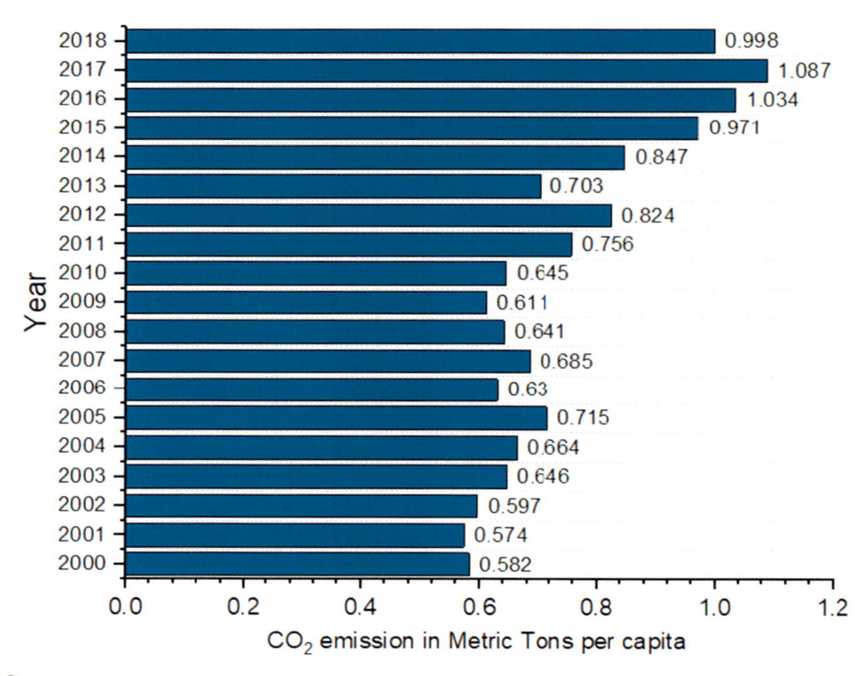

FIGURE 8.13

Sri Lanka: CO_2 emission in Metric Tons per capita (Illangasinghe, 2012).

Sri Lanka's CO_2 emission in Metric Tons per capita is depicted in Fig. 8.13 from 2000 to 2018.

1.5 Outlook for sustainable development in South Asia

Over the last decade, South Asian countries have made significant progress toward meeting the Millennium Development Goals (MDGs).

Fig. 8.14 illustrates that South Asia has already achieved targets for gender equality in primary education, increased forest cover and protected areas, poverty eradication, decreased carbon dioxide emissions per unit of GDP, tuberculosis reduction, and increased access to safe drinking water, and is on track to achieve the primary enrollment target and gender equality in secondary education. South Asia, like other subregions, has struggled to fulfill targets for maternal and child mortality, sanitation, and decreasing the proportion of underweight children. South Asia's accomplishments differ between aims and targets within and between countries (United Nations ESCAP, 2018).

South Asia has decreased severe poverty by 54.7% since 1990, exceeding the MDG objective of 50% reduction, thanks to bursts of acceleration in poverty reduction that occurred just before the global financial crisis and in recent years (Katekar and Deshmukh, 2020). South Asia has also met its MDGs for universal primary enrollment and primary completion. However, the subregion's net migration rate is 52.5% (United Nations ESCAP, 2018).

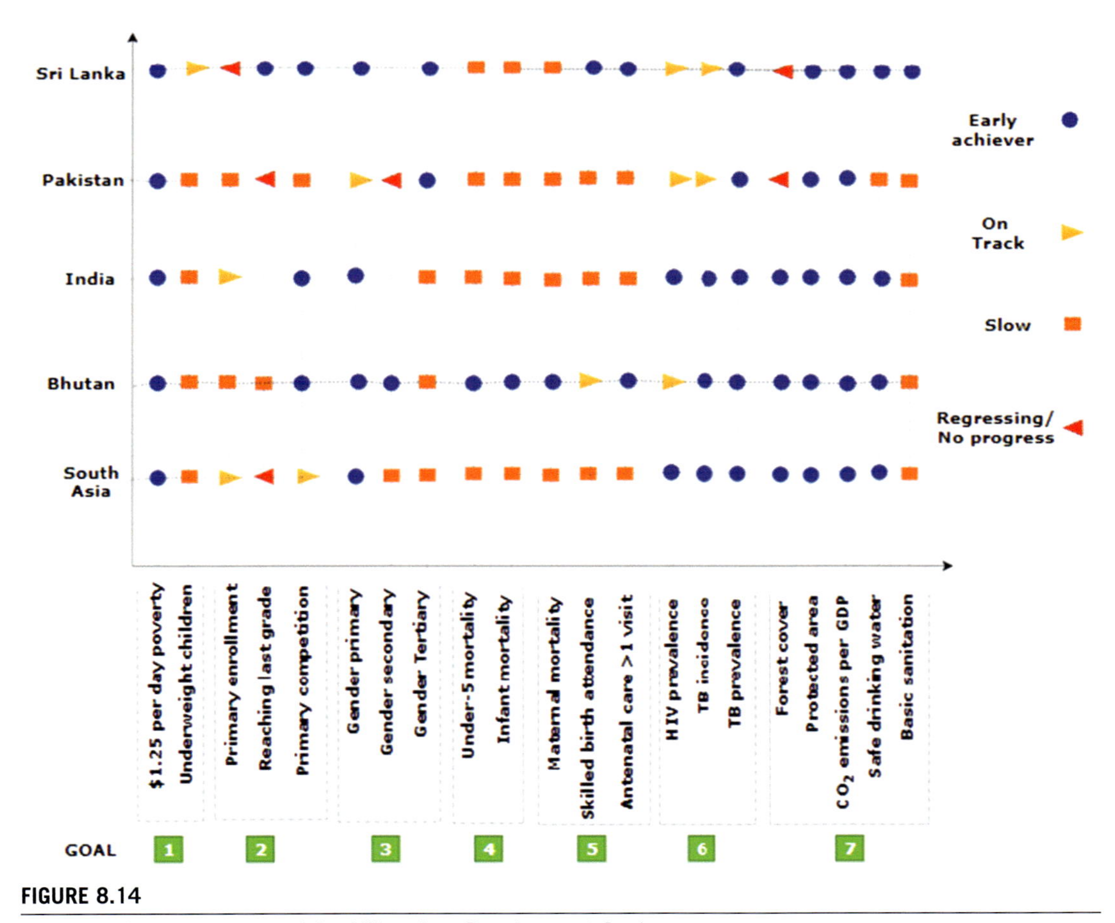

FIGURE 8.14

South Asia's progress toward the Millennium Development Goals.

Source: United Nations ESCAP (2018).

Table 8.5 Comparison of South Asian countries (Shukla, 2017).

South Asian country	Per capita electricity consumption (kWh)
Bhutan	430.2
India	644
Pakistan	457
Sri Lanka	636.3

As discussed in the above sections, energy supply and security are significant development challenges in South Asian countries. Tables 8.5 and 8.6 compare South Asian countries' per capita electricity consumption and electricity consumption and usage.

Table 8.7 demonstrates that several South Asian countries, like India (Coal—67.9%) and Sri Lanka (Oil—50.2%), rely on a single source for more than half of their total energy output.

Renewable energy sources provide enormous potential in South Asian countries. Table 8.8 below summarizes the renewable energy potential of solar power, hydropower, and wind power.

Table 8.6 Electricity consumption and uses: South Asian countries (Shukla, 2017).

Country	Energy consumed (million tons of oil equivalent)	Fossil fuels (% of total use)	Combustible renewable and waste (% of total use)	Alternative and nuclear energy (% of total use)	Energy produce (million tons of oil equivalent)	Energy use-energy production (Mtoe)
India	749.4	72.3	24.7	3	540.9	208
Sri Lanka	10.4	48.7	47.4	3.9	5.3	5.1
Bhutan	–	–	–	–	–	–
Pakistan	84.8	60.9	34.6	4.5	65.1	19.7

Table 8.7 Electricity generation in South Asian Countries (Shukla, 2017).

Country	Electricity production (kWh billion)	Coal (% of total)	Natural gas (% of total)	Oil (% of total)	Hydro power (% of total)	Renewable energy (% of total)	Nuclear power (% of total)
India	1052.3	67.9	10.3	1.2	12.4	12.4	5
Sri Lanka	11.6	8.9	0	50.2	39.7	39.7	1.2
Bhutan	–	–	–	–	–	–	–
Pakistan	95.3	0.1	29	35.4	29.9	29.9	0

Table 8.8 Renewable energy potential in South Asian Countries (Shukla, 2017).

Country	Solar power potential (kWh/m^2/day)	Hydropower potential (MW)	Wind power potential (MW)
India	5.0	150,000	102,778
Sri Lanka	5.0	2000	24,000
Bhutan	–	–	–
Pakistan	5.3	59,000	131,800

2. Energy and development in South Asia

Recent developments in South Asian energy security, such as the India–Nepal petroleum products pipeline and the India–Bhutan joint venture hydroelectric project, have reignited regional discussions about energy cooperation. While these are positive developments, they have encountered logistical, bureaucratic, and political setbacks like many others before them. Countries' growth can be assessed using the Human Development Index (HDI) and Happiness Index, covered in the following sections.

2.1 Human development index

HDI is a statistical measure tool that classifies countries based on statistics, lifespan, education, and income criteria to assign the four stages of human development. This index quantifies three characteristics of a country's well-being: its population's health, its educational attainment, and its standard of living based on the predicted number of years that its people can expect to live, as well as the country's GDP per capita. An HDI of 0.800 or above is considered "very high," while an HDI of 0.700 to 0.799 is defined as "high," and so on down to an HDI of 0.550 to 0.699, which is defined as "medium" (HDI of less than 0.550) (UNDP, 2020).

Table 8.9 represents the current HDI for Asian countries, as published in the Human Development Report of the United Nations Development Programme on December 15, 2020, and based on data collected in 2019.

2.2 Happiness index

The UN Sustainable Development Solutions Network publishes the World Happiness Report each year. It consists of articles and national satisfaction ratings based on respondents' ratings, also associated with various life variables.

Individuals from over 150 countries contributed data. Each variable is assigned a population-weighted average score on a scale of 0 to 10, then tracked over time and compared to similar variables in other countries. These variables include, but are not limited to, real GDP per capita, social support, healthy life expectancy, personal freedom, charity, and views of corruption (Helliwell et al., 2019).

In Table 8.10, the happiness score averaged over the years 2016–2018 is featured in the 2019 report. Finland is the happiest country on the planet, according to the 2019 Happiness Index. Denmark,

Table 8.9 HDI—South Asian countries (United Development Programme, 2019).

Country	HDI value (2019)	HDI rank in Asia region	HDI rank global
Sri Lanka	0.782	23	72
Bhutan	0.654	46	129
India	0.645	47	131
Pakistan	0.557	59	154

Table 8.10 Happiness index statistics of South Asian countries (Helliwell et al., 2019).

Country	Overall rank	GDP per capita	Social support	Healthy life expectancy	Freedom to make a life choices	Generosity	Perceptions of corruption
Pakistan	67	0.677	0.886	0.535	0.313	0.220	0.098
Bhutan	95	0.813	1.321	0.604	0.457	0.370	0.167
Sri Lanka	130	0.949	1.265	0.831	0.470	0.244	0.047
India	140	0.755	0.765	0.588	0.498	0.200	0.085

Norway, Iceland, and the Netherlands are the top four countries. The United Nations released the report on March 20, 2019. The World Happiness Report is a seminal study of happiness around the world.

3. Energy and social issues: South Asia

Worldwide, indoor air pollution (IAP) is a significant environmental problem. IAP refers to the contamination of internal air generated by solid fuel burning, contributing to indoor and outdoor air pollution. It is primarily a result of biomass (for example, wood, animal dung, agricultural wastes, and charcoal) and coal as cooking fuels (IHME, 2020). In actuality, interior air pollution is responsible for more than four million premature deaths every year, with children under the age of five accounting for half of all fatalities caused by IAP (The World Bank, 2019).

Cooking is a necessary aspect of everyday living. It is a family-friendly pastime that has cultural and social importance worldwide. However, solid fuels such as wood and coal are frequently utilized in traditional stoves for cooking in certain underdeveloped nations. Such harmful fuels and technology lead to IAP, leading to respiratory diseases, heart issues, and death (The World Bank, 2019).

As a result, the investigation into cooking fuels as a significant part of the energy and social issues utilized in various South Asian countries is required, which will be discussed in depth in the following sections.

3.1 Cooking fuel in Bhutan

The only Asian country with a net energy surplus is Bhutan. India has an energy surplus and exports energy, which accounted for 45% of its national net revenues and 19% of GDP in 2012–2013 and 99% of all power generated by hydropower (Dendup and Arimura, 2019).

Bhutan receives 99% of its hydroelectric energy, while it only uses 5% of its total potential. But net utilization is not viable due to the country's policy to retain forest cover of more than 60% at all times and to environmental and economic reasons. In addition, it is not advisable to rely on a single power source because sometimes it might lead to power outages. The hydroelectric power in Bhutan has a substantial sun power capacity. Bhutan pursues renewable energy resources, such as solar, wind, and biomass, and advances in the energy quality and the use of alternative transport fuels (Aravindh, 2016).

Primary cooking power sources and lightning sources in Bhutan are illustrated in Figs. 8.15 and 8.16, showing the fuel used in Bhutan.

3.2 Cooking fuel in India

According to a recent analysis on global and regional energy transition forecasts, India, Southeast Asia, and Sub-Saharan Africa will fall short of SDG target 7.1, impeding the global transition to a cleaner future.

As per the latest International Energy Outlook projections, 580 million people would rely on solid fuels for cooking in 2030, and India will fall short of the SDG 7.1 target. Despite India's population cutting its dependency on solid fuels by 15% (from 61% in 2016 to 46% in 2030) and despite anticipated advancements, over 580 million people will cook on solid fuels. By 2030, about 39% of India's population would still rely on solid fuels.

According to the CEEW research (Sunil et al., 2021), government measures have contributed to a significant increase in the use of LPG in the country. Thus, from 28.5% in 2011 to 71% in 2020, the proportion of families that use LPG as their primary cooking fuel has increased by more than twofold. Even though LPG/PNG is used as the principal fuel in 71% of houses, this number is still lower than

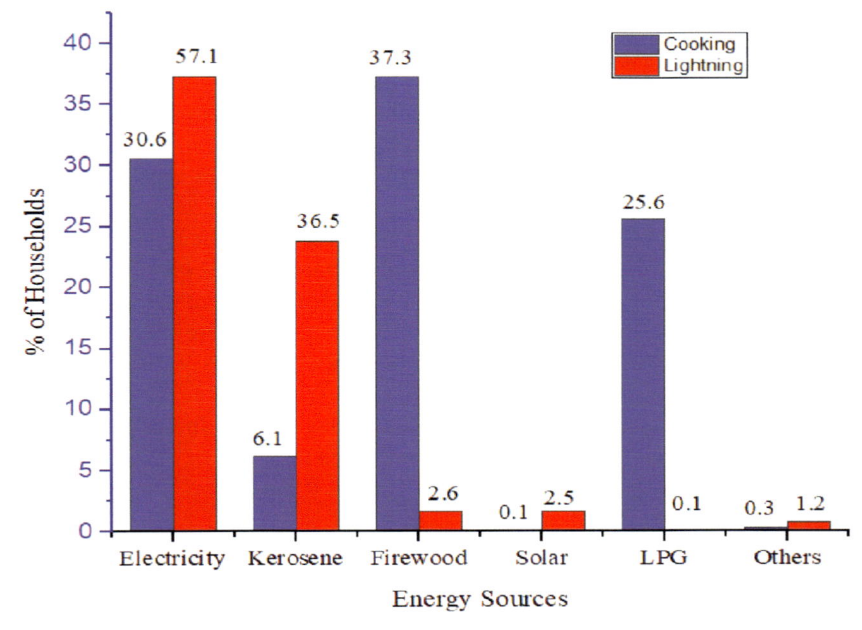

FIGURE 8.15

Primary energy sources for cooking and lightning in Bhutan (Aravindh, 2016).

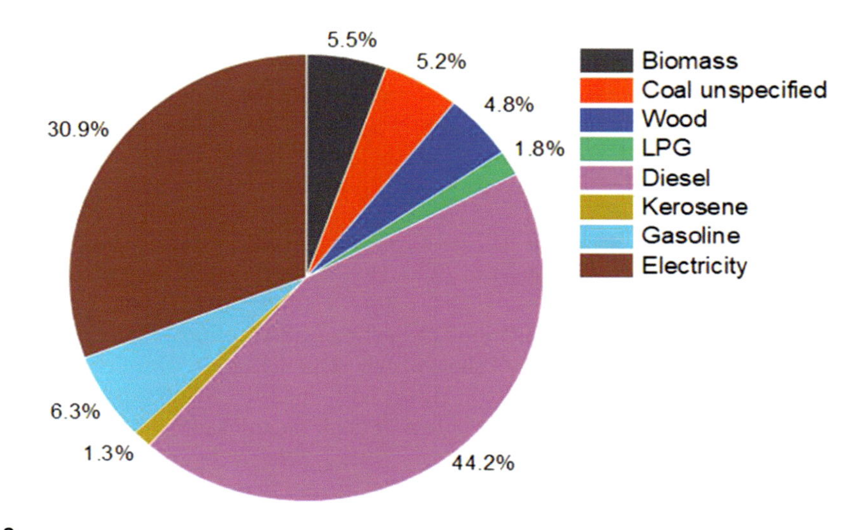

FIGURE 8.16

Cooking fuel mix of Bhutan (Aravindh, 2016).

the percentage of households having an LPG/PNG connection (85%). The penetration of LPG/PNG in urban areas (97.6%) is significantly higher than in rural areas (80%). Despite advancements in LPG technology, half of the Indian families continue to cook with firewood, and almost a fourth use dung cakes and other solid fuels for their cooking. Rural India has the highest proportion of solid fuel users; even though many families have LPG, firewood is used for at least a portion of their cooking needs. Stacking solid fuels with clean fuels is a source of concern since it exposes households to hazardous air pollutants continuously (Ambade et al., 2017).

Fig. 8.17 depicts the various types of cooking fuels utilized in Indian households. PNG and LPG are included because PNG adoption is shallow: 1.3% of urban homes and 0.4% of all Indian households use PNG. Coal, charcoal, lignite, and kerosene are all included in the "others" category.

While overall biomass feedstocks, such as firewood and crop waste, are strongly reliant on existing cooking fuels in India, accessible biomass feedstocks are subject to volatility and interruptions (Deshmukh et al., 2014).

Fig. 8.18 depicts the total percentage of household cooking fuels in India, indicating that the vast majority of cooking in India continues to be done using firewood.

3.3 Cooking fuel in Pakistan

Various studies have examined the way individuals in their homes have used fuel in the past (Williams, 2008). Energy consumption decisions among households are affected by household income, poverty, wealth, family size, and gender, according to certain studies from South Asia. Encouraged by the education of household heads and their family, modern and renewable fuel use is positively influenced, as these could be utilized for recreational activities or other profitable pursuits. When using modern fuels like propane, natural gas, or electricity, there is a positive correlation with

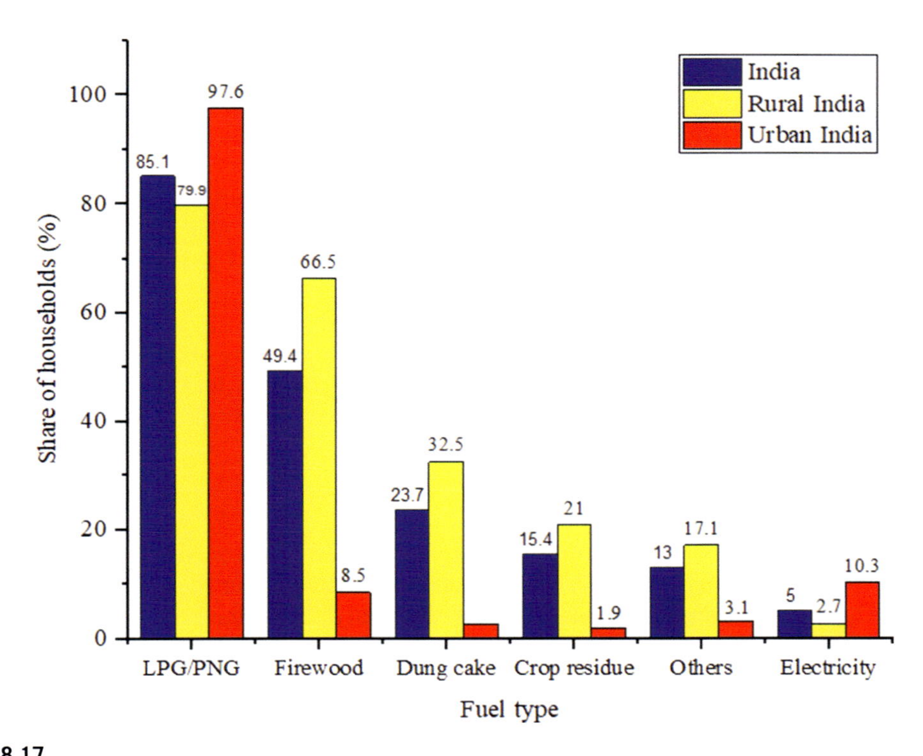

FIGURE 8.17

Share of Indian households in percentage by cooking fuel type (Sunil et al., 2021).

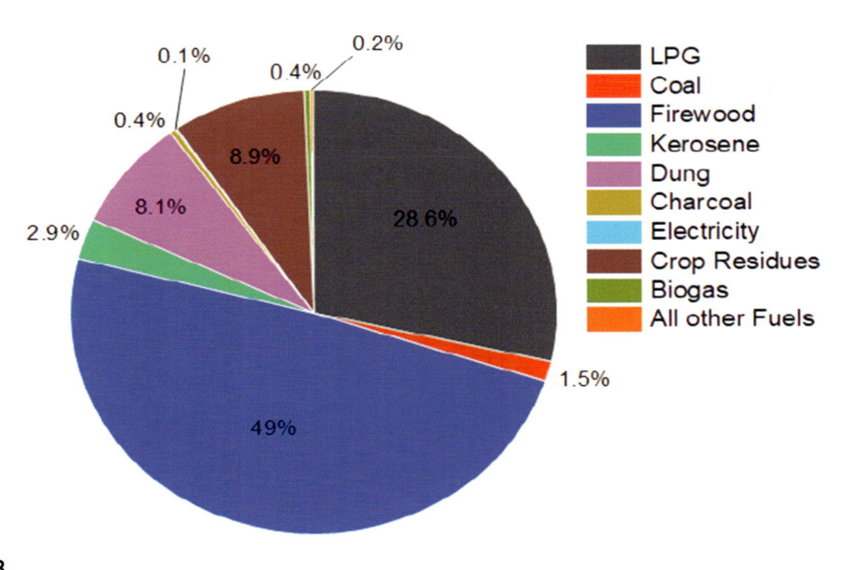

FIGURE 8.18

Cooking fuel mix in India (Katekar et al., 2021).

education levels. When using solid fuels like wood, coal, or charcoal, there is a negative correlation with education levels (Ur Rehman et al., 2019).

Reduced exposure to IAP, a primary cause of death in low-income homes, is brought about by having access to clean fuels or technologies such as clean cookstoves, presented in Fig. 8.19 (Rahut et al., 2019).

Fig. 8.20 indicates that 90% of households have access to the grid, 75.5% use fuelwood, 44.6% use LPG, 27.9% use agricultural residue, 16.2% use natural gas, 14.2% use solar, 10.3% use animal dung, and 3.4% use biogas. Biomass includes fuelwood, crop leftovers, and animal manure (Imran and Ozcatalbas, 2020).

Almost 56% of households that utilize fuelwood acquire it from their farms, 15.3% from other lands, and 28% buy it from the market. Their fields provided the majority of the crop residue. Almost 96% of households who use agricultural residue rely on their farms, with the remaining 3.6% collecting it from other property. Own farms were also a significant supply of animal manure (89.5%), with 10.5% collected from different farms (Table 8.11).

3.4 Cooking fuel in Sri Lanka

Sri Lanka's principal energy source is primarily oil and coal. Hydropower contributed approximately 40% of Sri Lanka's electricity in 2017, while coal has increased its power generation proportion since 2010. While Sri Lanka has made universal access to electricity, 15 million people relying on biomass

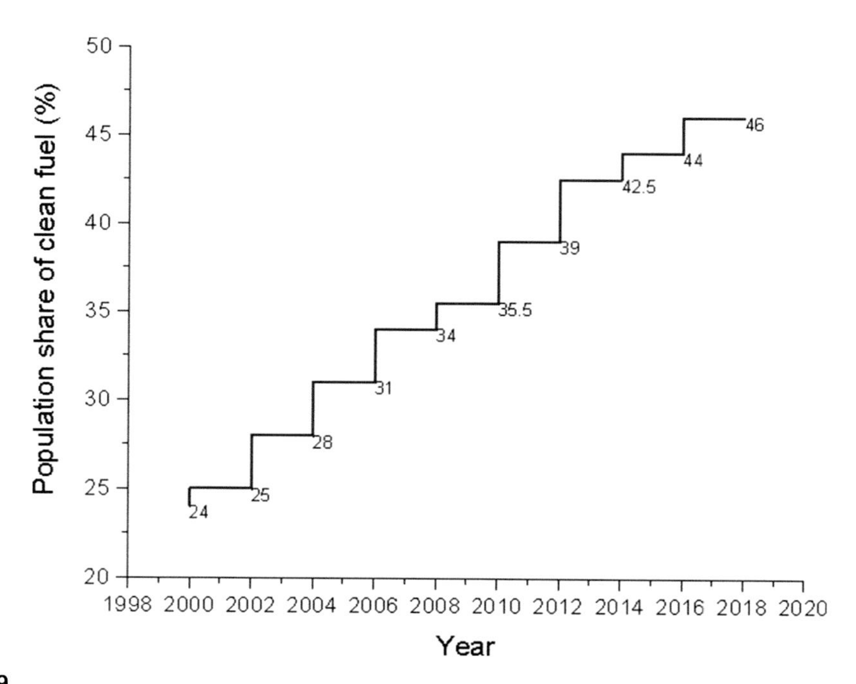

FIGURE 8.19

Population share of Pakistan having access to clean food fuel (Rahut et al., 2019).

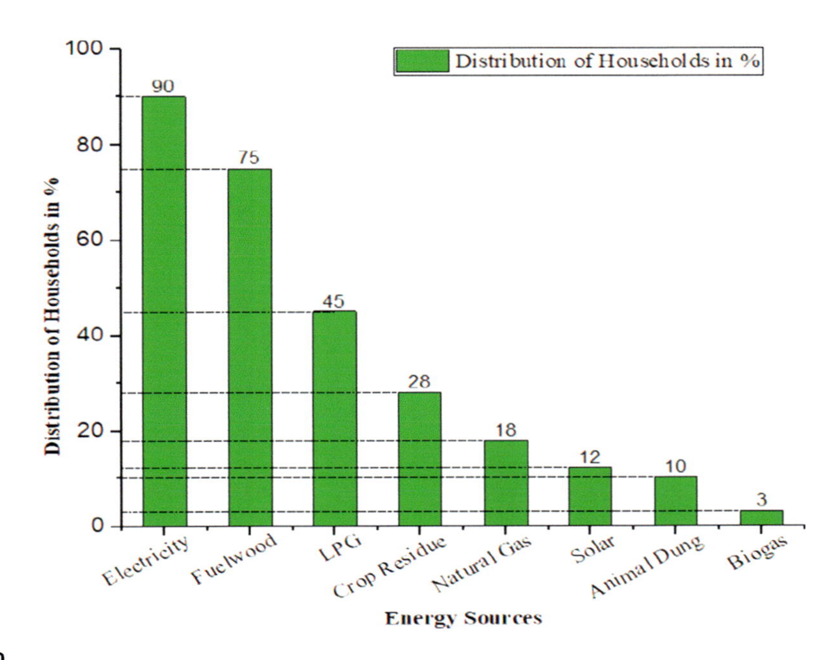

FIGURE 8.20

Household distribution based on access to various energy sources.

Data source: Imran and Ozcatalbas (2020).

Table 8.11 Sources of biomass (Imran and Ozcatalbas, 2020).

Source	Fuelwood (%)	Crop residues (%)	Animal dung (%)
Own farm	56.7	96.4	89.5
Collected from other farms	15.3	3.6	10.5
Purchased from the market	28	0	0

for cooking continue to face the challenge of clean cooking. The interactive diagram below illustrates the percentage of people that have access to clean cooking fuels (Eastern Research Group, 2016).

IAP exposures, the primary cause of fatality in low-income families, diminish the use of clean fuels or technology, such as clean cookstoves, presented in Fig. 8.21.

Lack of access to clean, modern fuels and cooking devices is one facet of energy poverty. By 2020, around 2.6 billion people frequently cook using carburant such as wood, animal dung, coal, or kerosene in developing countries (Lans and Vice, 2021). The cooking fuel mix in Sri Lanka is shown in Fig. 8.22.

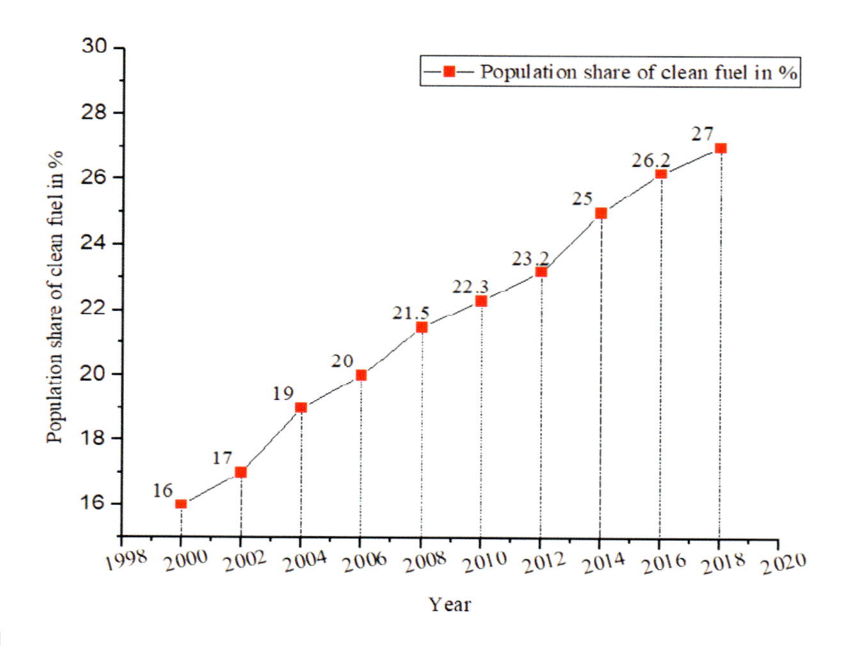

FIGURE 8.21

Share of the population of Sri Lanka with access to clean fuels for cooking.

Source: World Data Atlas—Sri Lanka (2019).

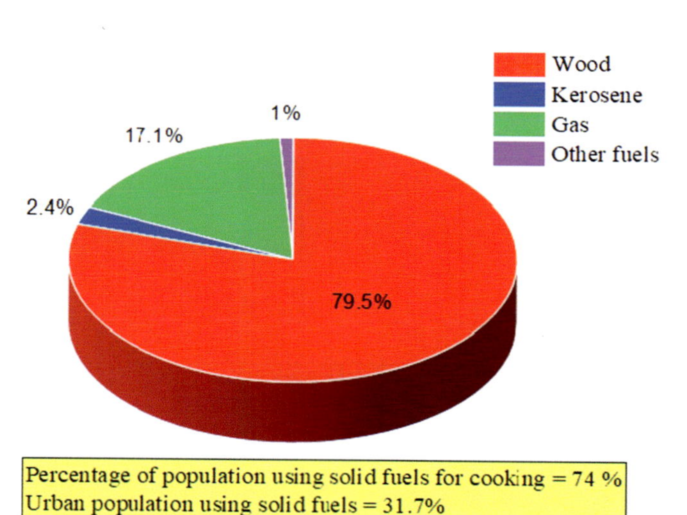

FIGURE 8.22

Cooking fuel mix in Sri Lanka.

Data source: Lans and Vice (2021).

4. Energy security: South Asia

Energy security is linked to national security and natural resources for energy use. In the workings of modern economies, the ability to get (relatively) cheap energy has become crucial (Ukey and Katekar, 2019). However, the uneven distribution of energy resources among countries has created considerable risks. With the consequent energy stability and insecurity, globalization has been aided by international energy linkages (Kumar Singh, 2013).

South Asia is at a crossroads with an expanding number of people, rising per capita revenues, emerging economies, and rising energy demand. The countries of South Asia rely heavily on imported fuels, particularly liquid carbon, for transportation. For example, with domestic sources, India, Pakistan, and Bangladesh fulfill their lower energy demand resulting in higher bills for energy imports. To satisfy their energy requirements, Maldives rely on oil goods. Both Nepal and Bhutan are mighty.

Meanwhile,energyconsumptionlevelsvarysubstantiallyacrossSouthAsiancountries.Theenormouspotentialofhydropowerin SouthAsiawillhelptheregionreduceitsdependenceonimportsoffossilfuelsandenhance thedemandforlocalresources(Anwar and Deshmukh, 2020). This makes for the need for producing and using locally accessible renewable energy and intraregional energy cooperation to assist minimize individual energy security risk countries in the face of climate change, essential for increased economic growth and poverty reduction in the region. Table 8.12 shows the economic indicators of South Asia.

4.1 Challenges in energy security and sustainable energy use

The growing population and income of the South and Southwest Asia subregion are increasingly putting pressure on natural resources (Katekar and Deshmukh, 2021b). Resource production and use are critical economic growth and development components and significant environmental consequences to growth. The aim of development plans should be to eliminate or minimize these ecological impacts from the

Table 8.12 Economic indicators of South Asia.

Country	Population (2018)	Population density (2018)	HDI ranking (2018)	Energy use*	Gross national income per capita (2018)	CO_2 emissions (2014)
Unit	Thousands	Person per km^2		Kilograms of oil equivalent per capita	US dollars	Metric tonnes per capita
Bhutan	754	20	134	367	3080	1.4
India	1,352,617	455	130	637	2020	1.7
Pakistan	212,215	275	150	460	1580	0.9
Sri Lanka	21,670	346	76	516	4060	0.9

*Represents the most recent statistics as of 2020.
Source: The World Bank Indicators (The World Bank et al., 2020).

start. Economic growth, environmental conservation, and social welfare are nonhierarchical pillars of sustainable development that must be viewed as equally important (Gupta, 2021).

4.1.1 South Asia's energy trade challenges

Regional marketplaces grow spontaneously when there are no constraints. Regional energy cooperation and integration are most feasible when nations support one another. Political conflicts, in contrast, make collaboration and trade in the region difficult. To put it another way, we aim for national self-sufficiency, which decreases our energy security. South Asian countries lack mutual trust. The issue of Pakistan and India getting independent dates back to 1947. Several attempts have been made to commence a peace process, including the "Lahore Declaration" in 2001. However, this has proven difficult, as significant problems remain, including deep-seated mistrust and unresolved conflicts (Kumar Singh, 2013).

Their suspicion of others at home influences the activities and voting behavior of South Asian populations. For example, during the 1980s, Sri Lanka was hit by ethnic strife, and as a result, mutual distrust developed between the Sri Lankan Sinhalese and the ethnic Tamils. The tension between the two countries has reduced over time, but suspicion lingers. Despite signing agreements on economic cooperation and a water-sharing deal, many issues remain unresolved between India and Bangladesh. The Nepalese rejection to renew the transit treaty between the two nations in 1989, which led to an economic blockade, was the direct effect of the SAARC's formation in 1985. Apart from India and Nepal, no other places on Earth have the infrastructure for electrical interconnections and gas pipelines crossing international borders. Commerce barriers in South Asia prevent regional trade, as illustrated in Table 8.13.

4.1.2 South Asia's energy cooperation challenges

For cross-border energy trade to be successful, all conditions must be met. These include the availability of energy resources, the economic potential of such help, the competitive cost benefits, adequate infrastructure, harmonious and capable policies, absence of trade barriers, a climate of mutual trust, and a solid political will to cooperate for the benefit of all parties (Sudan, 2021).

Table 8.13 Regional trade barriers in South Asia (Shrestha, 2012).

Barrier	Description
• Technical barriers	Grid synchronization and grid codes. Pipeline technology for natural gas
• Political barriers	Politicians that oppose electricity trade because of ongoing distrust, conflicts, and tensions across borders, particularly between India and Pakistan
• Cost barriers	Large-scale investments at the start
• Financial barriers	Inadequate finances and a significant risk
• Social barriers	Strong mistrust has an impact on how people act and vote
• Environmental barriers	Deteriorating water quality

The primary political hurdles to regional energy cooperation in South Asia are examined in this Table 8.14. It claims that investing in regional energy project planning can boost their viability while promoting integration and peacebuilding (Kumar Singh, 2013).

4.2 Energy policies for sustainable development

Several recent developments in South Asian energy security, like constructing the India–Nepal petroleum products pipeline and establishing a joint venture hydropower project between India and Bhutan, have reignited regional talks on energy cooperation. Even though these initiatives represent a positive step forward, they have been hampered by logistical, bureaucratic, and political setbacks (Deshmukh and Deshmukh, 2006).

Table 8.14 Energy cooperation challenges of South Asia.

Sr. no	Major challenges	Mechanism
1	Lack of competitive market power	• India is the first South Asian country to enable community-wide power trading, and it has just created an energy exchange for this purpose. • Other countries in the region do not have such alternatives for buyers and sellers of electricity.
2	Insufficient funding	• Power plants and transmission linkages, for example, demand a considerable investment in energy infrastructure. • South Asian governments lack the financial means to undertake such initiatives and rely on international donors and multilateral financing organizations for assistance.
3	Lack of coherence in energy strategy and related framework	• An essential requirement for cross-border commerce and investment is clear, coherent, and harmonized energy policies and a supportive legal and regulatory environment (Olhoff et al., 2011). • As a result, it is necessary to develop associated systems to enable and promote energy commerce between the countries of South Asia.
4	Inadequate infrastructure for establishing remote hydro projects	• Most hydroelectric plants are located in rural and hilly places without infrastructures such as high voltage or access roads (Dhurwey, 2019). • New infrastructures must be constructed and associated costs added to project costs to meet these needs. • This ultimately boosts the hydropower project's expenses and diminishes its competitiveness.

Continued

Table 8.14 Energy cooperation challenges of South Asia.—cont'd

Sr. no	Major challenges	Mechanism
5	Inadequate energy infrastructure	• Appropriate infrastructure for the transmission and transportation of energy resources must be present to have energy readily available. • Inability to provide infrastructure affected both the accessibility of resources and exploiting them. • The requirement for particular transmission lines within the region may be minimised by creating a solid energy infrastructure. Regional commerce could be conducted via interconnection points between adjacent member states (Jinturkar et al., 2014).
6	Institutional impediment	• Regional power exchange has trouble overcoming institutional constraints within the region. The current leadership has been unable to move the concept of power trade further. • There is no practical institutional framework between the countries that make up the South Asian Association for Regional Cooperation (SAARC).
7	A political frame of thinking	• Agreeing on national energy policies and approaches proved a pivotal obstacle to international cooperation. Trade-in power ought to be prevented (Anwar and Deshmukh, 2020).
8	Restriction of commercial and financial resources	• Tariff barriers restrict the capacity to detect and assess actual electricity costs. • The fundamental difficulty derives from the fact that power trade is typically viewed from a political rather than a business perspective (The World Bank et al., 2020).
9	Inadequate coherence and efficiency in the power sector	• State-owned enterprises dominate the region's power sector. • In contrast to that, the private sector has played a minimal role. Because of their poor financial performance, utilities have become reliant on public funds, which are limited and face a host of other competing demands (Manpreet and Keswani, 2017).

Source: Author's compilation.

4.2.1 South Asia should prioritize sustainable energy to close energy access gaps and speed up economic recovery

As 2020 concludes and the world continues to fight the COVID-19 pandemic, a new Sustainable Energy for All (SE for ALL) guide will show how Southern Asia can seize that chance of strengthening sustainable energy policy and finance, speeding up economic recovery, narrowing energy access gaps, and enhancing the lives of vulnerable populations within the region (Sudan, 2021). Energy policies to adopt in pandemic recovery and its measures for South Asia are given in Table 8.15.

Table 8.15 Energy policies to adopt in pandemic recovery and its measures for South Asia. To meet these pandemic recovery goals, governments will be provided with a set of enabling policies to implement, including.

Sr. no	Energy policy to adopt in pandemic recovery	Policy measures
1	Improvement of policy and institutional structures	• To remove structural impediments to renewable energy investments, regions must enhance their existing policies and institutions and improve their framework. • The region's progress has to be improved via strengthening capacity, regional and international collaboration, and business facilities.
2	Appreciating and growing your employees	• Underpinning this with investment in human resources and skills for sustainable energy-related jobs is necessary.
3	Investing in renewable energy rather than fossil fuels	• The region must focus on fossil fuels and renewable energy's "sunrise" sector. • Countries should move toward cost-reflective rates that do not disincentive renewable energy to attract additional investment (Deshmukh, 2016).

Source: Author's compilation.

If adopted, these enabling measures will result in lower energy costs, fewer environmental consequences, and better agricultural, gender, and health results. This reset can also inspire progress at the rate and scale required to meet SDG7 and help the global economy on the road consistent with the Paris Agreement and the SDGs (UNDP, 2020).

4.3 Energy security challenges in Bhutan

We need to acknowledge that energy is a crucial component of modern society and that energy resource availability and costs are vital to economic progress. Energy is not a goal in and of itself; it is a means to another goal: a healthy economy and environment. In terms of technology, skilled labor, and financial resources, Bhutan faces a complex problem in providing adequate, high-quality, and inexpensive energy to its inhabitants in a sustainable manner. The energy sector, particularly hydropower/energy investment planning and management, is highly complicated and demanding due to several factors.

Table 8.16 below shows some of the problems and constraints in the energy industry in Bhutan.

4.4 Energy security challenges in India

India, like China, is a rapidly developing superpower with a compelling need to fulfill rising energy demands. With over a billion people, or over a fifth of the world's population, India ranks sixth in energy

Table 8.16 Significant energy security challenges in Bhutan (Bhutan: Department of Energy, Ministry of Economic Affairs and Royal Government of Bhutan, 2009).

Sr. no	Major challenges	Reason/constraint
1	The price of electrification	• The expense of rural electrification has remained excessively high for the government and donor agencies due to the country's geographic position. • As a result, efforts to deliver clean, inexpensive, and easy "lifeline" energy to rural populations and impoverished cultures that cannot ultimately pay for it and its services continue to elude them.
2	Indoor pollution	• Rural populations, particularly women and children, bear a significant burden since they utilize traditional fuels like firewood and animal dung for cooking and heating, producing indoor pollution, and contributing to women's and children's health problems (Leach et al., 2014). • The quality of life must be improved and poverty reduced, especially in rural regions, by offering reliable and affordable energy as a stimulus for socioeconomic development.
3	Rising import dependence on oil	• With economic development, Bhutan's reliance on oil imports is increasing rapidly, raising serious concerns about the country's energy security (International Renewable Energy Agency, 2019). • As a result, it becomes important to expand the construction of oil refineries, prioritise fuel-efficient automobiles, and enhance the road conditions.
4	Strengthen energy efficiency and conservation actions	• Sustainable supply and end-use require more attention to energy efficiency and conservation methods. • The Government will support broad improvements in lighting, heating, ventilation, and insulation energy efficiency standards of buildings (Aravindh and Giri, 2016).
5	Enhance the electrical grid	• To achieve the national aim of "Electricity for All" by 2025 and operate and maintain the extensive transmission lines and consumer terminals, specialized labor and enormous resources would be required against insufficient revenue and budget (Anwar, 2020).
6	Increasing the base of hydropower resources	• Hydropower development demands a lot of investment, both in the tens of millions of dollars and billions of dollars. Thus, funding from many sources, both external and currently, is required. Due to the size and duration of energy investments, projects must be carefully planned and implemented rapidly (Sudan, 2021). • The Indian Government has supplied Bhutan with a 60% grant and a 40% loan for hydropower investment. An increasingly challenging situation has been set in motion in Bhutan because of the capital requirements of infrastructure development in India, making such financial assistance even if the Indian Government wanted to provide it difficult.

Table 8.16 Significant energy security challenges in Bhutan (Bhutan: Department of Energy, Ministry of Economic Affairs and Royal Government of Bhutan, 2009).—cont'd

Sr. no	Major challenges	Reason/constraint
7	Capitalizing the power sector through imports from the neighboring countries	• There is a significant electricity shortfall in our neighboring country, India. The Indian Ministry of Power has stated that 100,000 MW must be added by 2022, and the hydrothermal mix must be increased from 25:75 to 40:60. Bhutan is unquestionably in a position to seize this opportunity. • In addition, Nepal is interested in exporting electricity to India. The north-eastern states of India also have tremendous hydropower potential. Bhutan must act quickly and decisively to enhance its electricity exports to India (Kumar Singh, 2013).

Source: Author's compilation.

demand (Bhaisare et al., 2020). Its economy is predicted to grow by 7% to 8% over the next two decades, increasing the need for oil to power land, sea, and air transportation (Alagh, 2006).

Withashiftinfocusfromagriculturetoindustryandservices,India's rising population and economy have increased the energy intensity of its economy, resulting in the unprecedented need for energy supplies (Bhaisare et al., 2019). The problems faced in this respect are listed in Table 8.17 below.

India seeks to become a leading global economic force to fuel the energy needs for infrastructure, basic requirements, skill-building, job creation, and manufacturing. According to this, enhancing energy security can be accomplished by implementing the following methods. Table 8.18 illustrates possible measures to strengthen energy security in India.

4.5 Energy security challenges in Sri Lanka

Sri Lanka is a fascinating case study regarding energy and development issues. As a result of large-scale project implementation and the Mahaweli programme, the biomass for domestic consumption is currently excessive. However, in the following years, procuring wood fuel for houses will become increasingly complex, putting pressure on the commercial fuelwood sector.

The situation of Sri Lanka indicates that issues with satisfying people's energy needs and sustaining supplies exist even in countries that are not now considered to be in the midst of an energy crisis. Table 8.19 demonstrates significant energy security challenges in Sri Lanka.

4.6 Pakistan: energy policy and actions

Power generation in Pakistan is primarily a function of the public sector, with two vertically integrated utilities that control nearly all aspects of the supply chain. Since electricity demand fluctuates with the weather, the GOP turned to the private sector to create power.

Table 8.17 Significant energy security challenges in India.

Sr. no	Major energy security challenges	Mechanism
1	Obstacles to policy implementation	• A lack of capacity to attract foreign investment in domestic hydrocarbon exploration, as shown by NELP's inability to draw the attention of major global energy companies, among other things (Alagh, 2006). • Significant investments will be required to acquire hydrocarbon deposits in other countries. • In India, coal mining is hampered by regulatory and environmental approval delays. • India did not set up foreign-built nuclear reactors despite gaining access to essential technologies in vital areas through the Indo-US atomic pact (Sudan, 2021).
2	Accessibility challenge	• One of India's top energy consumers is the household sector. The entire primary energy use amounts to around 45%. Biomass accounts for 90% of total primary fuel usage in rural areas. For rural residents, this poses significant health risks. • There are currently 304 million Indians without power, and approximately 500 million Indians rely on solid cooking biomass.
3	Challenges in infrastructure and skills	• Scarce skilled labor and infrastructure for developing conventional and unconventional energy. • India does not have transit infrastructure to make energy accessible, and pipelines, for example, could be a beneficial approach to expand its overall gas supply. Because it can be used successfully in a wide range of demand sectors, gas will play an essential role in India's energy mix (Amin et al., 2020).
4	External challenges	• India is under great strain from its growing dependence on imported oil, regulatory uncertainties, international monopolies, and opaque policies on the price of natural gas India's precarious energy safety. • India aims for energy security by a range of partners, including the Indo-US nuclear pact, Middle East oil imports, etc. However, India's oil imports from Iran have been reduced in recent years, owing to disagreement between India's energy allies, such as the United States ands Iran (Eastern Research Group, 2016). • If a battle erupts between countries, China might gain a decisive edge by interfering with India's access to energy via the One belt, One Road plan. • Failure to involve all parties interested in the IPI gas pipeline and the TAPI gas pipeline, guaranteeing natural gas supplies for each country.

Source: Author's compilation.

Table 8.18 Measures to enhance energy security in India.

Sr. no	Measures enhancing energy security	Illustration
1	Expanding access to sustainable energy	• India has already pledged to provide energy to every home by 2022. A more audacious ambition would be to give electricity to every family 24 h a day. • To reach clean fuel in rural regions, a blend of the Pradhan Mantri Ujjwala Yojana and "efficient biomass chullahs" must be employed (Alagh, 2006). • On the agricultural front, NABARD and the Government must encourage the installation of solar irrigation pumps by using government and NABARD loans. • Potential unconventional energy sources need to be studied and investigated to make them technologically economic and accessible, such as geothermal power, tidal energy, etc.
2	Enhancing efficiency	• A complete cost-benefit study of energy-efficient technology and goods available to all industries, particularly agriculture, housing, and transport, should be conducted by the National Enhanced Energy Efficiencies Mission (NMEEE). • Institutionally, national and state-designated energy efficiency agencies should be enhanced. • To further improve vehicle fuel efficiency, auto gasoline quality should be enhanced to BS-VI standards before a statewide rollout in 2020 (Amin et al., 2020).
3	Changes in policy	• Coal-fired power plants generate around three-quarters of our electricity. India's domestic coal production must be increased to decrease its reliance on imports. It is necessary to expedite regulatory approvals, raise labor productivity, expand coal production, and improve distribution efficiency (Shukla et al., 2017). • Hydrocarbon Exploration and Licensing Policy (HELP) aims to minimize decision-making discretion by the Government, reduce conflicts, reduce administrative delays, and introduce the notion of income sharing, free marketing in the Indian oil and gas industry to boost growth. • The India Energy Security Scenarios, 2047 (ISS) was designed as an energy scenario construction tool. The main aim is to identify long-term energy demand and supply trends in 2047 (Sudan, 2021).
4	Diplomatic Energy of India	• India is establishing a network of energy links with Central Asian countries like Kazakhstan and the Gulf countries in the west on the broader area covering Myanmar, Vietnam in the east.

Continued

Table 8.18 Measures to enhance energy security in India.—cont'd

Sr. no	Measures enhancing energy security	Illustration
		• Indo-US Nuclear Deal has opened up new prospects for India concerning the facilitation of advanced nuclear fuel and nuclear technologies. India began to engage in nuclear fuel negotiations with China, Kazakhstan, and Australia. • India's SCO membership might now play a significant part in ensuring stronger energy cooperation by linking Central Asia to South Asia between energy producers and consumers.
5	Renewable Energy promotion	• It is estimated that to meet the goal of 175 GW of renewable energy capacity by 2022, 100 GW of renewable energy capacity needs to be in place by 2019 to fulfill that goal. • In three years, Solar Energy Company of India Limited (SECI) should create storage options to reduce pricing by adding residential and grid batteries to demand. • By 2022, a broad program will be started to exploit at least 50% of the country's biogas potential by promoting technology and enabling loans via NABARD.

Source: Author's compilation.

Table 8.19 Significant energy security challenges in Sri Lanka (Manpreet and Keswani, 2017).

Sr. no	Major difficulties	Illustration
1	A growing reliance on imported fossil fuels	• At the moment, fossil fuels (including coal) account for around 55% of Sri Lanka's energy mix, with the remaining 45% coming from a variety of alternative sources such as solar, wind, and hydropower. • Sri Lanka's energy requirements are primarily reliant on petroleum and coal imports, making for a sizable amount of the country's import spending. • If uncontrolled, it is believed that Sri Lanka would face a significant power crisis as early as 2022.
2	Transmission capacity is limited, particularly for the integration of renewable energy sources	• In 2016, the Sri Lankan government set an ambitious renewable penetration target of 100% renewable energy output by 2050. • By 2045, coal and oil production will be phased out, which comprise almost half of every generation.
3	Issues with improving present transmission and restoring the transmission network continuously	• Sri Lanka's transmission network system comprises a 200 kV trunk system functioning as a backbone and a 132 kV local system.

Table 8.19 Significant energy security challenges in Sri Lanka (Manpreet and Keswani, 2017).—cont'd

Sr. no	Major difficulties	Illustration
4	Introducing cost-reflective tariff	• As far as electrical demand is concerned, Colombo and its surrounding territories are the epicenters of 40% of the overall demand in the country. Additionally, the central region is supplied with hydropower, and the areas around Colombo are provided with thermal power generated by modest thermal power plants. • Sri Lanka, the only South Asian country with nearly 100% grid connectivity, has shifted its focus in recent months to improve the reliability and quality of electricity supply and revise its tariff structure. • The tariff is the most critical factor determining Sri Lanka's profitability as an entity. The ultimate objective of the tariffs methodology is to identify the appropriate cost to be charged from the consumer as tariffs (Asian Development Bank, 2019).

Source: Author's compilation.

Pakistan set out to satisfy its Vision 2030 criteria for secure and high-quality energy sources by implementing the *Energy Security Action Plan in 2005*, as illustrated in Tables 8.20 and 8.21. The strategy's focus is to utilize an optimal mix of all resources, including hydropower, oil, gas, coal, nuclear, and renewable energy such as wind and solar, to enhance the overall energy supply (Ur Rehman, 2019).

Table 8.20 Pakistan's energy policies summary, actions, and SEA status (Ur Rehman, 2019).

Energy policies and actions	Policies: • Policy for Power Generation 2002 • National Energy Conservation Policy • Policy for Development of Renewable Energy for Power Generation (small Hydro, Wind, and Solar Technologies) Actions: • Energy Security Action Plan (2005–30) • Assistance for hydropower projects • AEDB Programme of 100 solar homes per province • The commercialization of Wind Power Potential in Pakistan.
Guidance/legislations in energy	N/A

Table 8.21 Pakistan's policies and initiatives in the energy sector - an environmental assessment and the status of the SEA (Ur Rehman, 2019).

Type of assessment	SEA
Requirement mechanisms	Administrative
Legislations for Environmental evaluation/SEA	National Environmental Policy 2005
Applications	Policies, plans, and programs

5. Recommendations for South Asian countries to overcome energy security challenges

It is recommended that the governments of South Asian countries pursue the following guidelines to address the energy security challenges outlined above.

- International partnership in the manufacture and implementation of renewable energy products may give each participant novel technology, outstanding personnel, expertise, and management methods. International collaboration has a lot of promise for advancing renewable energy policies that maximize societal benefits.
- They are organizing public R&D initiatives to lower energy generating costs authentically. To decrease generating costs, centralizing governmental R&D programs might use intellectuals, research instruments, knowledge, and other research resources from around the country. It may paint a big picture of technical development in terms of lowering the cost of renewable energy generation, guiding R&D teams to work together toward a common objective.
- They coordinate local policies to manufacture renewable energy products at the central level. Policies promoting renewable energy development suited for regional economic, social, and resource availability are crucial and significant in South Asian countries. On the other hand, short-term return policies may result in a loss of social benefits at the national level, as the intervention may influence investors' decisions.
- Another crucial takeaway is that the Government cannot afford to bear the expense of guaranteeing secure and reliable electricity networks on its own. This is especially true in South Asia, where several state-owned utilities reported financial distress. As a result, drawing private capital into the energy sector is crucial.
- South Asia might benefit from capitalizing on regional possibilities for cooperation, given the region's inherent economic and geographic commonalities.
- Taking advantage of existing regional cooperation bodies, like the SAARC, could open additional investment and financing in the energy industry, especially when done on a national level.

6. Conclusions

South Asia is a critical geographic region due to its vast population and considerable energy and environmental challenges. About 24% of the world's population lives in the densely populated region in around 3.5% of the world's surface area. There is a wide range of energy difficulties in the region's countries, including, but not limited to, a lack of access to refined combustibles. The IEA estimates that South Asia's energy requirements will increase to more than double the world level over the next few decades. On the other side, there is a lack of resources and a robust policy definition and execution in the region. There are also significant environmental difficulties in the area, so these countries confront various problems for energy security, which are covered in the above sections of this book chapter. This study summarizes and concludes as follows:

- South Asian countries are improving their transmission and distribution networks to meet growing electrification, rising energy consumption, increased usage of renewable energy, distributed energy resources, and microgrids, among other new technologies.
- With over 1.2 billion people living in the region without a gas network, most South Asian countries rely on crude biofuels to meet their cooking and heating demands. While fossil fuels provide most of their energy, they also have a sizable renewable energy base.
- Bhutan is one of the countries with an energy surplus. While Bhutan's capacity utilization is below 5%, it has 99% of its electricity from hydropower. But, since the country mandates the retention of forest cover at 60% or more and due to environmental and economic considerations, net utilization is not feasible.
- Withatotalinstalledcapacityof36and28 GW, India has dramatically developed the wind and solar energy industries. In recent years, there has been a steady growth in adopting a more healthy, effective, and ecologically friendly cooking technique. South Asian countries must show that renewable energy has a place in their energy mix. Furthermore, regional energy trade alternatives must be explored.
- Economic growth in the South Asian region is unprecedented. However, expansion is becoming restricted by severe energy supply problems. South Asian countries have varying levels of energy endowment, and the region's governments might considerably benefit from enhancing the system of energy commerce through more excellent connectivity.
- At the moment, only India, Bhutan, and Nepal trade electricity. South Asian countries' national energy systems—gas and electricity networks—are mostly isolated. There are no intercontinental gas pipelines inside South Asia or between its neighbors.
- As a result, the region's domestic energy resources are hammered, and access to enormous energy resources in adjacent countries is prohibited, increasing the cost of energy supply and eroding the region's and individual countries' energy security.
- South Asian countries can benefit significantly from national and regional policies that promote renewable energy development and implementation. These policies can aid the region's developing countries establish market priorities and development paths for renewable energy.
- Additionally, it expands its renewable energy plan to include additional policies and actions to address anticipated problems associated with a large share of renewable energy in the energy portfolio and suitable mitigation measures such as smart grid technology.
- South Asian governments have implemented a range of initiatives and legislation to increase public understanding of the relevance of renewable energy sources. The country's contribution to GHG emission reduction and transformation into a green developing country is another essential aspect in launching renewable energy projects.

- National and international energy regulations suggest considerable effort in this direction; nonetheless, the South Asia region still has a lot of potential for entirely using renewable energy resources. As a result, greater public and government involvement are required to get a quantifiable effect.
- Renewable energy can attract investment, diversify energy sources, drive technical innovation, and support sustainable economic growth in South Asia. Additionally, increasing renewable energy's cost-effective penetration into the electrical grid requires considerable coordination among energy industry decision-makers.
- Energy security has been a national priority, from coal to oil. However, the concept of energy safety has developed less dependency on socioeconomic and environmental volatile oil supply. In addition, heavy dependence on imported energy would lead to large foreign-exchange expenditures, high energy expenses, low production costs, and reduced competitiveness.
- Reducing reliance on imported fuels and diversifying energy sources is an essential policy strategy for increasing energy security, reducing emissions, and encouraging economic growth.
- To achieve energy security, firm policies, efficient knowledge management, market and system reform, and market and system integration are required.

References

Alagh, Y.K., 2006. India 2020. Journal of Quantitative Economics 4 (1), 1−14. https://doi.org/10.1007/bf03404634.

Ambade, S., et al., 2017. Cram of novel designs of solar cooker. International Journal of Mechanical Engineering Research 7 (2), 109−117.

Amin, A., et al., 2020. How does energy poverty affect economic development? A panel data analysis of South Asian countries. Environmental Science and Pollution Research 27 (25), 31623−31635. https://doi.org/10.1007/s11356-020-09173-6.

Anwar, K., Deshmukh, S., 2020. Parametric study for the prediction of wind energy potential over the southern part of India using neural network and geographic information system approach. Proceedings of the Institution of Mechanical Engineers Part A Journal of Power and Energy 234 (1), 96−109. https://doi.org/10.1177/0957650919848960.

Aravindh, M., Giri, G.P., 2016. An overview on the solar energy utilisation in Bhutan. Concurrent Advances in Mechanical Engineering 2 (2), 1−7. https://doi.org/10.18831/came/2016021001.

Asian Development Bank, 2019. Sri Lanka: energy sector assessment, strategy, and road map. Asian Development Bank 1 (December), 1−109. Available at: https://www.adb.org/sites/default/files/institutional-document/547381/sri-lanka-energy-assessment-strategy-road-map.pdf.

Bhaisare, A., et al., 2019. Brackish water distillation for Gorewada water treatment Plant using solar energy-Case study. World Journal of Engineering Research and Technology 5 (3), 198−215. Available at: https://www.wjert.org/admin/assets/article_issue/32042019/1556619803.pdf.

Bhaisare, A., Katekar, V., Deshmukh, S., 2020. Novel energy efficient design of water-cooler for hot and dry climate. Test Engineering and Management 1 (1), 12523−12528.

Bhujade, S., et al., 2017. Biogas plant by using kitchen waste. International Journal of Civil, Mechanical and Energy Science (IJCMES) 1 (1), 64−69. https://doi.org/10.24001/ijcmes.icsesd2017.74.

Bhutan: Department of Energy, Ministry of Economic Affairs and Royal Government of Bhutan, 2009. Overview of Energy Policies of Bhutan (May), pp. 1−12. Available at: https://eneken.ieej.or.jp/data/2598.pdf.

Dendup, N., Arimura, T.H., 2019. Information leverage: the adoption of clean cooking fuel in Bhutan. Energy Policy 125 (March 2018), 181–195. https://doi.org/10.1016/j.enpol.2018.10.054.
Deshmukh, M.K., Deshmukh, S.S., 2006. System sizing for implementation of sustainable energy plan. Energy Education Science and Technology 18 (1/2), 1.
Deshmukh, S., 2016. Energy resource allocation in energy planning. In: Handbook of Renewable Energy Technology, pp. 801–846. https://doi.org/10.1142/9789814289078_0029.
Deshmukh, S., Jinturkar, A., Anwar, K., 2014. Determinants of Household Fuel Choice Behavior in Rural, vol. 64, pp. 128–133. https://doi.org/10.7763/IPCBEE.
Dhurwey, A.R., Katekar, V.P., Deshmukh, S.S., 2019. An experimental investigation of thermal performance of double basin, double slope, stepped solar distillation system. International Journal of Mechanical and Production Engineering Research and Development 9, 200–206.
Eastern Research Group, 2016. Comparative analysis of fuels for cooking: life cycle environmental impacts and economic and social considerations. Appendix A: detailed environmental, economic and social technical analyses. Global Alliance for Clean Cookstoves 1 (1), 1–272.
Group, U.N.S.D., 2021. The Sustainable Development Agenda, United Nations Sustainable Development Group. Available at: https://www.un.org/sustainabledevelopment/ (Accessed: 6 December 2021).
Group, W. bank, 2019. World Bank Data - Bhutan, World Bank Group. Available at: https://data.worldbank.org/country/BT (Accessed: 6 December 2021).
Group, T.W.B., 2019. The World Bank In Sri Lanka, The World Bank Group. Available at: https://www.worldbank.org/en/country/srilanka (Accessed: 6 December 2021).
Gupta, A., 2021. Energy security and reliance in South Asia. The National Bureau of Asian Research 1 (1), 1–8. Available at: https://www.nbr.org/publication/energy-security-and-resilience-in-south-asia/.
Helliwell, J.F., Layard, R., Sachs, J.D., 2019. World Happiness Report. Available at: https://happiness-report.s3.amazonaws.com/2021/WHR+21.pdf.
Hickman, L., Evans, S., Gabbatiss, J., 2020. Carbon Brief, Clear on Climate. European Climate Foundation. Available at: https://www.carbonbrief.org/category/energy (Accessed: 6 December 2021).
Hutfilter, U.F., et al., 2019. Decarbonising South and South East Asia - country profile - Pakistan. Climate Analytics (May). Available at: https://climateanalytics.org/publications/2019/decarbonising-south-and-south-east-asia/.
Illangasinghe, K., 2012. Sri Lanka: Sri Lanka. The Ecumenical Review 64 (2), 177–186. https://doi.org/10.1111/j.1758-6623.2012.00160.x.
Imran, M., Ozcatalbas, O., 2020. Determinants of Household Cooking Fuels and Their Impact on Women' S Health in Rural Pakistan, pp. 23849–23861.
India, G. of 2019. Ministry of New and Renewable Energy. Available at: https://mnre.gov.in/. (Accessed: 6 December 2021).
India, G. of 2021. Ministry of New and Renewable Energy. Available at: https://mnre.gov.in/. (Accessed: 6 December 2021).
The State of World Population Report, 2021: My Body Is My Own, 2021. United Nations in India. Available at: https://in.one.un.org/ (Accessed: 6 December 2021).
Institute for Health Metrics and Evaluation (IHME), 2020. GBD Compare Data Visualization. IHME, University of Washington, Seattle, WA.
International Renewable Energy Agency, 2019. Renewables Readiness Assessment: Kingdom of Bhutan, International Renewable Energy Agency. Available at: https://www.irena.org/-/media/Files/IRENA/Agency/Publication/2019/Dec/IRENA_RRA_Bhutan_2019.pdf.
Irfan, M., et al., 2020. 'Assessing the energy dynamics of Pakistan: prospects of biomass energy. Energy Reports 6, 80–93. https://doi.org/10.1016/j.egyr.2019.11.161.
Jaiswal, A., 2019. Transitioning India's Economy to Clean Energy. Natural Resources Defense Council. Available at: https://www.nrdc.org/experts/anjali-jaiswal/transitioning-indias-economy-clean-energy.

Jambhulkar, G., et al., 2015. Performance evaluation of cooking stove working on spent cooking oil. International Journal of Emerging Science and Engineering 3 (4), 26–31.
Jinturkar, A., et al., 2014. Fuzzy-AHP approach to improve effectiveness of supply chain. In: Supply Chain Management under Fuzziness, pp. 35–39.
Katekar, V.P., Deshmukh, S.S., 2020. Assessment and way forward for Bangladesh on SDG-7. Affordable and Clean Energy 20 (3), 421–438.
Katekar, V.P., Deshmukh, S.S., 2021a. Energy-drinking water-health nexus in developing countries. In: Asif, M. (Ed.), Energy and Environmental Security in Developing Countries. Advanced Sciences and Technologies for Security Applications. Springer, Cham. https://doi.org/10.1007/978-3-030-63654-8_17.
Katekar, V.P., Deshmukh, S.S., 2021b. Techno-economic review of solar distillation systems: a closer look at the recent developments for commercialisation. Journal of Cleaner Production 294, 126289. https://doi.org/10.1016/j.jclepro.2021.126289.
Katekar, V.P., Mohammed, A., Deshmukh, S.S., 2021. Energy and environmental Scenario of South Asia. In: Asif, M. (Ed.), Energy and Environmental Security in Developing Countries. Advanced Sciences and Technologies for Security Applications. Springer, Cham. https://doi.org/10.1007/978-3-030-63654-8_4.
Kumar Singh, B., 2013. South Asia energy security: challenges and opportunities. Energy Policy 63, 458–468. https://doi.org/10.1016/j.enpol.2013.07.128.
Lans, D. van der, Vice, W.M., 2021. Clean cooking industry snapshot. Clean Cooking Allianace 2 (1), 1–42. Available at: https://cleancooking.org/wp-content/uploads/2021/07/620-1-1.pdf.
Leach, M., Deshmukh, S., Ogunkunle, D., 2014. Pathways to decarbonising urban systems. Urban Retrofitting for Sustainability: Mapping the Transition to 2050 1 (1), 209–226.
Manoj Kumar, U., Ripunjaya Bansal, R.G., 2017. Energising India - Joint Project between NITI AAYOG and Institute of Energy Economics Japan (IEEJ).
Manpreet, S., Sandip Keswani, P.C., 2017. 100% Electricity Generation through Renewable Energy by 2050 - Assessment of Sri Lanka' S Power Sector. UNDP.
Olhoff, A., et al., 2011. Climate Impacts on Energy Systems: Key Issues for Energy Sector Adaptation. The World Bank, Washington, DC. https://doi.org/10.1596/978-0-8213-8697-2Library.
Pakistan, W., 2016. Worldometer Pakistan, Worldometer Pakistan. Available at: https://www.worldometers.info/world-population/pakistan-population/ (Accessed: 6 December 2021).
Portale, E., de Wit, J., 2014. Tracking progress toward sustainable energy for all in sub-Saharan Africa. World Bank Group 1 (1), 1–12. Available at: https://openknowledge.worldbank.org/bitstream/handle/10986/20254/908900BRI0Box300note0series02014032.pdf?sequence=6.
Rahut, D.B., et al., 2019. Wealth, education and cooking-fuel choices among rural households in Pakistan. Energy Strategy Reviews 24 (March), 236–243. https://doi.org/10.1016/j.esr.2019.03.005.
Sciences, U. of P. S. of A, 2021. Penn Arts and Sciences. Available at: https://www.sas.upenn.edu/ (Accessed: 6 December 2021).
Shrestha, R.M., et al., 2012. Economics of Reducing Greenhouse Gas Emissions in South Asia.
Shukla, A.K., Sudhakar, K., Baredar, P., 2017. Resource-Efficient Technologies Renewable Energy Resources in South Asian Countries: Challenges , Policy and Recommendations, vol. 3, pp. 342–346. https://doi.org/10.1016/j.reffit.2016.12.003.
Sudan, F.K., 2021. Addressing climate change and energy security through energy corporation: challenges and opportunities in South Asia. Achieving Energy Security in Asia 1 (1), 3–30. https://doi.org/10.1142/9789811204210_0001.
Sunil, M., Shalu, A., Abhishek, J., K. G, 2021. CEEW - State of Clean Cooking Energy Access in India - Insights from the India Residential Energy Survey (IRES) 2020 (September).

Susantono, B., et al., 2020. ADB Annual Report 2020: Operational Data. Asian Development Bank. https://doi.org/10.22617/FLS210109.

The World Bank, 2019. Clean Cooking: Why it Matters. The World Bank. Available at: https://www.worldbank.org/en/news/feature/2019/11/04/why-clean-cooking-matters (Accessed: 6 December 2021).

The World Bank, et al., 2020. The State of Access to Modern Energy Cooking Services, The State of Access to Modern Energy Cooking Services. Available at: http://documents1.worldbank.org/curated/en/937141600195758792/pdf/The-State-of-Access-to-Modern-Energy-Cooking-Services.pdf.

Ukey, A., Katekar, V.P., 2019. An experimental investigation of thermal performance of an octagonal box type solar cooker. In: Smart Technologies for Energy, Environment and Sustainable Development, LectureNotes on Multidisciplinary Industrial Engineering. Springer, Singapore, pp. 767—777. https://doi.org/10.1007/978-981-13-6148-7_73.

UNDP, 2020. Human Development Report 2020, The Next Frontier: Human Development and the Anthropocene, Briefing Note - Myanmar, pp. 1—7.

United Development Programme, 2019. Human Development Index and its Components, pp. 343—346.

United Nations ESCAP, 2018. Achieving the Sustainable Development Goals in South Asia: Key Policy Priorities and Implementation Challenges. Available at: https://www.unescap.org/sites/default/d8files/knowledge-products/UNESCAP-SRO-SSWASDGReport_Sep2018.pdf.

Ur Rehman, S.A., et al., 2019. Cleaner and sustainable energy production in Pakistan: lessons learnt from the Pak-times model. Energies 13 (1), 1—21. https://doi.org/10.3390/en13010111.

Vld, L.Q., Carlsson, H., 2010. Climate Change in South Asia: Strong Responses for Building a Sustainable Future. Asian Development Bank.

Williams, L., 2008. Hunger pains. Economist. https://doi.org/10.1097/00004836-198912000-00004.

World Energy Outlook, 2021, 2021. International Energy Agency. Available at: https://iea.blob.core.windows.net/assets/888004cf-1a38-4716-9e0c-3b0e3fdbf609/WorldEnergyOutlook2021.pdf (Accessed: 6 December 2021).

Worldometer, 2021. Current World Population. Worldometer. Available at: https://www.worldometers.info/world-population/.

CHAPTER

Buildings for sustainable energy future

Muhammad Asif

Department of Architectural Engineering, King Fahd University of Petroleum and Minerals, Dhahran, Saudi Arabia

Chapter outline

1. Introduction

Global warming is one of the biggest challenges faced by mankind. The human-induced warming of the atmosphere is reported to have reached approximately 1.2°C above preindustrial levels, increasing at 0.2°C per decade (Global Warming, 2019). Forecasts suggest that unless a serious collective effort is made at the global level, the average atmospheric temperature is expected to further rise by as much as 6°C by the end of this century (Ghosh, 2012). The rise in sea level due to the melting of glaciers is one of the most prominent implications of global warming. Estimates suggest that Antarctica is now annually losing around 160 billion tonnes of ice to the ocean, which is twice its previous findings (McMillan et al., 2014). Climate change as a result of global warming is resulting in wide-ranging problems such as seasonal disorder, a pattern of intense and more frequent weather-related events such as floods, droughts, storms, heat waves, and wild fires (Asif, 2021a,b; Qudratullah and Asif, 2020; Asif, 2013). It has been reported that since the advent of the 20th century, natural disasters such as floods, storms, earthquakes, and bushfires have resulted in an estimated loss of nearly eight million lives and over $7 trillion of economic loss (Karlsruhe Institute of Technology (KIT), 2016; Amos, 2016). Future projections suggest that by the year 2060, more than one billion people around the world might be living in areas at risk of devastating flooding due to climate change (Billion People, 2016). According to the US National Aeronautics and Space Administration (NASA), 2014–2018 have been the hottest five years on record (fourth Warmest, 2019). With the average atmospheric temperature in

Handbook of Energy and Environmental Security. https://doi.org/10.1016/B978-0-12-824084-7.00001-1
Copyright © 2022 Elsevier Inc. All rights reserved.

2018 recorded to be 0.83°C warmer than the 1951 to 1980 mean, almost 400 all-time high temperatures were reportedly set in the northern hemisphere over the 2019 summer (Stylianou and Guibourg, 2019). Realizing the need for a paradigm shift in human activities, countries across the world are placing an ever greater emphasis on sustainable development. Through the Paris Agreement, the world has adopted the first-ever universally legally binding global climate deal to avoid dangerous climate change by limiting global warming to well below 2°C. A recent Intergovernmental Panel on Climate Change (IPCC) report, however, has warned that the world is seriously overshooting this target, heading instead toward a higher temperature rise. It concludes that there is need for major changes in four big global systems: energy, land use, cities, and industry. The report warns that to limit warming to 1.5°C, world needs to invest around $2.4 trillion in renewable and energy efficiency initiatives every year through 2035 (Global Warming, 2018; Reed, 2018).

Buildings play an important role in the global ecocycle. Buildings' entire life cycle encompassing four crucial stages—raw materials production and supply, construction, use, and end of life—has significant ecological footprint. Buildings are important for their energy consumption as well as emission of greenhouse gases (GHGs) throughout their lifecycle. Globally, the building sector also significantly contributes to the gross development product and job market. Buildings are therefore crucial to not only the energy and environmental scenarios but also the broader sustainable development efforts in the world. Buildings offer a tremendous potential for improving their energy and environmental footprint through energy efficiency and renewable energy solutions.

2. Buildings and sustainable development

Buildings, accounting for over 36% of energy use, over 33% of GHG emissions, and around 40% of materials consumption at the global level, play an important role in the global energy and environmental outlook (UNEP, 2009; Asif, 2016a,b; Alrashed and Asif, 2015; Kannan and Strachan, 2009). Buildings consume energy over their entire life, i.e., from construction to decommissioning, as shown in Fig. 9.1. This use of energy during a building's life cycle is direct as well as indirect. The direct use of energy encompasses construction, operation, maintenance, renovation, and demolition of a building, whereas the indirect use of energy is related to the production of materials used in construction and installation of equipment (Cabeza et al., 2014; Sartori and Hestnes, 2007). The operational phase of buildings accounts for 40% and 90% depending on various factors including climatic conditions and user behavior (Hong et al., 2017; Guana et al., 2015).

Given its massive share in consumption of energy and natural resources, the building sector has a crucial contribution to make toward improving sustainability standards and addressing the energy and environmental challenges. In the fight against climate change, the world needs to make major and rapid changes across four major global systems: energy, land use, cities, and industry (McGrath, 2018). All these four systems are closely related to buildings. The United Nations' Sustainable Development Goals (SDGs) also regard buildings important—while SDG 11 covers "Sustainable Cities and Communities," a number of other SDGs are linked to the building sector. In the backdrop of the global drive for sustainability, buildings are experiencing significant improvement in their energy use patterns and efficiency standards especially in the developed countries. Efforts are being made both on the policy and technological fronts to improve the performance of buildings. The building sector is being provided wide-ranging support on multiple fronts, for example, financial and regulatory policies,

FIGURE 9.1

Life cycle of a building.

information and data, and training and capacity building. A combination of regulations, market-based instruments, incentives, capacity building, and information provision has a track record to deliver large-scale energy efficiency improvements, especially if backed by comprehensive national strategies and targets (IEA, 2019).

3. Energy efficiency in buildings

The building sector, compared with other sectors such as transport and industry, exhibits the largest potential for delivering long-term, significant, and cost-effective solution to improve energy efficiency and curtail GHG emissions. According to the United Nations Environment Program (UNEP), with proven and commercially available technologies, the energy consumption in both new and existing buildings can be reduced by 30%–80% with potential net profit during the building life span (UNEP, 2009; Asif, 2020). According to the International Energy Agency (IEA), with the emerging technologies and policy directives building sector in 2040 could be around 40% more efficient compared to their standing in 2017. Improvements can be made across all elements of building's energy consumption including space heating and cooling, water heating, lighting, appliances, and cooking (IEA, 2019). This potential for improvement is common to both the developed and the developing countries.

In the backdrop of the energy-saving potential in the building sector, developed nations have set out stringent targets to improve the energy performance of buildings. In 2009, the European Union (EU) adopted a 20-20-20 directive implying all member states to reduce their GHG emission by 20%, improve energy efficiency by 20%, and produce 20% of their energy from renewable resources. The revised EU climate goals aim to reduce GHG emissions by 80% by 2050, compared to the level of 1990 (European Commission, 2016). Similarly there have been such targets at national levels. The United Kingdom, for example, in June 2019 became the first major economy in the world to pass laws to end its contribution to global warming by 2050. The target will require the United Kingdom to bring all GHG emissions to net zero by 2050, compared with the previous target of at least 80% reduction from 1990 levels (REN21, 2020). The global drive toward sustainable buildings can be observed from the scale of investment being made to improve buildings. In 2018, the total investment in energy efficiency across buildings, transport, and industry sectors was US$240 billion of which buildings alone received 56%. Building envelope, heating, ventilation, and air conditioning (HVAC) systems, appliances, and lighting, respectively, received 30%, 15%, 7%, and 6% of the total energy efficiency investment (IEA, 2019).

Sustainable buildings can be accomplished through three key strategies: sustainable design measures, energy efficiency solutions, and renewable energy technologies. Building's energy and environmental performance can be best improved by incorporating energy efficient strategies at the early stages of the design process. Building's sustainability measures can be broadly classified into passive and active strategies.

Passive strategies contribute toward reducing the energy demand in buildings without the use of any mechanical or electrical system. These strategies are designed and constructed taking into account the climatic characteristics, such as wind direction, sun path, and solar angles, of the location. Some of the common passive design strategies include optimum orientation of building to reduce heat gain, shading to control natural sunlight allowing some light to enter for illumination, application of high performance windows and thermal insulation in walls to minimize solar radiation penetration, reducing the air infiltration of building to avoid excessive unnecessary load on the HVAC system, using wind and pressure differences to naturally ventilate the building, using solar wind towers for ventilation, coloring the building with light colors in hot climates to reflect solar radiation and vice versa, and to apply green roofs (Mahmoud et al., 2017). Active design strategies utilize mechanical and electrical systems either to reduce energy demand in buildings while maintaining the building's Indoor Environmental Quality and operation, or to generate energy on site using renewable energy systems. To maintain and operate the building actively while minimizing the energy consumption, the use of energy-efficient equipment such as energy-efficient HVAC systems, low energy-consuming LED light bulbs, and efficient and rated appliances and equipment is required. Moreover, the renewable energy technologies include using photovoltaic (PV) panels and other strategies to use the solar power, wind turbines to utilize wind for energy generation, and other processes such as the geothermal systems for heating (Chel and Kaushik, 2018). Renewable energy technologies are also vital to develop near-zero energy or zero-energy buildings (Alrashed and Asif, 2012a,b; Dehwah et al., 2018). Among these, rooftop PV panel is the most common technique to generate on-site energy on a building scale. For example, rooftop PV has the potential to contribute up to 16% of a university campus's annual energy requirements in Saudi Arabia (Asif, 2016a,b).

3.1 Building retrofitting

Existing buildings make up the bulk of building stock. It is therefore important to improve the energy performance of existing buildings as well. Energy efficiency of existing buildings can be improved through retrofitting. Building energy retrofitting can consist of wide-ranging energy efficiency measures such as improvement in envelope through incorporation of insulation, efficient buildings windows, air tightness, use of efficient appliances and lighting systems, upgraded HVAC system, and more effective operating regime. Energy retrofitting can be classified into different types depending upon the implemented energy efficiency measures and the involved cost. Energy retrofitting of building stock is one of the key priorities in developed countries. Globally, countries are investing in and developing successful energy retrofit projects to meet energy efficiency targets. A substantial proportion of energy saving can be accomplished through wide-ranging retrofitting measures.

4. Renewable energy in buildings

Renewable energy offers a potential solution to help address the global energy and environmental problems. Renewable resources—i.e., solar energy, wind power, hydropower, biomass, and geothermal—are abundant and inexhaustible offering environmentally clean energy. Over the last decade, renewables, especially solar energy and wind power, have made tremendous progress overtaking the supplies from fossil fuels and nuclear power both in terms of investment and capacity addition. Global renewable power capacity at the end of 2020 stood at 2838 GW. For the fourth year in a row, increment in the installed capacity of renewables has outpaced the net capacity addition of fossil fuel and nuclear power combined. Solar PV is one of the most promising and fast-growing renewable technologies. With 139 GW of new capacity addition in 2020, its total installed capacity stood at 760 GW at the end of the year (REN21, 2020).

Potential renewable technologies for application in buildings include solar PV, solar water heating, wind turbines, hydroelectric systems, and ground and air source heat pumps. Some of these technologies especially wind turbines and hydroelectric systems are heavily site dependent and are typically not applicable in urban environment. Solar PV offers great diversity in terms of scale and type of application. One of the most successful applications of solar PV has been in buildings as shown in Fig. 9.2 (Green). Although the utility scale projects are constituting the larger share of the solar market in most part of the world, small-scale applications are growing rapidly and are encouraged through incentives for the end users. It is estimated that rooftop solar PV can provide 25%–49% of the world's total electricity needs. Currently, rooftop application of PV accounts for 40% of the worldwide installed capacity of PV. Rooftop PV accounted for almost a quarter of total renewable capacity additions in 2018. In Germany, the share of installed rooftop PV is about 60% of the total PV market, with 35% installed on small to medium residential and commercial buildings. In the USA, PV systems on residential and nonresidential buildings have respective share of 20.5% and 19.1% of the total installations.

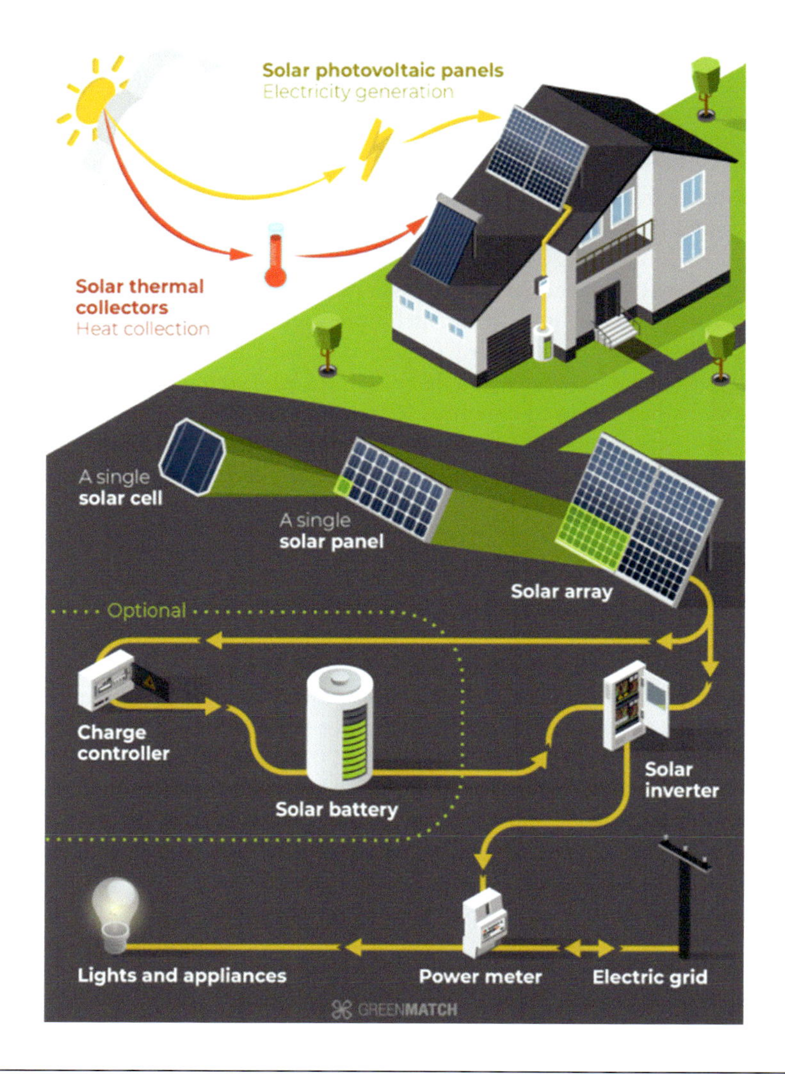

FIGURE 9.2

Solar systems for buildings.

5. Energy use in commercial buildings in the USA: case study

The building sector in the USA has the biggest share in the national energy consumption—in 2018, residential and commercial buildings accounted for 40% of the total energy use, a figure that mirrors the international trends (EIA; ASE). Commercial buildings have a wide variation in terms of types and functions. These include but are not limited to offices, hotels, restaurants, shopping malls, retail shops, hospitals, schools, and warehouses.

Commercial buildings vary in terms of their energy needs depending upon their use and the services they provide. As of 2018, the commercial building stock in the USA comprises 5.6 million buildings. The share of electricity in the fuel mix has steadily grown over the years (CBECS), presently making up 61% of the energy supplies followed by natural gas providing for 32% of the energy needs. Space heating is the greatest energy load, accounting for around 25% of the total energy used in commercial buildings (Energy). In terms of electricity consumption, as shown in Fig. 9.3, lighting has the largest share with refrigeration, cooling, and ventilation being in close proximity (Energy).

The top five energy-consuming categories of commercial buildings as classified by the US Energy Information Administration account for around 55% of the total consumption as shown in Fig. 9.4. Buildings categorized as "Mercantile and Services"—consisting of malls and stores, car dealerships, gas stations, and dry cleaners—have the greatest share being responsible for 15% of the total energy consumed in commercial buildings (Energy).

Renewable energy, especially solar PV, has significant potential in commercial buildings. Solar PV has experienced an exponential growth over the last couple of decades in terms of generation capacity (Growth, 2019). The growth in generation capacity has been helped by a sharp fall in the price of PV systems especially of PV modules. Commercial PV applications typically range from 10 kW to 2 MW in installed capacity, with 200 kW being the average size. The data from the National Renewable Energy Laboratory (NREL) suggest that the cost of PV systems in commercial buildings is around 25% cheaper than in residential buildings (US Solar, 2018). It also provides a breakdown of the cost for the average system in residential, commercial, and utility-scale applications, and the declining trend in cost of PV systems.

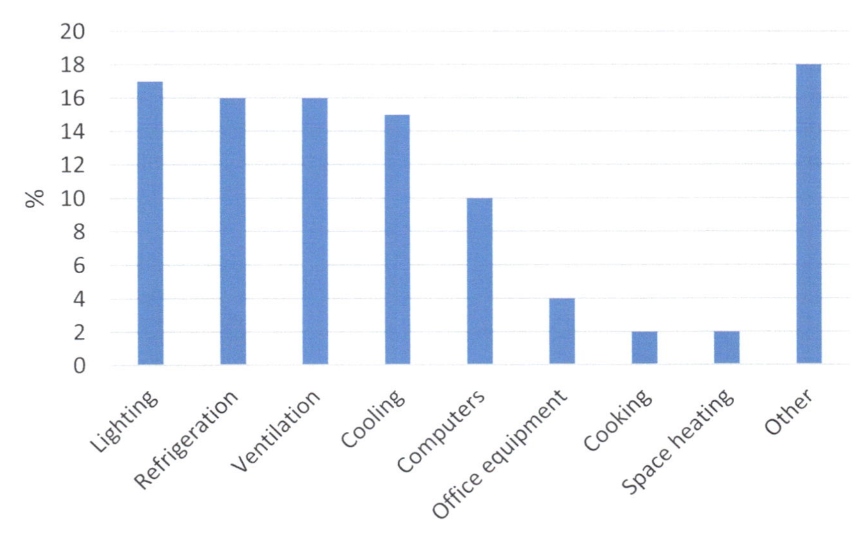

FIGURE 9.3

Breakdown of energy use in commercial buildings.

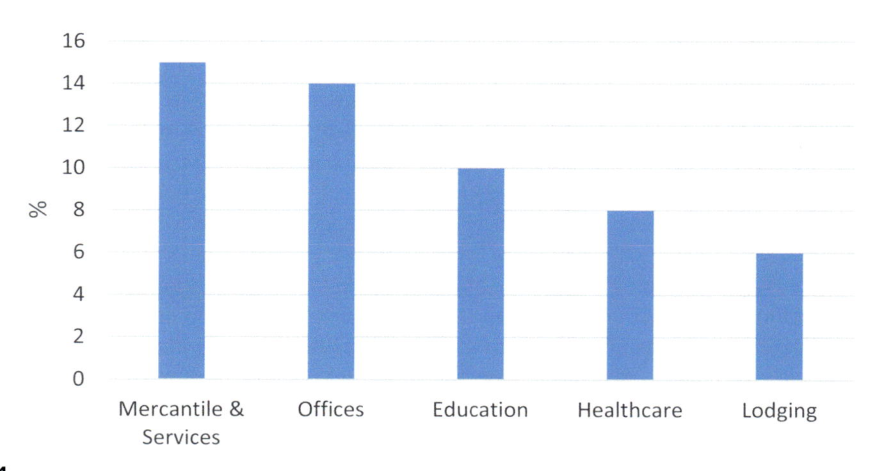

FIGURE 9.4

Five leading categories of commercial buildings.

5.1 Energy sustainability in shopping malls

In recent decades, the retail sector has experienced significant changes in terms of commercial, economic, and geographic characteristics. Shopping malls offer huge potential for energy savings due to the practice of regular rehabilitation and redesign of shopping centers. Efforts to improve energy efficiency and provide sustainable solutions for shopping centers needs to take this tendency into account. This state of constant flux offers the advantage of regular opportunities to improve the technical systems, such as lighting and ventilation, or the building envelope and monitoring systems. Consideration of these aspects along with the other drivers has the potential to achieve significant energy reductions and indoor environment quality improvement (Haase et al., 2015).

Commercial buildings account for 11% of the total final energy demand in Europe. The requirement for energy in commercial buildings can significantly vary depending upon their type and operations as also highlighted in Fig. 9.5 (Gokarakonda et al., 2017). Building energy management

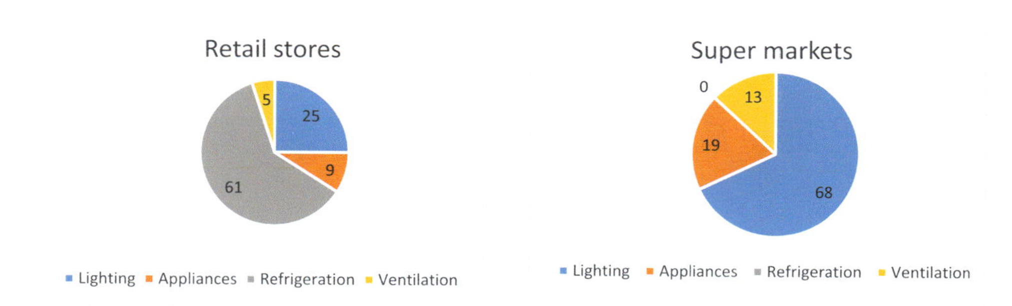

FIGURE 9.5

Distribution of energy in commercial buildings.

system, an important energy-saving technology, is reported to result in significant energy savings. In Europe, it is estimated that energy management systems bring on average 37% for space heating, water heating, and cooling/ventilation and 25% for lighting. However, there is a lack of awareness and understanding of the potential of energy management systems as in Europe, for example, only 25% of commercial buildings have properly installed energy management systems. In shopping malls, energy is primarily used for store lighting, ventilation, heating/air conditioning, and food refrigeration. In general, due to the high demand for refrigeration, food-driven stores, such as supermarkets, have significantly higher energy consumption than the others, ranging from 500 kWh/m^2 to 1000 kWh/m^2. Nonfood stores have energy consumption between 200 kWh/m^2 and 270 kWh/m^2. Among nonfood stores, energy consumption also varies depending on the sizes and functions. Besides, food stores and nonfood stores have very different energy consumption patterns (Gokarakonda et al., 2017). The important stakeholders potentially influencing the energy performance of shopping centers include shopping mall owners and investors, managers, internal and external facilities management companies, tenants, and customers. All these stakeholders have their own set of benefits and incentives associated with enhanced energy performance of shopping centers. At the same, they have respective sets of barriers to deal with.

Shopping centers are one of the major energy consumers among all types of buildings, owing to their huge lighting load, large and fluctuating number of shoppers, and long operating hours. This leads to significant demand for air-conditioning and lighting. Therefore, shopping malls are one of the major building types where building energy efficiency improvement is targeted. Shopping centers in Europe make up around 30% of the nonresidential building stock. There are several studies that have shown significant scope for energy saving in shopping centers (Curto et al., 2019).

One of the most prominent and recent study, the EU funded project CommONEnergy—that aimed to model current and future energy demand in European shopping centers and to derive policy recommendations—realized that there are many studies assessing buildings' energy efficiency; however, only few look at the nonresidential buildings and even less for commercial centers. The study found that energy demand for lighting makes up the highest share on the total final energy demand (33%) followed by space cooling (25%), appliances (16%), refrigeration (15%), ventilation (6%), and space heating (5%). It concludes that: customer satisfaction is the essential motivation to renovate; shopping centers are the only sector with high building renovation rates (4.4%); there is high potential to realize the energy-saving solutions along the planned esthetic renovations; to avoid lock-in effects and to achieve higher energy-saving potential; more ambitious energy efficiency measures have to be implemented; shopping centers are buildings regularly visited by the public, so they can be used to promote energy efficiency technologies. The study also concludes that energy-efficient retrofit offers a huge savings potential (0.85–2.81 TWh in EU 28 by 2030) (EC, 2018; Toleikyte, 2017).

6. Conclusions

The building sector is an important stakeholder in the global energy and environmental outlook. With the growing population and infrastructure development, especially in the developing regions, the role of buildings is set to become more important in future. Through energy efficiency and renewable energy measures, buildings can significantly become energy self-sufficient, also leading to reduced environmental footprint. Government policy support is critical to promote the development and adoption of low carbon, renewable and energy-efficient solutions in buildings. Recent trends indicate governments around the world are providing wide-ranging support (including enabling policies, information provision, and training and capacity building), and industry is making voluntary commitments to improve efficiency. A combination of measures like regulations, incentives, market-based instruments, and capacity building can improve the sustainability standards across the building sector.

References

Alrashed, F., Asif, M., 2012a. Prospects of renewable energy to promote zero-energy residential buildings in the KSA. Energy Procedia 18, 1096–1105. https://doi.org/10.1016/j.egypro.2012.05.124.

Alrashed, F., Asif, M., 2012b. Challenges Facing the Application of Zero-Energy Homes in Saudi Arabia: Construction Industry and User Perspective, in, pp. 391–398.

Alrashed, F., Asif, M., 2015. Climatic classifications of Saudi Arabia for building energy modelling. Energy Procedia 75, 1425–1430.

Amos, J., 2016. Economic Losses from Natural Disasters Counted. BBC.

ASE, Alliance to Save Energy, https://www.ase.org/about.

Asif, M., 2013. Energy and Environmental Security, Encyclopaedia of Environmental Management, vol. II. Taylor & Francis, New York, pp. 833–842.

Asif, M., 2016a. Urban scale Application of solar PV to improve sustainability in the building and the energy sectors of KSA. Sustainability 8, 1127.

Asif, M., 2016b. Growth and sustainability trends in the buildings sector in the GCC region with particular reference to the KSA and UAE. Renewable and Sustainable Energy Reviews 55, 1267–1273.

Asif, M., 2020. Role of energy conservation and management in the 4D sustainable energy transition. Sustainability 12, 10006. https://doi:10.3390/su122310006.

Asif, M., 2021a. Energy and Environmental Outlook for South Asia. CRC Press, USA, ISBN 978-0-367-67343-7.

Asif, M., 2021b. Energy and Environmental Security in Developing Countries. Springer, ISBN 978-3-030-63653-1.

Billion People Face Global Flooding Risk by 2060, Charity Warns, 2016. BBC.

Cabeza, F., Rincón, L., Vilariño, V., Pérez, G., Castell, A., 2014. Life cycle assessment (LCA) and life cycle energy analysis (LCEA) of buildings and the building sector: a review. Renewable and Sustainable Energy Reviews 29, 394–416.

CBECS, Commercial Buildings Energy Consumption survey, EIA, https://www.eia.gov/consumption/commercial/index.php (accessed on 10 November 2019).

Chel, A., Kaushik, G., 2018. Renewable energy technologies for sustainable development of energy efficient building. Alexandria Engineering Journal 57, 655–669. https://doi.org/10.1016/J.AEJ.2017.02.027.

Curto, D., Franzitta, V., Longo, S., Montana, F., Riva Sanseverino, E., 2019. Investigating energy saving potential in a big shopping center through ventilation control. Sustainable Cities and Society 49.

Dehwah, A.H.A., Asif, M., Rahman, M.T., 2018. Prospects of PV application in unregulated building rooftops in developing countries: a perspective from Saudi Arabia. Energy Build 171, 76–87. https://doi.org/10.1016/j.enbuild.2018.04.001.

EC, 2018. CommonEnergy, Project Report. European Commission.

EIA, US Energy Information Administration, https://www.eia.gov/tools/faqs/faq.php?id=86&t=1 (accessed on 10 November 2019).

Energy use in commercial buildings, EIA, https://www.eia.gov/energyexplained/use-of-energy/commercial-buildings.php (accessed on 10 November 2019).

European Commission, 2016. 2050 Long-Term Strategy. https://ec.europa.eu/clima/eu-action/climate-strategies-targets/2050-long-term-strategy_en.

fourth Warmest, 2019. 2018 Fourth Warmest Year in Continued Warming Trend, NASA, News. https://climate.nasa.gov/news/2841/2018-fourth-warmest-year-in-continued-warming-trend-according-to-nasa-noaa/.

Ghosh, P., 2012. Science Adviser Warns Climate Target 'Out the Window. BBC NEWS.. Available online. http://www.bbc.co.uk/news/science-environment-19348194. (Accessed 10 November 2019)

Global Warming of 1.5°C, 2018. Special Report for Policy Makers. Intergovernmental Panel on Climate Change.

Global Warming of 1.5°C, 2019. IPCC Special Report. Intergovernmental Panel on Climate Change.

Gokarakonda, S., Moore, C., Tholen, L., Xia-Bauer, C., 2017. Handbook: Building Energy Management in Large Shopping Malls and Medium-Sized Hotels. Wuppertal Institute.

Green Match, Solar Panels, www.greenmatch.co.uk.

Growth of Solar PV in USA, 2019. Statista.

Guana, L., Walmselya, M., Chen, G., 2015. Life cycle energy analysis of eight residential houses in brisbane, Australia. Procedia Engineering 121, 653–661.

Haase, M., stenerud Skeie, K., Woods, R., 2015. The Key Drivers for Energy retrofitting of European shopping centres. Energy Procedia 78, 2298–2303.

Hong, J., Zhang, X., Shen, Q., Zhang, W., Feng, Y., 2017. A multi-regional based hybrid method for assessing life cycle energy use of buildings: a case study. Journal of Cleaner Production 148, 760–772.

IEA, 2019. Energy Efficiency 2019. https://www.iea.org/reports/energy-efficiency-2019.

Kannan, R., Strachan, N., 2009. Modelling the UK residential energy sector under long term decarbonisation scenarios: comparison between energy systems and sectoral modelling approaches. Applied Energy 86, 416–428.

Karlsruhe Institute of Technology (KIT), 2016. Natural Disasters since 1900: Over 8 Million Deaths and 7 Trillion US Dollars Damage; Press Release 058/2016. Karlsruhe Institute of Technology, Karlsruhe, Germany.

Mahmoud, A.S., Asif, M., Hassanain, M.A., Babsail, M.O., Sanni-Anibire, M.O., 2017. Energy and economic evaluation of green roofs for residential buildings in hot-humid climates. Buildings 7, 30. https://doi.org/10.3390/buildings7020030.

McGrath, M., 2018. Final Call to Save the World from "Climate Catastrophe". BBC. https://www.bbc.com/news/science-environment-45775309.

McMillan, M., Shepherd, A., Sundal, A., Briggs, K., Muir, A., Ridout, A., Hogg, A., Wingham, D., 2014. Increased ice losses from Antarctica detected by CryoSat-2. American Geophysical Union. Geophys. Res. Lett. 41, 3899–3905.

Qudratullah, H., Asif, M., 2020. Dynamics of Energy, Environment and Economy: A Sustainability Perspective. Springer, ISBN 978-3-030-43578-3.

Reed, L., 2018. Climate Crisis Spurs UN Call for $2.4 Trillion Fossil Fuel Shift. Bloomberg.

REN21, 2020. Renewables 2020 Global Status Report. https://www.ren21.net/gsr-2020/.

Sartori, I., Hestnes, A.G., 2007. Energy use in the life cycle of conventional and low-energy buildings: a review article. Energy Build 3, 249–257.

Stylianou, N., Guibourg, C., October 9, 2019. Hundreds of Temperature Records Broken over Summer. BBC.

Toleikyte, A., June 2 , 2017. Technical Presentation, Vienna University of Technology, Energy Economics Group (EEG) (Vienna).

UNEP, 2009. Buildings and Climate Change: Summary for Decision Makers. United Nations Environment Programme, Paris, France.

US Solar PV System Cost Benchmark: Q1 2018, 2018. NREL.

CHAPTER

Wildfires, haze, and climate change

10

Maggie Chel Gee Ooi[1], Andy Chan[2], Mohd Talib Latif[3], Neng-huei Lin[4,5] and Li Li[6]

[1]*Institute of Climate Change, Universiti Kebangsaan Malaysia (UKM), Bangi, Selangor, Malaysia;* [2]*Department of Civil Engineering, University of Nottingham Malaysia, Semenyih, Selangor, Malaysia;* [3]*Department of Earth Sciences and Environment, Faculty of Science and Technology, Universiti Kebangsaan Malaysia (UKM), Bangi, Selangor, Malaysia;* [4]*Department of Atmospheric Sciences, National Central University, Zhongli District, Taoyuan, Taiwan;* [5]*Center for Environmental Monitoring Technology, National Central University, Zhongli District, Taoyuan, Taiwan;* [6]*School of Environmental and Chemical Engineering, Shanghai University, Baoshan, Shanghai, China*

Chapter outline

Handbook of Energy and Environmental Security. https://doi.org/10.1016/B978-0-12-824084-7.00013-8
Copyright © 2022 Elsevier Inc. All rights reserved.

1. Introduction

The future climate is expected to become hotter, with an anticipated rise of global temperature by 1.5 to 2°C comparing with preindustrial level (IPCC, 2014). This range naturally depends on the anthropogenic emission scenario: whether it is the representative concentration pathway (RCP) or the more recently introduced shared social pathway. The weather extremes will intensify and threaten the well-being of the people and exacerbate wildfire burning. The occurrence of wildfire has gradually become an annual occasion where the severity of extreme wildfires seems to have continuously escalated. The period of burning has also increased, and the fire season has been found to occur earlier than usual (Wuebbles et al., 2017). The climate change conditions with increased temperature, drought, and weather anomalies might create long-term change to the burning condition with higher fire risk.

In 2020 alone, the Southeast Asian fire, Amazon rainforest wildfires, Australia bushfires, and the California gigafire have clearly shown that the scales of destruction have tremendously intensified. The environmental and economic implications of biomass burning incidents are naturally detrimental. Wildfires damage the biodiversity-rich wilderness, threaten the habitat and livelihood of animals and plants, although it is still hard to quantify the actual statistics of the impact of wildfires on biodiversity to date (Michelle et al., 2020). The impact does not stop at the pristine forest loss, but also upsets the balance of the ecosystem, the carbon cycle, the hydrological cycle, and natural habitat to flora and fauna and their consequences (See Table 10.1). The postburning residues also increase the sedimentation of the river and stream, leading to flooding which would further degrade the soil fertility (Daniel et al., 2008). The economic damage is enormous from the disaster response stage which involve evacuation and relief effort, relocation, personnel mobilization, to postmortem recovery on destruction of properties and infrastructure, insurance claim, and the less tangible health and environmental cost. Table 10.1 tabulates information of wildfire coverage and implications.

When biomass burning haze (BBH) occurs, the mixture of burning gases and aerosol suspends in the atmosphere and reduces overall visibility. It is the largest source of primary fine carbonaceous aerosols and second largest source of trace gases (Akagi et al., 2011). The fire emissions released are mainly made up of burning aerosols, trace gases (CO_2, CO, NO_x, O_3, etc.) and toxins (persistent organic pollutants) that especially threaten human health (Chen et al., 2017). This becomes even more daunting when the burning occurs in the forest and on peatland, which are one of the largest natural terrestrial carbon reserves (Turetsky et al., 2015). The burning releases greenhouse gases (CO_2, O_3, and H_2O), which continue to warm up the ambient temperature and form a positive radiative forcing feedback loop to warm the atmosphere. On the other hand, the biomass burning also produces large amount of aerosol below the size of 2.5 μm ($PM_{2.5}$) and has the tendency to alter the radiative forcing, cloud properties, and precipitation formation (Andreae et al., 2004). Depending on circumstances in which immense amount of haze is produced, the changes on radiative balance and hydrometeorological processes would create regional climate-forcing effects.

The relationship of wildfires haze on weather and climate is discussed in Section 2. Section 3 guides through the basics of burning to understand the burning sources (burning types, fuel types, fire types, and fire spread) with relevant fire haze characteristics produced and weather that are conducive for the fire occurrence and propagation. Case study is provided in Section 4 to explain the uniqueness of wildfire burning regions and their corresponding haze formation. Mitigation and preventive

Table 10.1 Summary of the social, economic, and environmental implications of the recent wildfires.

Wildfire event	2015 Southeast Asian haze	2018 California wildfire	2019 Amazon rainforest wildfire	2019–20 Australia bushfires	2019–20 Arctic fire
Areas affected	Indonesia, Singapore, Malaysia	California	Amazon	Australia	Eastern Russia and Alaska
Land area destroyed	More than 2.6 million hectares	Up to 0.8 million hectares (Porter et al., 2020)	Up to 0.9 million hectares burnt (CBS News, 2019)	More than 12.6 million hectares burnt (Werner and Lyons, 2020)	More than 14.8 million hectares burnt (McCarty et al., 2020)
Loss of human lives	At least 19 fatalities (McKirdy, 2015)	At least 103 fatalities (Porter et al., 2020)	—	At least 33 fatalities (BBC News, 2020)	—
Loss of properties	—	22,868 structures destroyed (Porter et al., 2020)	Homes of indigenous tribe	More than 2000 houses destroyed (Werner and Lyons, 2020)	—
Other environmental implication	Loss of forest, peat, and other land; costs related to biodiversity may exceed US \$295 million for 2015 (World Bank Group, 2015)	—	Millions of animals and biodiversity	11.3 million Australians affected by smoke (Werner and Lyons, 2020); over one billion animals were killed (Werner and Lyons, 2020)	Melting of ice sheet, permafrost, etc. (McCarty et al., 2020)
Economic cost	Up to \$16.6 billion USD (World Bank Group, 2015)	More than \$150 billion USD (Wang et al., 2020)	More than \$957 billion USD (Gill, 2020)	\$75 billion USD (Quiggin, 2020)	—
CO_2 emitted	Up to 1150 million tonnes (Harris et al., 2015)	68 million tonnes (U.S. Department of the Interior, 2018)	154 million tonnes (Zuckoff, 2019)	434 million tonnes (Werner and Lyons, 2020)	Up to 244 megatonnes (Witze, 2020)

measures need to be taken to prevent and alleviate the potential wildfire risk and their haze implications. Section 5 provides an overview of available fire prediction models and is also discussed for their application and status. Section 6 discusses the future projection of the fire condition due to the changing climate condition and the potential implication and outlook.

2. Relationship between wildfires haze and climate

The Earth's climate is subjected to change by external physical factors known as climate forcing. These factors such as the changes in solar and anthropogenic activities do not preexist in the climate system, and hence their presence would induce changes to the existing system. One of the greatest factors which causes the climate change to take place is the change in radiative forcing, i.e., the energy budget distribution within the earth-atmosphere system. Apart from the direct radiative forcing, the burning aerosols emitted from biomass burning also exhibited a more complex role as an indirect effect of cloud interactions.

2.1 Direct radiative forcing

The emission from biomass burning contributes to both positive and negative radiative forcing. Among the components that contribute to the relative radiative forcing shown in Fig. 10.1, emission from

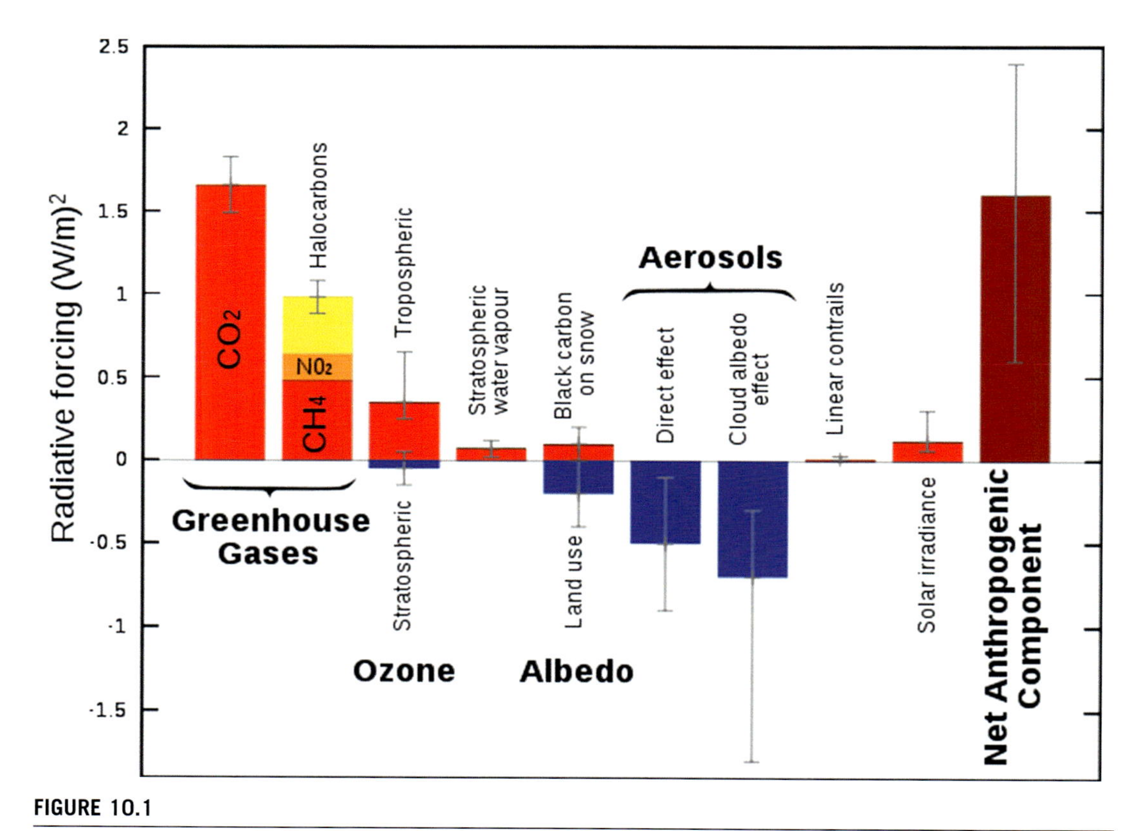

FIGURE 10.1

The relative radiative forcing of the Earth system contributed by different components.

Source: Leland McInnes (https://commons.wikimedia.org/wiki/File:Radiative-forcings.svg), "Radiative-forcings," https://creativecommons.org/licenses/by-sa/3.0/legalcode.

combustion has contributed to the positive forcing by greenhouse gases emission (CO_2), tropospheric ozone (O_3), black carbon (BC) deposited on snow as well as negative forcing by cleared land use and burning aerosols. CO_2 is known to be the main greenhouse gas that absorbs and retains heat from the sunlight and increases the atmospheric temperature. In this context, CO_2 is a product of complete combustion of the biomass. Considering the amount of CO_2 captured by plants during the photosynthesis process, the amount of CO_2 released back to the atmosphere is able to offset its carbon footprint, as well as its radiative forcing.

Biomass burning aerosol (BBA) pollutant has a more multifacet radiative forcing properties depending on the particle sizes distribution, optical properties, hygroscopicity, etc (Chen et al., 2017). Particulate matters ($PM_{2.5}$ and PM_{10}) make up the main compositions of the burning aerosol emission. The chemical constituents of the aerosols are complex, including the carbonaceous aerosol, ions, and organic compounds. The direct aerosol radiative forcing is mainly determined through the amount of aerosol loading and optical properties of aerosol including the single scattering albedo (reflectivity), specific extinction coefficient (absorption), and the scattering phase function (shape). The shadowing effect from thick biomass burning plumes with high aerosol optical depth (AOD) is able to scatter and attenuate light, and hence create a regional cooling effect on the surface beneath. The global BBA is found to contribute to -0.3 Wm^{-2} direct radiative cooling compared to the $+2.45$ Wm^{-2} from the anthropogenic greenhouse gases (Hobbs et al., 1997).

In general, the direct effect of BBA attenuates and negatively forces the solar radiation, while the carbonaceous component in the aerosol emitted contributes otherwise. Among which, the BC and brown carbon (BrC) absorb solar radiation as positive radiative forcing component (Ramanathan and Carmichael, 2008). Due to the short atmospheric lifetime of BBA, the effect of radiative heating might be short-term. When dark-coloured BC is deposited over surface of high albedo, such as snow and ice cover, it is able to absorb heat and accelerate the heating and melting of these surfaces (Kang et al., 2020). Nevertheless, anthropogenic emission also produce large amount of BC which cause the uncertainties on the exact amount of BC contributed by biomass burning. Similarly, the BrC, a yellowish/brownish light-absorbing organic carbon is also known for its heat-absorbing capability in the fresh BBH up to 30% of the total radiation absorption (Wang et al., 2018; Pani et al., 2021) (Fig. 10.2).

2.2 Indirect effect

Aerosol also serves as a cloud condensation nuclei (CCN) on the microphysical processes of cloud formation and its interaction with clouds, which are known as indirect effects. The incomplete combustion of biomass burning releases different organic compounds, the largest component of BBAs. The degradation of cellulose, lignin, and conifer through burning produces organics such as levoglucosan, methoxyphenols, and dehydroabietic acid, respectively (Simoneit, 2002). Majority of these organic aerosols are hygroscopic due to the presence of polar functional groups, which make them an efficient cloud droplet nucleate, and participant in the indirect effect. The presence of water-soluble ions such as sulfate, nitrate, and ammonium are also able to alter the hygroscopic properties of the aerosol particles. The aerosol particle size in accumulation mode (0.1 to 1 μm dry diameter) is known to be efficient CCN under humid conditions.

There is the first indirect effect, also known as cloud albedo effect in which aerosol affects the cloud optical properties and the second indirect effect, also known as the cloud lifetime effect in which aerosol is involved in the precipitation efficiency of the cloud. In the first indirect effect, the aerosol

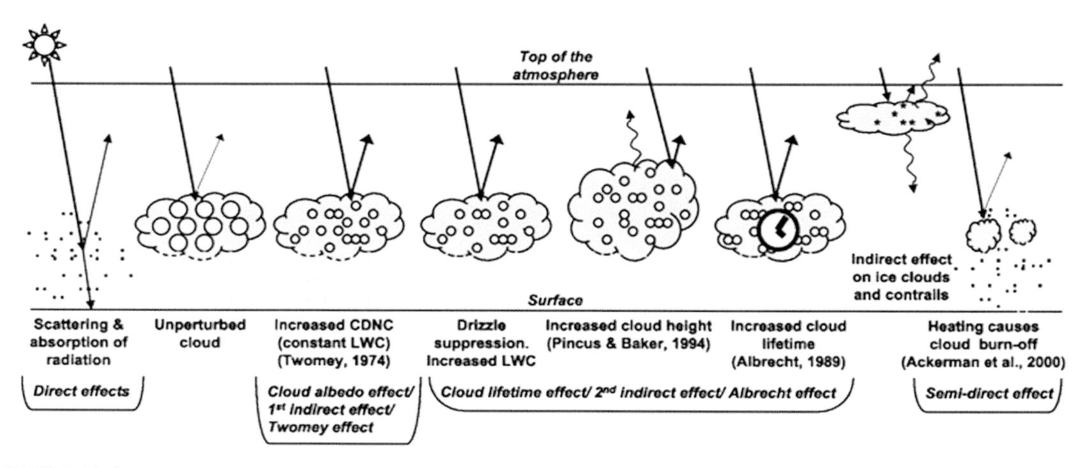

FIGURE 10.2

The aerosol radiative and climate forcing effect.

Source: IPCC Fourth Assessment Report: Climate Change 2007, Working Group I: The Physical Science Basis, Chapter 2: Aerosol, Figure 2.10.

acts as CCN with no change to the amount of liquid water content. Under such circumstances, the cloud droplet formed with BBA are much smaller than the natural aerosols exist. The larger amount of BBA as CCN increases the amount of cloud droplet form and hence increases the albedo of the cloud, from which the amount of radiation reduces reaching the atmosphere beneath.

In the second indirect effect, the precipitation efficiency and lifetime of the cloud are changed due to the change of the droplet number concentration. However, high concentration of BBA and smaller cloud droplets inhibit warm rain processes, i.e., reduce rainout at the lower level clouds and even ice cloud. The water droplet is unable to grow due to the abundance of available CCN. With the continuous accumulation of the small cloud water droplet, the cloud continues to grow into higher altitude and takes a much longer time to rain out. The inhibition invigorates the updraft of polluted air mass, where the large size and solid or mixed state hydrometeors are formed with greater release of latent heat to continue the updraft process. The rainout from the upper level then causes severe thunderstorms (Rosenfeld et al., 2008).

The influence of second indirect effect on radiative forcing is much smaller compared to the precipitation effect. Nevertheless, due to the large uncertainties in the distinctive properties of each aerosol components in the atmosphere, the understanding regarding the precipitation is still limited.

3. Wildfire burning and haze formation conditions

In this section, we would like to dissect the wildfire from the burning emission to their connection with weather conditions. The amount of BBH emission is dependent on the size of burnt area, fuel characteristics, fire types, and weather conditions. The collective of these factors greatly affects the emission species especially the organic compounds produced during the degradation of fuel.

3.1 Fuel types

Biomass fuel is high in organic content and is prone to sustain burnings, while the mass determines how much will be burnt. It involves different layers of biomass including the tree canopy (hereafter includes emergent layer, canopy, and understory), undergrowth, surface, and below surface biomass. Above surface level, there are live fuels such as trees, shrubs, and grass that consist of wooden and seasonal herbaceous components. While on the subsurface, there is a layer of dead fuel, mainly consisting of dry leaf mass that is easier to burn due to the availability of ignition source. Under intensive deforestation, drainage, and drought, the water table level in the soil drops and causes the deep soil burning more prone to occur (Turetsky et al., 2015).

The geographical location is hence a good guide for the vegetation cover types (fuel types), amount burnt, and burning cycles. The burning fuel is usually categorized according to the land cover and biome types, for example, savanna/grasslands, shrublands, tropical forest, temperate forest, agriculture waste, boreal forest, and peat, depending on the classification used. Although similarly falling under tropics, the burning over Africa is very different compared to the Maritime Continent (MC) and Amazonia fire due to the burning activities and types of land cover involved. The savannahs over Africa plain are burnt naturally during the dry season to replenish the soil nutrients, and the region is acclimated to recover from such drought and burning cycles every year, with most burning in the Northern Africa in December to January and the Southern Africa in July to August. On the other hand, the burning over MC and Amazon are mostly human-induced burning activities and propel into uncontrollable fire during the dry season. The former occurs during the slash-and-burnt land clearing, where the burning of agricultural waste spreads to the burning of the underlying peatland between August and October. The latter occurs in August to September as a result of intense deforestation and agricultural waste burning which subsequently spread more uncontrollably into the more pristine rainforest (Giglio et al., 2013).

The amount of air pollutants emitted from the vegetation cover can be determined from the emission factor and the relevant activity factors. Emission factor is a dimensionless value on the amount of air pollutant that will be produced based on the activity factors such as dry biomass burnt, heat produced, period of burning, and area burnt. It is usually determined from field experiment and empirical methods. There are several emission factor databases available (Akagi et al., 2011; Andreae and Merlet, 2001) for application of the atmospheric community and fire emission inventories. The emission factor might be slightly different among the fire emission inventory due to the land use classification, but its deviation is not the main factor causing the difference between the fire inventories but the algorithm each inventory adopted (Liu et al., 2020).

The emission factors for several main trace gases and carbonaceous aerosols from GFEDv4s are shown in Fig. 10.3. The air pollutants are arranged from left to right according to the emission factor from the highest to the lowest. CO_2 is a product of complete combustion; hence, the emission factor does not vary much among different vegetation covers. Notably, the peatland has a relatively high emission factor for CO and CH_4 due to the high amount of carbon content and its incomplete burning condition. While the tropical peatland contains a higher carbon content (56%) compared to the temperature and boreal peat (44.2%), the burning of tropical peatland produces larger amount of trace gases (CO_2, CO, and CH_4) but slightly smaller amount of particulate matters (Hu et al., 2018). The uncertainties of the burning dynamics are attributed as the deviation of the emission factors.

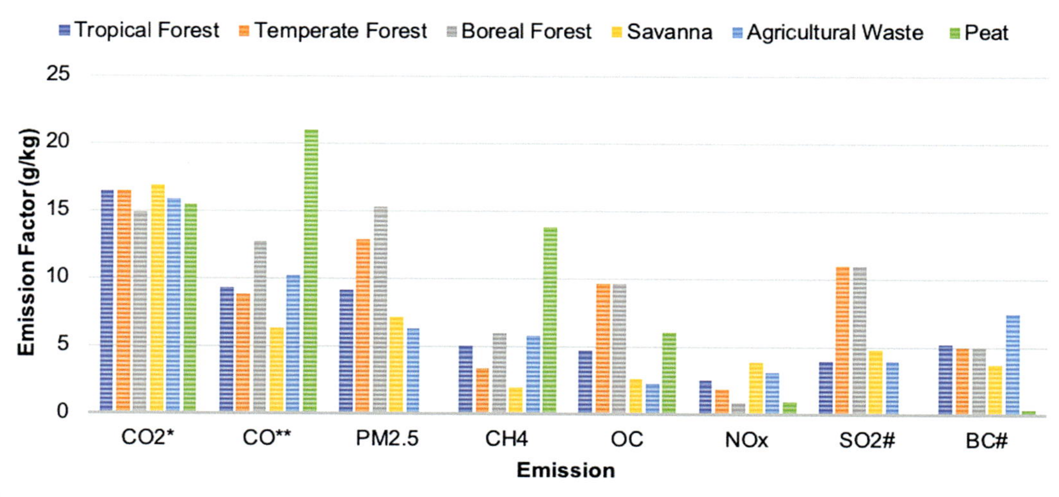

FIGURE 10.3

The emission factor of selected trace gases and aerosols extracted from GFEDv4.1s (Van Der Werf et al., 2017); symbols beside the pollutant name refer to scaled according to $^{*} = 10^{2}$, $^{**} = 10^{1}$, and $^{\#} = 10^{-1}$.

3.2 Fire types

There are two major types of fire involved in wildfire burning, the flaming and smoldering fire as shown in Fig. 10.4. The flaming fire when the burning occurs on and above the surface when combustion is complete. Due to the abundant oxygen supply, this fire occurs when the combustion is complete with high temperature and produces a large amount of heat. The heat difference between the burning plume and surrounding air and the atmospheric stability would determine the plumes rise. While the below-surface burning mainly consists of smoldering burning that occurs on the deeper soil layer. The smoldering fire can occur without any flames and also at much lower peak temperature of 450–700°C compared to flaming (1500–1800°C) with three times smaller heat fluxes required as ignition source (Santoso et al., 2019). While both fire types are distinct, they can lead to one another. Residual fire is the transition of flaming fire into smoldering fire where the remaining of the residue or below-ground biomass is burnt and the amount of heat generated is much lesser than direct flaming. On the other hand, the smoldering fire has the tendency to transform into flaming fire under unfavourable dry and strong wind conditions.

In areas like the Arctic and the MC, the peatland with high organic content can sustain for a tremendously long period due to the low temperature, causing these flameless burning conditions that are difficult to detect (Huang and Rein, 2019). Over the Arctic region, the burning happened during the boreal summer, but the burning emission has started earlier in the year 2019 and 2020 due to the early thawing and warming of temperature. Interestingly, the burning does not start on the surface but resurfaces from peatland and permafrost burning that has sustained underground over months and years (McCarty et al., 2020). It would be easier to detect the flaming and residue fire from the satellite observation due to the higher heat, while it is difficult to detect the smoldering fire directly from satellite observation. The aerosol retrieval method is generally used to identify the smoldering fire.

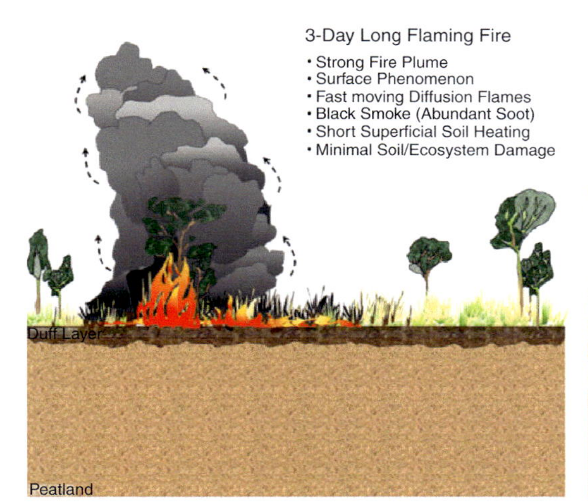

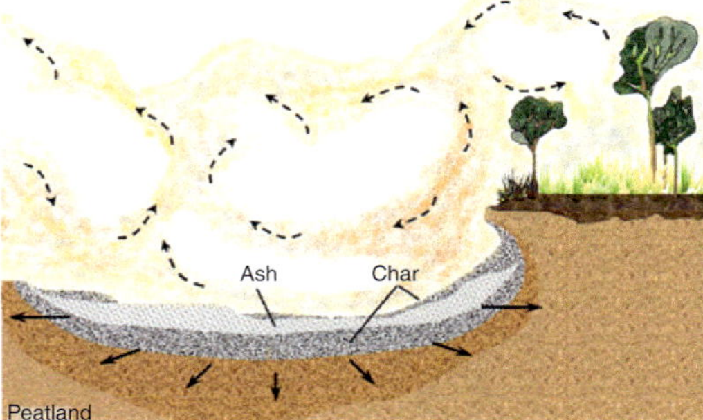

FIGURE 10.4

The emissions formed from flaming and smoldering fire.

Directly extracted from (Hu et al., 2018).

3.3 Fire spread

The burning fuel can change along the progress of the fire spread. The near surface and canopy wind determine the direction of the fire spread. The heat generated from the fire is able to heat and prepare the fuel to burn easily when the fire arrives. Under strong wind conditions, the burning residue or ember is able to be lifted and spread to a new burning spot, which is known as spotting. While the crowning is a process of fire spreading along the tree canopies, the fire spreading rate might be accelerated when there is incidence of strong wind gust or updraft, for example, near the mountain due to the availability of mountain-valley winds. Under circumstances where the spread has directed toward the forest region would cause great destruction to the pristine natural resources as occurred during the burning in MC peatland burning and Amazon forest fire. Conversely, when the fire spreads toward the populated residential or commercial regions, it can cause large economic properties and life loss.

The role of meteorological condition has previously contributed to a large-scale fire spread of burning seasons over Southern California. The strong and dry Santa Ana winds have contributed to the burning season in October to April. The dry katabatic winds that originate from the inland Great Basin have blown toward the coastal Southern California, a highly populated urban area (Jin et al., 2015). The wind intensified the fire burning condition and accelerated the spreading of fire by three times toward the populated coastal area. On a smaller scale, the strong storm updraft of the dry wind has evolved the fire into fire tornado that had happened during Woolsey and Carr fire in 2018. It has totaled up an estimate of 150 billion USD appreciable economical losses during the 2018 California fire (Wang et al., 2020).

3.4 Weather anomaly

The drought condition is the predominant factor leading to burning. It comprises different types of droughts, namely weather drought, agricultural drought, and hydrological drought. The weather drought has indirect influence on the agricultural drought and hydrological drought. Under weather drought conditions, the agricultural drought might exacerbate affecting the agricultural cultivation and vegetation to wilt and from which increases the fuel availability. Similarly, the weather drought causes the reduction of soil water moisture and water table that lead to the continuity of burning to spread. Precipitation has occurred less frequently to alleviate and stop the propagation of the fire burning. Weather anomaly is a condition where the weather is different from the long-term average baseline climate variability. Positive temperature anomaly is often related to the aforementioned weather drought condition that influences the temperature and precipitation. Such a natural regulatory phenomenon that is commonly related to biomass burning are the weather anomalies on different time scale: intraseasonal scale (e.g., Madden–Julian Oscillation (MJO) (Amirudin et al., 2020), Southern Annular Mode (SAM) (Harris and Lucas, 2019)), interannual scale (e.g., El-Niño Southern Oscillation (ENSO), Indian-Ocean Dipole (IOD)), and interdecadal (Atlantic Multidecadal Oscillation (AMO) (Brando et al., 2019)). Among which, the influence of the interannual weather anomalies are known to be closely related to burning in different regions.

The El-Niño condition is the warm phase in which the eastern and central tropical Pacific Ocean (ecTPO) experienced a higher sea surface temperature (SST) compared with the western tropical Pacific Ocean (wTPO). During ENSO, the warm air above the ecTPO rises, while the cold air over the northern south America and wTPO sinks as seen in Fig. 10.5. The former, where the Amazon is located and the latter, where the equatorial Southeast Asia (SEA) and Australia are located, hence experience a warm and dry weather condition due to the sinking dry air. Such anomaly weather conditions are found to be related to the extreme wildfire over the SEA peatland fire, East Australia bushfire, and Amazonian forest fire, creating a conducive environment for burning. IOD is also a condition of SST anomaly but happens in the Indian Ocean (IO). During the positive IOD, the western (eastern) IO experiences an above- (below-) averaged SST which leads to a rising (sinking) air above the western (eastern) IO. The eastern IO where the MC and the Western Australia is located has hence experienced a drier, warmer, and fire-conducive atmospheric condition.

ENSO is an interannual event that occurred every two to seven years. It usually lasts for a year from its development in June-July-August until the March-April-May of the following year. While the positive IOD is also an interannual event that occurred on a shorter span and lasted for 5–6 months starting in June to November. When both of the ENSO and positive IOD coincided with the peak burning season in the MC region from September to October, it caused large amount of fire burning and hence producing large amount of emission (Amirudin et al., 2020). Similarly, it occurred for the Southwestern Australia when the weather condition becomes drier for the land to burnt and the fire to sustain and propagate (Harris and Lucas, 2019; Chang et al., 2021).

4. Wildfires haze conditions

The prevailing weather condition is important on the dispersion and transport of the emission produced from fire burning. The structure of the emission plume is important from which the amount of heat produced will be a large consideration of the emission structure. The emission structure determines the dispersion of the emission and the affected area on the downwind. Fig. 10.6 shows the amount of $PM_{2.5}$ emitted from fire burning depending on the geographical area based on the GFEDv4s dataset.

(a) ENSO condition

(b) Positive IOD condition

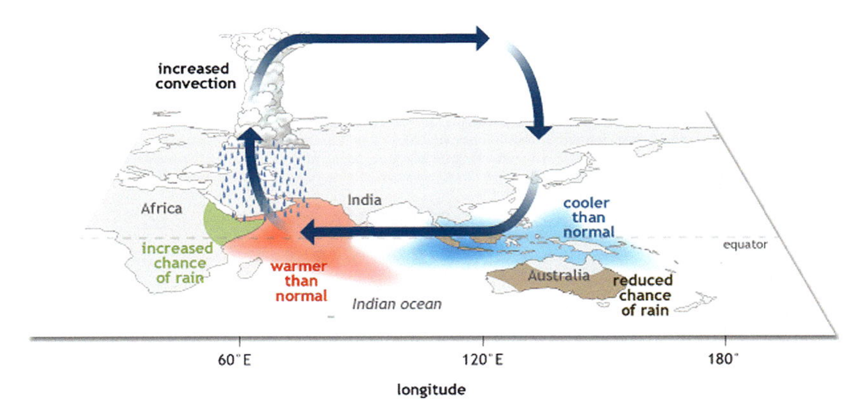

FIGURE 10.5

Illustration of atmospheric movement due to the (A) ENSO conditions (extracted from NOAA Climate.gov); (B) positive IOD condition.

Directly extracted from NOAA Climate.gov.

4.1 Fire emission inventories

There are two main approaches to develop the fire emission inventories: bottom-up and top-down approach. Bottom-up approach extracts fire emission information, either from satellite observations on land surface or local emission data, using the burned area method (e.g., FINN, GFED, and FLAMBE) or the fire radiative power (FRP) (e.g., GFAS and GBBEP-Geo). While top-down approach extracts fire emission information from the satellite observation based on the atmospheric structure, by using the FRP-based information and calibrate with satellite-derived dataset (e.g., QFED). The comparison of the selected four main fire emission inventories widely used is compiled in Table 10.2. These global fire emission inventories are mainly developed from the products extracted from Moderate Resolution Imaging Spectroradiometer (MODIS). Nevertheless, the amount of emission produced for different region can vary by several orders due to the interpretation and assumption of each emission inventories which can be easily compared through the FIRECAM google engine tool

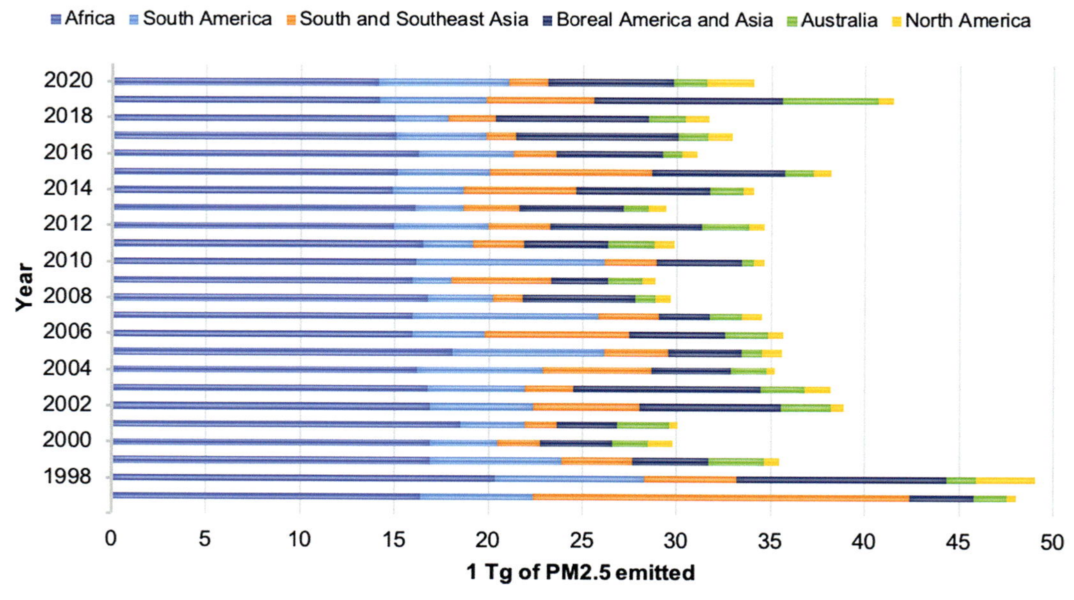

FIGURE 10.6

Global CO_2 emitted (in Tg) from fire burning according to Global Fire Emissions Database version 4.1 including small fire burned area (GFED4s).

(Liu et al., 2020). It has been attributed the difference between the emission inventories to few main factors including (i) the reliance on satellite fire products whether it is active fire or burnt area, (ii) missing satellite fire products due to cloud or haze cover, and (iii) difficulties in detection of small fires and others (Liu et al., 2020).

4.2 Case study: Peninsular Southeast Asia

The burning that occurs in the northern part of the SEA during the spring (December–April) annually has a much wider area of burnings across countries such as Myanmar, Thailand, Vietnam, and Laos. The burning is mainly initiated by human activities and the lack of soil moisture, while the dry weather condition spreads the burning further. The burning is escalated by the subsidence of dry air from higher latitude Tibetan Plateau. The strength of the dry air increases with the intensification of the India–Burma trough which is also connected to the ENSO of previous winter (Huang et al., 2016). The burning region over the western Myanmar and Thailand hence receives the inflow dry air to continue to fuel the burning.

The BBH from the spring burning over northern SEA can be lifted up to height of 700–800 hPa, approximately 2 km to 3 km due to the mountainous terrestrial condition and low-pressure trough present (Yen et al., 2013; Ooi et al., 2021). The strong lifting mechanism is able to carry the emission

Table 10.2 Comparison of fire emission inventories (GFEDv4s, QFED, FINNv1.5, and GFASv1).

	GFEDv4s	FINNv1.5	GFASv1.2	QFEDv2.5r1
Method	Bottom-up: burned area	Bottom-up: active fire	Top-down: FRP	Top-down: FRP with AOD data
Spatial Res	0.25° × 0.25°	1 km × 1 km	0.5° × 0.5°	0.1° × 0.1°/0.25° × 0.25°
Temporal resolution	Monthly (1995-NRT)	Daily (2002-NRT)	Daily (2001-NRT)	Daily (March 2000-NRT) (lag by six months)
Species	40 species of trace gases and aerosols	15 species plus NMOC that are speciated with MOZART (28), SAPRC99 (30), and GEOS-CHEM (11) chemical mechanism	42 species	17 species
Algorithm (see below for explanation)	BA × FL × CF × EF	BA × FL × CF × EF	FRE × beta × EF	FRP × beta × EF
Burned area estimate (BA); Fire radiative power (FRP) (MJ/km^2)	(BA) Active fire from MODIS fire observation; correction with MODIS burned area in some regions to include small fires	(BA) Active fires from MODIS Terra/ Aqua; 1 km^2 for burned area and 0.75 km^2 for grassland/savanna	(FRP) FRP from MODIS Terra/ Aqua	(FRP) FRP from the MODIS Level 2 fire products (MOD14 and MYD14) and the MODIS Geolocation products (MOD03 and MYD03)—according to biome types (4 LCs); cloud correction method by GFASv1 but more advanced treatment
Fuel loading (FL) [g/m^3]	CASA biogeochemical model (same with GFEDv3)	Land Cover map (MODIS LCT and VCF)—5 LCs (Hoelzemann et al., 2004)	—	
Combustion factor (CF) [-]/conversion factor (beta) [-]	(CF) Moisture condition for each fuel type (same with GFEDv3)	(CF) Function of tree cover (Ito and Penner, 2004; Wiedinmyer et al., 2006)	(beta) Link FRP in GFASv1.0 with dry matter combustion rate in GFEDv3/v4	(beta) Compared to GFED product
Emission factor (EF) (-)	(Akagi et al., 2011; Andreae and Merlet, 2001)	(Akagi et al., 2011; Andreae and Merlet, 2001) and others (Wiedinmyer et al., 2011)	(Andreae and Merlet, 2001) with updates (Kaiser et al., 2012)	GFEDv2 but calibrated with MODIS AOD data
Reference	Van Der Werf et al. (2017)	Wiedinmyer et al. (2011)	Kaiser et al. (2012)	Darmenov and da Silva (2015)

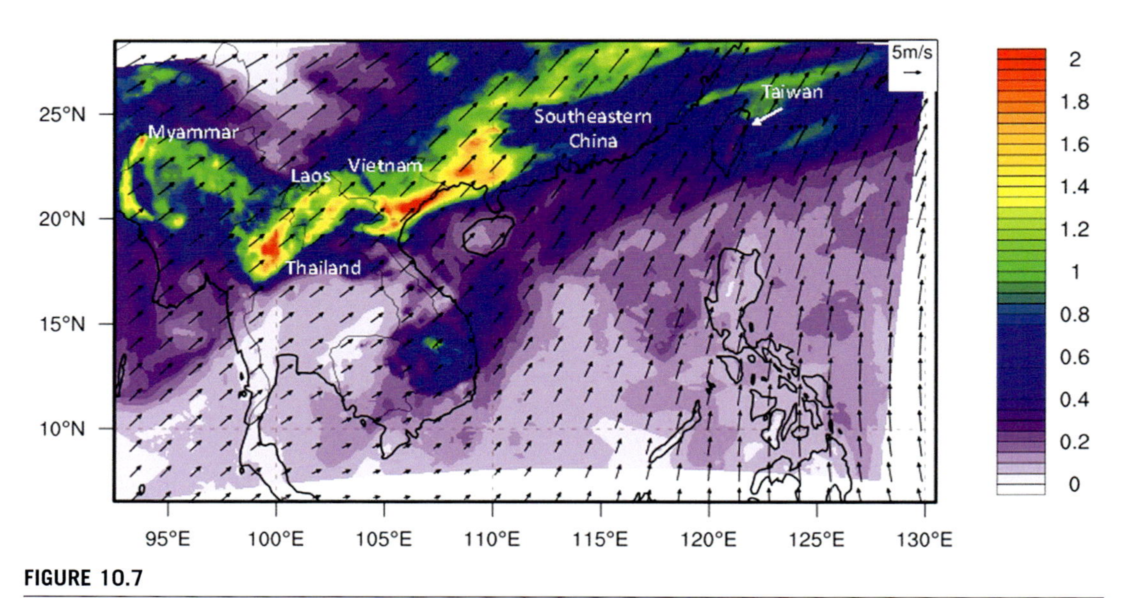

FIGURE 10.7

Modeled Aerosol Optical Depth from WRF-CMAQ output for the biomass burning over Indochina that transported eastwards toward western north Pacific Ocean during the case on March 20, 2013.

Adapted from (Ooi et al., 2021).

plumes into the subtropical westerlies jet and be carried over a larger distance toward the western North Pacific Ocean (Lin et al., 2013). Recently, it was found that the coexistence of the anticyclone over South China Sea is essential for the eastward transport of the emission plume (Huang et al., 2020). The burning pollutant from Indochina is found to be carried to western North Pacific Ocean and was detected at the Taiwan background receptor sites more than 1500 km away as shown in Fig. 10.7 (Chuang et al., 2014). The overhead of biomass burning plume might sink to the surface in circumstances that there are high-pressure system exiting the Asian continent. The physical and chemical properties of the BBA are also subjected to change to the coexistence of Asian dust storm that is frequent during the same season (Dong et al., 2018).

4.3 Case study: Maritime continent

The region experiences tropical wet climate condition according to the Köppen climate classification, while the heterogeneity of land and ocean has complicated the out-of-phase monsoon meteorology over different areas of the region (Ooi et al., 2017). The monsoon meteorology is an important factor of the burning and aerosol transport across the Southern Asian region (Reid et al., 2013). The burning in the MC usually occurs in the summer (June—October) over the Sumatra and Borneo islands, including Indonesia and Malaysia, and produces a large amount of burning emission (Oozeer et al., 2020). The burning is mainly initiated by agricultural activities, while it is sustained under the drier and warmer weather anomalies. The burning over the MC region produces large amounts of burning emission, mainly from the underground peatland fire. The burning in the region is highly susceptible by the dry weather anomaly, especially the eastern pacific ENSO, positive IOD, and dry phase MJO (Amirudin et al., 2020; Latif et al., 2018).

The movement of the haze is subjected to the southwest monsoon above equator and southeast monsoon below equator. In other words, the polluted airmass from the Southern Kalimantan (below equator) moves north-westwards and turns toward northeast direction after crossing the equator into the northern part to the Kalimantan. Similar condition applies for the biomass burning plumes from the Sumatra that travels north-eastwards to the Peninsular Malaysia and eventually entering the South China Sea. There are unusually early haze transport across the Peninsular Malaysia toward northern SEA in the June (e.g., 2013) that occurred due to the presence of tropical cyclones in the South China Sea (Oozeer et al., 2020; Cohen et al., 2017). The low-pressure system has prompted the movement of haze when Typhoon Bebinca has fully developed on 22 June off the coast of Vietnam as seen in Fig. 10.8. The BBA haze has also been found to negatively correlate to the precipitation through AOD and satellite products, reconfirming the role of BBA on climate forcing (Ng et al., 2017).

4.4 Case study: Australia bushfire

Under conducive weather condition, the fire could grow into extreme large fire where the moisture entrained into the burning plume would condense and release large amounts of heat and energy that continue to enhance the deep convection, known as pyroconvection. The clouds formed under such

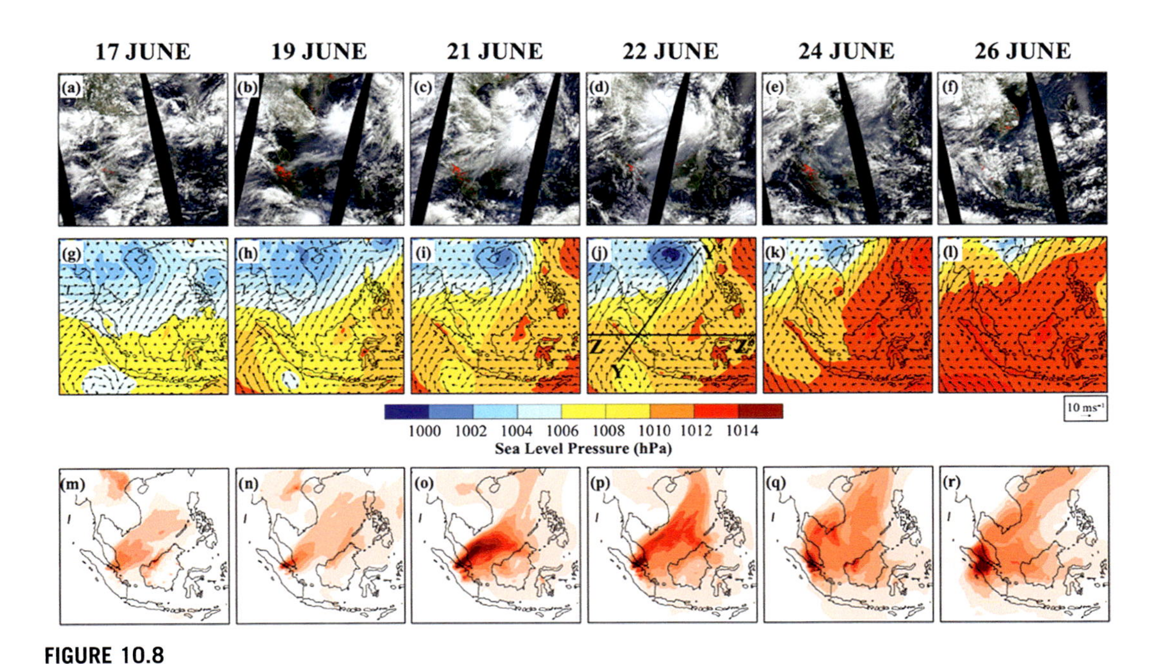

FIGURE 10.8

A case study (June 17, 2013–June 26, 2013) of $PM_{2.5}$ transport in the Maritime Continent from Sumatra toward the Southeast Sea due to the presence of tropical cyclones. First row is the MODIS satellite images, second row is the modeled sea level pressure and surface wind speed from WRF-Chem, third row is the surface $PM_{2.5}$ concentration.

Directly extracted from (Oozeer et al., 2020).

conditions are known as pyrocumulus (PyCu) and pyrocumulonimbus (PyCb) clouds. The deep convection can eventually lead to rainout on one side, while burning intensified on the other side due to the pressure gradient-driven wind gust. The large heat produced such positive feedback of the rainout and continuous release of heat for deep convection, which resembles a weather system itself (Bureau of Meteorology Australia, 2018). Under such an unstable atmospheric condition, the PyCb formed would evolve into thunderstorm and with possibility of lightning as ignition source for burning. The precipitation in the rain has subsequently evaporated due to the dry weather and creates strong downburst, further fueling the upstream burning.

The intense burning produced a large amount of heat that lifts the plume up above surface where the entrained water vapor has condensed to form clouds with the presence of the suspended aerosol as CCN. The thick carbonaceous aerosol particles would absorb solar radiation from reaching the surface and widening the temperature difference between surface and the environment, further enhancing the plume rise. Such situation has occurred during the 2019–20 Australia bushfire, also known as Black Summer. The plume injection height has reached approximately 16 km which can be well considered as lower stratosphere. It is also discovered that the PyCb has changed the stratospheric wind due to the heat absorbed by BBA as well as the stratospheric composition of CO_2, H_2O, O_3, and N_2O (Kablick et al., 2020). Also due to the height attained, the burning haze from the bushfire has been found to have drifted almost 10,000 km across the Pacific Ocean and arrived in Latin America at a height of 6 km (Kablick et al., 2020). Unfortunately, the future climate has found to increase the risk of extreme fire and pyroconvection in the southern Australia, inclusive of the highly populated southeast and southeast Australia (Dowdy et al., 2019).

5. Fire prediction model

Predictability of fire occurrence for emergency response plan requires the evaluation of the potential risk at targeted area and period, where action of prevention and remediation can be taken. This is able to reduce the chances of fire to grow out of hand. The early warning system for wildfire is hence a crucial guide for fire managers to understand the burning condition and plan ahead for the fire suppression effort. The wildfire danger rating tool is widely used to evaluate the fire risk based on the fuel availability, fuel characteristics, and weather condition. The integration of real-time monitoring data and forecasted weather into the existing fire danger rating tool has greatly improved the predictability of the fire risk.

5.1 Historical and continual monitoring

5.1.1 Fire-risk area mapping

Fire-prone area is contributed by several factors, but mainly the availability of the fuel. The mapping of the burning fuel needs to include information about the land cover and soil cover types. Land cover that is easily wilted under the dry weather and soil condition needs to be marked for high risk. Apart from fuel, mapping also involves the extent of destruction severity. Fire-prone areas that are close to human activities, residential, or infrastructure need to be designated with higher risk, for early detection. The risk mapping can finally be revised through the historical burning cases using the postburning scars and burned area product that is characterized by the vegetation cover change, burning ash, and charcoal deposits. Since most of the fires are man-made, tracing back recurring fire cases would be helpful to predict the possibility of next fire onsets.

5.1.2 Detection of fire burning

The common real-time detection of the fire burning is through witness reporting and ground sensors. The latter is rather passive and is usually based on a much larger scale to be seen and hence reported. Located over a fire-prone area, the terrestrial infrared or multimodal sensors mounted on elevated towers detect the immediate on-going fire. Complementing the small-scale coverage of the terrestrial sensor, the surveillance in fire-prone areas should be escalated through timely aerial monitoring and space-borne sensors to detect fire hotspot, burnt area, and fire emission. The real-time and near real-time detection keeps a lookout for thermal anomalies to identify burning hotspots. Technologies used involve small-scale aerial surveillance covered by drone or unmanned aerial vehicle monitoring up to regional and global coverage (Tsay et al., 2016). The Fire Information for Resource Management System (FIRMS) by NASA hosts global active fire data from MODIS sensors aboard NASA's Terra and Aqua satellites and VIIRS aboard Suomi NPP and NOAA-20 satellites. These polar orbiting satellites provide two datasets of global active fire daily, while the geostationary satellites provide continuous monitoring on a regional scale such as European Space Agency's (ESA) MSG-SEVIRI and NASA's GOES-16 ABI. These global sensors also detect fire smoke by retrieving the total column AOD to identify the potential fire emission. On top of the aerosol detection, ESA Copernicus Sentinel-5P's TROPOMI also provides global atmospheric composition for fire relevant emission. The AHI 8/9 sensor aboard Japanese Meteorology Agency's Himawari 8/9 satellite covers the East Asia and Western Pacific regions for aerosol detection efforts.

More specific sensors for analyzing the aerosol properties are available to identify the particle types and amount. The CALIOP sensor aboard the Cloud-Aerosol Lidar and Infrared Pathfinder Satellite Observation (CALIPSO) environmental satellite uses the top-down lidar system to identify the vertical profile of aerosol properties and diagnose the aerosol types including burning aerosols along the global swath. The Multi-angle Imaging SpectroRadiometer (MISR) onboard NASA Terra is able to detect the height of the biomass burning emission via the multiangle camera with different wavelengths. With the vertical dissection of aerosol distribution on different layers, the amount and types of aerosols could be identified diagnostically. More information on the satellite products is provided in Table 10.3.

5.2 Forest fire danger rating system

Fire danger indices have integrated meteorological and fuel availability to serve as warning indicator for potential fire risk over a large area and reference for fire managers on the fire containment effort. Several popularly used fire danger indices are the Canadian Forest Fire Danger Rating System, US National Fire Danger Rating System (NFDRS), and the forthcoming Australian Fire Danger Rating System (AFDRS). There are several components in the fire danger rating system, namely potential weather-associated fire danger and fire behavior rating that evaluate the existing fire behavior. The fire weather rating system that corresponds to the effect of climate change is discussed in this context. The rating system evaluates fire risk according to the weather condition. Weather conditions can influence the fire risk through the near-surface and vertical atmospheric stability.

5.2.1 Fire weather rating system

The near surface maximum temperature, low atmospheric moisture content, and maximum wind speed are critical in conditioning the fuel and soil for burning. There are several popular near-surface fire weather rating systems available, namely Fire Weather Index (FWI) by the Canadian Forest Service

Table 10.3 Available satellite products for near real-time fire burning detection with global and regional coverage.

	Product	Satellites	Source
MODIS active fire product	Active fire hotspot, aerosol detection and properties, and burned area product	NASA Terra/Aqua	https://earthdata.nasa.gov/earth-observation-data/near-real-time/firms/c6-mcd14dl
VIIRS active fire product (VNP14IMGTDL_NRT)	Active fire hotspot	Suomi NPP, NOAA-20	https://earthdata.nasa.gov/earth-observation-data/near-real-time/firms/viirs-i-band-active-fire-data
Advanced Very-High-Resolution Radiometer (AVHRR/3)	Active fire hotspot	NOAA POES, MetOp	https://www.ssd.noaa.gov/PS/FIRE/Layers/FIMMA/fimma.html
Spinning Enhanced Visible and Infrared Imager (SEVIRI)	Active fire hotspot, aerosol detection and properties, and vegetation cover	ESA Meteosat Second Generation (MSG)	https://eumetsat.int/seviri
Advanced Baseline Imager (ABI)	Active fire hotspot, aerosol detection and properties	NASA GOES-16	https://www.goes-r.gov/spacesegment/abi.html
Advanced Himawari Imager (AHI) 8/9	Aerosol optical thickness	JMA Himawari-8/9	https://www.data.jma.go.jp/mscweb/en/product/product_AOT.html
TROPOspheric Monitoring Instrument (TROPOMI)	Aerosol detection and atmospheric composition	ESA Copernicus Sentinel-5P	https//tropomi.eu
Cloud-Aerosol Lidar with Orthogonal Polarization (CALIOP)	Vertical column of aerosol characteristics	NASA CALIPSO	https://www-calipso.larc.nasa.gov/
Multi-angle Imaging SpectroRadiometer (MISR)	Vertical structure of aerosol characteristics	NASA Terra	https://misr.jpl.nasa.gov/
AErosol RObotic NETwork (AERONET)	Bottom-up AOD data	NASA ground network	https://aeronet.gsfc.nasa.gov/
NASA MicroPulse Lidar NETwork (MPLNET)	Bottom-up vertical aerosol characteristics	NASA ground network	https://mplnet.gsfc.nasa.gov/

and McArthur Forest Fire Danger Index (FFDI) by Australia. FWI has provided a general fire risk prediction by encompassing the weather information with three classes of fuel moisture code (Turner and Lawson, 1978). It is one of the most widely adopted fire risk index globally, including Europe, East Asia, and SEA. It includes Fine Fuel Moisture Code (FFMC) to indicate the surface fine fuel that can be easily ignited, Duff Moisture Code (DMC) as fuel made up of organic and woody with average moisture content, and Drought Code (DC) on the deep layer compact organic matter that is highly subjected for smoldering. The initial fire accounts the influence on wind speed on the surface fuels (FFMC), while the continually buildup of fire is determined by the fuels (DMC and DC) that sustain the burning. The detailed consideration of the fuel moisture code with time-lag difference includes the empirical representation of drought that takes longer time to be sufficiently ready for burning and more tough to contain. Nevertheless, the standard fuel moisture code in FWI is solely based on standard pine as burning fuel which needs to be made if applied to other forest regions.

The NFDRS developed by the US Department of Agriculture has covered additional information on the fuel and fire occurrence risk (Cohen and Deeming, 1985). The system has accounted for the dead and live fuel separately with a distinct moisture content algorithm. The fire ignition source such as human-caused and lightning-caused are also considered in the NFDRS, while the Canadian system has planned but yet to roll out in the Fire Occurrence Prediction system. With the details, the NFDRS is able to model different fire areas with higher accuracy, but the simplicity of FWI garnered greater application without compromising much of its accuracy. The McArthur FFDI used in AFDRS is a more straightforward empirical exponential function that integrates near-surface relative humidity, temperature, and wind speed with drought fraction. The latter is determined through fuel availability under circumstances of soil moisture deficit.

5.2.2 Fire weather stability index

Separate fire weather index is developed to account for the buildup and growth of the extreme or catastrophic fire due to the larger vertical scale atmospheric condition. The Haines Index (HI) developed and used in North America considers the atmospheric stability terms and moisture terms of the lower troposphere using temperature difference and dew point depression, respectively, over two vertical atmospheric layers. The HI is extended to over midtropospheric level to formulate Continuous Haines Index (CHI) as an indicator to determine the vertical stability and dry condition with the near surface condition that is relevant with the pyroconvective processes that often strike the southeast Australia. The CHI is found effective in determining the growth of large fires.

6. Conclusion and way forward

The interactions between wildfire, haze, and climate are interconnected. The influence of climate on biomass burning is clear. The intense heat and less precipitation during the dry season are expected to increase the availability of fuel moisture to burn and for fire to sustain. The weather condition is also decisive on the accumulation, dispersion, as well as the transport of the BBH. On the other hand, the influence of biomass burning and its haze on climate is also clear that the BBA is able to alter the precipitation pattern by inhibiting drizzle and form a larger scale thunderstorm through pyro-convection. From which, the emission from the fire can cause a small-scale weather or even regional displacement of atmospheric composition as well as affecting the stratospheric layer. Nevertheless, the knowledge on the aerosol-radiative cloud feedback processes to the climate still remains to be filled in.

The IPCC has used the RCP to represent radiation forcing of the substance within the atmosphere in the fifth assessment report (AR5), while in the sixth assessment report (AR6), shared socioeconomic pathways are adopted. In other words, the future climate projection is decided upon the socioeconomic activities and plan that governs the source of these radiative forcing. Improvement can be done through societal decision, so does the deterioration. For an example, the potential change imposed by the climate change in the fire burning is the conversion of land cover, especially on forest. The burning of the rainforest such as in the Amazonian forest would take a longer time to recover or might not even self-replenish, which eventually turn the forest into a shrubland region similar to Africa. This is definitely one of the undesired possibility, but not the only one.

Emergency response during burning cases is hence important, through the accumulation of knowledge and experience from historical cases. Immediate study and data collection on the burnt area, fire spread, fire type, and weather condition need to be documented and analyzed to develop the future remediation plan. A synergic work of ground-based network, remote sensing, and modeling for warning system on this subject shall be considered for the fire prevention, control, and containment effort.

Acknowledgment

This research was funded by the UKM-YSD Chair in Climate Change (code: ZF-2020-001). This work is also partly sponsored by the Shanghai International Science and Technology Cooperation Fund (No. 19230742500).

References

Akagi, S.K., et al., 2011. Emission factors for open and domestic biomass burning for use in atmospheric models. Atmospheric Chemistry and Physics 11 (9), 4039–4072.

Amirudin, A.A., Salimun, E., Tangang, F., Juneng, L., Zuhairi, M., 2020. Differential influences of teleconnections from the Indian and Pacific oceans on rainfall variability in Southeast Asia. Atmosphere (Basel) 11 (9), 13–15.

Andreae, M.O., et al., 2004. Smoking rain clouds over the Amazon. Science (80-) 303 (5662), 1337–1342.

Andreae, M.O., Merlet, P., 2001. Emission of trace gases and aerosols from biomass burning. Global Biogeochemical Cycles 15 (4), 955–966.

BBC News, 31-Jan. Australia Fires: A Visual Guide to the Bushfire Crisis. BBC News, London, England.

Brando, P.M., et al., 2019. Droughts, wildfires, and forest carbon cycling: a pantropical synthesis. Annual Review of Earth and Planetary Sciences 47, 555–581.

Bureau of Meteorology Australia, 2018. When Bushfires Make Their Own Weather. BOM [Online]. Available: https://media.bom.gov.au/social/blog/1618/when-bushfires-make-their-own-weather/. (Accessed 1 July 2021).

CBS News, 27-Aug. Brazil's Bolsonaro Says He Will Accept Aid to Fight Amazon Fires. CBS News, New York, US.

Chang, D.Y., et al., 2021. Direct radiative forcing of biomass burning aerosols from the extensive Australian wildfires in 2019-2020. Environmental Research Letters 16 (4).

Chen, J., et al., 2017a. Science of the Total Environment A review of biomass burning : emissions and impacts on air quality , health and climate in China. Science of the Total Environment 579 (November 2016), 1000–1034.

Chen, J., et al., 2017b. A review of biomass burning: emissions and impacts on air quality, health and climate in China. Science of the Total Environment 579 (November 2016), 1000–1034.

Chuang, M.T., et al., 2014. Carbonaceous aerosols in the air masses transported from Indochina to Taiwan: long-term observation at Mt. Lulin. Atmospheric Environment 89, 507–516.

Cohen, J.E., Deeming, J.D., 1985. The national fire-danger rating system: basic equations. General Technical Reports 16.

Cohen, J.B., Lecoeur, E., Ng, D.H.L., 2017. Decadal-scale relationship between measurements of aerosols, land-use change, and fire over Southeast Asia. Atmospheric Chemistry and Physics 17 (1), 721–743.

Daniel, N.G., Kevin, R.C., Leonard, D.F., 2008. Effects of Fire on Soil and Water. Ogden, UT.

Darmenov, A.S., da Silva, A., 2015. The quick fire emissions dataset (QFED) - documentation of versions 2.1, 2.2 and 2.4. Technical Report Series on Global Modeling and Data Assimilation 38 (September).

Dong, X., Fu, J.S., Huang, K., Lin, N.H., Wang, S.H., Yang, C.E., 2018. Analysis of the Co-existence of long-range transport biomass burning and dust in the subtropical west pacific region. Science Report 8 (1), 1–10.

Dowdy, A.J., et al., 2019. Future changes in extreme weather and pyroconvection risk factors for Australian wildfires. Science Report 9 (1), 1–11.

Giglio, L., Randerson, J.T., Van Der Werf, G.R., 2013. Analysis of daily, monthly, and annual burned area using the fourth-generation global fire emissions database (GFED4). Journal of Geophysical Research: Biogeosciences 118 (1), 317–328.

Gill, P., 06-Jan. Amazon Wildfires Will Cost Brazil Trillions of Dollars— Damage from Australia's Bushfires Maybe 5 Times Greater. Business Insider India, New Delhi, India.

Harris, S., Lucas, C., 2019. Understanding the variability of Australian fire weather between 1973 and 2017. PLoS One 14 (9).

Harris, N., Minnemeyer, S., Stolle, F., Payne, O., 16-Oct. Indonesia's Fire Outbreaks Producing More Daily Emissions than Entire US Economy. World Resrouces Institute, Washington, DC, USA.

Hobbs, P.V., Reid, J.S., Kotchenruther, R.A., Ferek, R.J., Weiss, R., 1997. Direct radiative forcing by smoke from biomass burning. Science (80-) 275 (5307), 1777–1778.

Hoelzemann, J.J., Schultz, M.G., Brasseur, G.P., Granier, C., 2004. Global Wildland Fire Emission Model (GWEM): Evaluating the use of global area burnt satellite data. Journal of Geophysical Research 109 (D14S04), 1–18. https://doi.org/10.1029/2003JD003666.

Hu, Y., Fernandez-Anez, N., Smith, T.E.L., Rein, G., 2018. Review of emissions from smouldering peat fires and their contribution to regional haze episodes. International Journal of Wildland Fire 27 (5), 293–312.

Huang, H.Y., et al., 2020. Influence of synoptic-dynamic meteorology on the long-range transport of Indochina biomass burning aerosols. Journal of Geophysical Research: Atmospheres 125 (3).

Huang, X., Rein, G., 2019. Upward-and-downward spread of smoldering peat fire. Proceedings of the Combustion Institute 37 (3), 4025–4033.

Huang, W.-R., Wang, S.-H., Yen, M.-C., Lin, N.-H., Promchote, P., 2016. Interannual variation of sprintime biomass burning in Indochina: regional differences, associated atmospheric dynamical changes, and downwind impacts. Journal of Geophysical Research: Atmospheres 121, 1–13.

IPCC, 2014. Climate Change 2014: Synthesis Report. Contribution of Working Groups I, II and III to the Fifth Assessment Report of the Intergovernmental Panel on Climate Change. Geneva, Switzerland.

Ito, A., Penner, J.E., 2004. Global Estimates of Biomass Burning Emissions Based on Satellite Imagery for the Year 2000. Journal of Geophysical Research 109 (D14S05), 1–18. https://doi.org/10.1029/2003jd004423.

Jin, Y., et al., 2015. Identification of two distinct fire regimes in Southern California: implications for economic impact and future change. Environmental Research Letters 10 (9).

Kablick, G.P., Allen, D.R., Fromm, M.D., Nedoluha, G.E., 2020. Australian PyroCb smoke generates synoptic-scale stratospheric anticyclones. Geophysical Research Letters 47 (13).

Kaiser, J.W., et al., 2012. Biomass burning emissions estimated with a global fire assimilation system based on observed fire radiative power. Biogeosciences 9 (1), 527–554.

Kang, S., Zhang, Y., Qian, Y., Wang, H., 2020. A review of black carbon in snow and ice and its impact on the cryosphere. Earth-Science Review 210 (August), 103346.

Latif, M.T., et al., 2018. Impact of regional haze towards air quality in Malaysia: a review. Atmospheric Environment 177, 28–44.

Lin, N.-H., et al., Oct. 2013. An overview of regional experiments on biomass burning aerosols and related pollutants in Southeast Asia: from BASE-ASIA and the Dongsha Experiment to 7-SEAS. Atmospheric Environment 78, 1–19.
Liu, T., et al., 2020. Diagnosing spatial biases and uncertainties in global fire emissions inventories: Indonesia as regional case study. Remote Sensor Environment 237 (November 2019), 111557.
McCarty, J.L., Smith, T.E.L., Turetsky, M.R., 2020. Arctic fire re-emerging. Nature Geoscience 13 (10), 658–660.
McKirdy, E., 29-Oct. Southeast Asia' Haze Crisis: A 'crime against Humanity. *CNN*, Atlanta, GA, US.
Michelle, W., et al., 2020. Impact of 2019-2020 mega-fires on Australian fauna habitat. Nature Ecology and Evolution 4, 1321–1326.
Ng, D.H.L., Li, R., Raghavan, S.V., Liong, S.Y., 2017. Investigating the relationship between aerosol optical depth and precipitation over Southeast Asia with relative humidity as an influencing factor. Science Report 7 (1), 1–13.
Ooi, M., et al., 2021. Improving prediction of trans-boundary biomass burning plume dispersion: from northern peninsular Southeast Asia to downwind western north Pacific Ocean. Atmospheric Chemistry and Physics 1–36.
Ooi, M.C.G., Chan, A., Subramaniam, K., Morris, K.I., Oozeer, M.Y., 2017. Interaction of urban heating and local winds during the calm inter-monsoon seasons in the tropics. Journal of Geophysical Research: Atmospheres 122, 1–25.
Oozeer, Y., et al., 2020. The uncharacteristic occurrence of the June 2013 biomass-burning haze event in Southeast Asia: effects of the Madden-Julian oscillation and tropical cyclone activity. Atmosphere (Basel) 11 (55), 1–24.
Pani, S.K., et al., 2021. Brown carbon light absorption over an urban environment in northern peninsular Southeast Asia. Environment Pollution 276, 116735, 2.
Porter, T.W., Crowfoot, W., Newsom, G., 2020. 2019 Wildfire Activity Statistics. California, US.
Quiggin, J., 10-Jan. Australia Is Promising $2 Billion for the Fires. I Estimate Recovery Will Cost $100 Billion. CNN Business, Hong Kong.
Ramanathan, V., Carmichael, G., 2008. Global and regional climate changes due to black carbon. Nature Geoscience 1 (4), 221–227.
Reid, J.S., et al., 2013. Observing and understanding the Southeast Asian aerosol system by remote sensing: an initial review and analysis for the Seven Southeast Asian Studies (7SEAS) program. Atmospheric Research 122, 403–468.
Rosenfeld, D., et al., 2008. Flood or drought: how do aerosols affect precipitation? Science (80-.) 321 (5894), 1309–1313.
Santoso, M.A., Christensen, E.G., Yang, J., Rein, G., 2019. Review of the transition from smouldering to flaming combustion in wildfires. Frontiers of Mechanical Engineering 5 (September).
Simoneit, B.R.T., 2002. Biomass Burning - A Review of Organic Tracers for Smoke from Incomplete Combustion, 17 no. 3.
Tsay, S.C., et al., 2016. Satellite-surface perspectives of air quality and aerosol-cloud effects on the environment: an overview of 7-SEAS/BASELInE. Aerosol and Air Quality Research 16 (11), 2581–2602.
Turetsky, M.R., Benscoter, B., Page, S., Rein, G., Van Der Werf, G.R., Watts, A., 2015. Global vulnerability of peatlands to fire and carbon loss. Nature Geoscience 8 (1), 11–14.
Turner, J.A., Lawson, B.D., 1978. Weather in the Canadian forest fire danger rating system: a user guide to national standards and practices. Rep. BC-X-177 40.
U.S. Department of the Interior, 2018. New Analysis Shows 2018 California Wildfires Emitted as Much Carbon Dioxide as an Entire Year's Worth of Electricity. Press Releases [Online]. Available at: https://www.doi.gov/pressreleases/new-analysis-shows-2018-california-wildfires-emitted-much-carbon-dioxide-entire-years [Accessed: 01-Jul-2021].

Van Der Werf, G.R., et al., 2017. Global fire emissions estimates during 1997-2016. Earth System Science Data 9 (2), 697–720.

Wang, J., et al., 2018. Light absorption of brown carbon in eastern China based on 3-year multi-wavelength aerosol optical property observations and an improved absorption Ångström exponent segregation method. Atmospheric Chemistry and Physics 18 (12), 9061–9074.

Wang, D., et al., 2020. Economic footprint of California wildfires in 2018. Nature Sustainability 4, 252–260.

Werner, J., Lyons, S., 05-Mar. The Size of Australia's Bushfire Crisis Captured in Five Big Numbers. ABC News, New York, US.

Wiedinmyer, C., et al., 2011. The Fire INventory from NCAR (FINN) – a high resolution global model to estimate the emissions from open burning. Geoscientific Model Development 4, 625–641.

Wiedinmyer, C., Quayle, B., Geron, C., Belote, A., McKenzie, D., Zhang, X., O'Neill, S., Wynne, K.K., 2006. Estimating emissions from fires in North America for air quality modeling. Atmospheric Environment 40 (19), 3419–3432. https://doi.org/10.1016/j.atmosenv.2006.02.010.

Witze, A., 2020. Why Arctic fires are bad news for climate change. Nature 585, 336–337.

World Bank Group, 25-Nov. Indonesia's Fire and Haze Crisis. The World Bank, Washington, DC, USA.

Wuebbles, D.J., Fahey, D.W., Hibbard, K.A., Dokken, D.J., Stewart, B.C., Maycock, T.K., 2017. Climate Science Special Report: Fourth National Climate Assessment, vol. I. Washington, DC, USA.

Yen, M.C., et al., 2013. Climate and weather characteristics in association with the active fires in northern Southeast Asia and spring air pollution in Taiwan during 2010 7-SEAS/Dongsha Experiment. Atmospheric Environment 78 (x), 35–50.

Zuckoff, E., 06-Sep. Carbon Dioxide Released by Amazon Fires Could Hasten Climate Change. CAI News, Woods Hole, MA, US.

CHAPTER

Environmental experience design research spectrum for energy and human well-being

11

Masa Noguchi[1], Li Lan[2], Sajal Chowdhury[1] and Wei Yang[1]

[1]*ZEMCH EXD Lab, Faculty of Architecture, Building and Planning, The University of Melbourne, Melbourne, VIC, Australia;* [2]*School of Design, Shanghai Jiao Tong University, Shanghai, China*

Chapter outline

1. Overview of energy and human health and well-being in economically booming countries

The population in the world's urban areas has steadily been on the rise since 2000 (Kharas, 2017). Middle-income groups contribute significantly to the world's socioeconomic growth, and they are increasing faster in Asia than anywhere else in other continents (ADB, 2010; Kharas, 2010). These middle-income groups are taking part of the driving forces behind the economic growth in booming countries (ADB, 2010). By 2030, the number of middle-income groups is estimated roughly at 35%–45% of the total world population (Kharas, 2010, 2017; OECD, 2019). In 2015, the United Nations' member states set Sustainable Development Goals (SDGs). The SDGs reflect a worldwide demand to take actions to protect the earth, end poverty, and secure the world peace and prosperity in view of their 2030 Agenda for Sustainable Development (United Nations Department of Economic and Social Affairs, 2021).

In Organisation for Economic Co-operation and Development (OECD) countries, enormous pressure on middle-class families is growing due to the increasing demand for energy use, housing affordability, human health and well-being, and quality education. According to the OECD report in 2019, over the last 25 years, housing cost has been ramped up by approximately 49% faster than the

Handbook of Energy and Environmental Security. https://doi.org/10.1016/B978-0-12-824084-7.00002-3
Copyright © 2022 Elsevier Inc. All rights reserved.

average household income increase. From the global perspective, building sectors are also accountable for nearly 40% of energy use and about one-third of greenhouse gas emission scenarios (Zhang et al., 2015). In the capitalist urbanization, the housing sector is embracing a crucial social concern as to human health and well-being globally (Tinson and Clair, 2020). The COVID-19 outbreak has made the masses realize the importance of human experiences and relationships to enhance physical and mental health and well-being within confined built environments.

Developing and emerging economies are mainly facing energy and health-related challenges in the 21st century. China and Bangladesh are among the Asian economic booming countries encompassing the large human resources, land, and GDP (BRAC, 2017; Salam et al., 2020; Zhang et al., 2015). Young generation from middle-income groups is being considered as the main driving force to the rapid development of these countries. Yet, the inadequate policies related to energy and human health and well-being are to some extent affecting the overall development of these countries negatively (Salam et al., 2020; Li et al., 2016).

1.1 Energy, housing, and urbanization in Bangladesh

Bangladesh is experiencing the second fastest-rising economy in South Asia according to the report of World Bank entitled "South Asia Economic Focus, Fall 2019: Making (De) centralization Work" (Beyer, 2019). Recently, the United Nations Committee for Development Policy (UN CDP) recommended Bangladesh as a developing nation from a Least Developed Country (LDC) after 45 years. According to the UN CDP recommendation, this country is scheduled to become a developing country officially in 2026 due to the impact of the COVID-19 on its economy and until 2026, the country will continue to enjoy the trade benefits as an LDC. Currently, the per capita income of Bangladesh is 1.7 times higher ($2,064) than the required threshold. According to Human Assets Index criterion, the country score is 75.3 points where the minimum requirement is 66. Moreover, the Economic Vulnerability Index (EVI) value of this country is estimated at 27.3, and this figure is less than the required EVI value of 32 points. This sustainable economic growth creates an enlarged demand for energy, housing, and urbanization (United Nations, 2021).

In Bangladesh, the young generation is the key contributor to the country's economy and has a significant influence on the socioeconomic growth of the country (ILO, 2020; Khatun and Saadat, 2020). Bangladesh is ranked as seventh in the global population with an estimated population of 164 million and will become the 24th world's largest economy by 2030 (Beyer, 2019). Around 47.6 million (approximately 29%) of the total population has been identified as the young generation of Bangladesh, whose age is between 18 and 35 years old (Khatun and Saadat, 2020; UNFPA, 2014). Hence, the government is formulating multiple strategies for this young generation's overall progress in Bangladesh (Khatun and Saadat, 2020). Millions of people from this young community in Bangladesh fall into middle-income families, and they are the core economic contributors to the country (BRAC, 2017; Chun, 2010; Kharas, 2010; Sadeque, 2013). According to the income level, there are mainly three subcategories of middle-income families in Bangladesh, i.e., lower-middle, middle, and upper-middle income groups. At present, people within these income groups are considered as well trained and educated, covering a substantial portion of the country's population. The young generation tends to migrate to urban areas (mainly Dhaka city) in Bangladesh (Sadeque, 2013; Tariq and Ahmed, 2020).

Dhaka is the capital and main urban business center of Bangladesh and the 11th fastest growing megacity globally where a population of about 21 million living in a land of 1,528 km^2 (Ahmed et al., 2018; Alam, 2018; Rajuk, 2015; Swapan et al., 2017). Dhaka's population has increased rapidly from 1.37 million to 21 million between 1970 and 2020. By 2030, Dhaka will have a population of about 28 million (Sarker, 2020; Satu and Chiu, 2019). This city is a dense urban area with a density of 49,182 persons per square kilometer (RAJUK, 2015; Swapan et al., 2017). To find new livelihood and business opportunities, 2000 to 2,500 people migrate to Dhaka city every day from different areas across the country (Ahmed et al., 2018; Alam, 2018; BRAC, 2017; Sarker, 2020; Swapan et al., 2017). Because of the rapid population growth, Dhaka is facing extreme challenges to grapple with the increasing housing demands (Alam, 2018; BRAC, 2017; Parveen, 2017; Satu and Chiu, 2019). In Dhaka, house prices are also increasing despite the urgent housing needs of middle-income families (BRAC, 2017; Sadeque, 2013). Therefore, almost 70%−78% of these middle-income families, particularly lower-middle and middle-income groups cannot afford their own houses or apartments in the current market price of housing in Dhaka city, Bangladesh (BRAC, 2017; Sadeque, 2013; Shams et al., 2014). Due to excessive house rents and apartment prices including energy costs, most middle-income families need to adjust their other daily expenditures to afford their dwelling unit. For example, they tend to reduce expenditures for their clothing, entertainment, food, and education to cope with their excessive housing cost and energy demand (BRAC, 2017). Therefore, it is increasingly becoming difficult for middle-income earners to buy or secure a decent residential dwelling unit in an urban area like Dhaka (BRAC, 2017; Sadeque, 2013).

Middle-income people are the main driving force of the country's economy. A significant portion of middle-income families is facing different health and well-being difficulties in Bangladesh due to a high level of energy demand, environmental pollution, living density, social uncertainty, and economic constraints in housing affordability (BRAC, 2017; Mridha and Moore, 2011; Satu and Chiu, 2019; TBS, 2019). Because of the socioeconomic restrictions, most middle-income families are living in small and congested dwelling spaces or apartments in the urban areas, where physical building design elements or components are the primary consideration of today's local high-density residential developments—not user experiences in the design decision-making process. These small living conditions create clumsiness leading to the deterioration of indoor environmental quality (IEQ) (e.g., no or less privacy) and indoor air quality (IAQ) (e.g., heavy CO_2 concentration and $PM_{2.5}$) (Chowdhury et al., 2020; Larcombe et al., 2019). Almost 55% of the total energy is consumed in the building sectors in Bangladesh because of the rising urbanization. Bangladesh had experienced an increase of approximately 30% in buildings' energy consumption (383 TWh as of 2017) since 2006 with an annual incremental rate of 3% (Salam et al., 2020; WEO, 2017). Residential electricity use in the urban areas is increasing rapidly with a rate of 48% annually. The households' energy consumption for active space cooling is also increasing drastically. These situations may negatively be affecting the people's health and well-being, as well as their productivity in the workplace driven by manpower that is a key booster of the local economy.

Urbanization is contributing to the drastic hike of energy demand in the city centers (Salam et al., 2020). It is expected that by 2050, about 50% of the total country population will be shifted to the urban centers in Bangladesh (Salam et al., 2020). Accordingly, the Government of Bangladesh is taking various initiatives including energy-saving building codes to alleviate energy demand in the building sectors (Salam et al., 2020; Hassan et al., 2012). To address this situation, both public and private housing sectors are now changing their target group from high and upper-middle income households to lower-middle and middle-income family groups through developing high-density small-sized apartments (Barua et al., 2010; BRAC, 2017; Kamruzzaman and Ogura, 2007). Consequently, in recent years, the demand for vertical expansion of high-density tall building developments in Dhaka has been ramped up exponentially, while the horizontal growth is weakened due to the buildable land shortage in the metropolitan areas (RAJUK, 2015; Seraj and Islam, 2013; Siddika et al., 2019; Swapan et al., 2017). Because of limited budget restrictions and comparatively high rents, the middle-income groups in Dhaka tend to live in small domestic environments. Notably, most lower-middle and middle-income families tend to live in tiny congested domestic spaces in high-density apartments due to housing affordability constraints. To accommodate the market trends, the builders only focus on exploring physical design elements (e.g., room numbers, sizes, building configurations, and floor layouts) in the architectural design decision-making process today (BRAC, 2017; Kamruzzaman and Ogura, 2007; Sadeque, 2013; Satu and Chiu, 2019). Nonetheless, these congested living spaces without consideration of user experiences may diminish the indoor environment and air quality that generates an impact on occupants' health and well-being (Fig. 11.1). However, because of middle-income families' socioeconomic limitations in Bangladesh, modifying their existing design elements or components that lead to some financial burdens may not be taken easily particularly by housing renters. Even so, the way of living in built environments can be changed through the adjustment of occupants' subjective perception and behavior including a sense of attachment to place in their domestic settings given—i.e., "domestic environmental experience" (Chowdhury et al., 2020).

1.2 Regional developments for climate mitigation in China

Urbanization has been accelerating in China since 1990, and about 40% of the population lives in the urban areas today. This equates to roughly 260 million population. The cities are growing rapidly with the aim of accommodating the expansion in the urban areas. Despite the several benefits of rapid urbanization, China continues to face severe resource and environmental degradation along with the speedy urbanization negatively affecting environmental problems and human health and well-being (Li et al., 2016). This rapid urbanization contributes to changing people's living standard or experiential quality that affects their health and well-being. Moreover, indoor air and environmental quality concerning temperature, humidity, lighting, noise, and pollutants in the built environment yield the complementary impact (Li et al., 2016). The building energy consumption in China has increased approximately 7% during the last decade. Therefore, energy efficiency in building sectors has become a key priority regarding energy security in reducing the demand for end users (Zhang et al., 2015).

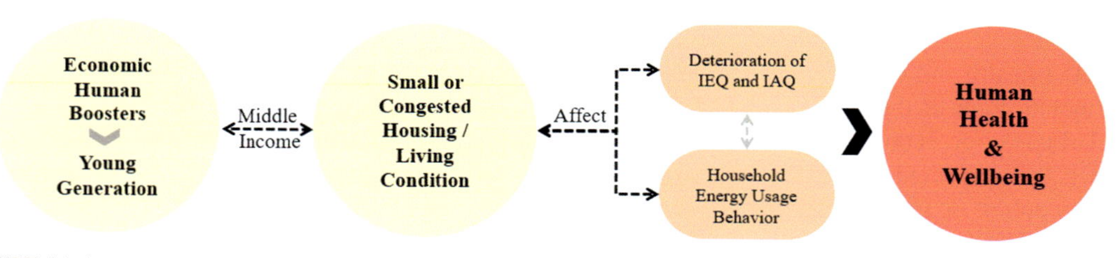

FIGURE 11.1

Living environment and human health scenario of middle-income families in Bangladesh.

Moreover, due to rapid urbanization and climate change, the country targets the development of high-density tall buildings and such high-rise constructions have grown exponentially over the last 10 years. Zhang et al. (2015) indicate some limitations as to a clear and consistent understanding of occupants' behavior and experiences around energy consumption in the buildings. Yet, the study explains that overall energy consumption in China's building sector was 350 Mtce in 2000, accounting for 27.5% of the total energy. By 2020, the energy usage was predicted to be 1089 Mtce (Tu, 2006). There is a growing need to grasp the more accurate energy consumption in Chinese building sectors and analyze occupants' behavior and experiences of the usage (Zhang et al., 2015).

As set out in the World Economic Forum on January 25, 2021, China will continue to work with the international community to actively implement the Paris Agreement and make greater contributions to the global response to climate change. Building energy efficiency is an important factor contributing to reducing carbon emissions in China in the coming years. It is forecasted that 35% of China's final energy will be used in the building sector by 2030 (Li, 2008). Within China, the regional difference in building energy-related carbon emissions is immense, with more emissions in the eastern region than those in the central and western regions. This is mainly because the economy in the eastern region is more developed, and people's living standards are higher with large living space and electricity-consuming appliances (Lin and Liu, 2015). To mitigate climate change, more efforts should be made in eastern regions to reduce building energy consumption and carbon emissions. To successfully achieve the targets set out in the Paris Agreement by the Chinese government, it is necessary to identify the innovative design and construction approaches for the implementation and integration of low-carbon building technologies. These building design and construction innovations can help mitigate climate change, bring down energy costs, create sustainable jobs, and lift economic productivity.

The building design strategies toward climate mitigation mainly focus on improving building envelope and ventilation systems and optimizing the use of renewable energy sources. By using highly insulated building envelopes and increasing building airtightness, the heating and cooling load of buildings can be reduced (Lin et al., 2020). Natural ventilation strategies like cross ventilation, stack ventilation, ventilated facade, and solar chimney are applied to buildings for the passive cooling while maintaining the IAQ. The selection of appropriate window size, opening position, and opening method are important design parameters for natural ventilation in buildings. Solar energy is the primary renewable energy source that has widely been applied to buildings for solar thermal and photovoltaic (PV) systems. Wang et al. (2016) proposed a new type of solar house structure using a heat collection and storage roof. It was found that flat and 45 degree slope roofs could increase the average room temperature by 5.0 and 8.3°C, respectively.

As for the building construction method, prefabricated buildings have been promoting in China in recent years as they are not only highly efficient, high quality, and low cost but also they effectively minimize construction waste (Kong et al., 2020). Although prefabricated construction has some potential benefits, the practice is still limited to an initial stage in China. A study on factors affecting prefabricated construction promotion in China shows that the policy plays a dominant role, while the management and market aspects are also significant for promotion of prefabricated buildings (Jiang et al., 2020). The urban action plan in China recognizes prefabrication as a key component of climate mitigation strategy. In a government report, it is mentioned that prefabricated buildings are expected to account for 30% of the new buildings over the next 10 years (Xu and Zhang, 2019).

China has a vast territory and complex topography, and therefore, the building design and construction approaches in different climatic regions vary with their characteristics and often take full advantage of natural ecological resources in local areas. For example, PV-integrated buildings are more widely used in China's central and southern provinces, where more solar radiation is received than the other regions, and nearly 40% of peak load is coming from the cooling demand in summertime (Li and Colombier, 2009). In Qinghai-Tibet Plateau where solar radiation is abundant, it is found that the buildings equipped with Trombe walls can store 52.6 MJ more energy during the day and achieve 72.8% energy-saving compared to the buildings with normal walls (Wang et al., 2013). Chinese quadrangles with a central courtyard that are often found in Beijing in the northern part of China can minimize solar radiation entry to the rooms during the summer, while maximizing solar radiation use for space heating and lighting as well as wind protection during the winter (Sun, 2013). Chinese quadrangles are typical vernacular dwellings that successfully reflect traditional Chinese architectural techniques and provide human thermal comfort throughout the whole year. To mitigate climate change, building design may need to integrate both modern technologies and traditional design methods. While the abovementioned building design and construction innovations can help mitigate climate change, they can also bring down energy costs, creates sustainable jobs, and lifts economic productivity. With the process of urbanization, some villages have become deserted due to the drastic and constant migration of rural residents to city areas in China. To revitalize the rural areas, the Chinese government has initiated some policies that aim to reduce the gap between urban and rural regional development. Housing is an important factor to narrow the gap between urban and rural living standards. The connotation of "housing" here includes the quality of the house, indoor environment, service life, impact on residents' lifestyle, and the surrounding environment. The residents in rural areas in China generally build houses by themselves, where the thermal comfort level of these self-built houses is very low. Houses in the southern part of China generally do not have heating facilities, and energy consumption of the building is at a low level. The residents in northern China tend to apply active heating methods. For example, the heating is powered using electricity, coal, firewood, or natural gas, resulting in various quality levels of the indoor environment. In recent years, there is a growing concern about human-centric passive design for houses in rural areas in China. The first passive house project in rural areas was developed in a village in Yanshou Town, Changping District, in Beijing in July 2017. The first floor of the house is equipped with a polystyrene board insulation system, and the outer wall uses 250 mm thick graphite polystyrene boards to form the insulation layer. The second layer adopts 40 mm thick vacuum insulation boards. The outer opening is a high-efficiency thermal insulation plastic steel window which can avoid forming the thermal bridge. Fresh air is maintained through the high-efficiency heat recovery system, which can reduce the supply of energy by recycling the heat in the exhaust air. In winter, a gas wall-hung furnace is used for floor heating. The furnace is equipped with a temperature controller to independently set the indoor temperature (Gao, 2019).

After the completion of the project, the Beijing Kangju Certification Center has been testing and tracking the indoor environment and energy consumption of passive houses. The field measurement results show that the residents choose the heating temperature in winter according to their own needs or preferences, and the indoor winter temperature is controlled between 18 and 25°C. The highest indoor temperature in summer is 27°C without the need of turning on air conditioning systems. The indoor relative humidity is kept in the range of 40%–60% throughout the year. A fresh air system delivers outdoor air to the room, and the indoor carbon dioxide concentration is controlled to maintain below 1,000 ppm (Gao, 2019).

The human-centric passive houses in rural areas not only improve the occupants' living environment and reduce energy consumption but also bring good economic and social benefits. Passive houses are an effective means of alleviating operational utility cost burdens while revitalizing the countryside's livelihood. The service life of ordinary houses in rural areas is estimated at about 15 to 20 years. Due to the poor performance of building materials and structural defects, the conventional houses tend to require for reconstruction after 15 years of the operation. On the other hand, the passive houses have a much longer service life, and the overhaul time is considered to be about 40 years after the construction. In comparison to the short-lived ordinary houses, durable passive houses may have the capacity for accommodating multiple generations.

Despite the benefits of passive houses, the promotion of passive design and relevant construction techniques is still very challenging in rural areas in China. First, most people in rural areas do not understand what a passive house is, and it is impossible to build a passive house for themselves. Some wealthy people can afford to conduct interior decoration upgrades or pursue an increase in building area. Nonetheless, they tend not to consider improving the building's thermal performance or the IEQ. Therefore, it is necessary to popularize the basic knowledge of passive house capacities in the rural areas and let the masses understand the benefits and values. Second, passive house construction requires a professional design and construction team. At present, there are only a few design and construction companies that have the expertise and experience for the delivery of passive houses in China. It is very common that construction workers do not grasp the passive house design and construction methods. Therefore, there is a clear and urgent need for professional passive house design and constriction training courses in China.

The built environment encompasses a system of energy and environment being occupied by the masses, whose behavior affect the consequences. To accommodate diverse needs and demands of individuals and societies, it needs to be customized or personalized. In parallel to studies on technological advancement, human-centric environmental experiences should be researched much further to ensure the delivery of socially, economically, environmentally, and humanly sustainable built environments that can be applied to privileged and unprivileged families, communities, and nations that are sharing our common future.

2. Indoor environmental quality on human well-being and productivity

To examine the effects on human well-being and productivity, IEQ will be explored in this section, yet focusing on the four selected factors: i.e., thermal, visual, and acoustic environment and IAQ. Poor IAQ accompanied by ineffective ventilation and thermal discomfort, as well as inadequate light and noise may yield negative effects on human well-being and productivity. Economic calculations indicate that the improved IEQ is cost-effective when the financial value of well-being and productivity benefits are considered (Fisk et al., 2011; Wargocki and Djukanovic, 2005). Improvement in occupants' well-being and productivity presumably leads to wider benefits; thus, it motivates building owners and tenants to pursue better IEQ. In this chapter, the term "well-being" is used to reflect accepted definitions of comfort and health—i.e., the "condition of mind that expresses satisfaction with the thermal environment and is assessed by subjective evaluation" (ASHRAE, 2017) and "a state of complete physical, mental, and social well-being and not merely the absence of disease or infirmity"

(WHO, 1946), respectively. The effects of the abovementioned selected IEQ factors on well-being will be summarized below.

Thermal environment: Thermal comfort is influenced by air temperature, mean radiant temperature, air velocity, humidity, personal metabolic rate, and clothing-induced thermal insulation. Elevated temperatures that caused thermal discomfort have been shown to produce acute subclinical health symptoms such as itchy, throat irritation, headache, fatigue, and difficulty in concentrating (Lan et al., 2010, 2011a; Fang et al., 2004). Raised temperature also can result in measurable changes in physiological responses including increased sympathetic nervous system activity (Lan et al., 2010). Warm and humid indoor environments encourage mold and fungus to grow (Spengler et al., 2001). Alternatively, low humidity and temperature alter the disease transmission of infectious disease particles, such as the influenza virus (Lowen et al., 2007). Low relative humidity (5%−30%) in office, for instance, increased the prevalence of complaints about perceived dry and stuffy air and sensory irritation of eyes, as well as aggravated eye tear film stability and the osmolarity of upper respiratory airways (Wolkoff, 2018).

Indoor air quality: IAQ refers to air quality within buildings, especially as it relates to the occupants' health and comfort. IAQ is influenced by pollutants generated indoors and outdoors, and the building systems that impact on ventilation. Depending on the presence of pollutants, the concentration levels, and the exposed time, poor IAQ leads to acute effects, such as asthma, fatigue, irritation, dizziness, fatigue, and headache, as well as chronic effects, such as cancer, some respiratory diseases, and heart disease (Cedeño-Laurent et al., 2018). Ventilation in buildings brings in fresh air from outside and dilutes pollutants generated indoors. Thus, it plays an important role in creating and maintaining healthy IAQ. The increase of ventilation levels helps decrease the percentage of subjects dissatisfied with the air quality and the intensity of odor, improve the perceived air freshness, and lower the intensity of subclinical health symptoms, such as dry mouth and throat, difficulty to think, and feeling bad (Wargocki et al., 2000).

Light and view: Nowadays artificial light can cover off visual needs despite the absence of natural light in buildings. Light not only provides visual function but also acts as a modulator of nonvisual functions, such as mood regulation, alertness, and work performance. Yet, the nonvisual function of light can be highlighted by the circadian rhythmic modulation. A normal synchronized circadian rhythm is essential for well-being, since the disrupted rhythm could lead to diseases, such as diabetes, obesity, depressive disorders, and Alzheimer in addition to tumor appearance (Stevens et al., 2007). Light controls a biological clock of the body through releasing a hormone (or melatonin). It induces sleep and regulates mood and mental abilities. Among all factors of lighting, illuminance and color temperature are the two factors considered to be most influential. A low or high color temperature corresponding to a low or high level of illuminance is empirically assessed as being pleasant or neutral (Kruithof, 1941). In general, the studies indicate that lower illuminance and color temperature (warm color) are more likely to enhance positive mood (Hsieh, 2015). However, elevated illumination suppressed the melatonin release significantly, which is linked to higher alertness.

Noise: Noise and its nonauditory effects are pervasive in the urban environment. The nonauditory effects including annoyance and psychological stress are widely suspected to cause some disability worldwide. Noise annoyance can result from sound interfering with daily activities, feeling, and thought, and it might lead to humans' negative emotional responses, such as anger, displeasure, exhaustion, and other stress-related symptoms (Basner et al., 2014). High exposure to environmental noise can play a role in cardiovascular disease. Noise can raise blood pressure, change heart rate, and release stress hormones. Consistent changes of these conditions can lead to risks for hypertension, arteriosclerosis, and even more serious events, such as a stroke or myocardial infarction (Basner et al., 2014; Münzel et al., 2018). Studies also suggest that traffic noise and air pollution exposure may interact with each other and with traditional risk factors, such as hypertension and type 2 diabetes (Münzel et al., 2017). Sleep deprivation is another aspect of health risk that is triggered by environmental noise. Noise exposures shorten sleep period, cause awakenings, and reduced two important stages of sleep, i.e., slow-wave and rapid eye movement sleep (Muzet, 2007).

2.1 IEQ effects on productivity

Since the cost of people in an office is an order of magnitude higher than the cost of maintaining and operating the building, spending money on improving the work environment may be considered as the cost-effective way to improve their productivity. IEQ was estimated to be more influential on productivity than job dissatisfaction or stress management (Roelofsen, 2002). The effects of the above-mentioned IEQ factors on productivity will be summarized below.

Thermal environment: In general, thermal environment is regarded as one of the important indoor environmental factors that affect human performance. Thermal discomfort caused by low or high ambiance air temperature reduced productivity (Lan et al., 2009, 2010). Seppänen et al. (2006a,b) established a relationship between air temperature and productivity based on the results of past studies that had used objective indicators of performance that were likely to be relevant to office work activities. The relationship indicates that the performance tends to increase with the room temperature up to 21–22°C, while decreased by the temperature above 23–24°C. The highest productivity was achieved at temperatures around 22°C (Seppänen et al., 2006). An index that integrates the effect of different thermal criteria would be considered as a useful tool in the assessment of productivity. Roelofsen (2002) used this approach and related the loss of performance with PMV. His relationship was created by regressing equal thermal sensations based on a Gagge's two-layer model against thermal comfort sensations in the thermal comfort model proposed by Fanger. Lan et al. (2011b) established a quantitative relationship between thermal environment characterized by thermal sensation votes and task performance. This relationship follows a bell-shaped curve centered around the conditions that are optimal for performance, and it suggests that a slightly cool environment and avoidance of moderately elevated temperatures will create conditions that are optimal for productivity (Fig. 11.2). This applies even to subjects, who are acclimatized to higher temperatures by living in tropical climates (Tham, 2004). Based on the relationship shown in Fig. 11.2, the impact of changes in clothing insulation during the summer and winter on the air temperature for optimum performance was shown in Table 11.1. When predicting the thermal sensation votes, the mean radiant temperature was assumed to be equal to air temperature (i.e., operative temperature equals the air temperature), in which the activity level was estimated at 1.2 Met, air velocity at 0.15 m/s, and the relative humidity was at 50%. In summer, the indoor air temperature for optimum performance can be increased from about 23.5 to 25.4°C, when people wear a short-sleeved shirt with walking shorts (whose clothing combination corresponds to 0.36 clo) instead of a short-sleeved shirt with trousers (0.61 clo). The indoor air temperature for optimum performance can be decreased from about 21.9 to 18.9°C in winter, when a long-sleeved shirt, thick long-sleeved sweater, T-shirt, suit jacket, and a long underwear bottoms, as well as trousers were worn (1.30 clo) instead of a long-sleeved shirt, and a thin long-sleeved sweater, as well as trousers (0.86 clo). A recent study suggests that moderately elevated

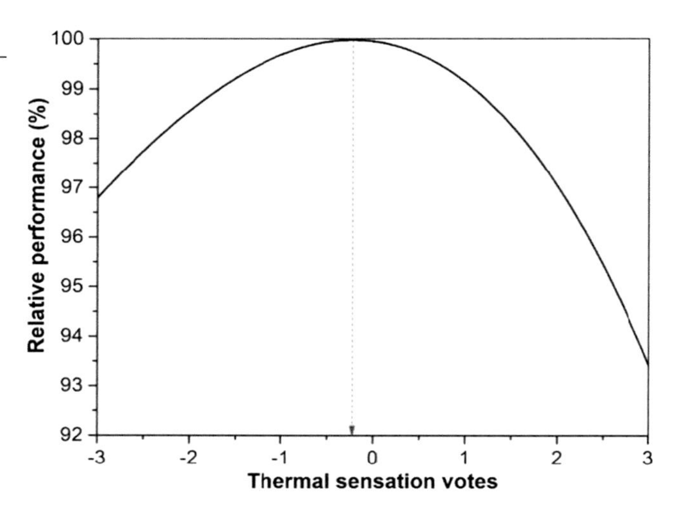

FIGURE 11.2

The relationships between thermal sensation and productivity.

Table 11.1 The air temperature for optimum performance at different clothing insulation levels in summer and winter.

Season	Clothing with estimated insulation level	Optimum air temperature (°C)
Summer	Short-sleeved shirt, walking shorts, 0.36 clo	25.4
	Typical clothing in summer, 0.5 clo	24.3
	Short-sleeved shirt, trousers, 0.57 clo	23.9
	Long-sleeved shirt, trousers, 0.61 clo	23.5
Winter	Long-sleeved shirt, thin long-sleeved sweater, trousers, 0.86 clo	21.9
	Typical clothing in winter, 1.0 clo	21.2
	Long-sleeved shirt, thick long-sleeved suit jacket, thick sleeveless vest, trousers, 1.19 clo	19.7
	Long-sleeved shirt, long-sleeved sweater, T-shirt, suit jacket, long underwear bottoms, trousers, 1.3 clo	18.9

temperatures despite the achievement of thermal comfort can still lessen user performance (Lan et al., 2020). Such effects deserve further studies.

Indoor air quality: Reducing the pollution load in indoor air proved to be an effective means of improving productivity of building occupants (Wargocki et al., 1999; Bakó-Biró et al., 2004). IAQ can also be modified by changing the ventilation rate. Seppänen et al. (2006a,b) reviewed the literatures relating productivity with ventilation rate, and most of the studies explored concluded that higher ventilation rates tend to contribute to increasing performance. They established a quantitative relationship between outdoor ventilation rate and productivity. It indicates that typically, a 10 l/s-person increase in outdoor air ventilation rate improves performance by 1%–3%. The performance improvement per unit through increasing the ventilation rate yet below 20 l/s-person is more significant than over 45 l/s-person.

Light and view: Lighting effect on productivity is determined by several light parameters. Among them mainly are the illuminance level (LL) and correlated color temperature (CCT). The effects of LL and CCT could be interacted, as shown in the Kruithof curve, which illustrates a region of LL and CCT that are often viewed as comfortable or pleasing to an observer (Kruithof, 1941). Nonetheless, in real terms, research outcomes relating to the effect of lighting environment on productivity tend not to be constant (Souman et al., 2018; Lok et al., 2018). Some studies observed beneficial effects of increased illuminance on cognitive performance, such as sustained attention, response inhibition, and working memory. The others did not find any significant improvement of sustained attention and working memory. For instance, Viola et al. (2008) concluded that daytime blue-enriched light (17,000K) could improve occupants' self-reported alertness and work productivity, while Vandewalle et al. (2007) and Smolders and de Kort (2017) admitted that such positive effects were not observed in their studies.

Noise: It is generally accepted that noise has negative effects on productivity. Yet, Smith (1989) reviewed the effects of noise on performance and concluded that noise effects are still not clear, and that beyond intensity issues, researchers need to analyze the questions of what type of noise at what intensity affects which type of task performance. Most researchers concerned high intensity of noise, such as above 80 dB, that is not commonly experienced in modern offices, where a 55–70 dB range of noise tends to be accepted. The limited research focusing on the effect of noise of such levels on performance has come to inconsistent results. For example, Witterseh et al. (2004) found that noise of low intensity decreased work performance by 3%. Delay and Mathey (1985) discovered that a subject's performance for a time estimation task increased as the noise intensity level was ramped up from 50 to 80 dB. Evans and Johnson (2000) did not find any effect of low intensity of office noise on typing performance during a well-designed 3 h experiment. There is still a clear need for further methodological research that helps examine a compelling quantitative relationship between productivity and ambiance noise.

The mechanism that mediates the effects of IEQ on human well-being and productivity has not been enough elucidated (Wargocki and Wyon, 2017). It is hypothesized that the IEQ affects human through complex interactions among physiological responses, psychological reactions, and cognitive function. Accordingly, the following section will focus mainly on exploring the human mental responses to the built environment and how the occupants' subjective needs and demands can be incorporated into the architectural design decision-making process.

3. Human psychological responses to built environments

The built environment triggers human perceptions (McClure et al., 2011). Every space creates opportunities for the users' everyday activities and experiences (Goldhagen, 2017; Kopec, 2018; Noguchi et al., 2018; Norman, 1988). Every single element in the built environment tends to contribute to linking human perceptions to their surrounding physical settings (Evans, 2003; Kopec, 2018; Ulrich et al., 1991). Therefore, the built environment has capacities for creating physical, biological, and psychological impacts on human health and well-being, directly or indirectly. The built environment embraces different ideologies, and it is a space where people live and connect to diverse sociocultural and environmental factors that affect human health and well-being (Evans, 2003; Goldhagen, 2017).

Humans perform in their living environment with diverse habits, and their perceptions differ from one another (Graham et al., 2015; Gosling et al., 2014). Humans tend to adjust or adapt to surrounding environments to achieve their needs and demands within different settings. Human satisfaction regarding feelings, moods, and emotions in indoor and outdoor living quality is essential (Amérigo and Aragones, 1997).

According to Mehrabian (1974), pleasure, arousal, and activation are the main human emotional responses to any spatial experience. Reflecting all those emotional responses through experiences resulting in satisfaction affects human behavior. Human mental responses to spatial settings are complex phenomena being difficult to define precisely (Pallasmaa, 2005; Ittelson et al., 1976). Even so, the identification of such perceptions and correlations seems to be desirable when spatial settings that affect humans' physical and mental responses are planned (Fig. 11.3).

Today's theories related to human emotion focusing mainly on physiological stimulation and activation characteristics define the fundamental responses of human subjective emotive states (Mehrabian and Russell, 1974). The emotions swing with peripheral trials in the built environment, and stimulation variations influence human biological aspects. Surrounding environments influence human emotional states even after they leave the space (Goldhagen, 2017; Mallgrave, 2018; Mehrabian and Russell, 1974). Seemingly, human emotional states stimulate cognitive memories and responses to mental situations in numerous ways (Mallgrave, 2018; Sussman and Hollander, 2021). In this respect, human sensory organs perform a core role that contributes to turning the recognition of physical settings into the notion of environmental experience (Fig. 11.4) (Mallgrave, 2018).

Moreover, human always perceives their environment involving six sensory attributes, i.e., vision, hearing, smell, skin-sense of air, haptic, and kinesthesia (Mallgrave, 2018; Goldhagen, 2017; Sussman and Hollander, 2021; Mehrabian and Russell, 1974). Any physical setting enhances human sensory

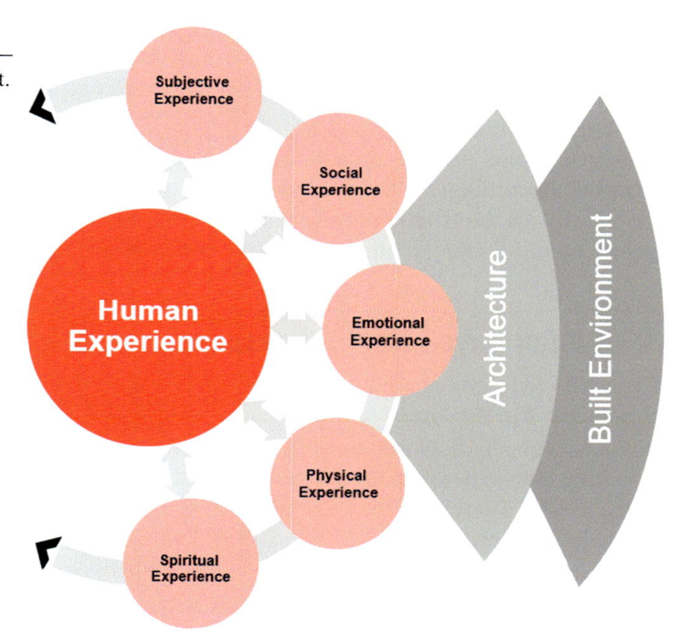

FIGURE 11.3

Human experiences in the built environment.

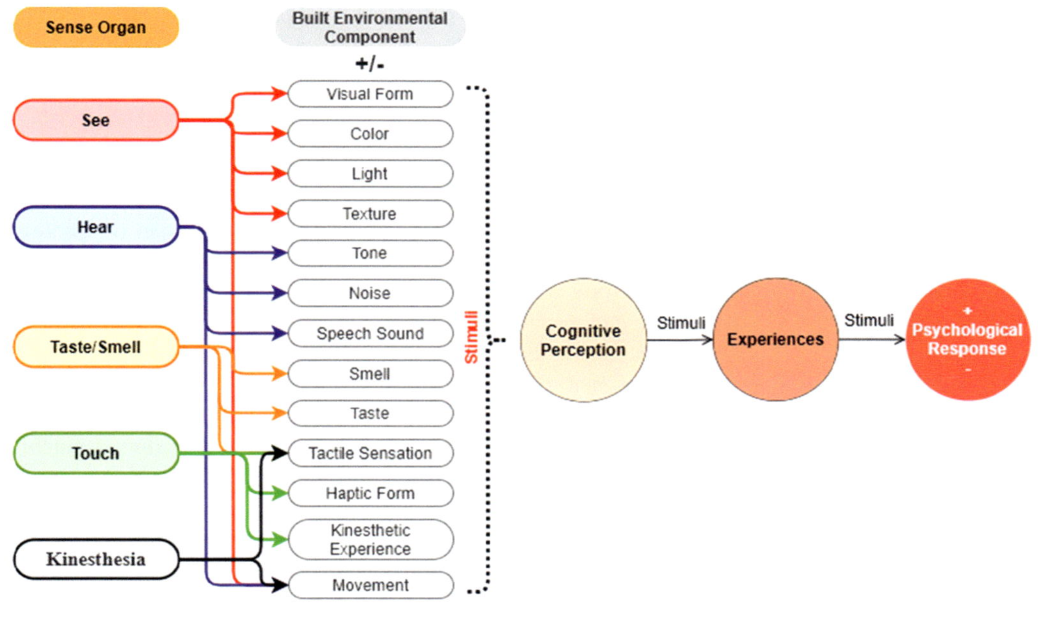

FIGURE 11.4

Relationship between the built environment and human response.

organs and promotes mental responses to whether they are auditory, tactile, or visual stimuli (Sussman and Hollander, 2021; Mehrabian and Russell, 1974). For example, lighting levels affect human moods and feelings that stimulate mental strain and alternate circadian rhythms (Ergan et al., 2018). Moreover, poor indoor lighting levels tend to reduce human psychological growth (Cooper et al., 2014; Lawrence, 1984). Noise affects individual privacy within living environments. Noise problems in the built environment may lead to a negative impact on human behavior (Evans, 2003; Ittelson et al., 1976). Human memory is also connected to the smell of any place whereas odors influence the human mind (Cooper et al., 2014). Several studies identify the psychological benefits of gardening and indoor–outdoor connectivity in the living environment (Kaplan, 1995). Nature tends to revive environmental quality as well as to provide visual satisfaction. Studies indicate that outdoor connectivity promotes environmental stimuli and human psychological relief from mental distress (Kaplan, 1995). Alongside this, spatial ergonomics related to shape, size, height, dimension, fixture, and density affect human psychological perceptions positively or negatively within living environments (Graham et al., 2015; Iavicoli et al., 2010).

Built environments stimulate human lives through their experiences (Goldhagen, 2017; Ittelson et al., 1976). In 1970, the publication entitled *Environmental Psychology: Man and His Physical Setting* by Proshansky, Ittelson, and Rivlin highlights the significance of relationships between human perceptions, behaviors, and built environments. This study concerns humans' physical, psychological, and sociocultural dimensions in built environments (Kopec, 2018; Maslow, 1971; Proshansky et al., 1970). Perhaps, these human factors can be considered as key variables in environmental design that influence individual mental responses, e.g., feelings, emotions, and moods, within the built environment. Several theories exist in the environmental psychology domain, and they study human responses to built environments (Graham et al., 2015; Mallett, 2004). For example, the theories include Brunswik's model of probabilistic lens, Gibson's model of affordance, Berlyne's model of aesthetics, social learning theory, integration theory, control theory, behavior setting theory, simulation theory, and attention restoration theory. Indeed, these theories articulate human–environment interactions (Goldhagen, 2017; Lawson, 2013; Proshansky et al., 1970).

Designing physical settings in the built environment may need to consider human psychological perception and well-being (Kopec, 2018; Goldhagen, 2017). Furthermore, its research approach specifies the built environment in numerous ways, such as independent variables of interpersonal effect, behavioral aspect, phenomena, and psychological context (Altman, 1992). Environmental psychology is also a core arena of understanding human relationships and connections associated with sociophysical settings in the built environment. It merges physical and social sciences and examines the relationship between humans' perceptions and their surrounding environments by making use of multidisciplinary theoretical models (Gifford et al., 2011). Recent research focuses on sustainability, and the understanding of everyday life experiences is becoming prominent, yet controversial (Sussman and Hollander, 2021; Ulrich et al., 2010). Environmental psychology is becoming more crucial than ever in light of user behavior that affects human health and well-being, as well as climate change mitigation and adaptation (Saegert, 2004). Wapner and Demick (2002) further interpreted this concept of contextualism considering six contexts: social, physical, psychological, natural, interpersonal, and cultural aspects of the built environment. Furthermore, Gifford et al. (2011) illustrated three dimensions, such as place, person, and psychological process in the built environment research. Goldhagen (2017) in the book entitled *Welcome to Your World: How the Built Environment Shapes Our Lives* reveals a need for the establishment of a conceptual framework that aims to understand human experiences according to individual needs and demands in built environments. In short, human factors in the architectural design process may need to be associated with users' sociocultural backgrounds, as well as their preferences and restrictions.

Today's architectural design approach experiences a gap between users' spatial needs and demands and psychological satisfaction (Chowdhury et al., 2020). The "environmental deterministic theory" describes the physical environment's impacts on human behavior (Vischer, 2008). This theory excludes or limits users' social and cultural contexts. On the other hand, the "social constructivism theory" engages with users' social and cultural perceptions as a challenge to measure the effects of built environments (Vischer, 2008). A human-centric design approach emerged positioning itself between the environmental deterministic and social constructivism theories, and it is to some extent addressing the effects of users' social, cultural, and environmental aspects on the architectural design process (Norman and Draper, 1986).

3.1 Environmental experience design research trajectory

Norberg-Schulz's "existential and architectural space" reflects the meaning of place and human perceptions and Relph's "place and placelessness" expresses cultural and emotional attachment within the built environment (Mallgrave, 2018). Kling (1977) coined the term "user-centered design (UCD)," which reflects a person-centric philosophical design approach that focused on human cognitive interaction with objects, products, and things. In 1986, the concept of UCD became widely popular as "user experience (UX) design" due to the publication entitled *User-Centered System Design: New Perspectives on Human-Computer Interaction* by Donald A. Norman at the University of California, San Diego (Norman and Draper, 1986). Human-centered design is a design philosophy starting with understanding and realizing human needs and demands where the design is projected to achieve through observations and user experiences (Norman, 1988).

Additionally, in a book entitled *The Design of Everyday Things*, Norman (2004) expanded the concept of "experience design" to the industrial design domain and elaborated the concept of human psychology within design thinking and articulated the importance in everyday human lives in consideration of product usability and usefulness. He also articulated four basic product design considerations, such as easy to determine, easy to evaluate, visible, and natural setting for experience design (Norman, 1988). As well, the author proposed seven design strategies, such as goal setting, planning, specification, performance, perception, interpretation, and comparison to systematize a product design decision-making process (Norman, 1988). Designing needs to interact between people and technology where discoverability and understanding are the two essential features of a good design (Norman, 1988). In the book entitled *Design for Experience: Where Technology Meets Design and Strategy*, Kim (2015) stated that the user is a focal point in experience design. The design incident is subjective and theories falling into humanities and social science may somewhat cover these issues. UX/UCD/UI concepts help form fundamental user experience design logic.

UX/UCD/UI definitions can be applied in numerous ways, yet all focus on exploring the users' perspective in the design process based on their needs and demands. In the ISO standard 9241-210, the UCD process has been considered as an interactive system based on people's perceptions and responses (Linden et al., 2019). According to the authors, the user-product interrelates with sociocultural factors in a precise context (Linden et al., 2019). UCD raised a philosophical agenda of users' expectations and experiences. Thüring and Mahlke draw attention to three design factors of user experiences, such as instrumental factors, noninstrumental factors, and emotional responses to form a complete decision and regulate user behavior (Linden et al., 2019). The authors also mentioned that user experiences emphasize on human perceptions, preferences, and emotional responses while a product or service is in use. Considering these factors, experience design can be regarded as an extension of the human-centered design domain. Pallasmaa (2005) stated that an architect's design requires to incorporate basic human needs of feelings and emotions where phenomenological analysis of these human factors is a prominent part of design decisions. According to Pallasmaa (2005), architectural phenomenology is a purely theoretical approach to interpreting human-environmental perceptions. McLellan (2000) mentioned that functionality, engagement, stimulation, enjoyment, and memory are the goals of experience design. The book entitled *The Handbook of Interior Design* also highlighted that reflection of user experiences enhances the quality of indoor living environment (Thompson, 2015). Reflecting the notion of experience design approaches, Ma et al. (2017) and Noguchi et al. (2018) introduced the term and concept of "environmental experience design (EXD)" (Fig. 11.5). Their EXD studies led to proposing a research framework applied to aged care facilities in Australia. Following the emergence of an EXD concept, Chowdhury et al. (2020) defined the term "domestic environmental experience" as "users' experiences of cognitive perceptions and physical responses to their domestic built environment." Furthermore, a conceptual correlation of the domestic environmental parameters (i.e., environmental design factor, spatial factor, and user context) was illustrated based on user perceptions.

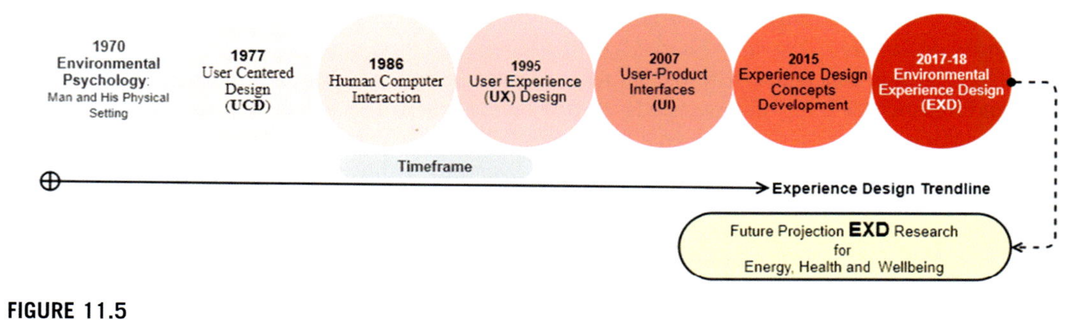

FIGURE 11.5

EXD research trajectory.

Today, architectural design research and practices tend to neither reflect multidimensional human factors nor encompass user experiences (Chowdhury et al., 2020). The design of built environments may be required to incorporate human-centered EXD approaches that aim to accommodate the occupants' individual needs and demands with the aim to enhance their health and well-being (Fig. 11.6) (Noguchi et al., 2018).

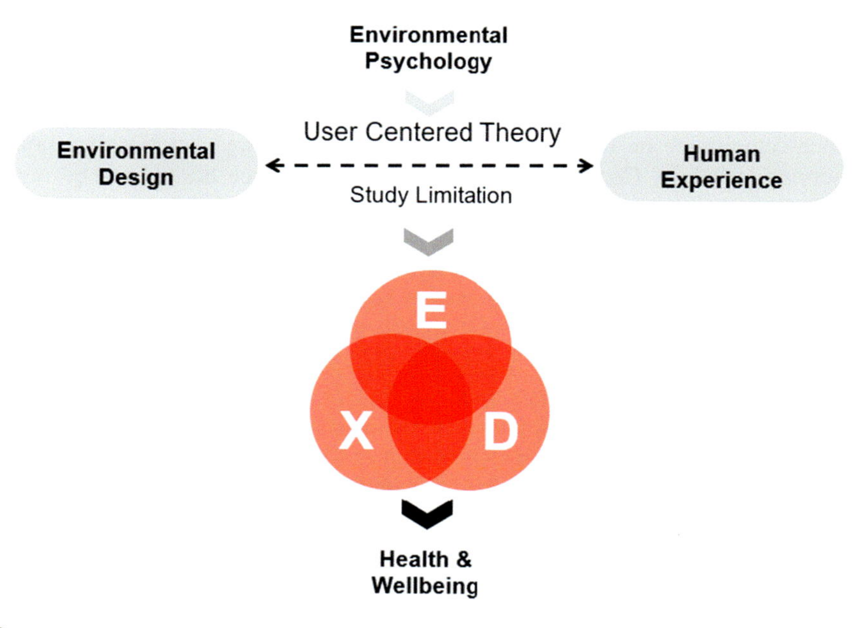

FIGURE 11.6

Environmental experience design disciplinary prospect.

4. Conclusions

Human behavior in built environments is stimulated not only by spatial settings but also by the users' perception within their sociocultural contexts. The user-centered design (UCD) may require an understanding of human perception and behavior within the built environment. Over the last years, environmental psychologists had a tendency to focus mainly on examining negative human responses (e.g., depression, anxiety, and stress) to built environments. Today, the study has begun exploring a culture of positive human thinking and feeling within defined spatial contexts. The research arena touches upon individual and societal well-being, such as pleasant, happiness, and meaningful life. Also, the new paradigm of human cognition is shifted to an image of human experiences in built environments where the users' mind and body interact. Human experiences generally indicate cognitive perceptions and physical responses to spatial settings where different types of individual experiences coexist. The built environment is indeed a system of energy and environment being occupied by the masses where their perception and behavior impact on the consequences. It can be enriched through the implementation of environmental experience design (EXD) that aims to accommodate users' needs and demands for enhancement of their health and well-being. EXD implementation may also have a significant potential for users' perceptual and behavioral changes toward energy savings. This chapter conceptualized an EXD research spectrum and the potential effects on energy and human well-being within the contexts of built environments. The demonstration and validation are required for further EXD studies.

References

Ahmed, S., Nahiduzzaman, K.M., Hasan, M.M., 2018. Dhaka, Bangladesh: unpacking challenges and reflecting on unjust transitions. Cities 77, 142–157.

Alam, M.J., 2018. Rapid urbanization and changing land values in mega cities: implications for housing development projects in Dhaka, Bangladesh. Bandung 5 (1), 1–19.

Altman, I., 1992. A transactional perspective on transitions to new environments. Environment and Behavior 24 (2), 268–280.

Amérigo, M., Aragones, J.I., 1997. A theoretical and methodological approach to the study of residential satisfaction. Journal of Environmental Psychology 17 (1), 47–57.

ASHRAE, 2017. Standard 55-Thermal Environmental Conditions for Human Occupancy. ASHRAE, Atlanta.

Asian Development Bank. Development Indi. Economics & Re, 2011. Key Indicators for Asia and the Pacific 2011. Asian Development Bank. https://www.adb.org/publications/key-indicators-asia-and-pacific-2011 (Accessed 7 February 2021).

Bakó-Biró, Z., Wargocki, P., Weschler, C., Fanger, P.O., 2004. Effects of pollution from personal computers on perceived air quality, SBS symptoms and productivity in offices. Indoor Air 14, 178–187.

Barua, S., Mridha, A.H.A.M., Khan, R.H., 2010. Housing real estate sector in Bangladesh present status and policies implications. ASA University Review 4 (1), 240–253.

Basner, M., Babisch, W., Davis, A., Brink, M., Clark, C., Janssen, S., Stansfeld, S., 2014. Auditory and non-auditory effects of noise on health. Lancet 383 (9925), 1325–1332.

Beyer, R.C.M., 2019. South Asia Economic Focus, Fall 2019: Making (De) Centralization Work. The World Bank, 142628: 1–88.

BRAC, 2017. The State of Cities 2017: Housing in Dhaka. Brac Institute of Governance and Development, Dhaka, Bangladesh. https://bigd.bracu.ac.bd/publications/the-state-of-cities-2017-housing-in-dhaka (Accessed 25 December 2020).

Cedeño-Laurent, J.G., Williams, A., MacNaughton, P., Cao, X., Eitland, E., Spengler, J., Allen, J., 2018. Building evidence for health: green buildings, current science, and future challenges. Annual Reviews of Public Health 39, 291–308.

Chowdhury, S., Noguchi, M., Doloi, H., 2020. Defining domestic environmental experience for occupants' mental health and wellbeing. Designs 4 (3), 26.

Chun, N., 2010. Middle Class Size in the Past, Present, and Future: A Description of Trends in Asia. Working Paper Series 217, Asian Development Bank (ADB). Asian Development Bank Economics, Manila. https://www.adb.org/publications/middle-class-size-past-present-and-future-description-trends-asia (Accessed 10 February 2021).

Cooper, R., Burton, E., Cooper, C. (Eds.), 2014. Wellbeing: A Complete Reference Guide, Wellbeing and the Environment, vol. 2. John Wiley & Sons.

Delay, E., Mathey, M., 1985. Effects of ambient noise on time estimation by humans. Perceptual and Motor Skill 61, 415–419.

Ergan, S., Shi, Z., Yu, X., 2018. Towards quantifying human experience in the built environment: a crowdsourcing based experiment to identify influential architectural design features. Journal of Building Engineering 20, 51–59.

Evans, G.W., 2003. The built environment and mental health. Journal of Urban Health 80 (4), 536–555.

Evans, G.W., Johnson, D., 2000. Stress and open-office noise. Journal of Applied Psychology 85, 779–783.

Fang, L., Wyon, D.P., Clausen, G., Fanger, P.O., 2004. Impact of indoor air temperature and humidity in an office on perceived air quality, SBS symptoms and performance. Indoor Air 14, 74–81.

Fisk, W.J., Black, D., Brunner, G., 2011. Benefits and costs of improved IEQ in U.S. offices. Indoor Air 21, 357–367.

Gao, Q., 2019. Energy consumption and cost analysis of rural passive housing demonstration project in Shaling new village of Changping District. Construction Science and Technology 380, 46–52 (in Chinese).

Gifford, R., Steg, L., Reser, J.P., 2011. Environmental Psychology. Wiley Blackwell.

Goldhagen, S.W., 2017. Welcome to Your World: How the Built Environment Shapes Our Lives. Harper, New York.

Gosling, S.D., Gifford, R., Mccuan, L., 2014. Environmental perception and interior design. The Body, Behavior, and Space 242, 278–290.

Graham, L.T., Gosling, S.D., Travis, C.K., 2015. The psychology of home environments: a call for research on residential space. Perspectives on Psychological Science 10 (3), 346–356.

Hassan, A., Rahman, M., Khan, F., Malik, M., Ali, M., 2012. Electricity challenge for sustainable future in Bangladesh. APCBEE Procedia 1, 346–350.

Hsieh, M., 2015. Effects of illuminance distribution, color temperature and illuminance level on positive and negative moods. Journal of Asian Architecture and Building Engineering 14 (3), 709–716.

Iavicoli, I., Leso, V., Fontana, L., Bergamaschi, A., 2010. Occupational exposure to urban airborne particulate matter: a review on environmental monitoring and health effects. In: Urban Airborne Particulate Matter, pp. 501–525.

ILO, 2020. Global Employment Trends for Youth 2020: Technology and the Future of Jobs. International Labour Organization. https://www.ilo.org/wcmsp5/groups/public/—-dgreports/—-dcomm/—-publ/documents/publication/wcms_737648.pdf (Accessed 8 December 2021).

Ittelson, W.H., Franck, K.A., O'Hanlon, T.J., 1976. The Nature of Environmental Experience. In: Wapner, S., Cohen, S.B., Kaplan, B. (Eds.), Experiencing the Environment. Springer, Boston, MA, USA, pp. 187–206.

Jiang, W., Huang, Z., Peng, Y., Fang, Y., Cao, Y., 2020. Factors affecting prefabricated construction promotion in China: a structural equation modeling approach. PLoS One 15 (1), e0227787.

Kamruzzaman, M., Ogura, N., 2007. Apartment housing in Dhaka City: past, present and characteristic outlook. In: Building Stock Activation. Tokyo, Japan. http://tmu-arch.sakura.ne.jp/pdf/26_proc_bsa_e/Proceedings_pdf/081-088 011SS_A1-4.pdf (Accessed 15 December 2021).

Kaplan, S., 1995. The restorative benefits of nature: toward an integrative framework. Journal of Environmental Psychology 15 (3), 169–182.

Kharas, H., 2010. The Emerging Middle Class in Developing Countries. OECD Development Centre Working Papers 285. OECD Publishing, Paris. https://www.oecd.org/dev/44457738.pdf (Accessed 5 February 2021).

Kharas, H., 2017. The Unprecedented Expansion of the Global Middle Class: An Update. Working Paper 100. Global Economy & Development, Washington, DC, p. 20036. https://www.brookings.edu/wp-content/uploads/2017/02/global_20170228_global-middle-class.pdf (Accessed 6 February 2021).

Khatun, F., Saadat, S.Y., 2020. Youth Employment in Bangladesh. Springer, Singapore.

Kim, J., 2015. Design for Experience: Where Technology Meets Design and Strategy. Springer.

Kling, R., 1977. The organizational context of user-centered software designs. MIS Quarterly 41–52.

Kong, A., Kang, H., He, S., Li, N., Wang, W., 2020. Study on the carbon emissions in the whole construction process of prefabricated floor slab. Applied Science 10 (7), 2326.

Kopec, D.A., 2018. Environmental Psychology for Design. Bloomsbury Publishing Inc., New York, p. 317.

Kruithof, A.A., 1941. Tubular luminescence lamps for general illumination. Philips Technical Review 6 (3), 65–73.

Lan, L., Lian, Z.W., Pan, L., Ye, Q., 2009. Neurobehavioral approach for evaluation of office workers' productivity: the effects of room temperature. Building and Environment 44 (8), 1578–1588.

Lan, L., Lian, Z.W., Pan, L., 2010. The effects of air temperature on office workers' well-being, workload and productivity-evaluated with subjective ratings. Applied Ergonomics 42 (1), 29–36.

Lan, L., Wargocki, P., Wyon, D., Lian, Z.W., 2011a. Effects of thermal discomfort in an office on perceived air quality, SBS symptoms, physiological responses, and human performance. Indoor Air 21, 376–390.

Lan, L., Wargocki, P., Lian, Z.W., 2011b. Quantitative measurement of productivity loss due to thermal discomfort. Energy and Buildings 43, 1057–1062.

Lan, L., Xia, L.L., Hejjo, R., Wyon, D.P., Wargocki, P., 2020. Perceived air quality and cognitive performance decrease at moderately raised indoor temperatures even when clothed for comfort. Indoor Air 30, 841–859.

Larcombe, D.L., van Etten, E., Logan, A., Prescott, S.L., Horwitz, P., 2019. High-rise apartments and urban mental health—historical and contemporary views. Challenges 10 (2), 34.

Lawlor, A., 1997. A Home for the Soul: A Guide for Dwelling with Spirit and Imagination. Clarkson Potter Publishers.

Lawrence, R.J., 1984. Transition spaces and dwelling design. Journal of Architectural and Planning Research 1, 261–271.

Lawson, B., 2013. Design and the evidence. Procedia-Social and Behavioral Sciences 105, 30–37.

Li, J., 2008. Towards a low-carbon future in China's building sector - a review of energy and climate models forecast. Energy Policy 36, 1736–1747.

Li, J., Colombier, M., 2009. Managing carbon emissions in China through building energy efficiency. Journal of Environmental Management 90 (8), 2436–2447.

Li, X., Song, J., Lin, T., Dixon, J., Zhang, G., Ye, H., 2016. Urbanization and health in China, thinking at the national, local and individual levels. Environmental Health 15 (1), 113–123.

Lin, B., Liu, H., 2015. CO_2 emissions of China's commercial and residential buildings: evidence and reduction policy. Building and Environment 92, 418–431.

Lin, Y., Zhong, S., Yang, W., Hao, X., Li, C., 2020. Towards zero-energy buildings in China: a systematic literature review. Journal of Cleaner Production 276, 123297.

Linden, J.V.D., Amadieu, F., Vayre, E., Leemput, C.V.D., 2019. User experience and social influence: a new perspective for UX theory. In: International Conference on Human-Computer Interaction. Springer, Cham, pp. 98–112.

Lok, R., Smolders, K.C.H.J., Beersma, D.G.M., de Kort, Y.A.W., 2018. Light, alertness, and alerting effects of white light: a literature overview. Journal of Biological Rhythms 33, 589–601.

Lowen, A., Mubareka, S., Steel, J., Palese, P., 2007. Influenza virus transmission is dependent on relative humidity and temperature. PLOS Pathogens 3, 1470–1476.

Ma, N., Chau, H.W., Zhou, J., Noguchi, M., 2017. Structuring the environmental experience design research framework through selected aged care facility data analyses in Victoria. Sustainability 9 (12), 2172.

Mallett, S., 2004. Understanding home: a critical review of the literature. The Sociological Review 52 (1), 62–89.

Mallgrave, H.F., 2018. From Object to Experience: The New Culture of Architectural Design. Bloomsbury Publishing.

Maslow, A.H., 1971. The Farther Reaches of Human Nature, vol. 19711. Viking Press, New York.

McClure, W.R., Bartuska, T.J. (Eds.), 2011. The Built Environment: A Collaborative Inquiry into Design and Planning. John Wiley & Sons.

McLellan, H., 2000. Experience design. Cyberpsychology and Behavior 3 (1), 59–69.

Meeting report: The role of environmental lighting and circadian disruption in cancer and other diseases. Environmental Health Perspectives 115 (9), 2007, 1357–1362.

Mehrabian, A., Russell, J.A., 1974. An Approach to Environmental Psychology. the MIT Press.

Mridha, A.M.M.H., Moore, G.T., 2011. The quality of life in Dhaka, Bangladesh: neighborhood quality as a major component of residential satisfaction. In: Investigating Quality of Urban Life. Springer, Dordrecht, pp. 251–272.

Münzel, T., Sørensen, M., Gori, T., Schmidt, F., Rao, X.Q., Brook, J., Chen, L.C., Brook, R.D., Rajagopalan, S., 2017. Environmental stressors and cardio-metabolic disease: part I—epidemiologic evidence supporting a role for noise and air pollution and effects of mitigation strategies. European Heart Journal 38, 550–556.

Münzel, T., Schmidt, F., Steven, S., Herzog, J., Daiber, A., Sørensen, M., 2018. Environmental noise and the cardiovascular system. Journal of the American College of Cardiology 71 (6), 688–697.

Muzet, A., 2007. Environmental noise, sleep and health. Sleep Medicine 11, 135–142.

Noguchi, M., Ma, N., Woo, C.M.M., Chau, H.W., Zhou, J., 2018. The usability study of a proposed environmental experience design framework for active ageing. Buildings 8 (12), 167.

Norman, D.A., 1988. The Psychology of Everyday Things. Basic books.

Norman, D.A., 2004. Emotional Design: Why We Love (Or Hate) Everyday Things. Basic Civitas Books.

Norman, D.A., Draper, S.W., 1986. User Centered System Design; New Perspectives on Human-Computer Interaction. L. Erlbaum Associates Inc.

OECD (Organisation for Economic Co-operation and Development), 2019. Under Pressure: The Squeezed Middle Class. OECD Publishing. https://www.oecd.org/social/under-pressure-the-squeezed-middle-class-689afed1-en.htm (Accessed 6 November 2020).

Pallasmaa, J., 2005. In: The Eyes of the Skin: Architecture and the Senses. John Wiley & Sons.

Parveen, R., 2017. Energy Independent Residential Development for Dhaka City, Bangladesh (Doctoral dissertation). University of Adelaide, Australia. Retrieved from: www.digital.library.adelaide.edu.au.

Proshansky, H.M., Ittelson, W.H., Rivlin, L.G. (Eds.), 1970. Environmental Psychology: Man and His Physical Setting. Rinehart and Winston, New York, Holt, pp. 21–26.

RAJUK (Rajdhani Unnayan Kartripakkha), 2015. Dhaka Structure Plan 2016–2035 (Dhaka, Bangladesh).

Roelofsen, P., 2002. The impact of office environments on employee performance: the design of the workplace as a strategy for productivity enhancement. Journal of Facilities Management 1 (3), 247–264.

Sadeque, C.M.Z., 2013. The Housing Affordability Problems of the Middle-Income Groups in Dhaka: A Policy Environment Analysis (Doctoral dissertation). The University of Hong Kong, Hong Kong. Retrieved from: https://hub.hku.hk/bitstream/10722/193500/2/FullText.pdf.

Saegert, S., 2004. In: Bechtel, R.B., Churchman, A. (Eds.), Handbook of Environmental Psychology. Wiley, New York (2002) 772 Journal of Environmental Psychology 2(24): 259–263.

Salam, R.A., Amber, K.P., Ratyal, N.I., Alam, M., Akram, N., Gómez Muñoz, C.Q., García Márquez, F.P., 2020. An overview on energy and development of energy integration in major South Asian countries: the building sector. Energies 13 (21), 5776.

Sarker, P., 2020. Analyzing urban sprawl and sustainable development in Dhaka, Bangladesh. Journal of Economics and Sustainable Development. ISSN 2222–1700.

Satu, S.A., Chiu, R.L., 2019. Livability in dense residential neighbourhoods of Dhaka. Housing Studies 34 (3), 538–559.

Seppänen, O., Fisk, W., Lei, Q.H., 2006a. Room Temperature and Productivity in Office Work, eScholarship Repository. Lawrence Berkeley National Laboratory, University of California, Berkeley, California. http://repositories.cdlib.org/lbnl/LBNL-60952.

Seppänen, O., Fisk, W.J., Lei, Q.H., 2006b. Ventilation and performance in office work. Indoor Air 16 (1), 28–36.

Seraj, T.M., Islam, M.A., 2013. Detailed areas plan: proposals to meet housing demand in Dhaka. In: Jahan, S., Kalam, A.K.M.A. (Eds.), Dhaka Metropolitan Development Area and its Planning Problems, Issues, and Policies. Bangladesh Institute of Planners, Dhaka, pp. 22–32.

Shams, S., Mahruf, M., Shohel, C., Ahsan, A., 2014. Housing problems for middle and low income people in Bangladesh: challenges of Dhaka Megacity. Environment and Urbanization Asia 5 (1), 175–184.

Siddika, A., Badhan, I.M., Wahid, N., 2019. Current trend of income and its impact on affordability of multi-ownership housing at Moghbazar, Dhaka. Journal of Housing and Advancement in Interior Designing 2 (1, 2).

Smith, A., 1989. A review of the effects of noise on human performance. Scandinavian Journal of Psychology 30, 185–206.

Smolders, K.C.H.J., de Kort, Y.A.W., 2017. Investigating daytime effects of correlated colour temperature on experiences, performance, and arousal. Journal of Environmental Psychology 50, 80–93.

Souman, J.L., Tinga, A.M., Te Pas, S.F., van Ee, R., Vlaskamp, B.N.S., 2018. Acute alerting effects of light: a systematic literature review. Behavioural Brain Research 337, 228–239.

Spengler, J.D., Samet, J.M., McCarthy, J.F., 2001. Indoor Air Quality Handbook. McGraw-Hill, New York.

Stevens, R.G., Blask, D.E., Brainard, G.C., Hansen, J., Lockley, S.W., Provencio, I., Rea, M.S., Reinlib, L., 2007. Meeting report: the role of environmental lighting and circadian disruption in cancer and other diseases. Environmental Health Perspectives 115 (9), 1357–1362.

Sun, F., 2013. Chinese climate and vernacular dwellings. Buildings 3, 143–172.

Sussman, A., Hollander, J.B., 2021. Cognitive Architecture: Designing for How We Respond to the Built Environment. Routledge.

Swapan, M.S.H., Zaman, A.U., Ahsan, T., Ahmed, F., 2017. Transforming urban dichotomies and challenges of South Asian megacities: rethinking sustainable growth of Dhaka, Bangladesh. Urban Science 1 (4), 31.

Tariq, S.H., Ahmed, Z.N., 2020. Effect of plan layout on electricity consumption to maintain thermal comfort in apartments of Dhaka. Energy Efficiency 13 (6), 1119–1133.

TBS (The Business Standards), 2019. Towards an Unliveable City. The Business Standard, City, Dhaka, Bangladesh. Retrieved from: https://tbsnews.net/bangladesh/towards-unliveable-city.

Tham, K.W., 2004. Effects of temperature and outdoor air supply rate on the performance of call center operators in the tropics. Indoor Air 14 (7), 119–125.

The Conversation, 2021. https://theconversation.com/china-just-stunned-the-world-with-its-step-up-on-climate-action-and-the-implications-for-australia-may-be-huge-147268 (Accessed 28 January 2021).

Thompson, J.A.A., Blossom, N.H. (Eds.), 2015. The Handbook of Interior Design. Wiley-Blackwell, Hoboken.

Tinson, A., Clair, A., 2020. Better housing is crucial for our health and the COVID-19 recovery. The Health Foundation, UK.
Tu, F., 2006. Situation and policy of building energy efficiency in China. Housing Industry 12, 29–32.
Ulrich, R.S., Simons, R.F., Losito, B.D., Fiorito, E., Miles, M.A., Zelson, M., 1991. Stress recovery during exposure to natural and urban environments. Journal of Environmental Psychology 11 (3), 201–230.
Ulrich, R.S., Berry, L.L., Quan, X., Parish, J.T., 2010. A conceptual framework for the domain of evidence-based design. Health Environments Research & Design Journal 4 (1), 95–114.
United Nations Population Fund (UNFPA), 2014. The State of the World Population 2014: The Power of 1.8 Billion: Adolescents, Youth and the Transformation of the Future. United Nations Population Fund, New York. Retrieved from: https://www.unfpa.org/sites/default/files/pub-pdf/EN-SWOP14-Report_FINAL-web.pdf (Accessed 21 March 2021).
United Nations, 2021. List of Least Developed Countries (As of 11 February 2021). Committee for Development Policy (CDP) United Nations, New York. https://www.un.org/development/desa/dpad/least-developed-country-category.html (Accessed 6 March 2021).
United Nations Department of Economic and Social Affairs, 2021. Sustainable Development. Retrieved from: https://sdgs.un.org/goals (Accessed 21 March 2021).
Vandewalle, G., Gais, S., Schabus, M., Balteau, E., Carrier, J., Darsaud, A., Sterpenich, V., Albouy, G., Dijk, D.J., Maquet, P., 2007. Wavelength-dependent modulation of brain responses to a working memory task by daytime light exposure. Cerebral Cortex 17, 2788–2795.
Viola, A.U., James, L.M., Schlangen, L.J.M., Dijk, D.J., 2008. Blue-enriched white light in the workplace improves self-reported alertness, performance and sleep quality. Scandinavian Journal of Work Environment & Health 34, 297–306.
Vischer, J.C., 2008. Towards a user-centred theory of the built environment. Building Research & Information 36 (3), 231–240.
Wang, D., Liu, Y., Liu, J., Ma, B., Chen, H., 2013. Measuring study of heating performance of passive solar house with Trombe wall in qinghai-tibet plateau. Acta Energiae Solaris Sin 34 (10), 1823–1828.
Wang, T., Liu, Y., Wang, D., Ma, C., 2016. Thermal process and optimization design for solar house with Trombe roof. Acta Energiae Solaris Sin 37 (9), 2286–2291.
Wapner, S., Demick, J., 2002. The increasing contexts of context in the study of environment behavior relations. In: Handbook of Environmental Psychology. Wiley, Hoboken, NJ, pp. 3–14.
Wargocki, P., Djukanovic, R., 2005. Simulations of the potential revenue from investment in improved indoor air quality in an office building. ASHRAE Transactions 111 (part 2), 699–711.
Wargocki, P., Wyon, D.P., 2017. Ten questions concerning thermal and indoor air quality effects on the performance of office work and schoolwork. Building and Environment 112, 359–366.
Wargocki, P., Wyon, D.P., Baik, Y.K., Clausen, G., Fanger, P.O., 1999. Perceived air quality, sick building syndrome (SBS) symptoms and productivity in an office with two different pollution loads. Indoor Air 9, 165–179.
Wargocki, P., Wyon, D.P., Sundell, J., Clausen, G., Fanger, P.O., 2000. The Effects of outdoor air supply rate in an office on perceived air quality, sick building syndrome (SBS) Symptoms and productivity. Indoor Air 10, 222–236.
WEO, 2017. Key Findings. International Energy Agency. Available online: https://www.iea.org/weo2017/ (Accessed 15 May 2020).
WHO, 1946. Preamble to the Constitution of the World Health Organization as Adopted by the International Health Conference. World Health Organization, New York, pp. 19–22.
Witterseh, T., Wyon, D.P., Clausen, G., 2004. The effects of moderate heat stress and open-plan office noise distraction on SBS symptoms and on the performance of office work. Indoor Air 14 (Suppl. 8), 30–40.

Wolkoff, P., 2018. Indoor air humidity, air quality, and health – an overview. International Journal of Hygiene and Environmental Health 221, 376–390.
Xu, C., Zhang, D., 2019. Understanding and thinking on the development of prefabricated buildings. Project Management 5, 5–8 (in Chinese).
Zhang, Y., He, C.Q., Tang, B.J., Wei, Y.M., 2015. China's energy consumption in the building sector: a life cycle approach. Energy and Buildings 94, 240–251.

CHAPTER

Environmental security in developing countries: a case study of South Asia

12

Mabroor Hassan[1,2], Muhammad Irfan Khan[1], Mazhar Hayat[3] and Ijaz Ahmad[4]

[1]*Department of Environmental Science, International Islamic University, Sector H-10, Islamabad, Pakistan;* [2]*Green Environ Sol (Private) Limited, Sector H-10, Islamabad, Pakistan;* [3]*National Adaptation Process, Ministry of Climate Change, Islamabad, Pakistan;* [4]*National Skills University, Islamabad, Pakistan*

Chapter outline

1. Highlights

- Environmental insecurity and climate change have threatened socioeconomic and environmental notions of South Asia.

Handbook of Energy and Environmental Security. https://doi.org/10.1016/B978-0-12-824084-7.00012-6
Copyright © 2022 Elsevier Inc. All rights reserved.

- The Paris Agreement and SDGs provide obligations to handle climate change.
- The MCDA was applied to climate change policies and plans in South Asia.
- Gaps in policies and their implementation are threatening the region.
- Environmental diplomacy could be pursuance to cooperation.

2. Dynamics of environmental security

The environmental challenges have shattered the dynamics of economics and security in the recent era. Environmental insecurity has been accepted worldwide as nonconventional ravage to economic development and security concerns. A paradigm of dynamics of environmental security has been diversified from ecosystem restoration and conservation, pollution prevention, illegal movement of hazard waste, waste management and access and availability of natural resources to political economy, human insecurity due to environmental migrations and disasters, scarcity of natural resources, social upheavals, equitable access to natural resources, climatic extremes, environmental sustainability, national security, and geopolitical concerns in a current decade (Hassan et al., 2021). The environmental stresses such as scarcity of natural resources, climatic extremes, inequitable access to natural resources, and climate change have preceded the environmental migrations, energy and water crises, social upheavals, human insecurity, unjust distribution of natural resources, violence, civil wars, political instability, regional conflicts, insufficient capacity to mitigate environmental challenges, economic instability, and national insecurity.

Burgeon population in South Asia is experiencing multiple forms of insecurities including geopolitical, energy, human, water, human, environmental, political, and economic at the same times (Matthew, 2011). These insecurities have complexed the dynamics of environmental security and their relationship with human well-being, social protection, political stability, economic development, national security, and regional cooperation in South Asia. It has endured economic lagging, political instability, high geopolitical tension, civil wars, and the highest vulnerability to impacts of climate change. Hence, a clear understanding of dynamics of environmental security, regional cooperation, environmental diplomacy, good governance, mitigation of climate change, and regional partnership on environmental security is way forward to economic development, eradicate poverty, human security, national security, geopolitical stability, and environmental sustainability.

3. Environmental security and sustainable development

The Brundtland Commission 1987 unwrapped the concept of sustainable development. The commission defines sustainable development as:

> ***Sustainable development is a development that meets the needs of the present without compromising the ability of future generations to meet their own needs.***

However, the researchers bestow the explanation of the concept of sustainable development. Circular Ecology (2021) denoted sustainable development as a pathway to environmental sustainability. Fischer (Fischer et al., 2017) describes sustainable development as a distinct research field that explores the understanding of robust interaction between social and natural systems. The concept also underpins ways forward to transform and influence social and natural systems for environmental sustainability. Murphy (2012) stressed to interlink social and environmental pillars of sustainable development. Murphy proposed a conceptual framework based on equity, public awareness, social cohesion, and participation to achieve sustainable development.

The World Bank has recognized environmental security as a cardinal constituent of sustainable development because of the synergy of environmental security with equity and justice among generations (Ide et al., 2021). The environmental crises due to deforestation, climate change, pollution, land degradation, energy and water insecurity, climatic extremes, natural resources scarcity, biodiversity loss, and conflicts have threatened and presented barriers to achieving sustainable development (Shokhnekh et al., 2020). The environmental stresses have the potential to affect the livelihood of people, equity, justice, and well-being of a sustainable society. It may lead to violation and conflicts due to ecological unbalances, damage to water and terrestrial resources, inequitable access to resources, scarcity of resources, economic losses, risks of morbidity, damage to health, low working capacity, and decrease in employment opportunities. Hence, vibrant environmental strategies, cooperation, effective planning, and prompt response to environmental insecurity have the potential to balance ecological and sustainability paradigm to achieve environmental security and sustainable development.

4. Climate change: vulnerable nations

Environmental security is delineated as a broader analytical integration of technical, economic, social, and environmental aspects of environmental challenges, and flattering its influence on nations to transform the traditional concept of national security. Water, energy, food, development, public health, ecosystem degradation, and climate change are the main security stressors (Grumbine, 2017). Among these stressors, anthropogenic-induced changes in climate are evident from rising global temperature, sea-level rise, a shift in the weather pattern, extreme weather events, more heatwaves, expansion in drylands, and glaciers melting across the globe (IPCC., 2014). South Asia is the home of most vulnerable countries to climate change, facing stripping impacts of climate change, and huge economic loss. Meanwhile, climate change seems to threaten environmental security due to shifts in the timing of ecological events, distribution of animal and plant species, upsurge scarcity of natural resources (water, energy, and food), public health concerns, social vulnerability, and its high economic cost (Spencer et al., 2017). Additionally, climate change might be the driver to national insecurity, regional instability, and conflict due to forced migration, and its impact on natural resource security (energy, water, and food) (James, 2016). Climate change hampers water scarcity, shocks to food production, violation of treaties, conflicts, and adverse impacts on livelihood and development in South Asia (Hassan et al., 2017). Since, climate change is affecting the communities at both local and regional levels in South Asia.

Singh et al. (2017) had identified shortcomings in climate change-induced vulnerability assessment in South Asia as well as the least attention to vulnerable systems, vulnerable communities, causes, environmental insecurity, funding, the adaptive capacity of systems, and implementation priority. The greenhouse gas (GHG) emissions (carbon dioxide, nitrous oxide, methane, and water vapors) from both natural and anthropogenic sources are major drivers of the greenhouse effect, global warming, and climate change in Pakistan (Ahmed et al., 2017; Hassan et al., 2013). Zaman et al. (2017) interlinked climate change with natural resources consumption, energy, GHG emissions, pollution, and environmental protection in Pakistan. Wilson et al. (2017) pointed out the grappling impact of climate change on water and environmental security including more frequent natural disasters, glaciers melting, damage to Hindu Kush-Himalaya (HKH) water tower, effects on headwater of major rivers of the continent, and transboundary water concerns and conflicts. According to Ur-Rehman et al. (2017), climate change has particularly affected the agrarian economy because they are dependent on agriculture and livestock for income generation. A decrease in milk production, crop failures, and effect on fisheries due to droughts and water scarcity have threatened sustainable access to natural resources, national economy, income, and employment opportunities. While Hassan et al. (2016) argued that natural disasters (flood, drought, heavy rainfall, storms, and cyclones), pests, plants diseases, water scarcity, and change in cropping season and patterns had reduced the productivity and profit as well as intimate threat to environmental and food security.

Spencer et al. (2017) contended that divesting impacts of climate change, its complex interactions with human, ecological, and physical systems will necessitate adaptive capacity, mitigation of climate change, and commute the future planning of states. Åhman et al. (2017) stressed developing nations to adopt the low-carbon development, decarbonization of the energy sector and industries to meet the targets of the Paris Agreement, energy efficiency, and effectively contribute to climate change mitigation. Hamilton et al. (Hamilton and Lubell, 2018) referred to analyze the regional or global dynamics, donor priorities, established governance structure, vibrant climate policy, and planning to mitigate the heterogeneous impacts of climate change. At the same time, Lawrence and Haasnoot (2017) consider climate uncertainty, socioeconomic conditions, and static adaptive decisions as challenges in the implementation of climate policy and plans. However, Underdal (2017) reasoned to evolve political feasibility, lags between environmental impacts and mitigation measures, divergence in views of polluters and vulnerable as well as conflicts between a region on shared natural resources to cope with the climate change governance challenges. It may be synthesized from the above discussions that the previous studies from South Asia focused on climate change-induced vulnerability, drivers of climate change, impacts, and significance of good climate change governance. But, comparative vulnerability assessment, analysis of climate change policies and plans, their coherence with the Paris Agreement, Sustainable Development Goals (SDGs), and Thimphu Statement on Climate Change, the relationship of climate change with environmental security, neutral negotiations for collective mitigation, and a quest for environmental diplomacy as pursuance to cooperation are seldom investigated. Additionally, climate change has endured divesting impacts on agriculture, energy, and water resources which are threatening national security (from hardcore conflicts and national security), human security, and food security. Hence, this article has addressed these particular knowledge gaps regarding climate change in a regional context to strengthen climate change governance, conflicts, environmental security, development, and collective mitigation efforts.

4.1 Shreds of evidence of climate change in South Asia

The Organization for Economic Cooperation and Development (OECD) reported 1%–3.3% of global annual gross domestic product (GDP) as a cost of climate change-induced impacts by 2060 due to an increase in temperature of 1.5 °C by the end of this century (OECD., 2015). However, a temperature increase of 3–6 °C above the baseline is projected in South Asia by 2100. The region has already evident 2 °C increase in Hindu Kush. Meanwhile, the rise in temperature and sea level, changes in precipitation, frequent extreme weather events, and glacier melting in the region is threatening energy, water, food, environmental security, infrastructure, public health, and development (Vinke et al., 2017). The comparison of climate change scenario and vulnerability in South Asia is presented in Table 12.1.

Hence, vulnerability assessment of climate change, judicious climate change policies and plans, coherence of state initiatives with SDGs and Paris Agreement, urgent practical mitigation measures, strong institutions, and effective regional efforts are required to respond and tackle climate change at a local, regional, and global level. However, there are knowledge gaps in comparative vulnerability assessment, quantitative analysis of climate change policies and plans, capacity mapping of institutions, and neutral negotiations for collective mitigation of climate change. This chapter has entailed climate change scenario in South Asia, the analysis of climate change policies and plans in South Asia, their coherence with the Paris Agreement, SDGs and Thimphu Statement on Climate Change, the relationship of climate change with environmental security, and unfolded quest for environmental diplomacy as pursuance to cooperation.

5. Approach toward the development of case study

5.1 Multi-criteria decision analysis framework

The climate change policy domain has endured high conflicts and controversies among political parties, states, government agencies, universities, and environmental organizations for decades. This disagreement among stakeholders on climate change policy domain prestige impact of climate change, conflicts, environmental damage, economic cost as well as damage the mitigation efforts (Jenkins-Smith et al., 2014). In this context, Multi-criteria Decision Analysis (MCDA) offers considerable support in the development of such complex and multidisciplinary climate change strategies involving complex interactions with natural systems, many stakeholders, immediate response, and cooperation (Michailidou et al., 2016). The experts of different academic and professional backgrounds (climate change experts, international relation experts, government representatives, environmentalists, NGO representatives, academia, independent observers, and economists) have been selected to elicit a valid response. The local experts were approached personally, but experts from other countries were approached via email or Skype through different platforms like South Asian Association Regional Cooperation (SAARC) Energy Center, South Asia Co-operative Environment Programme (SACEP), and Center of Excellence in Environmental Studies, King Abdulaziz University, Jeddah. The response rate of experts from India and Bangladesh was initially low when they were contacted directly, but they respond positively when they were contacted through different forums. The experts rated the significance and existing consideration of individual criteria in climate change policies, plans, regional commitments, and the Paris Agreement linked with environmental security and development in the context of Pakistan and South Asia. The compatibility of existing ranking was different in climate change policies, plans, and agreements due to differences in consideration, priorities, and goals of each state in the context of the Paris Agreement and environmental security. The data were interpreted using significance and existing ranking of each criterion and Pearson correlation between significance and existing ranking. The diligence of MCDA for climate change policies, plans, impacts, conflicts, mitigation, and adaptation is manifested from various studies (Iqbal et al., 2021; Barker, 2017; Daksiya et al., 2017; Runfola et al., 2017; Mi et al., 2017). Although South Asian countries had established and implemented climate change policies and plans (Table 12.2), Dhaka Declaration on Climate Change

Table 12.1 The climate change scenario and vulnerability in South Asia (GOB., 2005; GOB., 2009; GOI., 2008; GoIRA., 2009; GoM., 2005; GoM., 2007; GON., 2010; GON., 2011; GoP., 2015; GOSL., 2015; RGOB., 2006).

Country	Emissions	Temperature rise	Precipitation	Sea-level rise	Extreme weather events	Glacier	Focused/ Vulnerable sectors
Afghanistan	Greenhouse gas (GHG) emissions are increasing, but exact calculations are not available.	1–2.6 °C during 1976–2099 (Aich et al., 2017) 2 °C increase in Hindu Kush region	8%–10% increase is projected in Hindu Kush region in next decades.	Nil	Increase in frequency and intensity of floods, droughts, and frost events in different parts of the country. The country has faced more severe and longest droughts during 1998–2006.	Glaciers are melting, but exact calculations are not available.	Agriculture biodiversity Energy Forestry and rangelands Health Livelihood Water resources
Bangladesh	16.7–18.2 billion tons of CO_2 equivalent since 2005	0.03–0.5 °C during the last decades 1°C increase is expected by 2030	5% increase by 2030	4–6 mm/year during last decades 19 cm by 2030	Most vulnerable country to cyclones and sixth vulnerable country to floods (UNDP., 2004). Natural disasters eroded 73,552 ha. Floods hit one-quarter of the country every year and	Nil	Agriculture Disasters Environment Food security Health Industry Infrastructure Land and soil Livelihood Water resources Trade

					60% area once in each 4–5 yeas. Six severe floods in the last 25 years. More frequent cyclones and a severe cyclone in every three years. Floods and cyclones killed 41,000 and 46,000 people during 1970–98.		
Bhutan	Forest cover 72.5% of its total land area. Highest GHG emissions sequestration potential in the world.	Temperature is increasing, but exact calculations are not available.	An increase in precipitation has been observed, but exact details are not available.	Not available	More frequent flash flooding, landslides, and glacial lake outburst floods (GLOFs) are threatening hydropower generation (backbone of the economy). Five GLOF events in 1960, 1968, 1970, 1994, and 2015.	Glaciers are melting as wells resulting in glacial lakes. There are 2674 glacial lakes, among them, 562 are associated with glaciers. 24 lakes are dangerous for an outburst.	Agriculture biodiversity and forestry Health Livelihood natural disasters and infrastructure Water resources and energy
India	The GHG emissions are continuously increasing. Currently,	0.4 °C rise has been observed during the last decades. However, a	10%–12% precipitation increase in West Coast, Andhra	1.06–1.75 mm/year	More frequent, severe, and long floods, droughts, and	Glaciers are melting, but exact calculations	Agriculture Coastal area Disasters Ecosystems Energy

Continued

Table 12.1 The climate change scenario and vulnerability in South Asia (GOB., 2005; GOB., 2009; GOI., 2008; GoIRA., 2009; GoM., 2005; GoM., 2007; GON., 2010; GON., 2011; GoP., 2015; GOSL., 2015; RGOB., 2006).—cont'd

Country	Emissions	Temperature rise	Precipitation	Sea-level rise	Extreme weather events	Glacier	Focused/Vulnerable sectors
	1.02 metric tons of CO_2 per capita are emitted.	cooling trend has been reported in northwest India and some parts of South India.	Pradesh, and northwest India. A 6%–8% decrease in rainfall in North Madhya Pradesh, North Eastern India, some parts of Gujrat, and Kerala.		cyclones during the last 130 years. Increase in severe storms at the rate of 0.11 events per year. Increase in epidemics and diseases.	are not available.	Food security Forestry Health Livelihood Water resources
Maldives	Not available	1.5 °C increase by 2100 0.2–1.1 °C increase in sea surface temperature	No significant change in precipitation.	1.7 mm/year	Sea level is rising while 80% population is less than 1 m mean sea level. Increase in intense rainfall and cyclones. The tide height has been increased to 2.78–3.18m. Northern Maldives is being exposed to more frequent and severe storms.	Data are not available.	Agriculture Coastal zone management Critical infrastructure Fisheries Food security Human health Tourism Water

Nepal	0.25% contribution to global GHG emissions	0.06 °C increase during last decades 1.2 °C increase is projected by 2030	There is no clear trend about precipitation. However, a decline in premonsoon is observed in far and mid-Western Nepal. Meanwhile, there is a 5%–10% increase in precipitation in Eastern Nepal during winter as well as 15%–20% increase in summer month in the whole country.	Not available	Increase in frequency, intensity, and duration of floods, droughts, storms, soil erosion, and glacial lakes.	Glaciers are retaining at the rate of 10–60 m/year. The number of glaciers is increased by 9% in Nepal. There is a 20% decrease in glaciers area (Bajracharya et al., 2007). There are 26 dangerous glacial lakes.	Agriculture and food security Climate-induced disasters Forests and biodiversity Public health Urban settlement and infrastructure Water resources and energy
Pakistan	310 million tons of CO_2 equivalent CO_2 (54%) Methane (36%) Nitrous oxide (9%) Carbon monoxide (0.7%) VOCs (0.3%)	0.099 °C rise/decade Overall 0.47 °C increase from 1960 to 2010	Unpredictable trend. Decrease in rainfall during 1910–1950. Increase in rainfall during 1951–1962. Again, decrease in rainfall during 1963–1976. Increase in rainfall during 1977–1997. Decrease in precipitation during 1997-date.	1.1 mm/year Karachi coast (1.1 mm/year) Pasni coast (1.1 mm/year) Makran coast (1–2 mm/year)	Increase in frequency, intensity, and duration of floods, droughts, storms, heatwaves, severe cold wind, and glacial lakes. Consecutive and massive floods during 2010–2014. Recent severe drought in Thar. More heatwaves and	Glaciers are melting in Hindu Kush-Himalaya. A new 1100 m^2 lake associated with the Hinarchi glacier is clear evidence. There is an upward rise of 1 km in the snowline during the last 25 years. About 2500 glacial lakes, 52 of which	Agriculture and livestock The arid and hyperarid area Biodiversity and ecosystems Coastal areas Disasters Forestry Human health Livelihood Mountain areas Rangelands and pastures Water resources and energy Wetlands

Continued

Table 12.1 The climate change scenario and vulnerability in South Asia (GOB., 2005; GOB., 2009; GOI., 2008; GoIRA., 2009; GoM., 2005; GoM., 2007; GON., 2010; GON., 2011; GoP., 2015; GOSL., 2015; RGOB., 2006).—cont'd

Country	Emissions	Temperature rise	Precipitation	Sea-level rise	Extreme weather events	Glacier	Focused/ Vulnerable sectors
					high temperatures in the early summer season.	are dangerous. The country has experienced GLOFs such as Booni Gole Glacier outburst, Attabad Lake, and Passu lake outburst (Rasul et al., 2012).	
Sri Lanka	0.15–0.20 metric tons/ capita increase in CO_2	0.017–0.026 °C increase per year during 1996–2001 0.5 °C increase is projected from 2010 to 2039 2–3 °C increase is projected from 2070 to 2099	No change in North-East monsoon. Mean annual precipitation has been decreased by 7%.	1–3 mm/year	Increase in consecutive dry days and decrease in wet days. Expansion of dry zones. Increase in the number of thunder days and 10%–20% increase in cyclones. Increase in frequency and intensity of droughts, floods, and landslides.	Not available	Biodiversity and coastal resources Export Food security and water Health and human settlement Industry and energy Infrastructure Tourism

Table 12.2 Overview of climate change-related policies, plans, and institutional framework in South Asia.

Country	Climate change policies and plans	Institutional framework
Afghanistan	National Capacity Needs Self-assessment for Environmental Management (NCSA) National Adaptation Programme of Actions for Climate Change (NAPA) 2009	National Environmental Protection Agency (NEPA) Ministry of Agriculture, Irrigation, and Livestock (MAIL) Afghanistan National Disaster Management Authority Ministry of Water and Energy Department of Meteorology
Bangladesh	National Environmental Policy 1992 National Adaptation Programme of Actions (NAPA) 2005 Vision 2021 Bangladesh Climate Change Strategy and Action Plan 2009	Ministry of Environment and Forest National Disaster Management Council Meteorological Department
Bhutan	Bhutan 2020—A vision to peace Bhutan National Adaptation Programme of Actions (NAPA) 2006 National Environment Strategy 2015	The National Environment Commission Ministry of Industry and Trade Bhutan Water Cooperation Partnership Department of Agriculture Department of Energy Department of Hydro-met Services
India	National Environmental Policy 2006 National Adaptation Programme of Actions (NAPA) 2008 National Action Plan for Climate Change and Health 2016	Ministry of Environment, Forest and Climate Change Prime Minister's Council on Climate Change Department of Meteorology Ministry of Water Resources Ministry of Power Ministry of Agriculture and Farmers Welfare Ministry of Health
Maldives	National Adaptation Programme of Actions (NAPA) 2007 National Sustainable Development Strategy 2009 Maldives Climate Change Policy Framework 2015 Vision 2020 Environmental Protection and Preservation Acts Third Environmental Action Plan	Ministry of Environment, Energy and Water Ministry of Economic Development and Trade
Nepal	National Adaptation Programme of Actions (NAPA) 2010 Climate Change Policy 2011 Nepal Environmental Policy and Action Plan 1993	Ministry of Environment Climate Management Division Water and Energy Commission Ministry of Energy Ministry of Water Resources

Continued

Table 12.2 Overview of climate change-related policies, plans, and institutional framework in South Asia.—cont'd

Country	Climate change policies and plans	Institutional framework
Pakistan	National Environmental Policy 2005 Climate Change Policy 2012 Framework for the Implementation of Climate Change Policy 2014–2030 The Pakistan Climate Change Act 2017 Pakistan Vision 2025	Ministry of Climate Change Provincial Environmental Protection Agencies Ministry of Water Resources Ministry of Energy Planning Commission of Pakistan Ministry of Planning, Development and Reforms Ministry of National Food Security and Research Pakistan Meteorological Department
Sri Lanka	National Environmental Policy 2003 The National Climate Change Adaptation Strategy for Sri Lanka 2011–2016 The National Climate Change Policy of Sri Lanka 2012 National Adaptation Programme of Actions (NAPA) 2015	Climate Change Secretariat Ministry of Mahaweli Development and Environment Ministry of Power and Energy Ministry of Agriculture

2008, and Thimphu Statement on Climate Change 2010, they are not as palpable to cope with the challenge of climate change. Consequently, MCDA was applied to climate change policies and plans in South Asia and assess its relationship with environmental security and development. MCDA comprehending 32 technical, sustainability, and political criteria had been established in this chapter to highlight the knowledge gap in climate change policies and plans, coherence with SDGs and Paris Agreement, and effectiveness of institutions in South Asia. Additionally, the evaluation gauge (1–6) for each criterion has been established in this article to clarify understandable and meaningful ranking.

5.1.1 Identification of goals

Climate change has reshaped the social, economic, environmental, development, and political goals worldwide. It is considered a potential driver of change in energy, water, food, consumption, development, and stability dynamics. This study has set a goal to analyze climate change policies and plans in South Asia, assess gaps and their relationship with environmental security, recommend policy implications and rationalize environmental diplomacy.

5.1.2 Selection of criteria

The Intergovernmental Panel on Climate Change (IPCC) has stressed both developing and developed countries to ponder economic instruments, regulatory mechanisms, public health, international agreements, and government programs in their climate change policy for mitigation and adaptation of climate change (IPCC., 2014). Thirty-two different technical, economic, environmental, social, and political criteria (Fig. 12.1) were selected from the Fifth Assessment Report of IPCC (summary to policymaker) and various research studies (IPCC., 2014; Iqbal et al., 2021; Barker, 2017; Daksiya et al., 2017; Runfola et al., 2017; Mi et al., 2017).

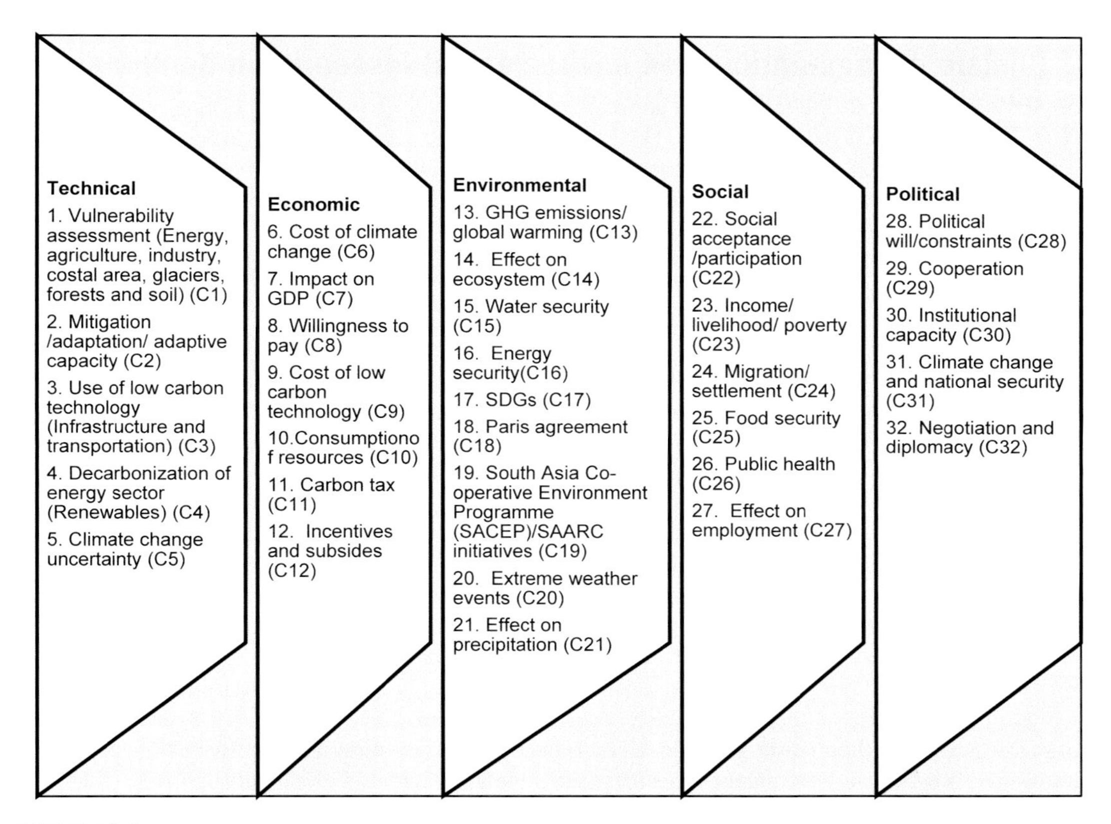

FIGURE 12.1

Evaluation criteria for analysis of climate change policies and plans in South Asia.

5.1.3 Weight determination and ranking

Each selected criterion has a different significance and contribution to climate change policies and their adaptive capacity. There is an essence to assign weight, rank, and normalize each criterion selected for climate change policies and plans (Senapati and Gupta, 2017). The study employed a simple multiattribute rating technique (SMART) to rank, weight, and normalize selected criteria. Different 40 experts have been selected from different backgrounds including national and regional climate change experts, international relation experts, government representatives, environmentalists, NGO representatives, independent observers, and economists. They were asked to weigh how well each country has done in performing discrete tasks or how important different facets regarding climate change and its divesting impacts were addressed in climate change policies and plans. They have to endorse significance ranking out of 100, begin with 10 (minimum rating) to least considered criterion, and increment the rank with increasing significance. The evaluation gauge (1. Absent (10 (minimum rating)), 2. Least considered (11–30), 3. Partially considered (31–50), 4. Moderately considered (51–70), 5. Strongly considered (71–90), 6. Very strongly considered (Above 90)) for each criterion has been established in this article.

6. Climate change-anticipated environmental insecurity in South Asia

The SDGs urge the states to take urgent action to combat climate change (Goal 13) and make climate change resilient and sustainable cities (Goal 11) (The United Nations (UN)., 2015). Similarly, the United Nations Framework Convention on Climate Change (UNFCCC) has established a global action framework on climate change to halt the increase in average global temperature below 2 °C above the preindustrial level in the form of Paris Agreement (UNFCCC., 2015; UFCCC., 2017). This chapter has assessed the relationship of climate change mitigation efforts in South Asia with both SDGs and Paris Agreement obligations.

6.1 Technical aspects

The technical aspects comprise vulnerability assessment, adaptive capacity, use of low-carbon technology (industry, infrastructure, and transport), decarbonization of the energy sector, and climate change uncertainty. Clause 3 of Thimphu Statement on Climate Change and article 2 of the Paris Agreement accented on member states to assess the climate change vulnerability (SAARC., 2010; SAARC., 2008). The experts reveal vulnerability assessment (C1) as an essence to mitigate and adapt climate change as well as identify threats to the environment. The National Climate Change Policy of Pakistan and Climate Change Framework for Implementation of Climate Change Policy had identified the agriculture, livestock, arid area, hyperarid area, biodiversity, ecosystems, coastal areas, forestry, livelihood, public health, mountain areas, wetlands, rangelands, water, and energy sectors more vulnerable to climate change. Similarly, all National Adaptation Programme of Actions and policies in the region had dispensed the vulnerability assessment (Table 12.3, and Fig. 12.2) and identified vulnerable sectors in their country (Table 12.1). It is notable to mention that biodiversity, ecosystems, agriculture, forests, soil, water, and energy are among the vulnerable sectors in all countries, which are apprehending threats to environmental security.

Article 2 (b) (increase inability to adapt), article 6 (mitigation and adaptation mechanism), and article 7 (mitigation, adaptation, and adaptive capacity) of the Paris Agreement provide comprehensive guidelines for mitigation, adaptation, and development of adaptive capacity (UFCCC., 2017). Hence, experts granted the significance of 80 and 75 to mitigation, adaptation, and adaptive capacity (C2) in national and regional contexts, respectively. The results had expressed the identification of mitigation and adaptation measures for vulnerable sectors in all policies, frameworks, and national adaptation plans in the region (Table 12.3, Fig. 12.2). However, all South Asian countries are struggling to adapt to climate change due to a lack of data, low institutional capacity, and financial concerns. The insufficient adaptive capacity is harnessing adverse impacts of climate change and environmental insecurity.

According to Paris Agreement, article 4(19) and article 10 broadly sermon the use of low-carbon technology, low-carbon development, decarbonization of energy sector, and transfer of low-carbon technology to developing countries and finical support (UFCCC., 2017). Similarly, clause 5 of Thimphu Statement on Climate Change 2010 and Dhaka Declaration on Climate Change 2008 accentuated the SAARC members to promote low-carbon technology (SAARC., 2010; SAARC., 2008). The experts support the use of low-carbon technology (C3) and decarbonization of the energy sector (C4) for climate change mitigation. The results exhibited that climate change policies and plans in the region except for Afghanistan and Bhutan (least considered) had moderately or strongly undertaken the promotion and use of low-carbon technology and energy generation (Table 12.3, Fig. 12.2). Meanwhile, all countries tinted lack of domestic manufacturing, skills, financial resources,

Table 12.3 An evaluation of current consideration of selected criteria climate change policies/plans in South Asia.

Criteria		AFG	BND	BTN	IND	MDV	NPL	PAK	SLK
Technical	C1	4	5	4	5	5	5	4	5
	C2	3	4	4	5	5	4	4	5
	C3	1	4	3	4	5	4	4	4
	C4	2	4	4	5	5	3	5	4
	C5	3	4	4	4	4	4	4	4
Economic	C6	1	4	3	2	4	2	2	3
	C7	2	4	4	3	5	4	3	4
	C8	1	3	4	3	3	3	1	2
	C9	1	2	1	1	1	1	1	1
	C10	1	4	3	3	4	4	3	3
	C11	1	1	1	1	1	1	1	1
	C12	1	3	1	3	3	2	2	3
Environmental	C13	2	4	4	5	5	5	5	5
	C14	4	4	4	5	5	4	5	5
	C15	4	4	5	5	5	4	5	5
	C16	3	4	4	5	4	4	5	4
	C17	4	4	4	4	4	5	4	4
	C18	1	2	2	4	4	4	3	3
	C19	1	2	2	3	2	3	3	3
	C20	2	5	4	4	5	4	4	5
	C21	4	4	4	4	4	4	1	4
Social	C22	4	4	4	4	4	4	4	4
	C23	4	5	4	4	5	4	4	5
	C24	1	4	3	4	5	5	3	5
	C25	3	5	4	4	5	4	4	5
	C26	2	4	3	5	4	4	4	4
	C27	1	3	3	3	4	4	3	4
Political	C28	4	4	4	4	5	5	4	4
	C29	2	3	3	3	5	3	3	3
	C30	3	4	4	4	4	4	4	4
	C31	1	2	3	4	4	3	1	3
	C32	2	4	4	4	4	4	4	4

(1. Absent (10), 2. Least considered (11–30), 3. Partially considered (31–50), 4. Moderately considered (51–70), 5. Strongly considered (71–90), 6. Very strongly considered (Above 90)).

research, and development as potential barriers to use low-carbon technology and meet the targets of the Paris Agreement. Energy-intensive industries are facing defy of innovations, technical concerns, meeting global marketing trends, high mitigation costs, and provisions of the Paris Agreement. The consideration of low-carbon development and decarbonization of the energy sector in climate change

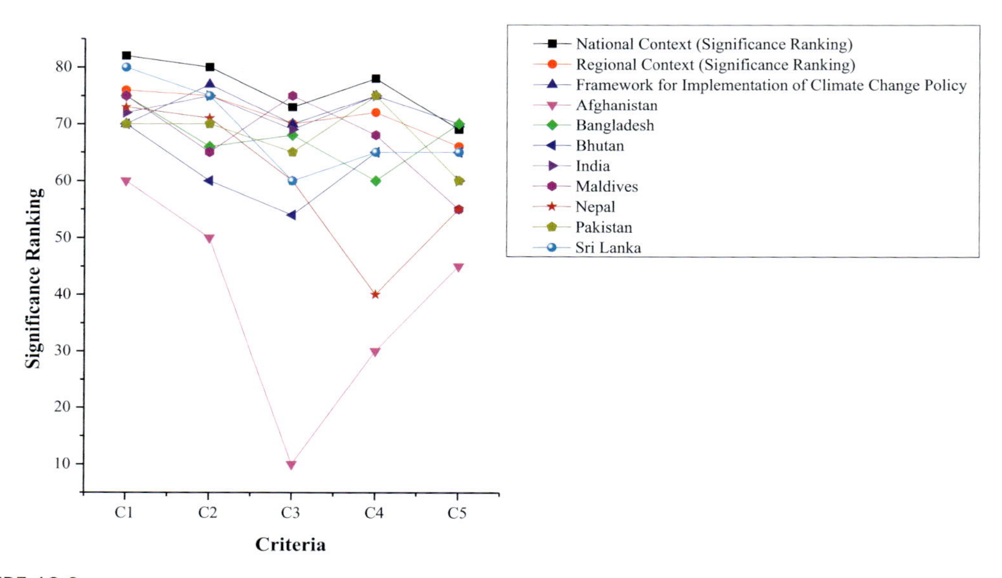

FIGURE 12.2

The ranking of technical criteria in climate change policies and plans in national and regional context.

policies and framework implicitly support the high reduction in GHG emissions and mitigation efforts. Perhaps, these barriers have slackened the shifting of the energy-intense system to low-carbon development and mitigation to climate change.

Sustainable economic development urges the establishment of adaptive plans to respond to uncertain environmental and socioeconomic changes. The experts suggested invoking the climate change uncertainty (C5) in their policies. The existing ranking of climate change policies and plans demonstrated that regional states had moderately considered the climate uncertainty (Table 12.3, Fig. 12.2). Although member states assert uncertainty in their adaptation plans, their initiatives and response to cope with these uncertainties were found to be insufficient, e.g., most of the countries in South Asia are being consistently hit by the flood for decades but still are unable to reduce the damages. Therefore, more evaluation of uncertainties and futuristic approach is obligatory to adapt to climate change and protect the environment. Overall, Maldives is more focused, but Afghanistan least contemplated on technical aspects of their climate change policies and plans among South Asian states (Table 12.3). The results represented a positive Pearson correlation between significance ranking and existing ranking of technical criteria in National Climate Change Policy (very strong) and Framework for Implementation of Climate Change Policy (moderate) (Table 12.4). Alike the national context, the climate change policies and plans had expressed a positive Pearson correlation between significance ranking and existing ranking of technical criteria in a regional context (Table 12.5). Climate change has invoked a threat to environmental and national security across the world, but the technical aspects confront an indispensable role in mitigation and adaptation. Therefore, profound attention to policies and plans is required to embark on mitigation and adaptation.

Table 12.4 The correlation between significance ranking and existing ranking of selected criteria in climate change policies and plans in a national context.

Criteria	National climate change policy	Framework for implementation of climate change policy
Technical	0.81*	0.46
Economic	0.67	0.67
Environmental	0.50	0.52
Social	0.71	0.55
Political/institutional	0.16	0.68

*(***P <.01, **P <.05, *P <.1) (Very weak = 0.00–0.19, Weak = 0.20–0.39, Moderate = 0.40–0.59, Strong = 0.60–0.79, Very strong = 0.80–1.0).*

Table 12.5 The correlation between significance ranking and existing ranking of selected criteria in climate change policies and plans in South Asia.

Criteria	Afghanistan	Bangladesh	Bhutan	India	Maldives	Nepal	Pakistan	Sri Lanka
Technical	0.44	0.12	0.25	0.85*	0.61	0.53	0.74	0.78
Economic	0.24	0.93***	0.43	0.53	0.65	0.62	0.71*	0.56
Environmental	0.36	0.35	0.67**	0.84***	0.64*	−0.20	0.58	0.54
Social	0.11	0.07	0.4	0.36	0.15	−0.01	0.50	0.38
Political	0.75	0.60	0.75*	0.56	0.5	0.83*	0.52	0.68

*(***P <.01, **P <.05, *P <.1) (Very weak = 0.00–0.19, Weak = 0.20–0.39, Moderate = 0.40–0.59, Strong = 0.60–0.79, Very strong = 0.80–1.0).*

6.2 Economic aspects

The economics of climate change is complex but cardinal in the design of climate change policy. As global environmental challenges particularly GHG emissions, climate change, and its adaptation robustly affected growth per capita income. The cost of climate change, impact on the GDP, cost of low-carbon technology, consumption of resources, carbon tax, incentives, and subsidies are major economic aspects of climate change. Climate change is causing the bulk of economic cost and damage. The economic cost of climate change includes the cost of damage and the cost of mitigation. Hsiang et al. (2017) estimated the increase in an annual loss by 0.6% GDP per 1 °C rise in temperature. The cost of damage resulted from the impact of extreme weather events on infrastructure, tourism, agriculture, glacier melting, energy, water, livelihood, health, and environment. Meanwhile, article 8 of the Paris Agreement evoked the parties to recognize the loss due to climate change (UFCCC., 2017). The experts also referred to mediate the cost of climate change (C6) and impact on GDP (C7) in both national regional contexts. The results had evinced that the National Change Policy and framework in Pakistan deem both criteria to a lesser extent. The framework referred to the loss of more than 9.6

billion US dollars in the flood of 2010 only (GoP., 2015). Bangladesh Climate Change Strategy and Action Plan 2009, Maldives Climate Change Policy Framework 2015, and their plans moderately cogitated the cost of climate change among regional states (Table 12.3). The government of Bangladesh has endured the loss of more than 10.88 billion US dollars during 1974–2007 by the only flood and invested more than 10 billion US dollars in climate-resilient infrastructure (GOB., 2009). Climate change has induced the cost of more than one billion US dollars to the tourism sector and 300 million dollars combined cost during Tsunami in the Maldives (GoM., 2007). However, other countries in the region undermine the cost of climate change (Table 12.3, Fig. 12.3). Furthermore, the climate change policies and plans have identified the impact of climate change on GDP except for Afghanistan, India, and Pakistan (less attention) (Table 12.3, Fig. 12.3). The experts stressed to behold willingness to pay (C8) in climate change policy, but pointed out a pathetic response in the current socioeconomic condition of South Asia. The results confirmed that South Asian countries merely or partially reflected the willingness to pay in their climate change policies (Table 12.3, Fig. 12.3). There is a dire need to rationalize climate change damage and economic management of climate as an integral part of climate change policy. The climate change mitigation and adaptation efforts are anticipating participation and willingness to pay by states and as an emitter. Despite implementation of climate change policies, the developing countries are struggling to control increasing GHG emissions due to a lack of low-carbon technologies.

Low-carbon technologies involve high costs, but developing countries are unable to afford the cost without support due to limited financial resources. For this purpose, the developed nations will voluntarily assist the developing countries under article 9 of the Paris Agreement (UFCCC., 2017). Similarly, the SAARC members decided to develop a mutual fund under the Dhaka Declaration on Climate Change 2008. The experts visualized to ponder the cost of low-carbon technology (C9).

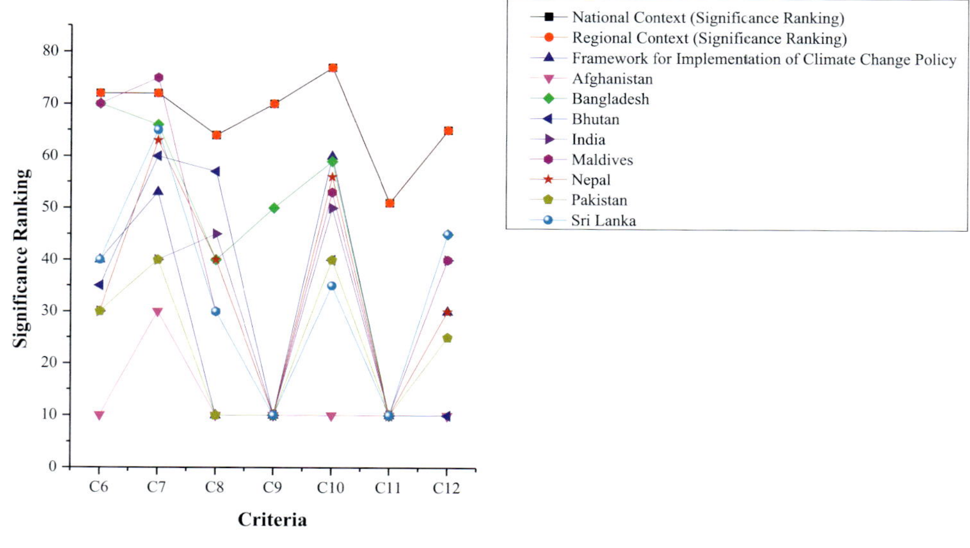

FIGURE 12.3

The ranking of economic criteria in climate change policies and plans in national and regional context.

The results had shown that the South Asian countries did not estimate the cost of low-carbon technologies and esteem low significance (Table 12.3 and Fig. 12.3). However, all regional states had affirmed the lack of financial resources and looking for financial assistance from Clean Development Mechanism (CDM), Global Environment Facility (GEF), Green Climate Fund (GCF), and support under the Paris Agreement. However, the lack of financial resources will affect the efforts to mitigate climate change and threaten environmental security.

The prices of fossil fuels, energy demand, GHG emissions, and natural resources consumption are key players of policy agenda and environmental sustainability. The experts granted significance ranking (77) to consumption patterns (C10) in both national and regional contexts. The climate change policies and plans in the region except Afghanistan had conceived the modestly (Bhutan, Pakistan, and Sri Lanka) or moderately (Bangladesh, India, Maldives, and Nepal) consumption pattern (Table 12.3 and Fig. 12.3).

Carbon taxes and prices promote carbon-efficient technologies, adjust relative prices, internalize global warming, reduce abatement cost, ensure energy conservation and efficiency, increase cooperation, thematic criteria in climate change policy, harmonize regulations and environmental concerns. While other argued that carbon taxes to reduce GHG emissions will put the burden on low-income groups as well as increase the price of electricity unless the state rationalizes redistribution of income, transfers payments and variation in tax, or introduces incentives and subsidies. The experts supported the second argument and rated low significance to a carbon tax (C11) as compared to incentives and subsidies (C12). Besides, they argued that almost 50% of people in South Asia are living below the poverty line and less aware of climate change; therefore, they will resist paying the carbon tax and be affected by energy prices. Meanwhile, governments are unable to rationalize redistribution of income due to fragile governance, ineffective policies, and regulations. However, they suggested the phasing out of GHG emissions by introducing transition time, incentives, and subsidies. The results had revealed the dearth of the carbon tax but modest consideration of incentives and subsidies in climate change policies and plans in South Asia (Table 12.3 and Fig. 12.3). The immaterial contemplation of both criteria will affect the implementation of policies and plans. In sum, Maldives had more focus, but Afghanistan had least pondered economic aspects in their climate change policies and plans (Table 12.3). The results conferred diverse positive Pearson correlation between significance ranking and existing ranking of economic criteria of climate change policies and plans in both national and regional context (Tables 11.4 and 11.5). Overall, the economic criteria are partially considered in climate change policies and in both national and regional contexts which will potentially threaten their effectiveness, sustainability, and environmental security.

6.3 Environmental criteria

Climate change has intensified the global environmental changes including GHG emissions, rainfall, scarcity of natural resources (energy, water, and food), biodiversity, habitat, and ecosystems loss. Therefore, SAARC members agreed to establish a regional CDM in Dhaka Declaration on Climate Change 2008 (SAARC., 2010). Alike, articles 4 and 5 of the Paris Agreement adjudicate the parties to focus on assessment and mechanism for abatement of GHG emissions. Meanwhile, parties declared to hold mean global temperature rise below 2 °C above preindustrial level (UFCCC., 2017). The experts signaled the GHG emissions as a potential threat to environmental security and connoted to insinuate GHG emissions, their abetment, and global warming (C13). The results conferred the climate change

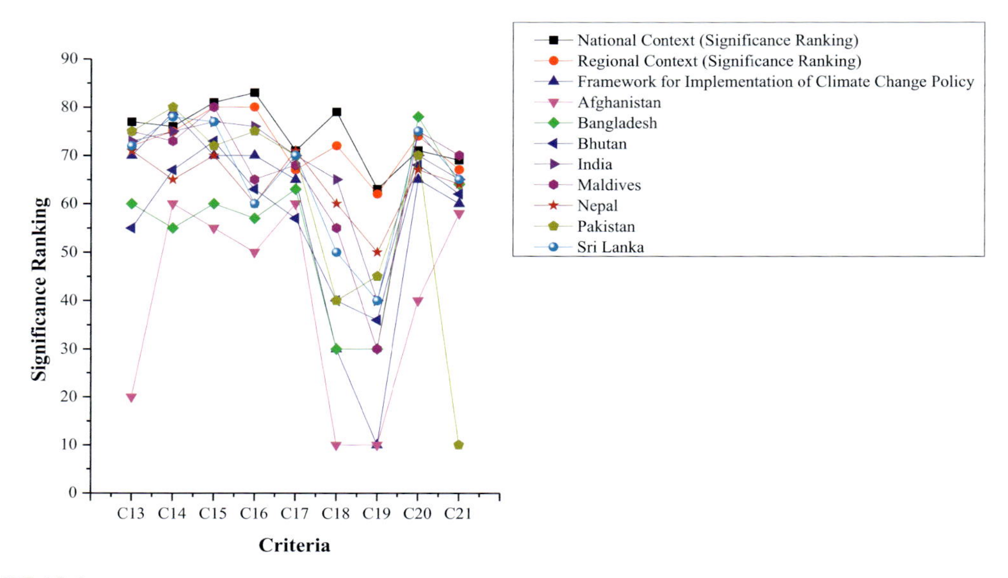

FIGURE 12.4

The ranking of environmental criteria in climate change policies and plans in national and regional context.

policies and plans in the region except for Afghanistan moderately or strongly dictated GHG emissions, temperature rise (Table 12.2), and their abatement measures (Table 12.3 and Fig. 12.4). However, their effectiveness and implementation are confronting various problems including governance issues, human, technical, and financial limitations.

Clauses 10, 11, and 12 of the Thimphu Statement on Climate Change and article 7 (2) of the Paris Agreement particularly focused on biodiversity and ecosystems under a change in climate context (UFCCC., 2017; SAARC., 2010). The experts had granted a high significance ranking to the effects of climate change on ecosystems (C14). The results had demonstrated that the effect of climate change on the ecosystem had been thoroughly addressed in all climate change policies and plans in the region (Fig. 12.4). However, the implementation of mitigation and adaptive measures to hold the impact of climate change on biodiversity and ecosystems are still awaited, consistent decrease, extinction of many species, and habitat loss deposed climate-induced environmental insecurity.

The energy-water-climate change nexus has reshaped the discourse of decision-making, planning, management of natural resources, consumption patterns, and economic development. The experts had assigned an outstanding significance ranking to water security (C15) and energy security (C16) in climate change policies and planning in both national and regional contexts. They proposed to integrate the energy, water, and climate change nexus. The results exhibited the moderate or strong consideration of water security in climate change policies and plans in the region (Table 12.3). However, energy security was found to be mince in climate change policies and plans in South Asia (Fig. 12.4). The experts also criticized the governments in the region for straggling efforts to water and energy security as presented in climate change policies and plans.

The SDGs provide a pathway to global sustainable development during 2015—30. The experts are valued to discuss the SDGs (C17) in climate change policies and plans. Although, all climate change policies and plans in the region except the National Action Plan for Climate Change and Health 2016 of India (encompass SDGs) were drafted and implemented before SDGs. Hence, SDGs were not discussed (Table 12.3), but they had esteemed the urgent measures to combat climate change (Fig. 12.4) because it was also a part of Millennium Development Goals. They had also focused on interlinked sectors in their objectives including agriculture, energy, water, livelihood, infrastructure, biodiversity, and ecosystems. Meanwhile, the success of goal 13 (urgent measures to combat climate change) can assure the achievement of other goals including goal 2 (zero hunger by reducing risk to agriculture), goal 3 of good health and well-being (reducing the negative health effects of air contamination), goal 6 of availability and sustainable water management (reducing the risk to water resources), goal 7 of affordable and clean energy for all (ensure water supply for power generation, reducing the risk of climatic variation and structural changes), goal 11 (sustainable cities), goal 14 (conservation of marine resources), and goal 15 (conservation of terrestrial ecosystems). In the current scenario of climate change, Paris Agreement is assumed as a cumulative positive step to reduce global GHG emissions, address climate change, and be more relevant in the implementation of climate change policy.

The Paris Agreement demonstrated the common global ambition to combat climate change, its adaptation, and more support to developing countries. The agreement aims to clasp average global temperature below 2 °C above the preindustrial level, intensify the capability of developing countries, suitable fiscal support, transfer of technology, robust transparent framework, and capacity building of developing as well as vulnerable countries to deal with climate change (UFCCC., 2017). The experts dispensed high significance ranking to adjure the Paris Agreement (C18) in climate change policies and plans. However, the results indicated the climate change policies and plans in the region had partially mediated the Paris Agreement except for Afghanistan (did not consider) in terms of global support to combat climate change (Table 12.3 and Fig. 12.4), perhaps, the climate changes policies and plans in region were formulated and implemented before the Paris Agreement. Furthermore, the assessment of “nationally determined contributions (NDCs)” to climate change mitigation is still awaited in some countries in the region. This will affect the harmonization of their efforts to global mitigation of climate change.

The South Asia Cooperative Environment Programme (SACEP) is an intergovernmental organization working in collaboration with the SAARC to promote and assist the protection and management of the environment in South Asia (SACEP., 2021). The SAARC has introduced SAARC Environment Action Pan 1997. Dhaka Declaration and Action Plan on Climate Change 2008, Delhi Statement on Cooperation in Environment 2010, Thimphu Statement on Climate Change 2010, SAARC Convention on Cooperation on Environment 2010, and SAARC Agreement on Rapid Response to Natural Disasters 2011 to protect the environment, abate climate change, and regional cooperation on the environment (SAARC., 2021). The experts supported regional environmental protection and cooperation through SACEP and SAARC (C19). Meanwhile, they pointed out the inadequacy, low technical capacity, and regional geopolitical narratives as barriers. The results revealed that the climate change policies and plans in the region had imparted low significance to SACEP and SAARC initiatives (Table 12.3 and Fig. 12.4). It was observed that Dhaka Declaration and Action Plan 2008 and Thimphu Statement on Climate Change are too much brief and inadequate to cope with the horrendous challenge of climate change. They are unable to develop diplomatic pressure to revisit and transform them into a comprehensive agreement.

The experts also evoked extreme weather events (C20) and change in precipitation (C21) as core criteria to reduce the consequences of climate change. The results had manifested the moderate consideration of extreme weather events in climate change policies and plans in the region (Table 12.3). However, the effects on precipitation were on climate change policies and plans except for the National Climate Change Policy of Pakistan (Fig. 12.4). The detail of effects of climate change on precipitation in South Asia is furnished in Table 12.1. Overall, the Government of India has more contemplated the environmental aspects among regional countries in the climate change context (Table 12.3). The results bestow the moderate positive Pearson correlation between significance ranking and existing ranking of environmental criteria in the national context (Table 12.4). Similarly, the positive Pearson correlation (weaker to very strong) between significance ranking and existing ranking of environmental criteria was exhibited in regional context except for Nepal (negative correlation) (Table 12.5). The modest attention to some environmental criteria and inappropriate implementation of environmental risks-related mitigation measures in climate change policies and plans seem to favor the climate change to escalate environmental degradation as well as threatens human and environmental security.

6.4 Social criteria

The social criteria such as public participation, social acceptance, impacts on income, livelihood, and poverty, migration, settlement, food security, public health, and effect of employment incorporate multifold trust and stakeholder concerns in climate change policy. The Thimphu Statement on Climate Change 2010 (clause 4 and 6) and the Paris Agreement (article 7(5) and article 12) emphasized the involvement of stakeholders and social acceptance (UFCCC., 2017; SAARC., 2010). The experts recommended considering social acceptance and participation (C22). The climate change policy and framework in Pakistan had moderately mediated social acceptance and participation. Alike, other climate change policies and plans in the region had also moderately addressed social acceptance and participation (Table 12.3 and Fig. 12.5).

Climate change has affected the livelihood of many people in developing countries. The Thimphu Statement on Climate Change 2010 (clause 12) and the Paris Agreement (article 7(2)) have visualized the impacts of climate change on livelihood, income, and poverty (UFCCC., 2017; SAARC., 2010). The experts granted significance to livelihood, income, and poverty (C23), migration and settlement (C24), and effect on employment (C25). The climate change policy and framework of Pakistan had moderately dealt with the livelihood, income, poverty, and effect on employment. However, they had paid modest attention to climate-induced migration and settlement. While, other climate policies and plans in the region had reflected moderate (Afghanistan, Bhutan, India, and Nepal) or strong (Bangladesh, Maldives, and Sri Lanka) attention to income, livelihood, and poverty. The climate policies and plans in the region except Afghanistan had entertained the migration and settlement, but all countries had addressed the effect of clime change on employment (Table 12.3 and Fig. 12.5). However, exact data of evaluation of impacts on livelihood, income, poverty, migration, settlement, and employment are not available in South Asian countries.

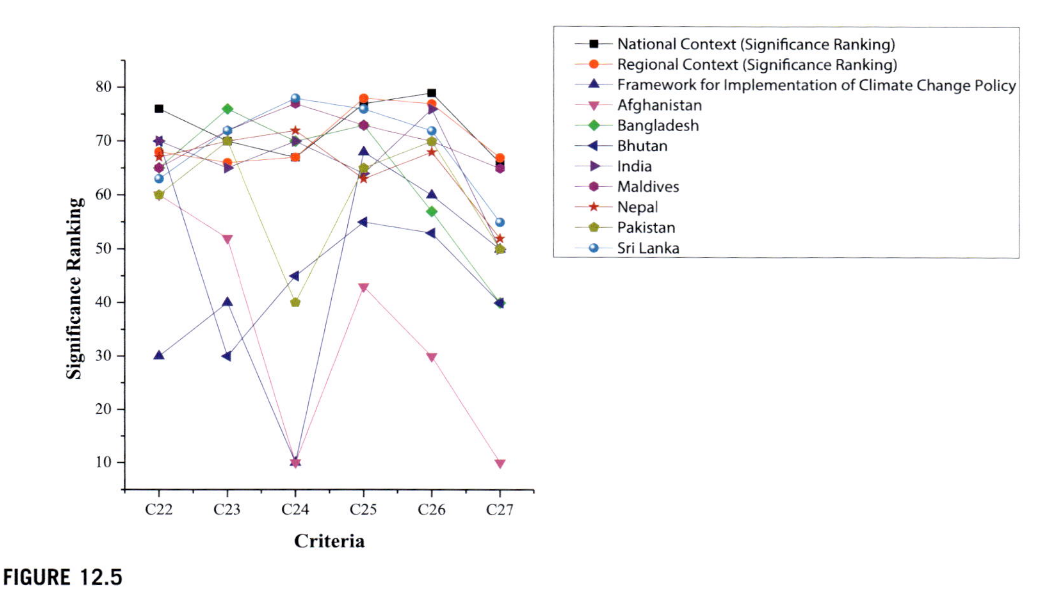

FIGURE 12.5

The ranking of social criteria in climate change policies and plans in national and regional context.

Agricultural productivity, crop, and food security have been adversely affected by climate change in many countries across the world. The experts administered high significance to food security (C25) in the climate change context of South Asia. The results expressed that South Asian countries are aware of the threat to agriculture and food security due to climate change (Fig. 12.5). The South Asian countries had partially (Afghanistan), moderately (Bhutan, India, Nepal, and Pakistan), or strongly (Bangladesh, Maldives, and Sri Lanka) considered climate change as a threat to food security in their policies and plans (Table 12.3). However, they have to implement the proposed mitigation measures for agriculture and food security as soon as possible.

Climate change has intensified the infectious disease dynamics, spread, distribution, seasonality, intensity, and transmission due to variation in temperature and precipitation. The experts rated high significance to public health (C26) in climate change policies. The results had indicated that public health was moderately or strongly pondered expect Afghanistan (least) and Bhutan (partial) as climate change-related concerns in climate change policies and plans (Table 12.3 and Fig. 12.5). However, the implementation of proposed mitigation measures is urgently required to handle public health. Overall, Maldives has more stress on the social aspects among regional countries in the climate change context (Table 12.3). The results divulged diverse positive Pearson except for climate change policy and plans of Nepal (negative) correlation between significance ranking and existing ranking of social criteria in the context of South Asia (Tables 11.4 and 11.5). The social dimension of climate change policies and plans in the region driving the lunge to contemplate social structure demographic groups, vulnerable locations, displacement, health, food security, and well-being to avoid the climate-induced conflicts for long-term socioeconomic and environmental sustainability.

6.5 Political criteria

The imminent threat of climate change has reshaped the political, public, and policy narratives as a discourse of environmental security. The goals of the Paris Agreement are not possible to achieve without cooperation from the aslant political spectrum. More focus on different political narratives such as political will, cooperation, international obligations, negotiation, and diplomacy is likely to frame better adaptation measures and achieve the climate goals. Articles 3 and 7(9) of the Paris Agreement postulated the parties to demonstrate the political will to combat climate change (UFCCC., 2017). The experts allocated an outstanding significance to political will (C28) and revealed it as an imperative climate change adaptation. The results evinced the moderate or strong demonstration of political will in the policies and plans in the region (Fig. 12.6), but experts portrayed the contradictory scenario for implementation of plans and secure committed obligations. They had criticized footling practical measures except for India, Maldives, and Nepal. These are eroding political will toward environmental concerns, climate change, and environmental security.

The Paris Agreement has opened a new arena of cooperation on climate change. Articles 3, 7(6), and 11 have particularly focused on cooperation between developed and developing nations on climate change in the form of capacity building, technical support, technology transfer, and financial cooperation (UFCCC., 2017). Meanwhile, clauses 2 and 9 of the Thimphu Statement on Climate have also riveted the regional cooperation on climate change (SAARC., 2010). The experts allotted an exceptional significance to cooperation (C29) for the success of climate change policies and plans in both national and regional contexts. The climate change policies and plans had considered regional cooperation on climate as nidus criterion for adaptation of climate change policy (Fig. 12.6). However, the governments are promoting cooperation on climate change to some extent at the national level, but

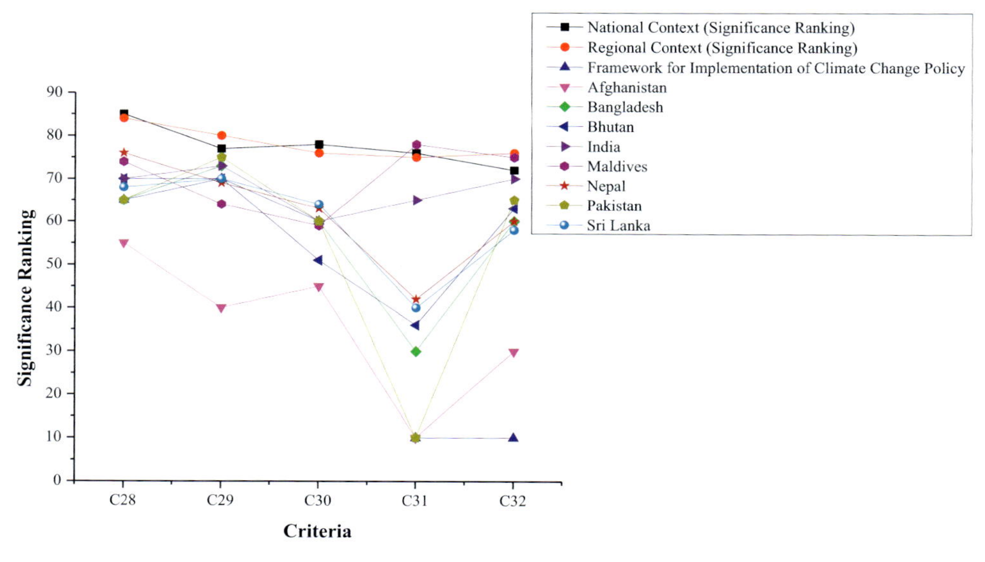

FIGURE 12.6

The ranking of political criteria in climate change policies and plans in national and regional context.

no significant proceeding was observed after Thimphu Statement on Climate Change 2010, SAARC Convention on Cooperation on Environment 2010, and SAARC Agreement on Rapid Response to Natural Disasters 2011 at the regional level (Table 12.3). The lack of cooperation and climate change at the regional level has threatened environmental security and development.

The Thimphu Statement on Change (clause 9) refers to develop institutional capacity (SAARC., 2010). In addition, the Paris Agreement (articles 6, 8 (c) and 11) (UFCCC., 2017) proposed to develop institutional capacity (C30) in developing countries, technical and financial support by developed countries. The results evident the moderate existence of an institutional framework for climate change in all countries except Afghanistan (weak institutional framework) (Table 12.3 and Fig. 12.6). However, the technical capacity was found low because many employees were from irrelevant academic backgrounds and experience, having limited knowledge of climate modeling, socioeconomic vulnerability assessment, climate variability and uncertainty, adaptation measure, fundamentals of climate change policy, sustainable natural resource management, and international agreement. This limited capacity will potentially threaten the implementation of policies, adaptation, and national and environmental security.

Climate change is directly linked with national security, and failure to manage climate can impair the global economy as well as destabilize the security landscape. Additionally, climate will have exacerbated the social, economic, political, and environmental instability with growing conflicts due to stripping impacts of climate change. The experts assimilated climate change as a matter of national security (C31). The climate change policy and plans in Afghanistan and Pakistan snubbed the relationship between climate change and national security. However, other climate change policies and plans in the region had partly or moderately excogitated the relationship between climate and national security (Table 12.3 and Fig. 12.6). The less implication of this criterion might endanger national sovereignty and environmental security in the region. The experts suggested utilizing negotiation and diplomacy (C32) to strengthening clime change policies, plans, and cooperation between national and regional entities. The results had shown that the climate change plans of the region had moderately mediated the negation and diplomacy except for Afghanistan (least considered) (Table 12.3 and Fig. 12.6). However, there is a need to initiate diplomacy and define a mechanism for negotiation at both national and regional to combat climate change in the long haul. Alike social aspects, Maldives has more emphasized the political aspects among South Asia in climate change policies and plans (Table 12.3). The results expressed a variable positive Pearson correlation between significance ranking and existing ranking of political criteria in the context of South Asia (Tables 11.4 and 11.5). The notion of political aspects develops concord among actors on mitigation measure, increases the effectiveness of policies and cooperation, and protects the environment.

7. Environmental diplomacy as pursuance to cooperation

Climate change has more threatened the developing countries, complexed their socioeconomic and environmental problems as well as exacerbated the conflicts. For instance, climate change is a federal subject, while formulation, implementation, and monitoring of compliance with environmental regulations is now a provincial subject after the 18th amendment in Pakistan. Many conflicts are mounting due to interposition in the jurisdiction of each other and unclear directions. Meanwhile, implementation of mitigation measures is a joint venture of the Ministry of Climate Change, provinces, and many ministries including Ministry of Water Resources, Ministry of Energy, Ministry of Food Security, Planning and Development Division, Provincial Environmental Protection Agencies, Ministry of Foreign Affairs, and Ministry of Industries along with some institutions including National Disaster Management Authority (NDMA), Water and Power Development Authority, National Energy Efficiency and Conservation Authority (NEECA), Pakistan Metrological Department (PMD), etc. There are numerous conflicts about jurisdiction, engagement of human resources, data availability, capacity, regulatory measures, set objectives, and financial resources.

Alike, the limited regional cooperation on climate change has been observed during the study. The regional declarations (Dhaka Declaration on Climate Change and Thimphu Statement on Climate Change) are incomplete, insufficient, and peanut capacity in a regional context. Furthermore, an intermittent debate on the consequences of climate change was noticed on regional forums. At the same time, the impacts of climate are more frequent and severely damaging the region (Table 12.1). Therefore, strong cooperation, negotiation, and diplomacy beyond game theory and territoriality can remedy the divesting impact of climate change in the context of Pakistan and South Asia. Thus, environmental diplomacy can cater to neutral, nonaggressive, and sustainability-based mechanisms to negotiate, settle disputes, and frame cooperation on climate change in both national and regional contexts. It is different and effective than traditional diplomacy in current geopolitical and historical narratives of South Asia due to its neutral characteristics, engagement of a variety of actors, and negotiation capacity based on sustainability to balance social, economic, political, and environmental narratives. Hence, environmental diplomacy can negotiate broader agenda of climate change, revisit regulatory and institutional frameworks, provide a pathway to comprehensive regional agreement or strategy, settle disputes, explicate national and regional cooperation on climate change to reduce poverty, sustainable development, and environmental security.

7.1 Institutional capacity for environmental diplomacy on climate change

The SAARC and SACEP can entertain the region for negotiation and cooperation on climate change. Alike, the global institutions such as the World Bank, the United Nations, the UNFCCC, the World Meteorological Organization, the United Nations Environment Programme (UNEP), International Union for Conservation of Nature (IUCN), and regional institutions like South Asian Network for Development and Environmental Economics (SANDEE), International Center for Integrated Mountain Development (ICIMOD), Asian Development Bank (ADB) in association with forums, the Economic Cooperation Organization (ECO), Association of South East Nations (ASEAN), and Shanghai Cooperation Organization (SCO) can diplomatically involve parties to put their efforts to discuss a concern, negotiation, settle conflicts, and harness long-term cooperation on climate change for sustainable future of South Asia.

8. Conclusions

A paradigm of dynamics of environmental security has been diversified from ecosystem restoration and conservation, pollution prevention, illegal movement of hazard waste, waste management, and access and availability of natural resources to political economy, human insecurity due to environmental migrations and disasters, scarcity of natural resources, social upheavals, equitable access to natural resources, climatic extremes, environmental sustainability, national security, and geopolitical concerns. The South Asian countries had initiated the climate change governance and implementation of regulatory mechanisms after the Kyoto Protocol, but a positive response is quite limited. The substantial technical (low-carbon technology and energy sector), economic (cost evaluation of climate change, impact on GDP, less willingness, cost of low carbon, carbon tax, incentives, and subsidies), environmental (Paris Agreement, less SACEP/SAARC initiatives, extreme weather events, and change in precipitation), social (migration, settlement, food security, public health, and employment), and political (political will, low institutional capacity, and national security) gaps had dwindled the effectiveness of climate policies, plans and mitigation, and adaptive measures in the region. The substantial distortion with obligations of the Paris Agreement and SDGs were noted that include adoption of low-carbon technology, decarbonization of energy sector, enhance the swift share of renewable energy in the energy mix, low institutional capacity, and lack of cooperation and partnership. The ineffectiveness of climate change policies in addition to low institutional capacity, absence of data, no NDCs, immaterial research, poor planning, and conflicts had high potential to mutilate human, national, and environmental security discourse. Based on the results of this study, it concluded that the Government of Maldives has more comprehended the selected criteria in climate change policies and plans. But, the climate change mitigation guidelines and the National Adaptation Programme of Actions for Climate Change (NAPA) of Afghanistan have least pondered the selected criteria and echoed more vulnerability and threats to environmental security and development. The current scenario of climate change reflected the quest for environmental diplomacy as pursuance to cooperation for sustainable development, viable national and environmental security. The following policy implications are recommended based on this study:

- The Government of Pakistan in coordination with the Ministry of Climate Change, Ministry of Inter-provincial Coordination, and Council of Common Interests (CCI) should redefine the climate change governance structure in Pakistan according to the 18th amendment and review the climate change policies and plans according to provisions of SDGs and Paris Agreement (assess the NDCs).
- The Governments in South Asia in collaboration with the Ministries of Foreign Affairs should:
- Revisit the Dhaka Declaration on Climate Change and Thimphu Statement on Climate Change to develop comprehensive regional guidelines for climate change mitigations.
- Conduct detailed climate vulnerability assessment, develop climate change inventory for each country, and integrate environmental challenges (energy, water, and climate change) in national policies and foreign policy.
- Rationalize the technical, human, financial, and administrative disagreement to mitigate climate change in Pakistan as well as South Asia because it has been observed low priority of the governments, insignificant budget, inadequate technical and human resources to combat climate change in South Asia.
- Initiate the diplomatic negotiations to enhance the capacity of SAARC to mitigate climate change and practice environmental diplomacy to render cooperation between provincial entities, ministries, and regional cooperation.
- Utilize the nonstate stakeholders such as business, industry, media, and NGOs to ensure environmental sustainability and security.
- Initiate the environmental diplomacy for trust-building, dispute settlement for long-term cooperation, sustainable development, and environmental security.

Acknowledgments

This research article is part of Dr. Hassan's PhD dissertation entitled, "A Study on the Relationship of Development with Environmental Security in National and Regional Context: The Case of Environmental Diplomacy for Pakistan" conducted at the Department of Environmental Science (DES), International Islamic University (IIU), Islamabad.

References

Åhman, M., Nilsson, L.J., Johansson, B., 2017. Global climate policy and deep decarbonization of energy-intensive industries. Climate Policy 17 (5), 634–649.

Ahmed, K., Rehman, M.U., Ozturk, I., 2017. What drives carbon dioxide emissions in the long-run? Evidence from selected South Asian countries. Renewable and Sustainable Energy Reviews 70, 1142–1153.

Aich, V., Akhundzadah, N.A., Knuerr, A., Khoshbeen, A.J., Hattermann, F., Paeth, H., Paton, E.N., 2017. Climate change in Afghanistan deduced from reanalysis and coordinated regional climate downscaling experiment (CORDEX)—South Asia simulations. Climate 5 (2), 38–48.

Bajracharya, S.R., Mool, P.K., Shrestha, B.R., 2007. Impact of Climate Change on Himalayan Glaciers and Glacial Lakes: Case Studies on GLOF and Associated Hazards in Nepal and Bhutan. ICIMOD, Kathmandu, Nepal.

Barker, T., 2017. The economics of avoiding dangerous climate change. In: Green Economy Reader. Springer International Publishing, pp. 237–263.

Circular Ecology, 2021. What Is the Difference between Sustainability and Sustainable Development? https://circularecology.com/introduction-to-sustainability-guide.html. (Accessed 10 June 2021).

Daksiya, V., Su, H.T., Chang, Y.H., Lo, E.Y., 2017. Incorporating socio-economic effects and uncertain rainfall in flood mitigation decision using MCDA. Natural Hazards 87 (1), 515–531.

Fischer, D., Haucke, F., Sundermann, A., 2017. What does the media mean by 'sustainability 'or 'sustainable development'? An empirical analysis of sustainability terminology in German newspapers over two decades. Sustainable Development 25 (6), 610–624.

GOB., 2005. National Adaptation Programme of Actions (NAPA) 2005. Ministry of Environment and Forest, The Government of Bangladesh, Dhaka, Bangladesh.

GOB., 2009. Bangladesh Climate Change Strategy and Action Plan 2009. Ministry of Environment and Forest, The Government of Bangladesh, Dhaka, Bangladesh.

GOI., 2008. National Adaptation Programme of Actions (NAPA) 2008. Ministry of Environment, Forest and Climate Change, The Government of India, New Delhi, India.

GoIRA., 2009. National Adaptation Programme of Actions for Climate Change (NAPA) 2009. National Environmental Protection Agency, The Government of Islamic Republic Afghanistan, Kabul, Afghanistan.

GoM., 2005. Maldives Climate Change Policy Framework 2015. Ministry of Environment, Energy and Water, The Government of Maldives, Male', Maldives.

GoM., 2007. National Adaptation Programme of Actions (NAPA) 2007. Ministry of Environment, Energy and Water, The Government of Maldives, Male', Maldives.

GON., 2010. National Adaptation Programme of Actions (NAPA) 2010. Ministry of Environment, The Government of Nepal, Kathmandu, Nepal.

GON., 2011. Climate Change Policy 2011. Ministry of Environment, The Government of Nepal, Kathmandu, Nepal.

GoP., 2015. Framework for the Implementation of Climate Change Policy 2014-30. Ministry of Climate Change, The Government of Pakistan, Islamabad, Pakistan.

GOSL., 2015. National Adaptation Programme of Actions (NAPA) 2015. Ministry of Mahaweli Development and Environment, The Government of Sri Lanka, Sri Jayawardenepura Kotte, Sri Lanka.

Grumbine, R.E., 2017. Using trans boundary environmental security to manage the Mekong river: China and South-East Asian countries. International Journal of Water Resources Development 1–20.

Hamilton, M., Lubell, M., 2018. Collaborative governance of climate change adaptation across spatial and institutional scales. Policy Studies Journal 46 (2), 222–247.

Hassan, M., Mumtaz, W., Raza, I., Syed, W.A., Ali, S.S., 2013. Application of air dispersion model for the estimation of air pollutants from coal-fired brick-kilns samples in Gujrat. Science International 25 (1), 141–145.

Hassan, M., Ali, K.W., Amin, F.R., Ahmad, I., Khan, M.I., Abbas, M., 2016. Economical perspective of sustainability in agriculture and environment to achieve Pakistan Vision 2025. International Journal of Agriculture and Environmental Research 2 (1), 129–142.

Hassan, M., Afridi, M.K., Khan, M.I., 2017. Environmental diplomacy in South Asia: considering the environmental security, conflict and development nexus. Geoforum 82, 127–130.

Hassan, M., Khan, M.I., Mumtaz, M.W., Mukhtar, H., 2021. Energy and environmental security nexus in Pakistan. Energy and Environmental Security in Developing Countries. Springer, pp. 147–172.

Hsiang, S., Kopp, R., Jina, A., Rising, J., Delgado, M., Mohan, S., Rasmussen, D.J., Muir-Wood, R., Wilson, P., Oppenheimer, M., Larsen, K., 2017. Estimating economic damage from climate change in the United States. Science 356 (6345), 1362–1369.

Ide, T., Bruch, C., Carius, A., Conca, K., Dabelko, G.D., Matthew, R., Weinthal, E., 2021. The past and future (s) of environmental peacebuilding. International Affairs 97 (1), 1–16.

IPCC., 2014. Summary for policymakers. In: Field, C.B., Barros, V.R., Dokken, D.J., Mach, K.J., Mastrandrea, M.D., Bilir, T.E., Chatterjee, M., et al. (Eds.), Climate Change 2014: Impacts, Adaptation, and Vulnerability. Part A: Global and Sectoral Aspects. Contribution of Working Group II to the Fifth Assessment Report of the Intergovernmental Panel on Climate Change. Cambridge University Press, Cambridge, p. 132.

Iqbal, K.M.J., Akhtar, N., Amir, S., Naseer, H.M., Khan, M.I., 2021. A governance index modelling of public sector institutional capacities for modern agriculture and climate compatible development. International Journal of Modern Agriculture 10 (2), 4242–4257.

James, J., 2016. Climate Change and National Security. Department of Strategic Studies, Victoria University of Wellington, Wellington, New Zealand.

Jenkins-Smith, H., Nohrstedt, D., Weible, C., Sabatier, P., 2014. The advocacy coalition framework: foundations, evolution, and ongoing research. In: Sabatier, P., Weible, C. (Eds.), Theories of the Policy Process. Westview Press, Boulder, CO, pp. 183–224.

Lawrence, J., Haasnoot, M., 2017. What it took to catalyse uptake of dynamic adaptive pathways planning to address climate change uncertainty. Environmental Science and Policy 68, 47–57.

Matthew, R., 2011. Regional dynamics of environment and security in post-conflict states. Procedia Social and Behavioral Sciences 14, 28–30.

Mi, Z.F., Wei, Y.M., He, C.Q., Li, H.N., Yuan, X.C., Liao, H., 2017. Regional efforts to mitigate climate change in China: a multi-criteria assessment approach. Mitigation and Adaptation Strategies for Global Change 22 (1), 45–66.

Michailidou, A.V., Vlachokostas, C., Moussiopoulos, N., 2016. Interactions between climate change and the tourism sector: multiple-criteria decision analysis to assess mitigation and adaptation options in tourism areas. Tour Manager 55, 1–12.

Murphy, K., 2012. The social pillar of sustainable development: a literature review and framework for policy analysis. Sustainability: Science, Practice, & Policy 8 (1), 15–29.

OECD., 2015. The Economic Consequences of Climate Change. The Organization for Economic Cooperation and Development Publishing, Paris. https://doi.org/10.1787/9789264235410-en.

Rasul, G., Mahmood, A.S., Khan, A., 2012. Vulnerability of the indus delta to climate change in Pakistan. Pakistan Journal of Meteorology 8 (16), 89–97.

RGOB., 2006. Bhutan National Adaptation Programme of Actions (NAPA) 2006. The Royal Government of Bhutan, Thimphu, Bhutan.

Runfola, D.M., Ratick, S., Blue, J., Machado, E.A., Hiremath, N., Giner, N., White, K., Arnold, J., 2017. A multi-criteria geographic information systems approach for the measurement of vulnerability to climate change. Mitigation and Adaptation Strategies for Global Change 22 (3), 349–368.

SAARC., 2008. Dhaka Declaration on Climate Change 2008. The South Asian Association for Regional Cooperation, Dhaka, Bangladesh.

SAARC., 2010. Thimphu Statement on Climate Change 2010. The South Asian Association for Regional Cooperation, Thimphu, Nepal.

SAARC., 2021. Environment, Climate Change and Natural Disasters. Thee South Asian Association for Regional Cooperation. http://saarc-sec.org/areas_of_cooperation/area detail/environment-natural-disasters-and-biotechnology/click-for-details_6. (Accessed 10 July 2021).

SACEP., 2021. Introduction to South Asia Cooperative Environment Programme. The South Asia Cooperative Environment Programme. http://www.sacep.org/. (Accessed 10 July 2021).

Senapati, S., Gupta, V., 2017. Socio-economic vulnerability due to climate change: deriving indicators for fishing communities in Mumbai. Marine Policy 76, 90–97.

Shokhnekh, A.V., Telyatnikova, V.S., Mushketova, N.S., Panova, N.S., 2020. Ensuring environmental security strategies in social and economic development on the platform of responsible consumption system. In: E3S Web of Conferences, vol. 220. EDP Sciences, p. 01088.

Singh, C., Deshpande, T., Basu, R., 2017. How do we assess vulnerability to climate change in India? A systematic review of literature. Regional Environmental Change 17 (2), 527–538.

Spencer, B., Lawler, J., Lowe, C., Thompson, L., Hinckley, T., Kim, S.H., Bolton, S., Meschke, S., Olden, J.D., Voss, J., 2017. Case studies in co-benefits approaches to climate change mitigation and adaptation. Journal of Environmental Planning and Management 60 (4), 647–667.

The United Nations (UN)., 2015. Transforming Our World: The 2030 Agenda for Sustainable Development. United Nations, New York. Available at: https://sustainabledevelopment.un.org/post2015/transformingourworld/publication.

UFCCC., 2017. The Paris Agreement: Essential Elements. The United nations Framework Convention on Climate Change. http://unfccc.int/paris_agreement/items/9485.php (accessed 20-8-21).

Underdal, A., 2017. Climate change and international relations (after Kyoto). Annual Review of Political Science 20, 169–188.

UNDP., 2004. A Global Report: Reducing Disaster Risk: A Challenge for Development. The United Nations Development Programme. http://www.undp.org/bcpr. (Accessed 19 July 2021).

UNFCCC., 2015. The adoption of Paris agreement 2015. In: The United Nations Framework Convention on Climate Change (UNFCCC), Conference of Parities (COP21), (Paris, France).

Ur-Rehman, M.A., Hashmi, N., Siddiqui, B.N., Afzal, A., Zaffar, A., Masud, K., Khan, M.R.A., Dawood, K.M., Shah, S.A.A., 2017. Climate change and its effect on crop and livestock productivity: farmers' perception of Rajanpur, Pakistan. International Journal of Advanced Research in Biological Sciences 4 (4), 30–36.

Vinke, K., Martin, M.A., Adams, S., Baarsch, F., Bondeau, A., Coumou, D., Donner, R.V., Menon, A., Perrette, M., Rehfeld, K., Robinson, A., 2017. Climatic risks and impacts in South Asia: extremes of water scarcity and excess. Regional Environmental Change 17 (6), 1569–1583.

Wilson, A.M., Gladfelter, S., Williams, M.W., Shahi, S., Baral, P., Armstrong, R., Racoviteanu, A., 2017. High Asia: the international dynamics of climate change and water security. Journal of Asian Studies 1–24.

Zaman, K., Abdullah, I., Ali, M., 2017. Decomposing the linkages between energy consumption, air pollution, climate change, and natural resource depletion in Pakistan. Environmental Progress and Sustainable Energy 36 (2), 638–648.

CHAPTER

13 How Japan's international cooperation contributes to climate change

Nobuhiro Sawamura

Asia Pacific Energy Research Centre, Tokyo, Japan

Chapter outline

1. Introduction

International society and public opinion have focused on climate change due to abnormal climate conditions such as extremely large typhoons and hurricanes worldwide. In response to mounting public opinion and the international community's interests, the United Nations Framework Convention

Handbook of Energy and Environmental Security. https://doi.org/10.1016/B978-0-12-824084-7.00026-6
Copyright © 2022 Elsevier Inc. All rights reserved.

on Climate Change (UNFCCC) was signed in Rio in 1992 during the United Nations Conference on Environment and Development (UNCED), and the Conference of the Parties (COP) is held every year. As a result of accumulated efforts, the Kyoto Protocol was adopted at COP3 held in Kyoto in 1997, and the Paris Agreement was adopted at COP21 held in Paris in 2015.

The government of Japan and other Japanese organizations (e.g., NGOs and private companies) have contributed more to international cooperation on climate change since the international society discussed and took actions on climate change. The government of Japan has aided developing countries in reducing greenhouse gas (GHG) emissions by its Official Development Aid (ODA) and other measures. However, as critics point out, there are challenges to Japan's international cooperation on climate change. Furthermore, due to the long-term economic stagnation of Japan and its limited budget, its ODA disbursement has almost leveled off during the past two decades (as you can see in Figs.13.1 and 13.2). Therefore, how efficiently its limited and precious resources should be allocated is important regarding Japan's ODA.

This chapter analyzes how Japan's international cooperation on climate change can be improved by referring to its past efforts and experiences. Mainly, its policies and activities are discussed from the viewpoint of mitigation and adaptation. Then, when it comes to climate change, mitigation tends to be more focused on than adaptation. However, adaptation is as important as mitigation.

2. Mitigation

2.1 Official development aid

The Japan International Cooperation Agency (JICA) has provided ODAs to developing countries on behalf of the government of Japan. In terms of international cooperation on climate change, JICA's primary cooperation schemes include technical cooperation, finance, investment cooperation, and

FIGURE 13.1

The gross disbursement of Japan's official development aid (MOFA, 2018).

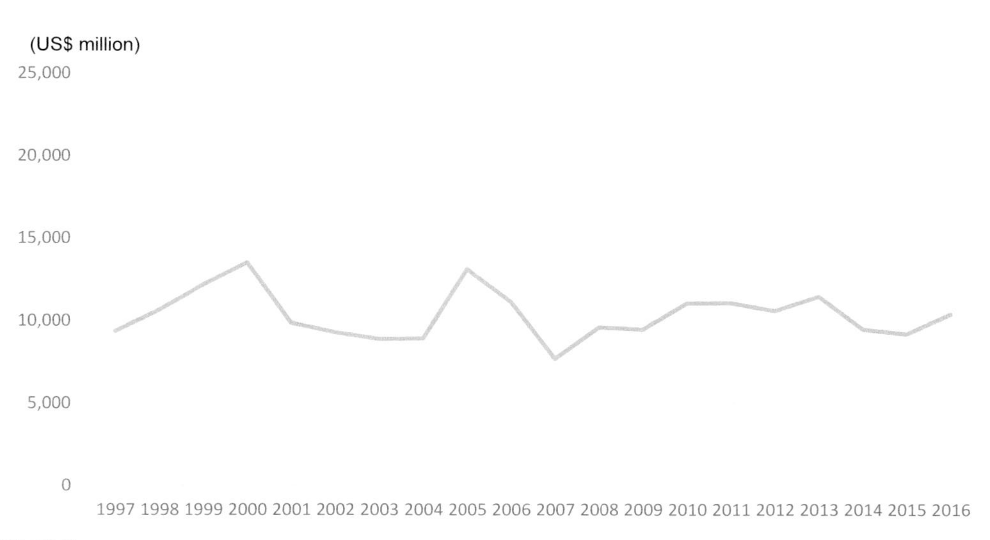

FIGURE 13.2

The net disbursement of Japan's official development aid (MOFA, 2018).

grants (JICA, 2021a). Then, three central climate change cooperation, namely, reducing emissions from Deforestation and Forest Degradation in developing countries (REDD+), Clean Development Mechanism (CDM), and Joint Crediting Mechanism (JCM), are examined and analyzed later. Furthermore, Japan Bank for International Cooperation (JBIC) and New Energy and Industrial Technology Development Organization (NEDO) have also participated in CDM projects in cooperation with the government of Japan, Japanese private companies, and hosts.

2.2 Reducing emissions from deforestation and forest degradation in developing countries

2.2.1 Background

For global warming prevention, forests such as absorbing carbon dioxide have been focused on, especially since the 1990s. At COP13 held in Bali in 2007, the international society decided to adopt REDD as climate change. REDD targets only to restrain deforestation and forest degradation in developing countries. Then, from COP14 held in Poznan in 2008, in addition to REDD, REDD+ has spread widely.

According to UN-REDD Program (2021):

> REDD+ is a mechanism developed by Parties to the United Nations Framework Convention on Climate Change (UNFCCC). It creates a financial value for the carbon stored in forests by offering incentives for developing countries to reduce emissions from forested lands and invest in low-carbon paths to sustainable development.

REDD+ includes forest conservation, sustainable forest management, and conservation of forest carbon stocks. International organizations and developed countries have provided REDD+ projects for developing countries by both multilateral and bilateral approaches.

2.2.2 Japan's REDD+ activities

Japan has been implementing REDD+ projects by two methods. First, it has been financially supporting REDD+ activities through bilateral partnerships with countries in Asia. Second, it has provided transfers of REDD+ technology and methodology for capacity building (Park et al., 2013, 219–221).

Japan's REDD+ financing projects include grants, subsidies, and loans, while other countries offer mostly grants (Park et al., 2013, 219). One of the reasons Japan provides loans is its principle, "self-help," can be raised. Supporting the "self-help" efforts of governments, other organizations, and people in developing countries is an essential policy based on Japan's ODA philosophy. Then, if projects are run only by grants, hosts tend to make relatively fewer efforts (because hosts can get finances for free and do not need to repay loans). However, if loans run projects, hosts make much more effort (because hosts must repay loans).

JICA has managed several REDD+ projects in Asia, such as Lao PDR and Viet Nam, regarding capacity building. In Viet Nam at a local level, some provincial REDD+ Action Plans (PRAPs) have started with support from JICA. In Dien Bien Province, the REDD+ project, "Sustainable Forest Management in the Northwest Watershed Area" (SUSFORM-NOW), produced a fruitful PRAP facilitating the local government and people to participate in forest protection activities (Anastasia Yang et al., 2016, 37). Through capacity building in developing countries, local governments, other organizations, and residents can raise awareness about forest protection activities and continue their projects even after supports by Japan's ODA end. Therefore, Japan emphasizes capacity building based on its principle, the "self-help." If hosts get more experience and know-how under their responsibilities, they do not need more assistance from others. As a result, they can manage projects such as REDD+ by themselves.

2.2.3 Challenges regarding REDD+

REDD+ has faced several challenges though its roles contributing to climate change have been highly evaluated.

First, there are concerns that REDD+ may damage biological diversity if only gaining carbon credits from REDD+ is focused on (Hyakuta and Yokota, 2021, 22). If planting a specific kind of trees and excessive forest management and conservation efforts are taken, balance in an ecosystem might be broken, leading to extinction crisis of endangered species. For both promoting REDD+ and preserving biological diversity, both host and recipient countries need to study and consider characteristics in object areas with the local community and the third-party organizations before starting projects. Furthermore, REDD+ projects need to be continuously monitored to preserve biological diversity after projects start and carbon credits are issued.

Second, some argue that REDD+ may have adverse influences on inhabitants living in recipient countries. Governments in recipient countries try to find and preserve the most suitable forests and lands for better REDD+ projects. As a result, they restrict the inhabitants' access rights to these areas. Then, inhabitants are driven out and cannot use these areas for their living (Harada, 2010, 5–6). Inhabitants tend to be engaged in forestry and agriculture in protected areas and coexist with nature generation after generation. If restrictions on these areas are too strict, habitants lose their living, and traditional forestry and agriculture vanish there. It is vital to establish new REDD+ mechanisms and systems to protect the inhabitants' access rights and living. Same with other mechanisms and systems on climate change, how impartially precious resources such as forests are allocated among stakeholders such as governments and inhabitants is one of the most difficult challenges.

2.3 Clean development mechanism

2.3.1 Background

The Kyoto Protocol obliges developed countries (Annex I Parties) that signed this treaty to reduce GHG emissions following each target's agreement. However, developing countries (Non-Annex I Parties) are exempt from the GHG emission duty under this treaty. As developed countries have less room for reductions and face more expensive marginal reduction costs than developing countries, the Kyoto mechanisms were introduced for developed countries to achieve their targets. The Kyoto mechanisms were designed based on market mechanisms. It has three measures, CDM, Emission Trading, and Joint Implementation (JI). Among them, Japan had put stress on CDM.

According to UNFCCC (2021a), "*the Clean Development Mechanism (CDM), defined in Article 12 of the Protocol, allows a country with an emission-reduction or emission-limitation commitment under the Kyoto Protocol (Annex B Party) to implement an emission-reduction project in developing countries*."

CDM projects produce certified emission reduction (CER) credits sold to developed countries to achieve their reduction targets. However, to use CER credits to meet its goal, the government of Japan needs to approve by inspecting whether each CDM project satisfies approval requirements and purchases CER credits from each business company in charge of each project. Furthermore, unless each host and CDM Executive Board (CDM EB) approve each project, CER credits cannot be earned. Besides these, there are more requirements for CDM projects to be approved and registered by CDM EB.

Based on the Marrakesh Accords, ODA cannot be diverted to CDM projects (the Marrakesh Accords). However, the Ministry of Foreign Affairs of Japan (MOFA, 2007) recognizes that "*ODA can be used for CDM projects if both donor and recipient countries confirm that it does not lead to the diversion of ODA*." Indeed, JBIC and Japan Carbon Finance, Ltd. (JCF) conducted the Zafarana Wind Power Plant Project in Egypt to reduce its dependency on fossil fuels (JPY13,497 million were distributed through loan aid). As a result, this project was approved by CDM EB and registered as Japan's first ODA project qualified as CDM (MOFA, 2007). Furthermore, the JICA Assisted Delhi Metro Phase-II Project in India (ODA loan assistance of JPY 212 billion) and the Coconut Shell Charcoaling and Power Generation Project in Sri Lanka (ODA loan assistance amounting to a total of JPY7.97 billion) were approved and registered as CDM projects (JICA, 2009; 2011). However, it is difficult for ODA-backed CDM projects to be approved and registered as official CDM projects since there are challenges regarding CDM.

2.3.2 Challenges regarding CDM

As one of the challenges regarding CDM, there were projects that CDM EB disapproved as CDM projects.

In response to a rising trend about energy saving in Viet Nam in the 2000s, NEDO started the energy-saving model project at Thanh Hoa Beer Joint Stock Company of Hanoi Alcohol Beer Beverage Corporation in 2004. The primary purpose of this project is to verify and disseminate Japan's energy-saving and petroleum alternative energy technologies in the Asia-Pacific region and ultimately secure its stable energy supply (NEDO, 2012). Therefore, it is expected that reducing GHG emissions was not the primary objective.

However, this project was not approved and registered as a CDM project by CDM EB. As reasons to dismiss this project, CDM EB explained:

> The DOE (DNV) and project participant failed to substantiate that the emission reductions calculations are in line with the methodology, as the discount emission factor of 10% to the ex-ante baseline emissions, which was proposed by the project participants, is not in line with the methodology's requirements. The Board further noted that the historical data used to calculate the baseline emissions was not provided either in response to the request for review or the review (UNFCCC, 2018).

As NEDO recognized that the ODA fund for energy-saving and petroleum alternative energy technologies and the additional government spending for reducing GHG emissions (CDM) were divided in this project, it expected that its additional spending would not be regarded as the ODA conversion to this CDM project. On the other hand, the Ministry of Industry of Viet Nam insisted that it was impossible to distinguish between the ODA fund and the additional expenditure, although it recognized that the ODA was not diverted to this project (Nakashima, 2012, 164−165). From the perspective of CDM ED, it is also difficult to judge whether additional government spending corresponds to ODA conversion or not.

Furthermore, "additionality" was one of the most severe challenges regarding CDM. A business operator in charge of a CDM project was responsible for proving that this project would "additionally" decrease GHG emissions on the condition that CDM is used. If this "additionality" was not certified, CDM ED did not approve this project as a CDM project issuing CER credits by CDM EB (Motoda, 2016, 126). Therefore, business operators hesitated to be actively involved in CDM projects. As a result, sometimes, developed countries could not support CDM projects positively, especially from financial aspects.

2.4 Joint crediting mechanism

2.4.1 Background

As discussed earlier, Japan and other countries faced difficulties about whether their projects were approved by CDM EB and registered as projects qualified as CDM. Especially, "additionality" was one of the strictest requirements for them, as mentioned previously. Then Arimura (2015) points out other problems. First, "*it took enormous costs and time for inspections and registrations of each project and issuances of CERs*." Second, "*compared with CDM projects concerning wind power and hydropower, the number of energy-saving projects is considerably small*." Third, "*CDM may not satisfy sufficiently what developing countries need*" (55).

As the new scheme replacing the Kyoto mechanisms, the government of Japan has established and implemented the JCM and has suggested it as one of the new systems under UNFCCC.

2.4.2 The JCM system

JCM is the new mechanism that Japan contributes to reducing GHG emissions in developing countries by disseminating its energy-saving and low-carbon technologies and products. This mechanism adopts a bilateral agreement. Credits gained from JCM projects are shared between Japan and a contracting country of its bilateral agreement. Japan signed its first JCM agreement with Mongolia in January 2013. As of January 2017, Japan has established JCM systems with 17 partner countries (Mongolia, Bangladesh, Ethiopia, Kenya, Maldives, Viet Nam, Lao PDR, Indonesia, Costa Rica, Palau, Cambodia, Mexico, Saudi Arabia, Chile, Myanmar, Thailand, and the Philippines) (MOE, 2021).

One of JCM's advantages is that JCM may satisfy more sufficiently what developing countries need than CDM. Unlike CDM, under JCM, the government of Japan can decide the management of its system and the eligibility of each project. Therefore, it becomes more flexible to reflect what a partner country needs in each JCM project (Motoda, 2016, 125).

As another merit of JCM, this system may reduce more costs for approvals and registrations than CDM. Under JCM, since a bilateral joint committee composed of Japan and its partner country approves and registers each project, it is assumed that each process would be more simplified (Arimura, 2015, 56).

2.4.3 Challenges regarding JCM

Although the government of Japan and its related organizations have made many efforts to establish and implement JCM as an internationally approved system, JCM still has various challenges.

First, there is uncertainty about whether international society accepts JCM's certification method of GHG reduction amount or not. As Japan and its partner country judge and decide GHG reduction amount, there is a possibility that international society has doubts about JCM's certification method and cannot accept JCM (Arimura, 2015, 56). Concerning this challenge, the government of Japan needs to make the JCM system even more transparent and more diligently release information about its system to the public.

Second, how enthusiastically and voluntarily Japanese companies are involved in JCM is essential. Under the circumstances that they face severe global competition, it is tough for them to set and achieve their GHG reduction targets at a higher level. In the case of CDM, many Japanese companies utilized CDM because they had to achieve their targets set by the "Voluntary Action Plan on the Environment" of KEIDANREN (Japan Business Federation) (Arimura, 2015, 57). Therefore, Japanese companies' incentives to be involved in JCM will depend on what scheme replacing "Voluntary Action Plan on the Environment" KEIDANREN will make. Although decarbonization is a global trend in business, it is crucial to facilitate discussions among the government of Japan, KEIDANREN, and Japanese companies and improve the JCM system for the Japanese industry community to use JCM actively. Then if Japanese companies work on its government to make JCM more match with the Sustainable Development Goals (SDGs), they can obtain more supports from public opinion and more investments from individual investors and institutional investors at home and abroad. Furthermore, if JCM is enhanced to have a more affinity with Environmental, Social and Governance (ESG), Japanese corporations will have high incentives to participate in this new mechanism. Then these can be a good opportunity for them to develop their businesses and earn more profits in the future.

Third, as the number of JCM projects has increased since 2014, their expected GHG reduction amounts and issued credits are small. This trend has continued because most JCM projects adopt energy-saving technologies such as high-efficiency air conditioners and small-scale renewable technologies such as solar power and microhydro. For increasing GHG reduction amounts and amounts of issued credits, it is recommended to increase the number of projects introducing the waste heat recovery system into factories as the waste heat recovery system is expected to obtain many GHG reductions. Then, most energy-saving projects aim to reduce carbon dioxide emissions. If these projects target to reduce other GHG emissions with global warming potential (GWP) such as methane, it believes in achieving more projects with a large amount of GHG reductions (IEEJ, 2019, 19–22). To reduce methane emissions and other GHGs with GWP, partnerships between public and private sectors need to be stronger, and investments in research and development should increase to invent new and advanced technology. Regarding carbon dioxide, technologies such as Carbon dioxide Capture and Storage (CCS) and Carbon dioxide Capture, Utilization, and Storage (CCUS) have improved gradually. If the same and similar technologies for other GHGs with GWP are invented, JCM projects may increase.

Fourth, although the government of Japan has made enormous efforts to increase partner countries, significant emitting emerging and developing countries such as China and India are not included in the JCM scheme. To make JCM agreements with China and India, Japan needs to give more satisfactory results with existing partner countries and improve the JCM system. If the JCM system is judged to be more attractive than other market mechanisms, there will be an opportunity to involve China and India in the future.

3. Adaptation

3.1 Background

As countermeasures against climate change, international society has also focused on adaptation. According to UNFCCC's interpretation (UNFCCC, 2021b), "*adaptation refers to adjustments in ecological, social, or economic systems in response to actual or expected climatic stimuli and their effects or impacts*." While mitigation aims to reduce global warming by reducing GHG emissions as a preventive measure, adaptation goals prepare for dealing with effects caused by future global warming. Unlike mitigation, the Kyoto Protocol and the Paris Agreement have not set the numerical target with a fixed time for adaptation. As no numerical target with a fixed time, each member country faces difficulties incorporating adaptation into its ODA policy. It is challenging to plan and implement adaptation policies because policymakers face uncertainties about future influences caused by climate change.

Furthermore, the fact that climate change influences various fields makes it difficult to make adaptation policies. Since adaptation does not have market mechanisms such as CDM issuing credits that developed countries can use for achieving their reduction obligations and goals, they are relatively less reluctant to make efforts on adaptation than mitigation. Additionally, it is more challenging to show the effects of adaptation by money than mitigation.

However, adaptation is as important as mitigation in terms of climate change countermeasures. Japan has contributed to international cooperation on climate change through its ODA and the international scheme in terms of adaptation. JICA has realized the importance of adaptation because adaptation policies give multifunctional benefits. Furthermore, ODA's adaptation policies provide not only their direct but also indirect effects in the future.

3.2 Official development aid

As adaptation, Japan has provided its ODA projects for agriculture, disaster prevention, water supply in developing countries. However, as previously mentioned, indicating the amount of money of adaptation's effects is more complicated than mitigation's effects. Therefore, adaptation has been relatively less focused on than mitigation. However, adaptation connotes a win-win policy and co-benefit that may cause potential merits. Therefore, there are cases in its ODA projects that lead to win-win policies and have co-benefit in the end. Hereafter, two kinds of cases are analyzed.

3.2.1 The project for the seawall construction in Malé Island in the Republic of Maldives

The Maldives is a small island country comprised of approximately 1190 islands and has a small population of 320,000 people (2010) (JICA, 2021b). It is vulnerable to the sea-level rise caused by climate change. According to the UNFCCC's report, "*over 80% of the land area in the Maldives is less than 1 m above mean sea level*" (MHAHE, 2001, 5). Therefore, based on UNDP (2021) data, "*a sea-level rise of even a meter would cause the loss of the entire land area of Maldives.*"

The island of Malé, where the capital of the Maldives locates, suffered from abnormal flooding by seawater in 1987. The total amount of damages caused by this calamity supposably came up to about US$ six million. As the Maldivian government requested urgent foreign aid to construct shore protection facilities, the government of Japan started an aid cooperation project to construct shore protection walls in Malé Island in 1987 (JICA, 2000, 1). The government of Japan provided the four phases of its project for seawall construction as grant aid. However, according to Nakashima (2012), "*as of 1987, this flood's damages had not been recognized to be caused by adverse effects of climate change*" (201). After that, international society started to focus on and take action on climate change. Small Island Developing States (SIDS) gathered in Malé to discuss a sea-level rise issue in 1989. "The Malé Declaration on Global Warming and Sea Level Rise" was signed at this meeting (UN, 2017, 2). In 1999, the Ministry of Home Affairs, Housing, and Environment (MHAHE), Republic of Maldives, released the Second National Environmental Action Plan (NEAP-II) and listed climate change and sea-level rise as critical issues on it (MHAHE, 1999).

Initially, this project started to mitigate damages caused by high tides. Then, its expected direct effect was disaster prevention. However, it recognized that shore protection facilities constructed for preventing disasters at that time could be beneficial for responding to future calamities such as floods caused by climate change and sea-level rise. Namely, this project produced adaptation functions to climate change as the indirect effect at the end (Nakashima, 2012, 211). In the field of disaster prevention, constructing banks such as a seawall is a win-win policy. According to Takemoto and Mimura (2007), "*the win-win policy refers to a policy for both current solving issues and being an adaptation to the future problems caused by climate change*" (361). There are many fields that the win-win policy can apply in terms of climate change. If Japan's adaptation projects based on the win-win policy are more adopted, more benefits can obtain at fewer expenses under its limited ODA budget.

Along with the win-win policy, the roles of Mainstreaming of Adaptation are highly evaluated. Some argue that adaptation should not be taken independently because its effect is limited. They add that adaptation should be positioned at the center of other policies such as agricultural and food policies. If adaptation is taken with other key policies, the effects of other policies will be sustainable and resilient against future risks. This concept is defined as Mainstreaming of Adaptation (Mimura,

2006, 108). By adopting more Mainstreaming of Adaptation, Japan can provide various projects with robustness and multifunction and contribute to solving various problems and climate change.

3.2.2 The project for rural water supply in the Republic of the Gambia

The Republic of the Gambia locates in the western part of Africa. It has a small population of 2347.71 thousand (2019) (The World Bank, 2021). Its primary industry is agriculture, such as groundnuts, rice, and maize (UN, 2021). However, the Gambia is not rich in oil. In the Gambia, infrastructures of water and sewage, and electricity have been undeveloped nationwide. According to JICA (2004):

> The Government of the Gambia has constructed water supply facilities with boreholes and provided the population with technical support for operation and maintenance. However, the nationwide water supply coverage rate is as low as 62% and 53% for rural areas (i).

Therefore, the remaining population cannot access the nationwide water supply and are forced to use dirty water. As the government of Japan received requests from the government of Gambia, it started the project for rural water supply (JICA, 2004a, i). JICA has provided the projects for rural water supply three times (phase 1−3) (JICA, 1990, 2004, I, 2010) and signed a grant agreement with the government of the Republic of the Gambia for the new project for rural water supply (phase 4) (JICA, 2020a). In the second and third projects, diesel generators were converted into solar pumping systems for lifting power at the water supply facilities (JICA, 2004, ii, 2010). Furthermore, JICA will adopt solar pumping systems in the fourth project (JICA, 2020b, ii).

These projects include not only a win-win policy but also a co-benefit. Co-benefit is defined as "*in addition to the attainment of the original goal and effects caused by adaptation on climate change, obtaining other effects contributing to achieving sustainable development such as environmental improvement and energy saving at the same time*" (Fujimori et al., 2008, 30−31). These projects brought about a stable supply of hygienic water as the direct effect and adaptation to reduce vulnerability to water supply caused by climate change as the indirect effect (win-win policy). Based on Fujimori et al. (2008), "*as a co-benefit, cost reduction of purchasing fossil fuels for diesel generators and reduction of GHG emissions can be expected*" (31).

Same as the project for the seawall construction in Malé Island in the Republic of Maldives, if the number of projects connotating co-benefit increases, more benefits can be gained in each project.

4. Conclusion

Climate change has been one of the most severe issues globally because it is assumed to cause natural disasters such as catastrophic floods and forest fires threatening human beings, other creatures, and the Earth. In this situation, the government of Japan and its other related organizations have worked on international cooperation on climate change both in mitigation and adaptation.

Regarding mitigation, they have mainly utilized ODAs, REDD+, CDM, and JCM for international cooperation on climate change.

REDD+ is one of the most prominent efforts to climate change since it can contribute to environmental conservation, such as preserving forests and addressing climate change impacts. For instance, Japan's efforts, through REDD+ projects making financial plans and enhancing capacity building, governments and local people in recipient countries can more facilitate awareness about climate change and forest preservation activities. On the other hand, there are potential challenges such as damaging biological diversity and inhabitants living in recipient countries regarding REDD+. However, if keeping a good balance among them, the number of REDD+ projects with better quality will increase.

Furthermore, though the government of Japan, other Japanese organizations, and business operators faced several severe challenges when taking on CDM projects, they have reasonably utilized CDM. However, Japan is a member country of the Kyoto first commitment period (2008–2012), but it is not a member country of the Kyoto second commitment period (2013–2020). Therefore, although Japan can transfer CER credits gained from its CDM projects registered during the Kyoto first commitment period, it cannot transfer credits registered during the Kyoto second commitment period. The Conference of the Parties serving as the Meeting of the Parties to the Kyoto Protocol (CMP) had been planned to be held with COP26 in Glasgow, UK, in November 2020. However, CMP and COP26 were postponed in November 2021 due to the COVID-19 pandemic (UNFCCC, 2020). Therefore, the new scheme replacing CDM will be established after the Kyoto second commitment period (2013–2020). Negotiations regarding Article 6 of the Paris Agreement, especially, new market mechanism, have continued.

Considering how Japan aims to contribute to international cooperation on climate change through market mechanisms, the government of Japan has established and tried to disseminate JCM, which can be a new scheme replacing CDM. Although international society has not agreed on the new market mechanism superseding CDM, the government of Japan has actively and enthusiastically built the JCM system and increased its partner countries. Still, there are many challenges, but JCM has high potentials backed by Japan's experiences, knowledge, and technologies about the environment.

Regarding adaptation, Japan and JICA have provided ODA projects to developing countries suffering from disasters such as floods.

In this chapter, the importance of adaptation was analyzed, although it tends to be less focused on than mitigation for reasons mentioned previously. After all, compared with mitigation, it is more difficult to clearly show when and how adaptation will produce results in the future due to a lack of the numerical target with a fixed time agreed by international schemes and uncertainties of its effects. However, discussions about adaptation have become livelier at COP because severe natural disasters assumed to be caused by climate change happen every year.

This chapter analyzed two adaptation projects conducted by JICA. Through these two projects, JICA illustrated that it could provide ODAs with relatively higher cost-effectiveness. Namely, these projects connoted the win-win policy and co-benefit that can deliver more benefits and effects simultaneously. Also, if the Mainstreaming of Adaptation advances more, adaptation projects can solve other issues such as poverty, famine, and infectious diseases in developing countries.

International society tends to focus on more mitigation than adaptation because there is no specific numerical target with a fixed time of adaptation at both the Kyoto Protocol and the Paris Agreement. In addition, it is uncertain and difficult to predict future natural disasters assumed to be caused by climate change and effects produced by adaptation. However, adaptation is as important as mitigation regarding climate change. Therefore, Japan can be more vigorously involved in international cooperation on climate change by making new standards and systems highly evaluated globally. Furthermore, Japan needs to cooperate more with international organizations and distribute more resources and finances to adaptation.

5. Glossary list

CCS	Carbon dioxide Capture and Storage
CCUS	Carbon dioxide Capture, Utilization, and Storage
CDM	Clean Development Mechanism
CDM EB	CDM Executive Board
CER	Certified Emission Reduction
CMP	Conference of the Parties serving as the Meeting of the Parties to the Kyoto Protocol
COP	Conference of the Parties
DAC	Development Assistance Committee
DNV	Det Norske Veritas
DOE	Designated Operational Entity
ESG	Environmental, Social and Governance
GEC	Global Environment Centre Foundation
GHG	Green House Gas
GWP	Global Warming Potential
IEEJ	Institute of Energy Economics, Japan
JBIC	Japan Bank for International Cooperation
JCF	Japan Carbon Finance, Ltd.
JCM	Joint Crediting Mechanism
JI	Joint Implementation
JICA	Japan International Cooperation Agency
KEIDANREN	Japan Business Federation
MHAHE	Ministry of Home Affairs, Housing, and Environment
MOE	Ministry of Environment Government of Japan
MOFA	Ministry of Foreign Affairs of Japan
NEAP-II	Second National Environmental Action Plan
NEDO	New Energy and Industrial Technology Development Organization
ODA	Official Development Aid
PRAPs	Provincial REDD+ Action Plans
REDD+	Reducing emissions from Deforestation and Forest Degradation in developing countries
SDGs	Sustainable Development Goals
SIDS	Small island developing states
SUSFORM-NOW	Sustainable Forest Management in the Northwest Watershed Area
UN	United Nations
UNCED	United Nations Conference on Environment and Development
UNDP	United Nations Development Programme
UNFCCC	United Nations Framework Convention on Climate Change

References

Arimura, T., 2015. [International Link of the Emission Trading and Trend and Outlook of the Related Economic Analysis] Kokunai haishutsu ryou torihiki no kokusai rinku oyobi kanren keizai bunseki no doukou to tenbou (in Japanese). Kankyo Keizai and Seisaku Kenkyu 8 (1), 2015.3. https://www.jstage.jst.go.jp/article/reeps/8/1/8_50/_pdf/-char/ja.

Fujimori, M., Kawanishi, M., Mimura, N., 2008. Proposal of adaptation function assessment to integrate adaptation to climate change into ODA projects. Environmental Systems Research 36. https://www.jstage.jst.go.jp/article/proer2000/36/0/36_0_27/_pdf/-char/ja.

Harada, K., 2010. [From Biological Diversity Preservation to Climate Change Mitigation: a Study about REDD's Influences on Protected Areas] Seibutsu tayousei hozen kara kikou hendou kanwa he-REDD ga hogo chiiki ni ataeru eikyou ni kansuru kousatsu- (in Japanese). Forest Economic Research Institute 62 (10), 5–6. https://www.jstage.jst.go.jp/article/rinrin/62/10/62_KJ00008619876/_pdf/-char/ja.

Hyakuta, K., Yokota, Y., 2021. [REDD+ System and Policy] REDD+ no seido seisaku (in Japanese). [New endeavors, REDD+ to preserve tropical forests] REDD+ nettairin wo hozen suru aratana torikumi (in Japanese). Shinrin Kagakku 60, 22. https://www.forestry.jp/publish/ForSci/BackNo/sk60/60.pdf.

The Institute of Energy Economics, Japan (IEEJ 2019), 2019. [the FY2018 Research Project about Infrastructure Development for Obtaining Joint Credits, etc.: The Report on International Trend about Negotiations, Etc. About Market Mechanisms] Heisei 30 Nendo (2018 Nendo) Nikokukan Kurejitto Shutoku Tou No Tameno Infura Seibi Chosa Jigyou: Shijo Mekanizumu Kousho Tou Ni Kakaru Kokusai Doukou Chosa Houkoku Sho (In Japanese), pp. 19–22. March 2019.

JICA, 1990. Basic Design Study Report on Integrated Water Use Project in the Republic of the Gambia. https://libopac.jica.go.jp/images/report/10865103_01.pdf.

JICA, 2000. Basic Design Study Report on the Project for the Seawall Construction in Malé Island (Phase IV) in the Republic of Maldives, 1. https://libopac.jica.go.jp/images/report/11586492_01.pdf.

JICA, 2004. Basic Design Study Report on the Integrated Water Use Project (Phase 2) in the Republic of the Gambia". https://libopac.jica.go.jp/images/report/11761434_01.pdf.

JICA, 2009. First ODA-Backed Private Sector Project Registered as CDM, -Coconut Shell Charcoaling and Power Generation Project Funded by Sri Lankan Environmentally Friendly Solutions Fund. https://www.jica.go.jp/english/news/press/2009/090420.html.

JICA, 2010. Preparatory Survey Report on the Project for Rural Water Supply Phase III in the Republic of the Gambia. https://libopac.jica.go.jp/images/report/12004297_01.pdf.

JICA, 2011. Completion of JICA Assisted Delhi Metro Phase- II Project in India. https://www.jica.go.jp/india/english/office/topics/110915.html.

JICA, 2020a. Signing of Grant Agreement with the Gambia: Contributing to the Improvement of Public Health and Living Standards through Constructing Water Supply Facilities in Rural Areas. https://www.jica.go.jp/english/news/press/2020/20200806_41.html.

JICA, 2020b. Preparatory Survey Report on the Project for Rural Water Supply Phase IV in the Republic of the Gambia ii. https://libopac.jica.go.jp/images/report/12354437_01.pdf.

JICA, 2021a. JICA's Cooperation on Climate Change: Towards a Sustainable and Zero-Carbon Society. https://www.jica.go.jp/english/our_work/climate_change/index.html. https://www.jica.go.jp/english/our_work/climate_change/c8h0vm00005rzelb-att/cooperation_01_en.pdf.

JICA, 2021b. Maldives. https://www.jica.go.jp/maldives/english/index.html.

MHAHE, 2001. First National Communication of Maldives to the UNFCCC, vol. 5. https://unfccc.int/resource/docs/natc/maldnc1.pdf.

MOFA, 2018. Trends in the ODA of major DAC countries (gross) and (net). In: White Paper on Development Cooperation 2017. https://www.mofa.go.jp/policy/oda/page22e_000864.html.

Mimura, N., 2006. [The Position and Challenges of Adaptation regarding Global Warming Countermeasures] Chikyu ondanka niokeru taiousaku no ichiduke to kadai (in Japanese). Chikyu Kankyo 11 (1), 108.

Ministry of Home Affairs, Housing, and Environment, 1999. Republic of Maldives (MHAHE 1999) NEAP-II (the Second National Environmental Action Plan). https://thimaaveshi.files.wordpress.com/2009/09/neap2.pdf.

Ministry of Environment Government of Japan (MOE, 2021. Joint Crediting Mechanism. Carbon Markets Express. https://www.carbon-markets.go.jp/eng/jcm/index.html.

Ministry of Foreign Affairs of Japan (MOFA, 2007. Box 1. Using ODA to Promote the Clean Development Mechanism (CDM), Japan's Official Development Assistance White Paper 2007. https://www.mofa.go.jp/policy/oda/white/2007/ODA2007/html/box/bx01001.htm.
Motoda, T., 2016. Introducing the International Cooperation Mechanism as a Means of Tackling Climate Change in Developing Countries: Joint Crediting Mechanism (JCM) and City-To-City Cooperative Framework. Global Environment Centre Foundation (GEC), Japan. https://www.jstage.jst.go.jp/article/mcwmr/27/2/27_123/_pdf.
Nakashima, K., 2012. Evaluation Method Concerning International Cooperation on Climate Change. Hokkaido University Press, Japan.
NEDO, 2012. [FY2006 Progress Report: The Project of Dissemination of the Results about the Model Project of Energy-Saving at a Beer Factory in Viet Nam] Heisei 18 Nendo (2006 Nendo) Seika Hokokusho: Biiru Kojo Shoenerugiika Moderu Zigyo Nikakaru Seika Fukyu Zigyo (Viet Nam) (In Japanese). https://www.nedo.go.jp/library/seika/shosai_201202/20110000001696.html.
Park, M.S., Choi, E.S., Youn, Y.-C., 2013. REDD+ as an International Cooperation Strategy under the Global Climate Change Regime: Forest Science and Technology: Taylor&Francis, pp. 219−221. https://www.tandfonline.com/doi/pdf/10.1080/21580103.2013.846875.
Takemoto, A., Mimura, N., 2007. Study on international Framework on adaptation to climate change in developing countries. Environmental Systems Research 35, 361. https://www.jstage.jst.go.jp/article/proer2000/35/0/35_0_355/_pdf/-char/en.
The Marrakesh Accords, "3. Modalities and procedures for a clean development mechanism, as defined in article 12 of the Kyoto Protocol" and "J. Work programme on mechanisms" (decisions 7/CP.4 and 14/CP.5). https://unfccc.int/cop7/documents/accords_draft.pdf.
The World Bank, 2021. Population, the Data. https://data.worldbank.org/indicator/SP.POP.TOTL?locations=GM.
United Nations (UN, 2017. HE ambassador marlene moses ambassador/permanent representative of Nauru to the united Nations, "climate change impacts on small island developing states", united Nations open-ended informal consultative process on oceans and law of the sea, Eighteenth meeting: "the effects of climate change on oceans,". Oceans and Law of the Sea 2. In: https://www.un.org/depts/los/consultative_process/icp18_presentations/moses.pdf.
UN, 2021. Country facts. In: Permanent Mission of the Republic of the Gambia to the United Nations. https://www.un.int/gambia/gambia/country-facts.
UN-REDD Programme Collaborative Online Workspace, 2021. ABOUT REDD+. https://www.unredd.net/about/what-is-redd-plus.html.
United Nations Development Programme (UNDP, 2021. Climate Change Adaptation, Maldives. https://www.adaptation-undp.org/explore/maldives.
United Nations Framework Convention on Climate Change (UNFCCC, 2018. CDM Executive board review of the project activity. In: The Model Project for Renovation to Increase the Efficient Use of Energy in Brewery (1516). https://cdm.unfccc.int/Projects/DB/DNV-CUK1200406374.33/Rejection/IGUMWP9U8NUCM0PAN2018FLS1SDDHN.
UNFCCC, 2020. The CDM Executive Board Considers CDM beyond 2020. Article/06 Oct, 2020. https://unfccc.int/news/the-cdm-executive-board-considers-cdm-beyond-2020.
UNFCCC, 2021a. The Clean Development Mechanism. In: https://unfccc.int/process-and-meetings/the-kyoto-protocol/mechanisms-under-the-kyoto-protocol/the-clean-development-mechanism.
UNFCCC, 2021b. What Do Adaptation to Climate Change and Climate Resilience Mean? https://unfccc.int/topics/adaptation-and-resilience/the-big-picture/what-do-adaptation-to-climate-change-and-climate-resilience-mean.
Yang, A., Nguyen, D.T., Vu, T.P., Trung Le, Q., Pham, T.T., Larson, A.M., Ravikumar, A., 2016. REDD+ Policy and Practice: Analyzing Multilevel Governance in Vietnam, p. 37. http://www.jstor.com/stable/resrep16292.11.

CHAPTER

Environmental sustainability in Asia: insights from a multidimensional approach

14

Thai-Ha Le[1,3], Ha-Chi Le[2] and Canh Phuc Nguyen[3]

[1]*Natural Capital Management Program, Fulbright School of Public Policy and Management, Fulbright University Vietnam, Ho Chi Minh City, Vietnam;* [2]*Monash University, Melbourne, VIC, Australia;* [3]*University of Economics Ho Chi Minh City, Ho Chi Minh City, Vietnam*

Chapter outline

1. Introduction

Over the past decades, Asian countries have experienced tremendous economic growth but along with that, natural resources and ecosystems have also been severely damaged, and lack of water resources and the amount of hazardous waste is also increasing (Le et al., 2019a). These environmental issues have had a negative impact on public health and social welfare. Extreme weather, air pollution, groundwater pollution, microplastic waste, and marine plastic are currently emerging issues that can reverse the recent progress of the region. Countries in the region have put in place many policies and measures to protect and manage the environment, although the effectiveness can be brought to varying degrees. This study attempts to compute a composite index of environmental sustainability to evaluate the trend for selected countries in Asia and hence expects to recommend regional policy directions.

For the past decades, environmental sustainability has emerged as a well-researched yet controversial issue. Specifically, given the currently parlous state of our planet, it has been established to be of utmost importance and relevance to humanity's existence, and thus has received extensive coverage in both media and academia. However, despite the considerable coverage, sustainability itself remains

Handbook of Energy and Environmental Security. https://doi.org/10.1016/B978-0-12-824084-7.00017-5
Copyright © 2022 Elsevier Inc. All rights reserved.

a complicated concept to precisely define. After years of being used so commonly and ubiquitously, sustainability appears to have become a corporate buzzword, more often associated with everyday products (Hopkins et al., 2009), which makes it appear less imperative and vital than how the concept should be perceived. Historically, sustainability has been formally defined by professionals in the field, under the guidance of the United Nations' Brundtland Commission (Brundtland, 1987), as activity or development that meets the needs of the present while preserving resources to ensure sufficient remaining resources for future generations. This definition is accompanied by the notion that long-term sustainability is only achieved through harmonious integration of environmental protection, economic development, and social equity across the globe. Prevalently, the core of sustainability centers around the three interconnected pillars of environment, economics, and social equity (Basiago, 1998; Boyer et al., 2016).

Out of the aforementioned three, environmental sustainability stands out as the most pressing matter. While economic development seems to accelerate and social equity reinforced as time goes by, natural systems have been deteriorating and at a breaking point in many parts of the world. Observable changes and degradation in the ecosystems require policymakers to adopt sustainable development (Gigliotti et al., 2019), especially given that one essential requirement for attainment of economic and social sustainability is, for the most part, dependent on a healthy environment (Pearce and Barbier, 2000). Ensuring environmental sustainability has been identified as one of the eight Millennium Development Goals by the United Nations in the 2000, and later disaggregated into more specific goals, such as climate action and affordable and clean energy, in the succeeding 17 Sustainable Development Goals in 2016. The principles of environmental sustainability are critical for policy-makers in designing modern management strategies to protect and safely consume natural resources in order to secure an environmentally sound and life-sustaining global ecosystem.

It is important to enforce environmental laws and regulations to secure ecological sustainability because the environment plays a critical role in the life of human beings by directly affecting their health status and well-being. Contaminated environment due to human activities could cause malnutrition and diseases, morbidity, and shortening of life span (Iwejingi, 2011). A report by the United Nations, Global environment outlook GEO 6 (2019), claimed that one in four premature deaths and diseases worldwide is due to manmade pollution and environmental damage, causing approximately nine million deaths in 2015 alone. Deadly smog-inducing emissions, chemicals polluting drinking water, and the accelerating destruction of ecosystems crucial to the livelihoods of billions of people are driving a worldwide epidemic that hampers the global economy. Environmental sustainability is further threatened by increasing globalization which encourages economic activities, i.e., increased farming, mining, fishing, consumption of fossil fuels, and water usage, which lead to resource depletion of several types: deforestation, aquifer depletion, mining for fossil fuels and minerals, soil erosion, and pollution or contamination of resources—all of which may cause disastrous impacts on human beings. Climate change stimulated by these anthropogenic factors may boost the possibility and intensity of natural disasters, most commonly droughts, storms, and tornados, which damage cities on a large scale.

In a nutshell, environmental sustainability is crucial because biological diversity is an overarching perquisite for most of life-supporting services, and necessary services can only be supported by a healthy global ecological system. Ultimately, before the needs for economic development and social equity, humans must consume resources in order to survive. Humanity requires oxygen, land, water, food, and other goods and services provided by nature for survival, all of which fall under the concept of environmental sustainability. Because of its significance, along with its interconnectedness with economic sustainability and social sustainability, it is important to be able to critically assess and accurately evaluate the suitability of sustainable environmental policies by adopting and applying appropriate environmental indicators for long-term monitoring.

Nevertheless, very few studies that have evaluated multiple dimensions of environmental sustainability for Asia. To extend this line of research, we construct a comprehensive environmental sustainability index (ESI) based on seven selected indicators in order to examine the trend across 31 Asian countries over the period 1992–2016. The countries included in the study sample are categorized into three distinct groups, i.e., low and lower-middle income countries, upper-middle income countries, and high-income countries, based on the latest country classification by income level (World Bank, 2020) for a thorough analysis. The results indicate that the trend of environmental sustainability varies even across countries belonging to the same income group. For the majority of the countries included in this study sample, we depict the trend of decreasing environmental sustainability. The results of this study expect to generate significant policy implications for Asia to improve its overall environmental sustainability.

The rest of this study will be structured as follows. Section 2 reviews the related literature on the concept of environmental sustainability and its measurement. Section 3 presents the variables, data sources, and methodology. Section 4 discusses the results and implications. In this section, the performance of environmental sustainability across countries in Asia would be compared over time. Section 5 concludes with recommendations on how to improve environmental sustainability in Asia.

2. Literature review

2.1 The concept of environmental sustainability

First conceptualized in 1992 by scientists at the World Bank and then developed in 1995 by Goodland, environmental sustainability is broadly defined as the idea that despite resource depletion through human consumption and activities, the environment can retain its ability to support human life and sustain existing biological systems into the future (Brinkmann, 2020). Goodland (1995) identified the term as a set of constraints to regulate major economic activities including the use of nonrenewable and renewable resources and the assimilation of pollution and waste. At the same time, Holdren et al. (1995) focused on biogeophysical aspects to conceptualize environmental sustainability, defining it as a continued effort to maintain and improve the integrity of the Earth's ecological systems and sustain the biosphere while providing sufficiently for the current and future generations to achieve economic and social development. Since then, the term has gradually become commonly established. Sutton (2004) defined it as "the ability to maintain the qualities that are valued in the physical environment." Environmental sustainability later also encompassed the goal of promoting life-supporting sustainably engineered systems without harming the natural ecosystems by the US National Science Foundation in 2009.

Fundamentally, environmental sustainability is a multifaceted concept. In order to comprehensively account for it, one must acknowledge the basic principles embodied in the term. The Current Opinion in Environmental Sustainability—the first scholarly journal to synthesize and review research on environmental change and sustainability—disintegrates environmental sustainability into six major areas including climate systems, human settlements and habitats, energy systems, terrestrial systems, carbon and nitrogen cycles, and aquatic systems. In 2001, the OECD Environmental Strategy for the First Decade of the 21st century stated four specific criteria for environmental sustainability comprising regeneration and substitutability with regards to renewable and nonrenewable resources, as well as assimilation and avoiding irreversibility regarding hazardous pollutants. Later, Moldan (2012) added more primary principles of the concept including long-term perspective, flexibility, regard for different scales, respect for nature and biological diversity, account for feedbacks, significance of local conditions, and understanding of the nonlinear evolution of complex systems. It is important to recognize that the ecosystems and natural services are interlinked to human well-being, and that environmental sustainability is, in essence, to maintain those services at a suitable level. Thus, construction of a complete indicator for environmental sustainability requires acknowledgment of the ecosystem services, both quantitatively and qualitatively (Moldan, 2012).

2.2 Constructing a composite index for environmental sustainability

Numerous attempts to quantify and evaluate a country's level of environmental sustainability attainment have been made by international organizations since the late 1990s (Tambouratzis, 2016). The primary objective of constructing an indicator is to provide a general and broad-based "snapshot" evaluation of the country's quality and sustainability of the environment (Halsnæs and Shukla, 2008; Chen, 2009). The indicator may serve as guidance to critically assess and accurately measure the suitability of sustainable environmental policies for long-term monitoring. The indicator, ideally, should take into account the state of natural resources and impacts of sustainability issues on global environmental change, including but not limited to air, land, water, urban life, and human behaviors and activities (Parmar and Bhardwaj, 2013; Agrawal et al., 2014; Kashem et al., 2014; Yigitcanlar and Teriman, 2015; Ramadass et al., 2015). The focus of building a composite index for environmental sustainability is to demonstrate the performance of the living planet along with its resources, fauna and flora, and human population and habitat with regards to the continuous process of attaining survival and development into the future (Huang and Shih, 2009). Thus, the index facilitates the understanding of the country in how well in terms of environmental sustainability they are faring compared to other nations, and how they can achieve higher level of sustainability by adopting approaches of countries with similar parameter values but has a higher sustainability score (Morse, 2016).

The last two decades have witnessed a considerable appearance of indices for environmental sustainability (De Sherbinin et al., 2013) as well as guidelines on how to construct the indices optimally (Zhou et al., 2006). The indices include, but are not limited to, the ecological footprint (Wackernagel and Rees, 1998), the living planet index (WWF, 1998), the empirical ESI (Sutton, 2003), the environmental vulnerability index (Kaly et al., 2004), the ESI (Samuel-Johnson and Esty, 2000; Esty et al., 2005), sustainable transport indicators (Rassafi and Vaziri, 2005), the composite environmental index (Zhou et al., 2006), the watershed sustainability index (Chaves and Alipaz, 2007), the environmental performance index (Hsu et al., 2016), and the living planet index (McRae et al., 2017). The majority of these indices aims to provide extensive data on global metrics for environment and sustainability-related matters.

Although an extensive collection of environmental sustainability-related indices exist, many researchers prefer to build their own composite index that is more relevant and meaningful to their study. For instance, to measure the low-carbon performance in Chinese provinces, Zhou et al. (2015) adopted various different carbon-related indicators. The process typically requires multiple steps, i.e., selection of indicators, normalization of data, weighting, and aggregation that aligns with the methodological development of composite indicators. A composite indicator or index refers to a mathematical combination with specified weightages of individual indicators to measure and summarize a multifaceted concept. Generally, in order to construct a composite index, a comprehensive indicator framework and an appropriate allocation of suitable weights for separate indicators are needed (Romero et al., 2018).

In the economic literature, several attempts at constructing a composite index for environmental sustainability have been made. Two most widely employed methods are the simple additive weighting (SAW) method and the weighted product (WP) method. For example, Hahn (2000) constructed an ESI for 76 countries to assess the economic impact of environmental policies using the SAW method, while Ebert and Welsch (2004) applied the WP method. Singh et al. adopted the noncompensatory multiple criteria decision analysis to build an index for environmental performance. Khanna and Kumar (2011) debated that a composite index measuring environmental sustainability performance can be better constructed using the weighted displaced ideal. Ultimately, the selection of the set of indicators and weighting method should be based on and relevant to the user's perspective, should contain the specific study-related characteristics, and should meet the requirements and priorities of the study.

3. Methodologies

Based on the existing literature and subject to data availability for Asia, we will select a range of indicators for constructing our composite ESI.

Since the selected indicators rely on different units and different scales, we first normalize them through z-transformation before aggregating them into a composite index. In the standardization procedure, scaling is performed based on deviation from the mean.

Normalization using standardized Z-score is performed as follows:

$$Zee = \frac{X_i - \overline{X}}{\sigma} \tag{14.1}$$

where:

$\overline{X}$ = group average.

σ = standard deviation.

Next step, the data are processed by conducting principal component analysis (PCA). The PCA analysis enables evaluating the impact of fluctuations in certain variables on the final outcome. PCA is commonly used to investigate strong predictors in a dataset since it reveals the data structure and explains the variations (Le et al., 2019b; Radovanović et al., 2018). In the existing literature, PCA has rarely been used to quantify the level of environmental sustainability (see Table A1 in the Appendix). However, since this approach is widely employed in the analysis of the phenomena affected by a group of indicators, the use of PCA in this study to measure the impacts of different components of energy security on its level is justified.

Following Jolliffe (2002), we construct one PCA as follows. Let X be a random vector whose values are taken in $\Re^m$, has a mean μ_X and covariance matrix of $\sum_X$. Eigenvalues of $\sum_X$ in descending order are as follows $\lambda_1 > \lambda_2 > \cdots > \lambda_m > 0$, so that the i-th eigenvalue of $\sum_X$ is the i-th largest of them. Similarly, a vector α_i that corresponds to the i-th eigenvalue of $\sum_X$ is the i-th eigenvector of $\sum_X$. Next, we derive the form of principal components by maximizing $\text{var}\left[\alpha_1^T X\right] = \alpha_1^T \sum_X \alpha_1$, subject to

$\alpha_1^T\alpha_1 = 1$. The Lagrange multiplier method is then applied to solve this optimization problem, as follows:

$$L(\alpha_1, \phi_1) = \alpha_1^T \sum\nolimits_X \alpha_1 + \phi_1\left(\alpha_1^T\alpha_1 - 1\right) \tag{14.2}$$

$$\frac{\partial L}{\partial \alpha_1} = 2\sum\nolimits_X \alpha_1 + 2\phi_1\alpha_1 = 0 \Rightarrow \sum\nolimits_X \alpha_1 = -\phi_1\alpha_1 \Rightarrow \mathrm{var}\left[\alpha_1^T X\right] = -\phi_1\alpha_1^T\alpha_1 = -\phi_1.$$

Since $-\phi_1$ is the eigenvalue of $\sum_X$, with α_1 being the corresponding normalized eigenvector, $\mathrm{var}\left[\alpha_1^T X\right]$ is maximized by choosing α_1 to be the first eigenvector of $\sum_X$. In this case, $z_1 = \alpha_1^T X$ is the first principal component of X, while α_1 is the vector of coefficients for z_1, and $\mathrm{var}(z_1) = \lambda_1$.

Next, we find the second principal component, $z_2 = \alpha_2^T X$, by solving the optimization problem of maximizing $\mathrm{var}\left[\alpha_2^T X\right] = \alpha_2^T \sum_X \alpha_2$ subject to z_2 being uncorrelated with z_1. Because $\mathrm{cov}\left(\alpha_1^T X, \alpha_2^T X\right) = 0 \Rightarrow \alpha_1^T \sum_X \alpha_2 = 0 \Rightarrow \alpha_1^T\alpha_2 = 0$, solving this problem is equivalent to solving maximization of $\alpha_2^T \sum_X \alpha_2$, subject to $\alpha_1^T\alpha_2 = 0$, and $\alpha_2^T\alpha_2 = 1$.

The Lagrange multiplier method is also applied as follows:

$$L(\alpha_2, \phi_1, \phi_2) = \alpha_2^T \sum\nolimits_X \alpha_2 + \phi_1\alpha_1^T\alpha_2 + \phi_2\left(\alpha_2^T\alpha_2 - 1\right) \tag{14.3}$$

$$\frac{\partial L}{\partial \alpha_2} = 2\sum\nolimits_X \alpha_2 + \phi_1\alpha_1 + 2\phi_2\alpha_2 = 0$$

$$\Rightarrow \alpha_1^T\left(2\sum\nolimits_X \alpha_2 + \phi_1\alpha_1 + 2\phi_2\alpha_2\right) = 0 \Rightarrow \phi_1 = 0$$

$$\Rightarrow \sum\nolimits_X \alpha_2 = -\phi_2\alpha_2 \Rightarrow \alpha_2^T \sum\nolimits_X \alpha_2 = -\phi_2.$$

Because $-\phi_2$ is the eigenvalue of $\sum_X$, and α_2 is the corresponding normalized eigenvector, $\mathrm{var}\left[\alpha_2^T X\right]$ is maximized by selecting α_2 to be the second eigenvector of $\sum_X$. In this case, $z_2 = \alpha_2^T X$ is the second principal component of X, while α_2 is the vector of coefficients for z_2, and $\mathrm{var}(z_2) = \lambda_2$. Following the similar procedure, the i-th principal component $z_i = \alpha_i^T X$ could be constructed by choosing α_i to be the i-th eigenvector of $\sum_X$, and has variance of λ_i. The main outcome of PCA is the principal components which are a set of values of linearly uncorrelated variables and have orthogonal vectors of coefficients.

In short, the *ESI* is constructed as follows:

$$ESI_{it} = a_1E_1 + a_2E_2 + \ldots + a_nE_n \tag{14.4}$$

where E denotes the selected input indicators for constructing the composite index and $a_1, a_2, \ldots$ are the coefficients (the Eigenvectors) derived from the PCA analysis for the first principal component after standardizing the variables.

In order to determine the outcome of the PCA, we assign weights to each variable included in the construction of the composite index. The first principal component is regarded as the best representative of the value of the composite index, which is newly established based on the input variables. The resulting weights indicate the correlation degree between an input factor and the composite index. The weights enable us to identify the variables that account for a significant part of the composite index.

The standardization approach leads to all principal components having zero mean. The standard deviation for each component is the square root of the eigenvalue.

Before the PCA is performed, we will conduct two tests, namely, Bartlett's test of sphericity and Kaiser–Meyer–Olkin (KMO) test of the sampling adequacy, in order to examine whether these data are suitable for factor analysis (Hair et al., 1998; Tabachnick et al., 2007). The Bartlett's test of sphericity is conducted to check if the correlation matrix used in the PCA is an identity matrix. The significance of the test statistics is necessary to justify the use of factor analysis (Hair et al., 1998; Tabachnick et al., 2007). Furthermore, we perform the KMO test to assess the sampling adequacy of factor analytic data matrices. The statistic indicates the proportion of common variance among variables that might be caused by underlying factors. The value of the KMO index ranges from 0 to 1, with being greater than 0.5 generally suggesting the suitability for factor analysis (Hair et al., 1998; Tabachnick et al., 2007).

The empirical PCA involves two steps. At the first stage, the factors with the lowest correlation pairwise are identified, and the total variance of the variable contributed by these factors is calculated. This procedure enables us to extract the factors that contribute the largest part of the variation in the original variables. The first factor explains the highest part of the total variation. The second factor that contributes the largest portion of the remaining unexplained variance is then extracted. The first factor and the second factor have no correlation. This procedure is continued until the number of identified components equals the number of variables originally included in the formation of the composite index. Afterward, we can extract the components that account for a portion of variance above a certain threshold level, which is generally set at one (Mundfrom et al., 2005). Based on the constructed composite index for environmental sustainability, we are able to identity which components are the critical drivers of environmental sustainability in each country during the sample period. We can also depict the trend of environmental sustainability across countries over time for comparison. The results of this study are thus expected to generate significant policy implications for governments in Asia to take action on various environmental issues.

4. Empirical results

Table 14.1 provides the list of 31 countries included in the study sample categorized into three distinct groups, i.e., low and lower-middle income countries, upper-middle income countries, and high-income countries, based on the latest country classification by income level (World Bank, 2020). The environmental indicators and their units considered for this study are listed in Table 14.2, with one air quality and pollution, one forest and biodiversity, two land use and agriculture, one human health and disaster, and two population indicators, forming a total of seven comprehensive indexes for a specified Asian sample. Subject to the data availability, we consider the period spanning from 1992 to 2016. The summary statistics for these variables are then presented in Table 14.3. It can be observed from Tables 13.2 and 13.3 that the variables have different units and are on different scale, and there exists substantial variance in some variables compared to others. This justifies the necessity of normalization of variables prior to the adoption of PCA.

Generally, the variances for most of the variables are large, indicating a significant disparity in the environmental sustainability performance of the countries in the sample. CO_2 emissions, which indicates the air quality and pollution theme in the environmental sustainability composite index, have a minimal value of 0.066 metric tons per capita and a maximal value of 24.627 metric tons per capita,

Table 14.1 List of countries in the study sample (31 countries).

Low and lower-middle income countries (16)
Bangladesh, Bhutan, Cambodia, India, Korea Dem. People's Rep., Kyrgyz Rep., Lao PDR, Mongolia, Myanmar, Nepal, Pakistan, Philippines, Sri Lanka, Tajikistan, Uzbekistan, Vietnam
Upper-middle income countries (10)
Azerbaijan, China, Georgia, Indonesia, Kazakhstan, Malaysia, Maldives, Thailand, Turkey, Turkmenistan
High-income countries (5)
Brunei Darussalam, Cyprus, Japan, Korea Rep., Singapore

Table 14.2 Asia's Environmental Sustainability indicators considered (1992–2016).

Variable	Theme	Indicator	Unit
POLLUTION	Air quality and pollution	CO_2 emissions (metric tons per capita)	
FOREST	Forest and biodiversity	Forest area (% of land area)	%
LAND1	Land use and agriculture	Arable land (% of land area)	%
LAND2	Land use and agriculture	Agricultural land (% of land area)	%
HEATH	Human health and disaster	Mortality rate, infant (per 1000 live births)	
POP1	Population pressures on ecosystem services	Fertility rate, total (births per woman)	
POP2	Population pressures on ecosystem services	Population growth (annual %)	%

Table 14.3 Summary on statistical descriptions of the selected variables.

Variable	Obs	Mean	Std. Dev.	Min	Max
POLLUTION	775	4.011	4.521	0.066	24.627
FOREST	775	32.087	24.661	1.226	82.108
LAND1	775	16.094	14.339	0.365	66.137
LAND2	775	37.514	22.078	0.931	83.981
HEALTH	775	34.345	24.337	2.000	102.700
POP1	775	2.590	0.953	1.076	6.007
POP2	775	1.349	1.044	−9.081	7.786

showing a mean of 4.011 and a standard deviation of 4.521 metric tons per capita. Forest area for each country in the study sample ranges from 1.226% to 82.108%, with a mean of approximately 32% and a standard deviation of 24.6%. Regarding arable land and agricultural land, the range also varies widely from around 0.37% to 66% of land area for the former and from 1% to 84% of land area for the latter. Infant mortality rate, which indicates human health and disaster, has arguably the largest variance among the variables. For the theme of population pressures on ecosystem services, fertility rate and population growth have a seemingly small variance, with means of approximately 2.6 births per woman and 1.3% growth, respectively.

The results for Bartlett's test of sphericity and KMO Measure of Sampling Adequacy presented in Table 14.4 suggest the suitability of these data for factor analysis as the *P*-values for Bartlett test are 0.00, i.e., smaller than the significance level alpha = 0.01, and the KMO values are greater than 0.5 for all cases. The estimated factors and their eigen values are reported in Table 14.5. From the results, the first three factors are considered for the PCA on Asia's ESI, as together they constitute roughly 80%, which is an acceptably large percentage, of the total variance of the index.

Table 14.6 displays the estimated magnitudes of the coefficients, which denote each chosen variables' impacts to that component. The magnitudes of the coefficients differ depending on the corresponding variables' variances. Due to standardization prior to the PCA process, all variables have zero mean and there exists zero correlation between the principal components. Identification of the variables that are most strongly correlated with each component is important in order to interpret the principal components. For this study, we consider a correlation of 0.5 and above—in either direction from zero—to be statistically significant, and reckon it as the baseline for interpretation of the results.

For normalized variables using standardized Z-score, the first principal component is significantly correlated with two of the normalized variables, namely, LAND2 and HEALTH. The first principal component increases with increasing LAND2, which is agricultural land (% of land area), and HEALTH, which denotes infant mortality rate. The finding hints at a more positive view to agriculture in terms of environmental impact. Traditionally, the expansion of agricultural land is associated with anthropogenic activities that cause widespread and severe soil degradation and contamination by pollutant substances (Hou et al., 2018). This problem has been prominent in Asia, especially in rice-growing countries of East and Southeast Asia such as China, Thailand, Korea, Vietnam, Myanmar, etc.

Table 14.4 Results of Bartlett test of sphericity and Kaiser–Meyer–Olkin measure of sampling adequacy.

	Bartlett test of sphericity			Kaiser–Meyer–Olkin measure of sampling adequacy
	Chi-square	Degrees of freedom	*P*-value	
Z-score normalization	26462.8[a]	21	0.000	0.623
Min-max normalization	2229.0[a]	21	0.000	0.548
Softmax normalization	7684.2[a]	21	0.000	0.552

Note: Bartlett test of sphericity: H_0: variables are not intercorrelated.
[a]Indicates statistical significance at 1% level.
Source: Authors' calculations.

Table 14.5 Total variance explained.

	Component	Eigenvalue	% of variance	Cumulative variance (%)
Normalized variables using standardized Z-score	1	2.90	41.44	41.44
	2	1.64	23.39	64.83
	3	1.08	15.43	80.26
	4	0.68	9.78	90.04
	5	0.44	6.31	96.35
	6	0.26	3.65	100.00
	7	0.00	0.00	100.00
Normalized variables using min-max normalization	1	2.68	38.35	38.35
	2	1.61	23.02	61.37
	3	1.08	15.45	76.81
	4	0.75	10.74	87.55
	5	0.52	7.39	94.94
	6	0.21	2.94	97.88
	7	0.15	2.12	100.00
Normalized variables using softmax normalization	1	2.90	41.48	41.48
	2	1.75	25.02	66.50
	3	1.13	16.20	82.70
	4	0.63	9.03	91.73
	5	0.38	5.49	97.22
	6	0.19	2.78	100.00
	7	0.00	0.00	100.00

Recently, more attention has been paid to the alleviation and remedy of agricultural expansion's negative influences on the ecosystems of the region. For example, the Chinese government has adopted various soil governance policies and soil restoration programs, as well as publishing a Soil Pollution Prevention and Control Action Plan in May 2016 (Wang et al., 2016; Coulon et al., 2016). Apart from soil, programs and policies to address other aspects of the environment such as agricultural species and genetics conservation, more eco-friendly cropping systems, and integration of rice-fishery systems have been introduced (Yu and Wu, 2018).

Our results suggest that the efforts to preserve and restore the ecosystems while expanding agricultural land and activities have been effective and have the potential to enhance environmental sustainability in selected Asian countries. On the other hand, the finding that environmental sustainability and infant mortality rate are positively correlated is an unprecedented one. This could possibly be explained by the divergence of governmental investment and public expenditure from healthcare infrastructures to environmental protection and restoration. A low healthcare coverage would

Table 14.6 Impact assessment of selected indicators on the principal components of the composite environmental sustainability index in 31 Asian countries (1992–2016).

	Normalized variables using standardized Z-score			Normalized variables using min-max normalization			Normalized variables using softmax normalization		
	Principal component (80%)			Principal component (76%)			Principal component (82%)		
Variable	**1**	**2**	**3**	**1**	**2**	**3**	**1**	**2**	**3**
POLLUTION	−0.187	−0.466	0.560	−0.391	−0.161	0.494	−0.178	−0.462	0.545
FOREST	−0.480	0.069	−0.284	−0.326	0.507	−0.377	−0.482	0.081	−0.293
LAND1	0.318	0.174	−0.483	0.336	−0.221	−0.367	0.306	0.113	−0.576
LAND2	0.550	−0.206	0.076	0.335	−0.583	0.126	0.547	−0.215	0.070
HEALTH	0.550	−0.206	0.076	0.508	0.234	−0.106	0.548	−0.211	0.060
POP1	0.169	0.624	0.147	0.466	0.393	0.159	0.197	0.618	0.153
POP2	0.023	0.523	0.582	0.201	0.345	0.653	0.061	0.543	0.504

negatively affect the country's healthcare sustainability, thus adversely influencing infant mortality rate (Saleem et al., 2019). The finding indicates that while maintaining environmental sustainability is important, governments in selected countries should not neglect healthcare services and other vital aspects that would affect the citizens' standard of living. Moreover, the two indicators LAND2 and HEALTH are implied to vary together. In layman's term, if agricultural land expands, infant mortality rate also rises, and vice versa. The correlation of the first principal component with these normalized variables are roughly the same, at 0.55. Interestingly, despite being closely related to LAND2, the coefficient for LAND1, while also positively correlated with the first principal component, is not statistically significant.

The second principal component increases with POP1 and POP2, hence can be viewed as a measure of population. The third component increases with increasing POLLUTION and POP2. This suggests that rising CO_2 emissions tend to move in tandem with escalating population growth. This is in line with myriad existing researches, including the most recent and relevant ones. Although the coefficient is not statistically significant at -0.48, which is quite close to our benchmark correlation of ± 0.5 and above, the finding suggests a slight possible negative correlation between the first principal component and FOREST. Similarly, for the third principal component, the coefficient for LAND1 is -0.483, hinting at a potential though not statistically significant negative correlation between the two. The results are relatively robust to different techniques of normalization including softmax and min-max standardization. However, the results obtained from softmax normalization is more similar to the main results of this study, while min-max normalization produces some slight differences.

The plots of ESI as measured by PCA for each country in the study sample are presented in Figs. 14.1−14.3. The plots for 31 selected countries are classified into three different income groups, with Fig. 14.1 showing low and lower-middle income countries, Fig. 14.2 showing upper-middle income countries, and Fig. 14.3 showing the last five high-income countries. As discussed in the methodology section, three techniques including Z-score (denoted by the suffix zze in the graphs), min-max (mmx), and softmax standardizations are employed to produce normalized variables to construct the principal components. In order to facilitate comparison between the results from different techniques, three plots of principal component scores based on these three techniques of normalization are displayed on the same graph for each of the 31 Asian countries.

In a nutshell, we find that even for countries belonging to the same income group, the trend of environmental sustainability varies. The trend of decreasing environmental sustainability is observed for the majority of the countries included in this study sample. For low and lower-middle income countries, the trend of decreasing environmental sustainability is the most common and is found for most of the countries in this income group. The exceptions are Cambodia and Tajikistan with a relatively stable and unchanging PCA score, and North Korea, Myanmar, Kyrgyz Republic, Sri Lanka, and Vietnam with an increasing environmental sustainability score. Our findings are in line with those of Shah et al. (2019), where Sri Lanka is found to have substantially increased its environmental sustainability performance over time, and of World Bank (2019) for Myanmar. For the upper-middle income group, improved environmental sustainability is found in only Azerbaijan and Indonesia out of 10 countries. For the five high-income countries in the study sample, a trend of decreasing environmental sustainability is observed, although the trend for Japan is quite unclear, and there have been

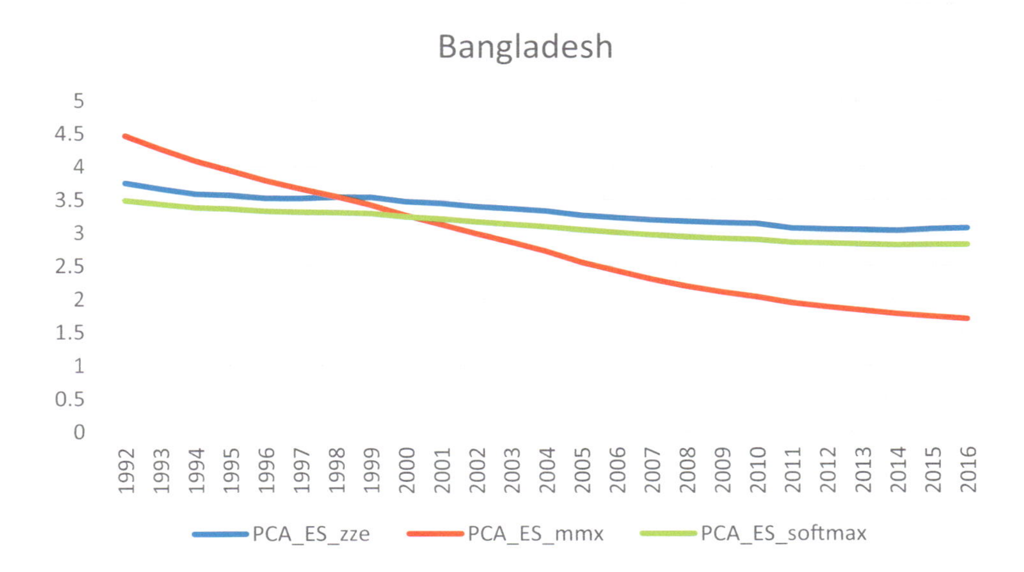

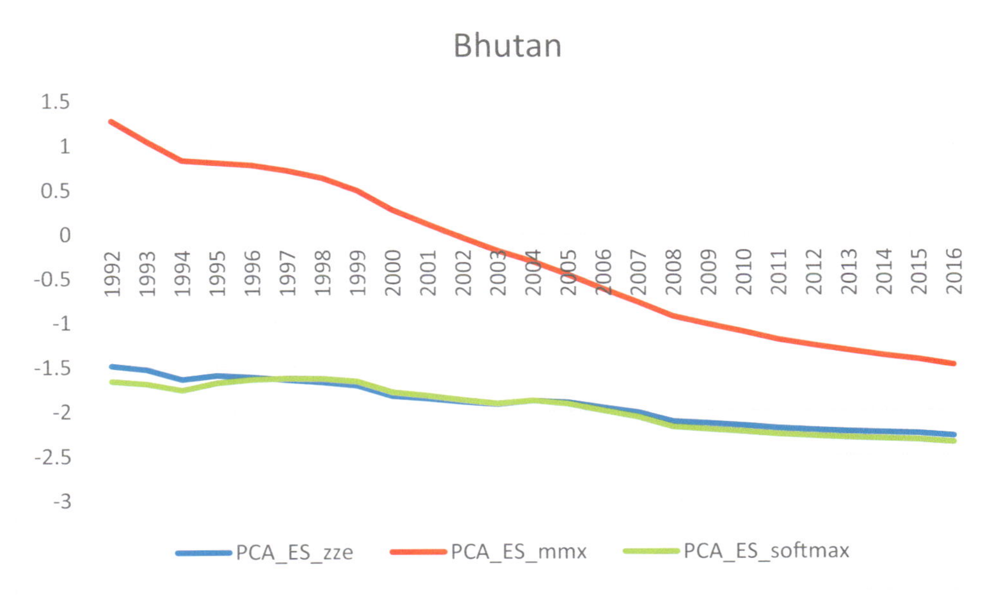

FIGURE 14.1

Environmental Sustainability Index in selected Asian countries.

Low and lower-middle income countries

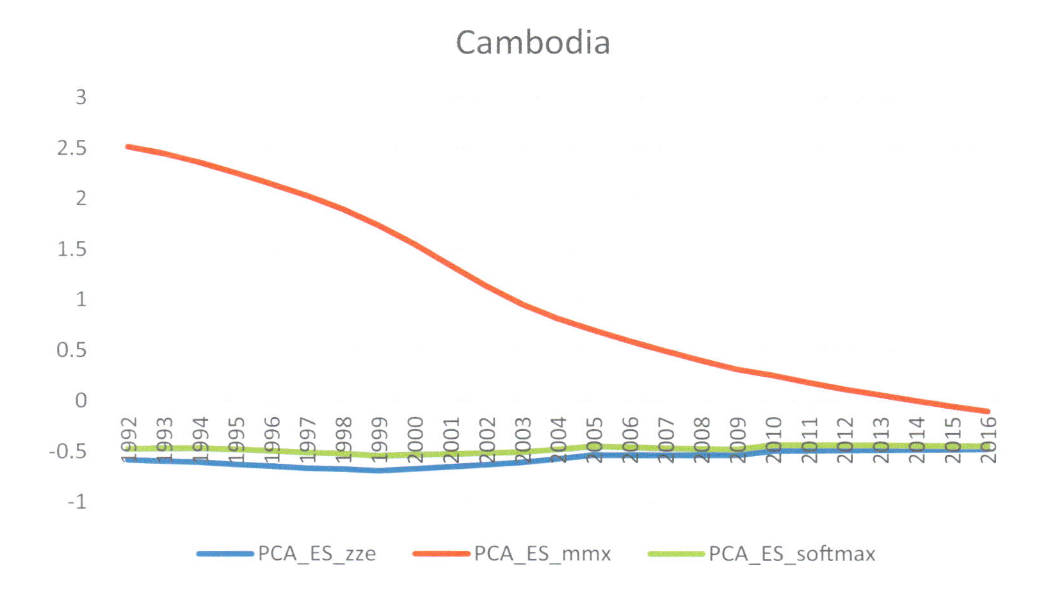

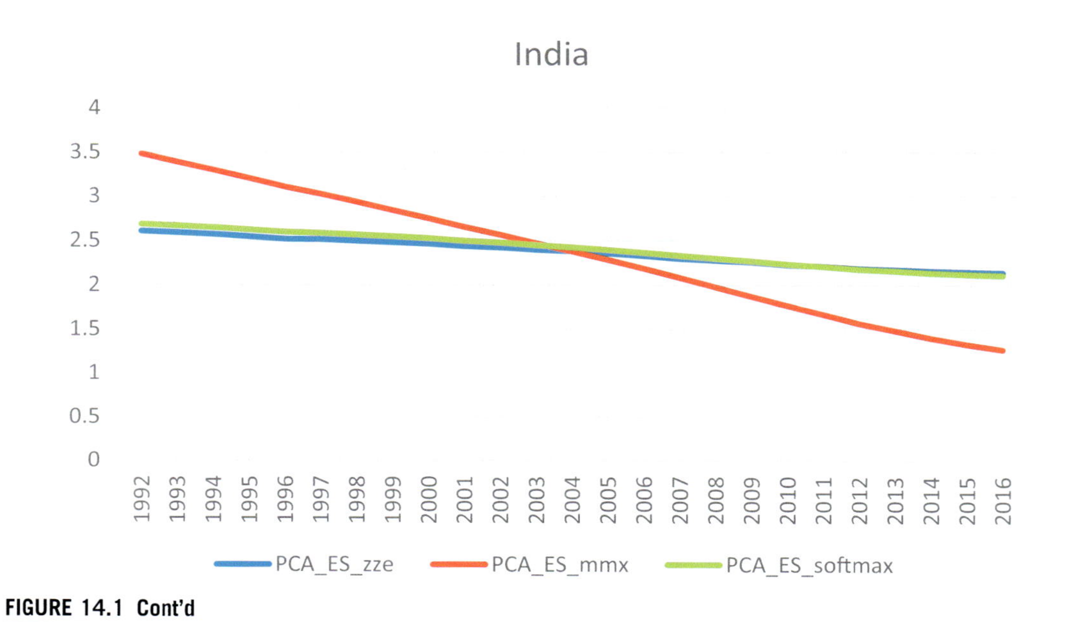

FIGURE 14.1 Cont'd

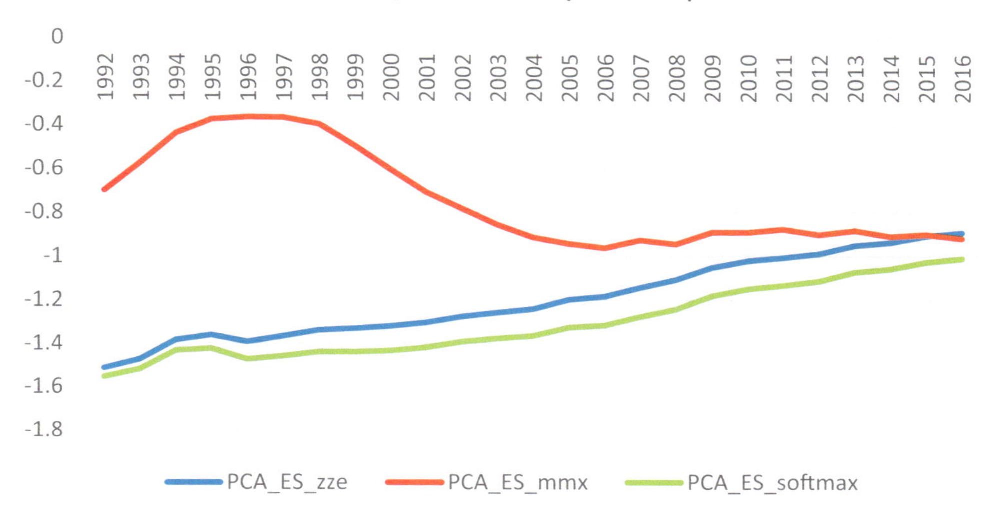

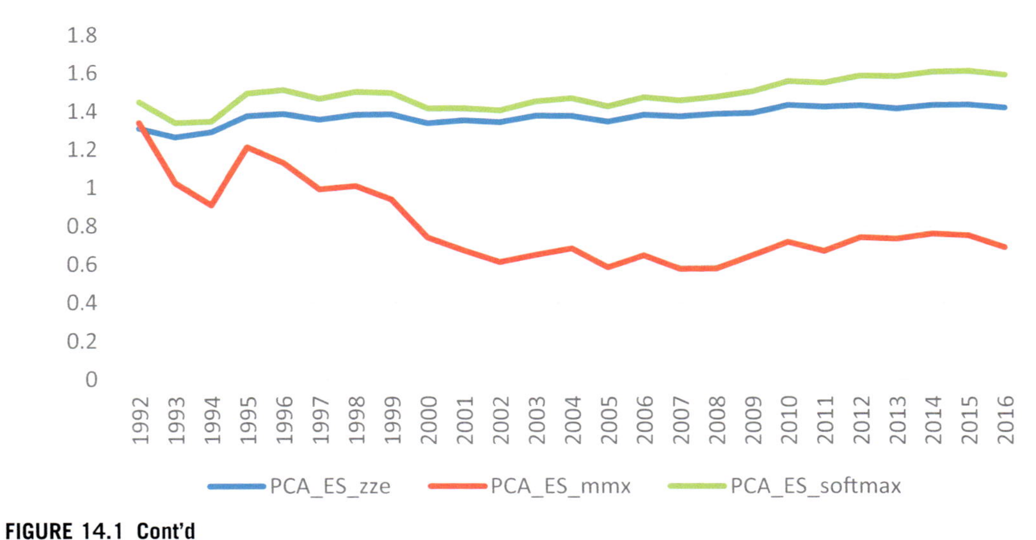

FIGURE 14.1 **Cont'd**

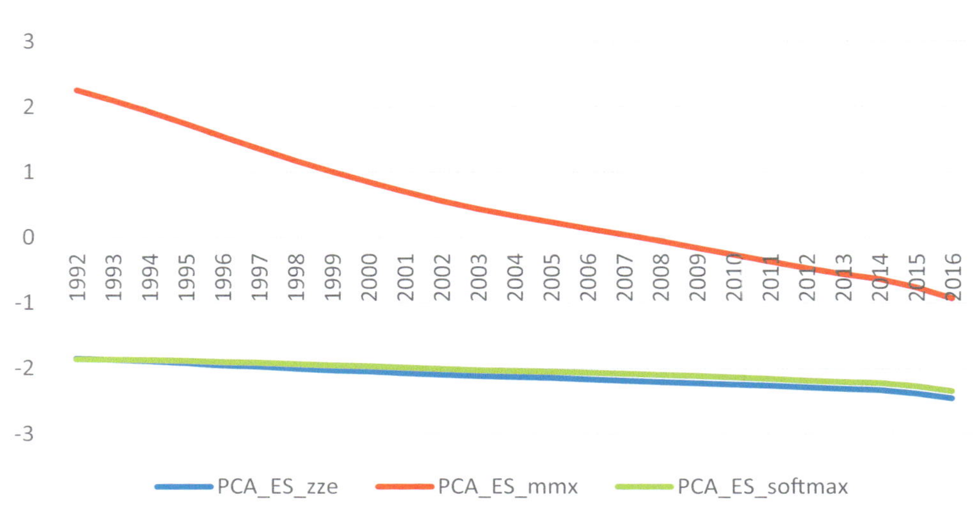

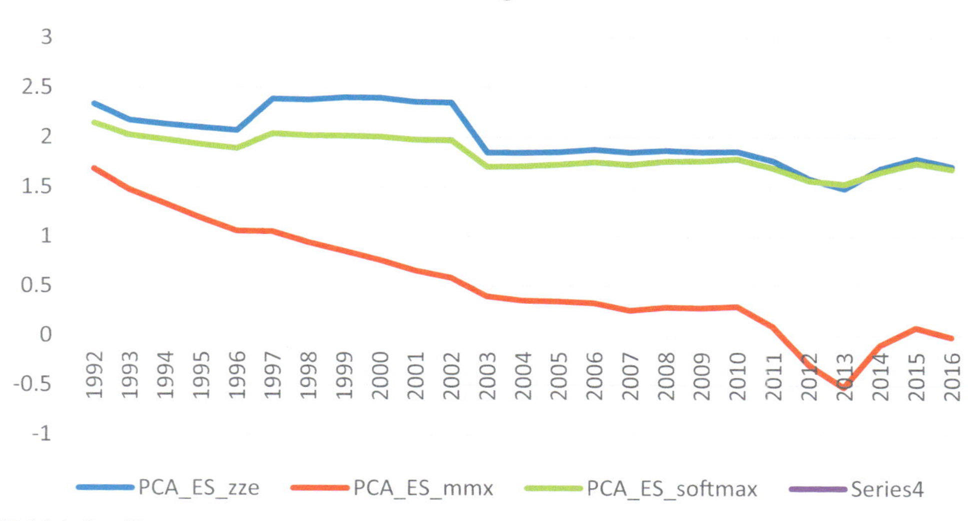

FIGURE 14.1 Cont'd

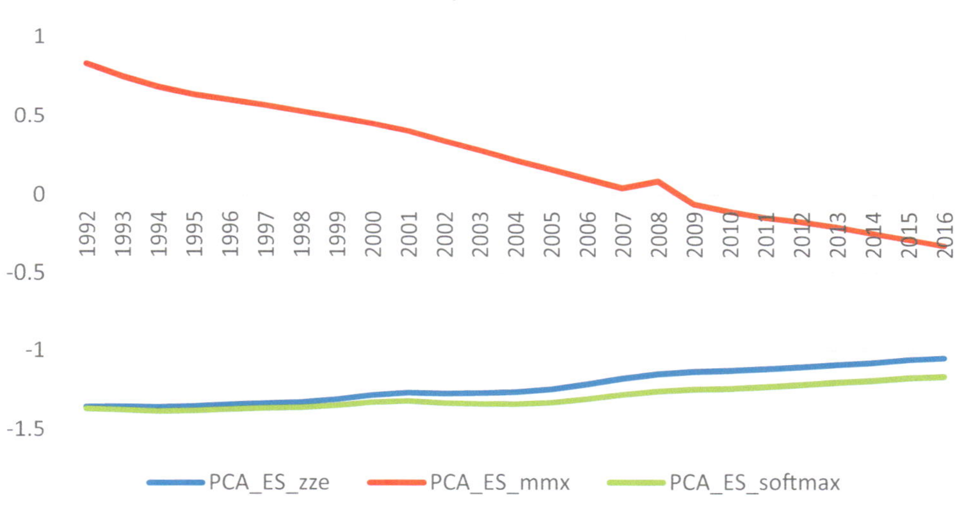

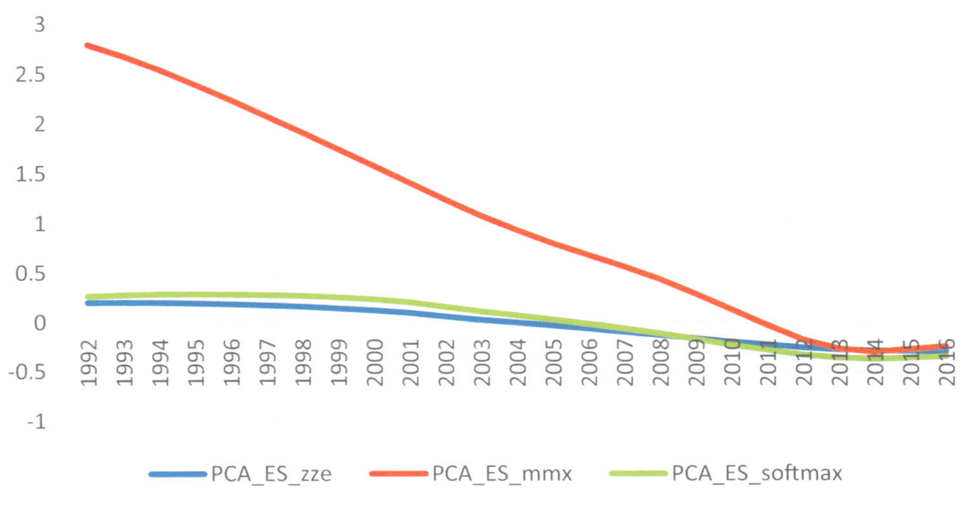

FIGURE 14.1 Cont'd

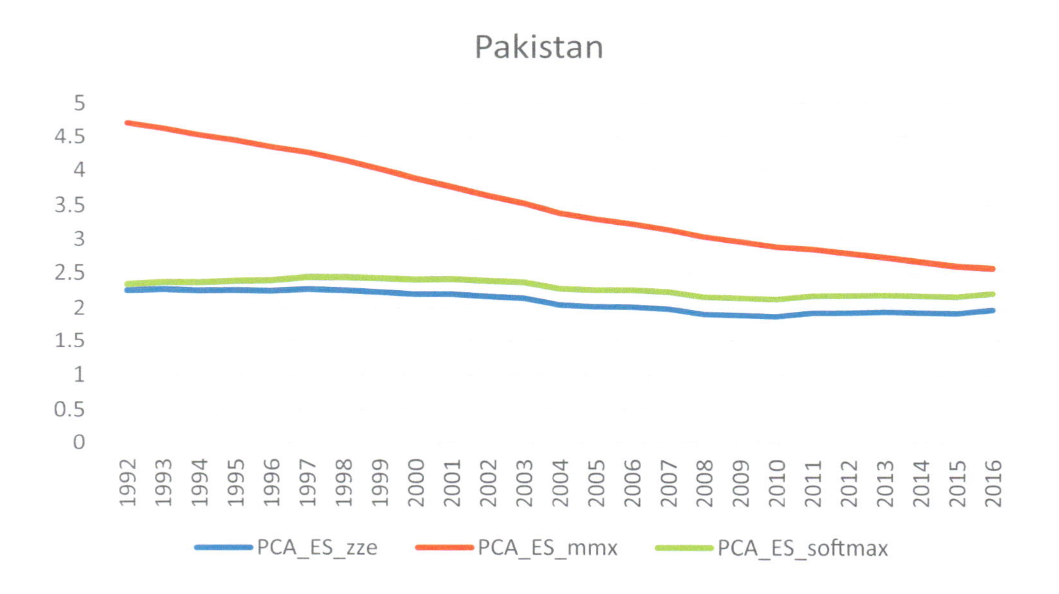

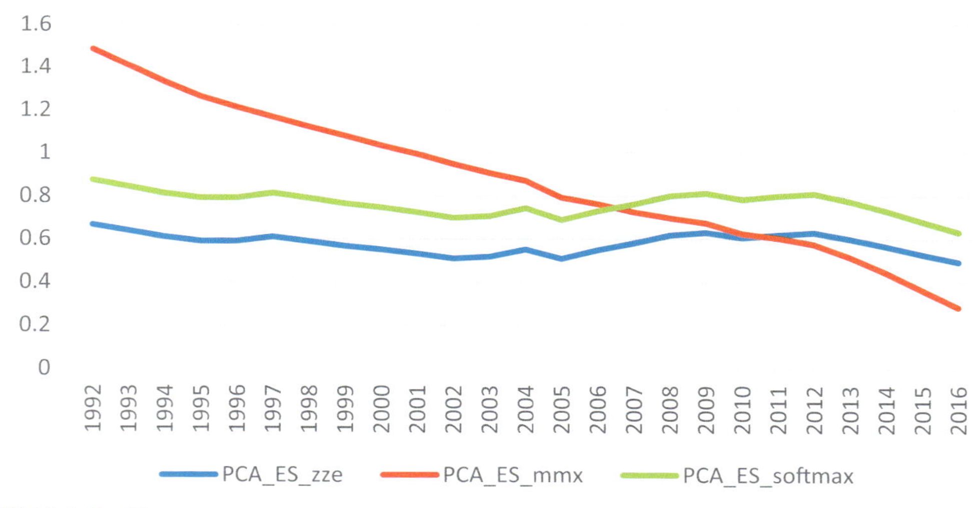

FIGURE 14.1 Cont'd

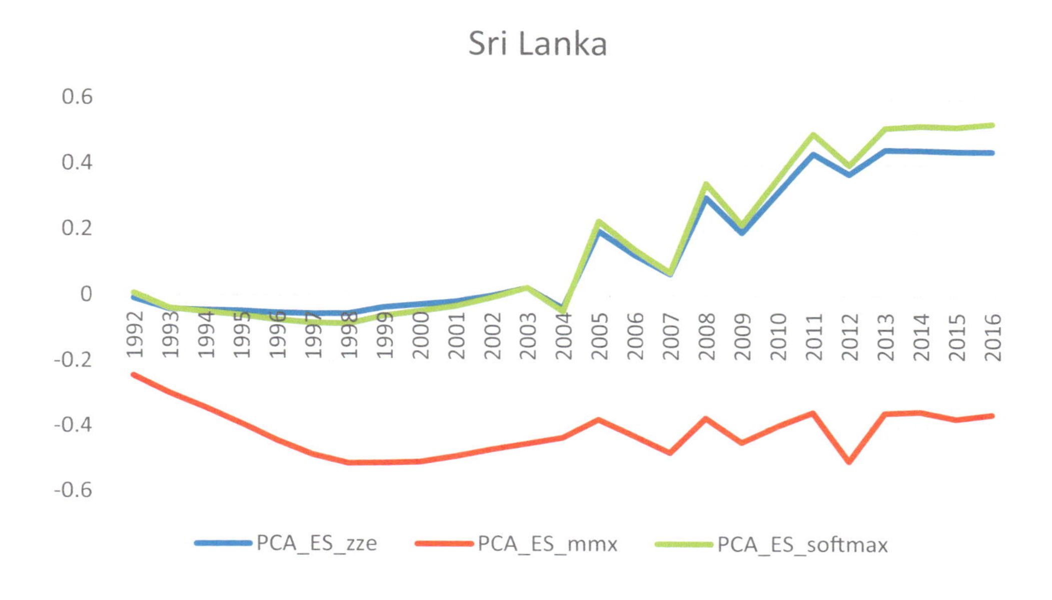

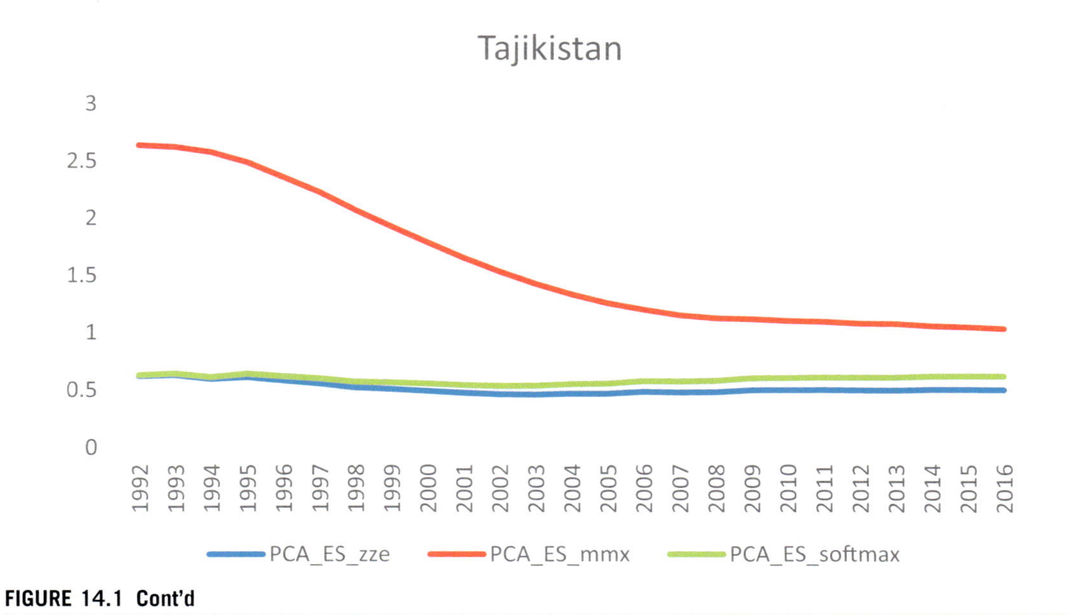

FIGURE 14.1 Cont'd

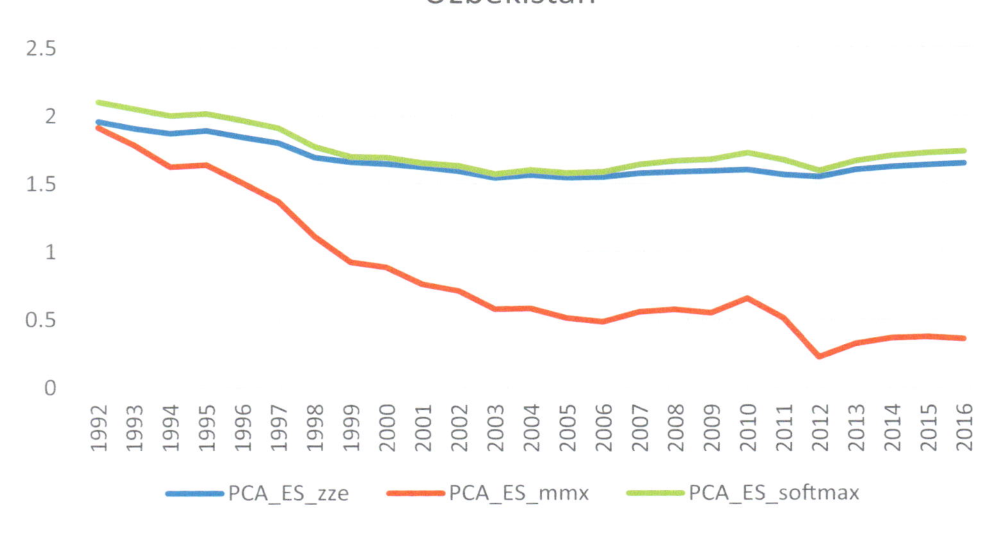

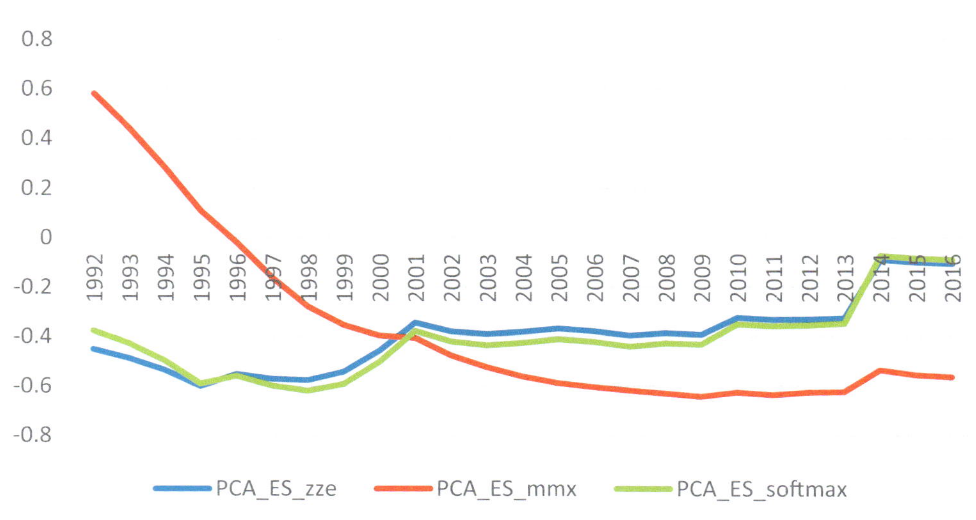

FIGURE 14.1 Cont'd

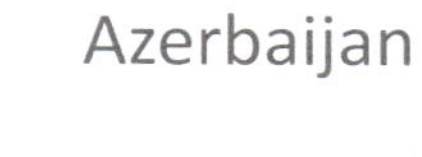

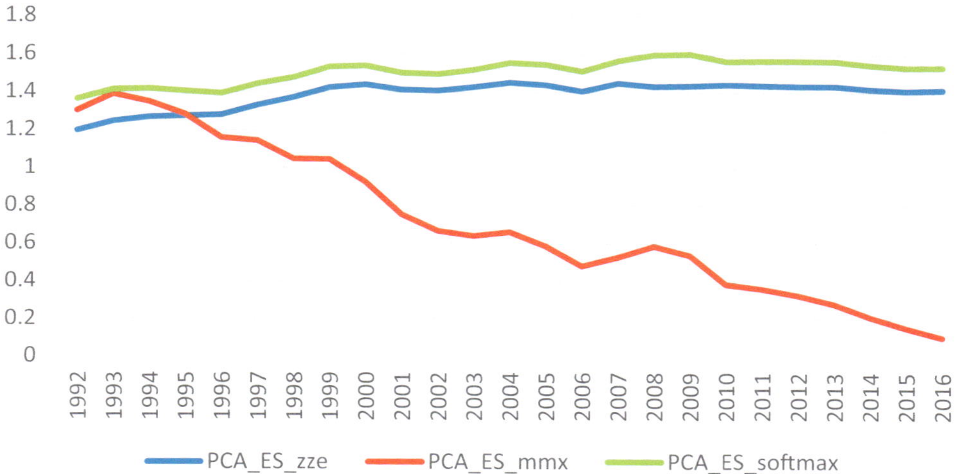

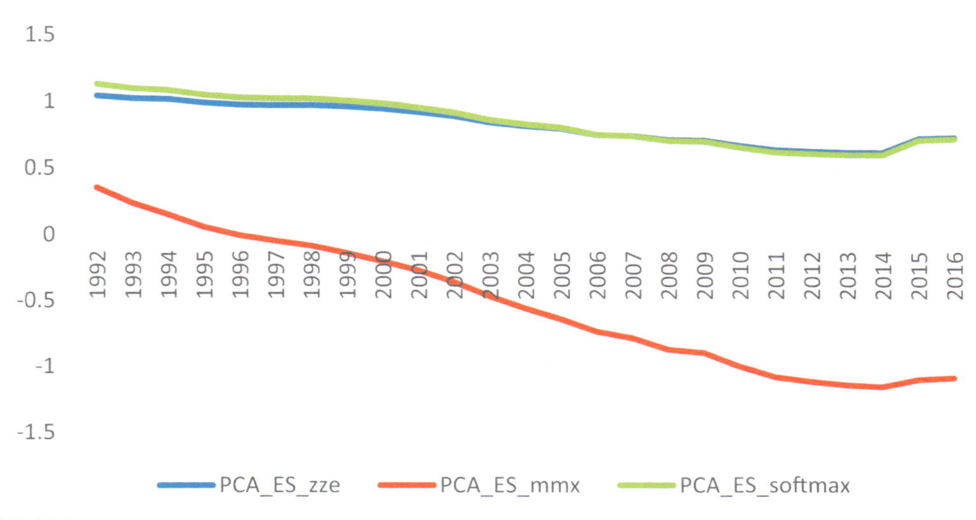

FIGURE 14.2

Environmental Sustainability Index in selected Asian countries.
Upper-middle income countries

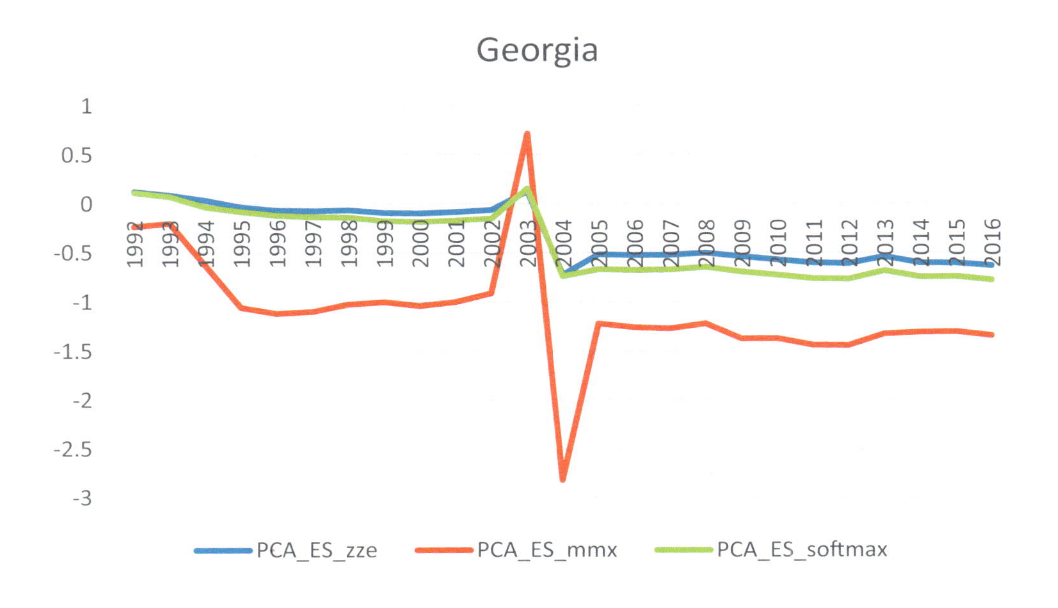

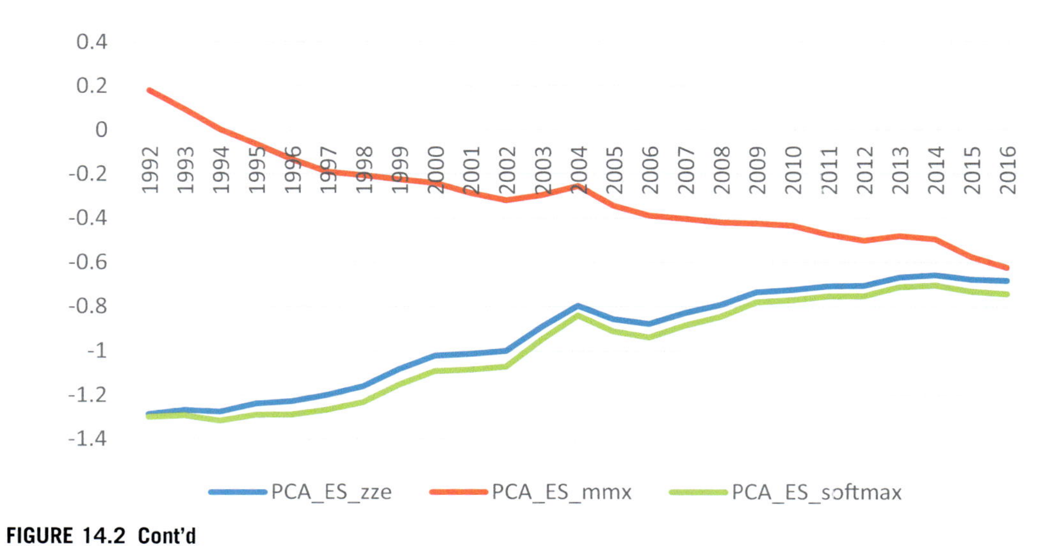

FIGURE 14.2 Cont'd

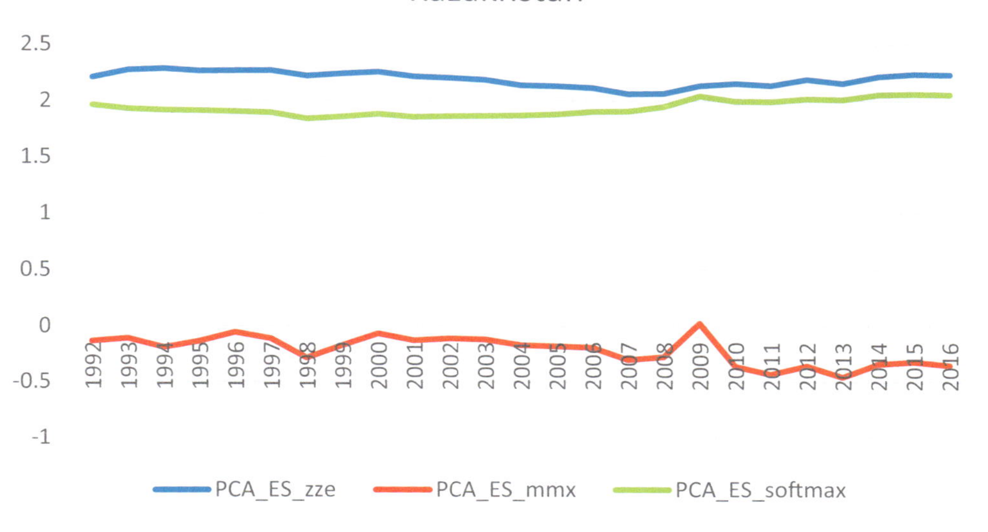

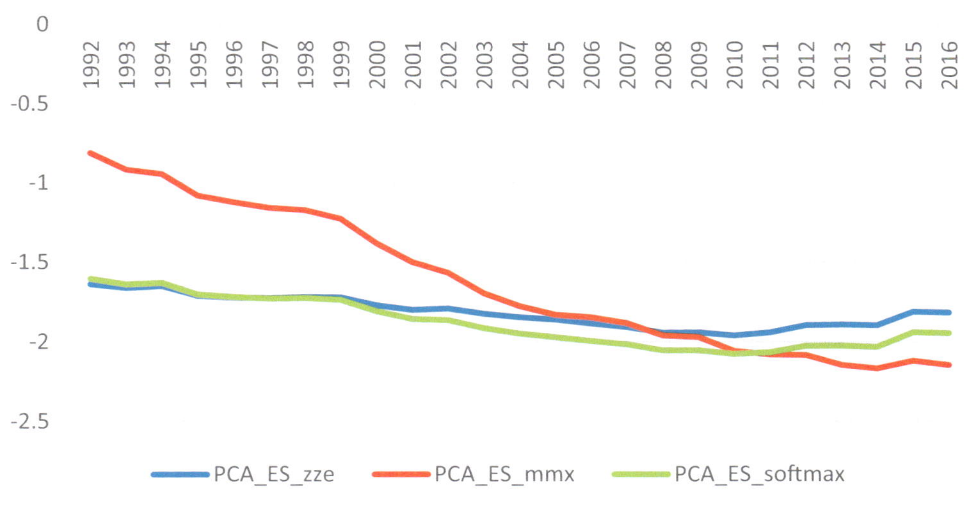

FIGURE 14.2 Cont'd

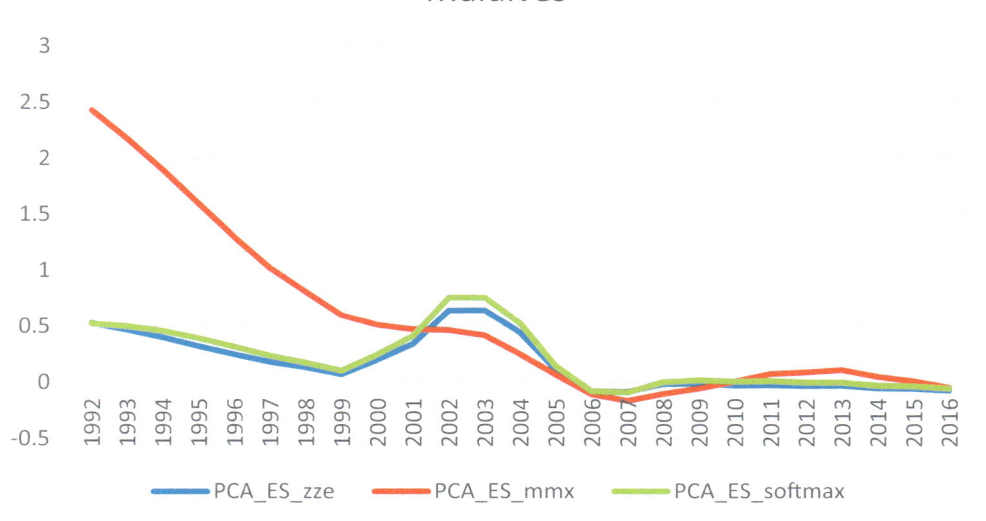

Thailand

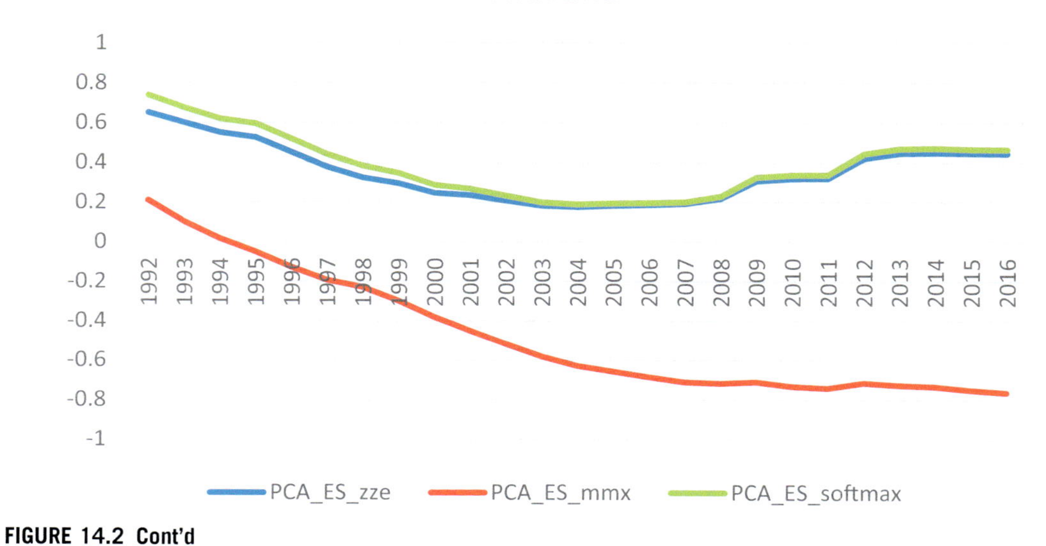

FIGURE 14.2 Cont'd

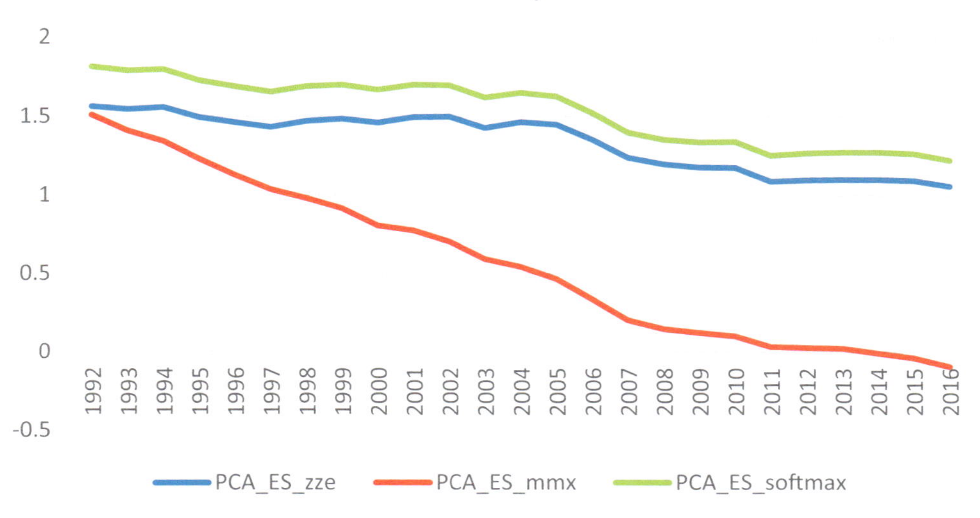

Turkmenistan

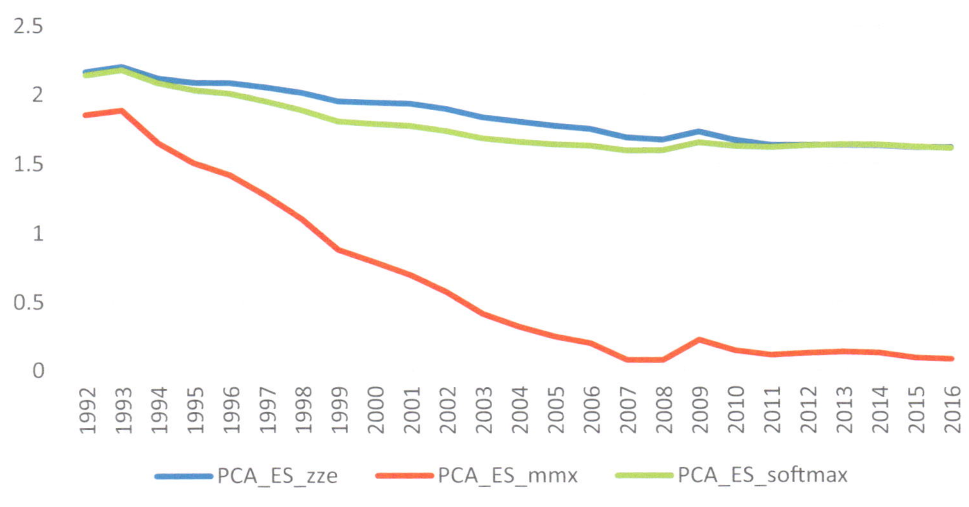

FIGURE 14.2 Cont'd

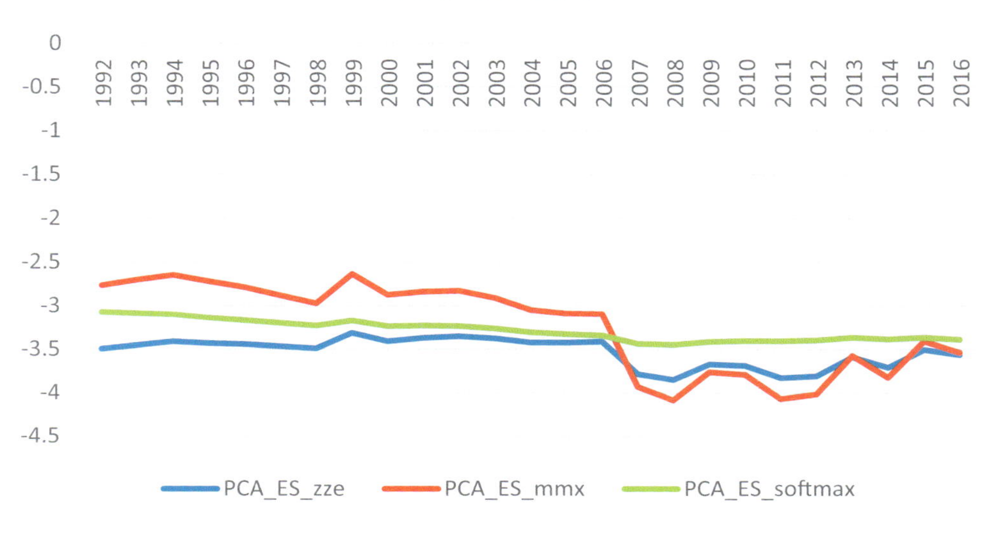

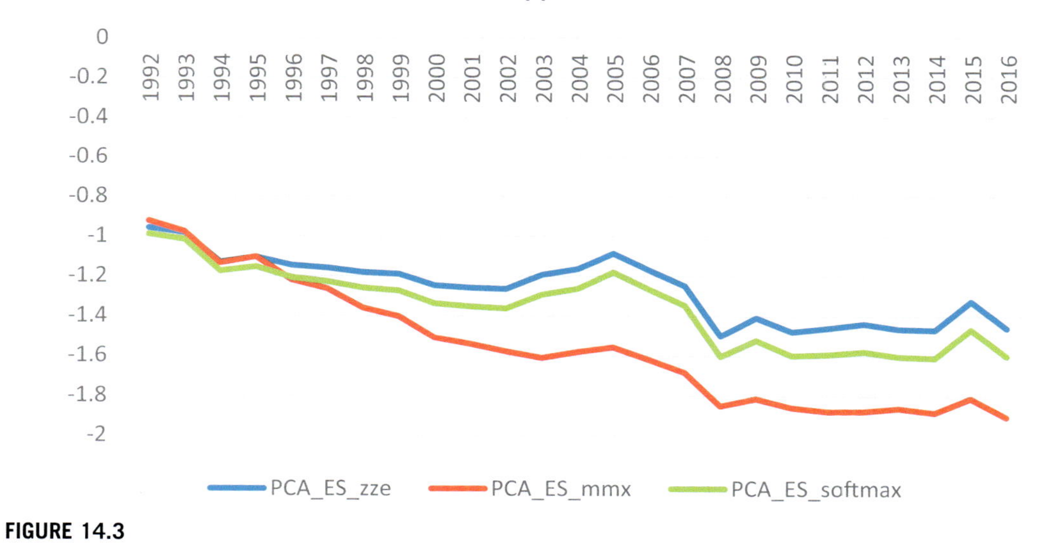

FIGURE 14.3

Environmental Sustainability Index in selected Asian countries.
High-income countries

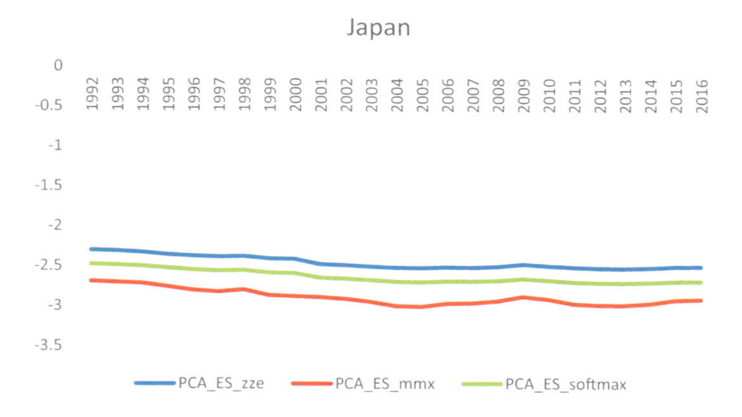

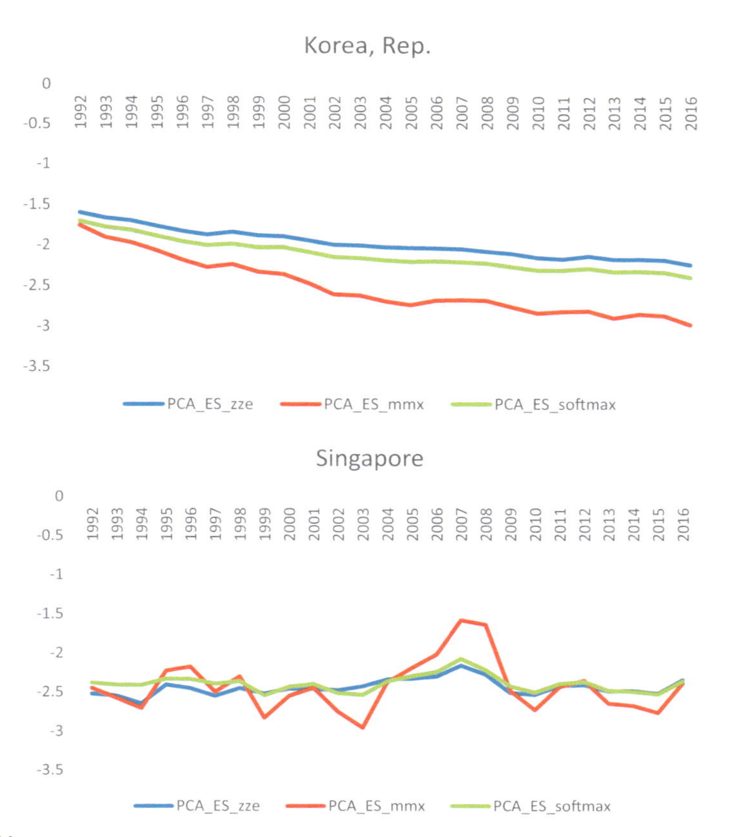

FIGURE 14.3 Cont'd

some slight improvement in the PCA score for Singapore. The results are robust to different techniques of standardization. This finding could be attributable to the differences in institutional quality as well as the diverse political will and governmental actions taken since 1992 to facilitate environmental sustainability of these countries.

5. Concluding remarks

In spite of impressive economic growth and urbanization rates that have lifted millions out of poverty, the environmental degradation in Asia is alarming and has negatively affected human life. However, very few studies have analyzed environmental sustainability issues for the case of Asia from a multiple-dimensional perspective. To fill this gap, this study employs a multivariate quantitative evaluation framework as the mainstream of analysis. We constructed a composite ESI based on seven selected indicators to depict the trend across 31 Asian countries over the period 1992–2016. For a thorough analysis, the countries included in the study sample are categorized into three distinct groups, i.e., low and lower-middle income countries, upper-middle income countries, and high-income countries, based on the latest country classification by income level by the World Bank.

Overall, we find a varying trend of environmental sustainability even across countries belonging to the same income group. For the majority of the countries included in this study sample, the trend of decreasing environmental sustainability is observed. Despite the proximity in income levels, the environmental regimes and policies adopted will affect the state and the sustainability of the environment in the country more directly and significantly than mere income. For instance, Singapore is known as a smart country/city with leading environmental and geospatial innovation that greatly promote urban livability and environmental sustainability (Lim et al., 2021). Green building, effective urban planning with sustainable construction perceptions, proenvironmental behaviors and practices, as well as the adoption of renewable electricity are possible contributors to Singapore's increasing trend of environmental sustainability (Zhang and Tu, 2021). Another example would be Indonesia with various policies and regulations to promote environmental sustainability that are rigorously enacted and reinforced, as well as the adoption of community-based waste management, the implementation of electronic waste management, and ecotourism (Darchen and Searle, 2018). Our findings thus recommend a clear-cut environmental legislative framework, along with effective and efficient urban planning and adoption of high technology electronic devices to both satisfy the increasingly complex human needs and maintain the sustainability of the environment.

Appendix

Table A1 Existing studies on the environmental sustainability index.

No	Authors	Country case	Data period (frequency)	Methodologies	Main variables	Other variables	Findings
1	Sun et al. (2020)	South Asian countries	2001–2015	Common weight data envelopment analysis (DEA)-like mathematical model.	Environment (CO_2 intensity, carbon emissions per capita, CO_2 from electricity generation, ratio of forest area, proportion of public green space, CH_4 emission), energy (energy consumption per capita, carbon-based energy imports, renewable energy sources, diversity in imported energy supply), economic (energy intensity, R&D expenditure per capita, loan volume to GDP).		Produced a ranking for environmental sustainability performance of South Asian countries.
2	Wendling et al. (2018)	180 countries	1995–2020	Once all indicators have been constructed on the 0–100 point scale, they are aggregated at each level of the framework hierarchy. At each level of the aggregation, a simple weighted arithmetic average is calculated.	32 performance indicators across 11 issue categories (air quality, sanitation and drinking water, heavy metals, waste management, biodiversity and habitat, ecosystem services, fisheries, climate change, pollution emissions, agriculture, water resources) covering environmental health and ecosystem vitality.		Developed an environmental performance index to rank 180 countries on environmental health and ecosystem vitality. These indicators provide a gauge at a national scale of how close countries are to established environmental policy targets.
3	Shah et al. (2019)	Eight South Asian countries	2006–2017	Multiplicative data envelopment analysis (MDEA) model.	Environmental sustainability (energy intensity, CO_2 intensity, per capita CO_2 emission, CO_2 emission from electric	Energy security (energy dependency, energy consumption per capita, GDP per capita, share of	Developed an Energy Security and Environmental Sustainability Index (ESESI). Produced a ranking based on ESESI scores for South Asian countries.

Continued

Table A1 Existing studies on the environmental sustainability index.—cont'd

No	Authors	Country case	Data period (frequency)	Methodologies	Main variables	Other variables	Findings
					generation, ratio of change in forestry).	renewable resources in electricity generation, diversity in TPES, electrification ratio).	
4	Dias (2017)	141 countries	2015	Exploratory factor analysis (EFA), confirmation factor analysis (CFA).	10 indicators: Stringency of environmental regulations, enforcement of environmental regulations, sustainability of travel and tourism industry development, particulate matter (2.5) concentration, environmental treaty ratification, baseline water stress, threatened species, forest cover change, wastewater treatment, costal shelf fishing pressure.		The environmental sustainability index of the Travel & Tourism Competitiveness Reports has consistency problems. Developed a modified Travel & Tourism Environmental Sustainability Index (TTESI) which is more consistent and reliable.
5	Singh et al. (2019)	22 Asian economies	1990–2012	Composite Z-score technique	Seven main components with 25 subindicators: Air quality and pollution (CO_2 emissions from transport (% of total fuel combustion), CO_2 emissions from manufacturing, CO_2 emissions from electricity and heat production, per capita CO_2 emissions, PM2.5 air pollution), energy management (electric power consumption, access to electricity, combustible renewables and waste, electricity production from natural gas sources, fossil fuel energy consumption, renewable electricity output, energy use), forest and biodiversity (protected areas domestic, forest area), land use and agriculture (% of arable land equipped for irrigation, agricultural land), human health and disaster (incidence of tuberculosis, infant mortality rate, improved		Created the environmental sustainability index (ESI) for selected countries. Economic growth, economic development, human development, social development, and environmental development are expressively related.

					sanitation facilities), population pressures on ecosystem services (fertility rate, annual population growth), water generation management (improved water source, renewable internal freshwater resources per capita).		
6	Pravitasari et al. (2018)	Indonesia	2014	Factor analysis (FA), Local Indicator of Spatial Association (LISA) analysis.	Regional Sustainability Index (RSI), where environmental index comprises number of drought events, flood events, landslide events, percentage of household living along the river, percentage of household living in the slum area, number of people suffering from malaria, respiratory tract infection, diarrhea and vomit, percentage of village curing water pollution, land conversion from agricultural land (excluding rice field) to nonagricultural land	Economic index (10 indicators) and social index (10 indicators).	Developed an RSI to evaluate and report economic, social, and environmental sustainability conditions at regional level.

Continued

Table A1 Existing studies on the environmental sustainability index.—cont'd

No	Authors	Country case	Data period (frequency)	Methodologies	Main variables	Other variables	Findings
7	World Economic Forum (WWF, 2002).	142 countries	1980–2000	Averaging the Z-scores for each variable in the indicator. Each variable received equal weight, missing variables are simply excluded from calculation. Composite index was calculated by taking the unweighted average of the values of the 20 indicators.	20 core indicators (air quality, water quantity, water quality, biodiversity, land, reducing pollution, reducing water stress, reducing ecosystem stresses, reducing waste and consumption pressures, reducing population growth basic human sustenance, environmental health, science/tech, capacity for debate environmental governance, private sector responsiveness, eco-efficiency, participation in International collaborative efforts, reducing greenhouse gas emissions, reducing transboundary environmental pressures) comprising 68 underlying variables.		
8	Esty et al. (2005)	146 countries	N/A	The ESI uses equal weights at both the indicator and the variable level. The ESI is the equally weighted average of the 21 indicators.	Five components (environmental systems, reducing environmental stress, reducing human vulnerability, social and institutional capacity, global stewardship) embracing 21 core indicators and 76 underlying variables.		- Environmental sustainability is a fundamentally multidimensional concept. Some environmental challenges arise from development and industrialization, while other challenges are a function of underdevelopment and poverty-induced short-term thinking. - Significant differences exist across countries.

Acknowledgment

Thai-Ha Le and Canh Phuc Nguyen receive funding from the University of Economics Ho Chi Minh City, Vietnam.

References

Agrawal, K.K., Jain, S., Kr Jain, A., Dahiya, S., 2014. Assessment of greenhouse gas emissions from coal and natural gas thermal power plants using life cycle approach. International Journal of Environmental Science and Technology 11 (4), 1157–1164.

Basiago, A.D., 1998. Economic, social, and environmental sustainability in development theory and urban planning practice. Environmentalist 19 (2), 145–161.

Boyer, R.H.W., Peterson, N.D., Arora, P., Caldwell, K., 2016. Five approaches to social sustainability and an integrated way forward. Sustainability 8 (9), 878.

Brinkmann, R., 2020. Megaconnections in environmental sustainability through the twenty-first century. In: Environmental Sustainability in a Time of Change. Palgrave Macmillan, Cham, pp. 279–287.

Brundtland, G.H., 1987. Report of the World Commission on Environment and Development: Our Common Future. New York, p. 8.

Chaves, H.M.L., Alipaz, S., 2007. An integrated indicator based on basin hydrology, environment, life, and policy: the watershed sustainability index. Water Resources Management 21 (5), 883–895.

Chen, C.C., 2009. Environmental impact assessment framework by integrating scientific analysis and subjective perception. International Journal of Environmental Science & Technology 6 (4), 605–618.

Coulon, F., Jones, K., Hong, L., Hu, Q., Gao, J., Li, F., Chen, M., et al., 2016. China's soil and groundwater management challenges: lessons from the UK's experience and opportunities for China. Environment International 91, 196–200.

Darchen, S., Glen Searle (Eds.), 2018. Global Planning Innovations for Urban Sustainability. Routledge.

De Sherbinin, A., Reuben, A., Levy, M.A., Johnson, L., 2013. Indicators in Practice: How Environmental Indicators Are Being Used in Policy and Management Contexts. Yale and Columbia Universities, New Haven and New York.

Dias, J.G., 2017. Environmental sustainability measurement in the Travel & Tourism Competitiveness Index: an empirical analysis of its reliability. Ecological Indicators 73, 589–596.

Ebert, U., Welsch, H., 2004. Meaningful environmental indices: a social choice approach. Journal of Environmental Economics and Management 47 (2), 270–283.

Esty, D.C., Levy, M., Srebotnjak, T., De Sherbinin, A., 2005. Environmental Sustainability Index: Benchmarking National Environmental Stewardship. Yale Center for Environmental Law & Policy, New Haven, pp. 47–60.

Gigliotti, M., Schmidt-Traub, G., Bastianoni, S., 2019. The sustainable development goals. Encyclopedia of Ecology 4, 426–431.

Goodland, R., 1995. The concept of environmental sustainability. Annual review of ecology and systematics 26 (1), 1–24.

Hahn, R.W., 2000. The impact of economics on environmental policy. Journal of Environmental Economics and Management 39 (3), 375–399.

Hair, J.F., Anderson, R.E., Tatham, R.L., William, C., 1998. Multivariate Data Analysis. Prentice Hall, Upper Saddle River, NJ.

Halsnæs, K., Shukla, P., 2008. Sustainable development as a framework for developing country participation in international climate change policies. Mitigation and Adaptation Strategies for Global Change 13 (2), 105–130.

Holdren, J.P., Daily, G.C., Ehrlich, P.R., 1995. The meaning of sustainability: biogeophysical aspects. In: Defining and Measuring Sustainability: The Biogeophysical Foundations, pp. 3–17.

Hopkins, M.S., Townend, A., Khayat, Z., Balagopal, B., Reeves, M., Berns, M., 2009. The business of sustainability: what it means to managers now. MIT Sloan Management Review 51 (1), 20.

Hou, D., Ding, Z., Li, G., Wu, L., Hu, P., Guo, G., Wang, X., Ma, Y., O'Connor, D., Wang, X., 2018. A sustainability assessment framework for agricultural land remediation in China. Land Degradation & Development 29 (4), 1005–1018.

Hsu, A., Esty, D., Levy, M., de Sherbinin, A., 2016. The 2016 Environmental Performance Index Report. Yale Center for Environmental Law and Policy, New Haven, CT. https://doi.org/10.13140/RG.2.2.19868.90249.

Huang, P.-S., Shih, L.-H., 2009. Effective environmental management through environmental knowledge management. International Journal of Environmental Science & Technology 6 (1), 35–50.

Iwejingi, S.F., 2011. Population growth, environmental degradation and human health in Nigeria. Pakistan Journal of Social Sciences 8 (4), 187–191.

Jolliffe, I.T., 2002. "Graphical Representation of Data Using Principal components." Principal Component Analysis, pp. 78–110.

Kaly, U., Pratt, C., Mitchell, J., 2004. The Environmental Vulnerability Index 2004. South Pacific Applied Geoscience Commission (SOPAC), Suva.

Kashem, S.B., Irawan, A., Wilson, B., 2014. Evaluating the dynamic impacts of urban form on transportation and environmental outcomes in US cities. International Journal of Environmental Science and Technology 11 (8), 2233–2244.

Khanna, M., Kumar, S., 2011. Corporate environmental management and environmental efficiency. Environmental and Resource Economics 50 (2), 227–242.

Le, T.-H., Chang, Y., Park, D., 2019a. Economic development and environmental sustainability: evidence from Asia. Empirical Economics 57 (4), 1129–1156.

Le, T.-H., Chang, Y., Taghizadeh-hesary, F., Yoshino, N., 2019b. Energy insecurity in Asia: a multi-dimensional analysis. Economic Modelling 83, 84–95.

Lim, T.K., Rajabifard, A., Khoo, V., Sabri, S., Chen, Y., 2021. The smart city in Singapore: how environmental and geospatial innovation lead to urban livability and environmental sustainability. In: Smart Cities for Technological and Social Innovation. Academic Press, pp. 29–49.

McRae, L., Deinet, S., Freeman, R., 2017. The diversity-weighted living planet index: controlling for taxonomic bias in a global biodiversity indicator. PLoS One 12 (1), e0169156.

Moldan, B., Janoušková, S., Hák, T., 2012. How to understand and measure environmental sustainability: Indicators and targets. Ecological Indicators 17, 4–13.

Morse, S., 2016. Measuring the success of sustainable development indices in terms of reporting by the global press. Social Indicators Research 125 (2), 359–375.

Mundfrom, D.J., Shaw, D.G., Lu Ke, T., 2005. Minimum sample size recommendations for conducting factor analyses. International Journal of Testing 5 (2), 159–168.

Organisation for Economic Co-operation and Development, 2001. OECD Environmental Strategy for the First Decade of the 21st Century: Adopted by OECD Environmental Ministers. OECD.

Parmar, K.S., Bhardwaj, R., 2013. Water quality index and fractal dimension analysis of water parameters. International Journal of Environmental Science and Technology 10 (1), 151–164.

Pearce, D.W., Barbier, E., 2000. Blueprint for a Sustainable Economy. Earthscan.

Pravitasari, A.E., Rustiadi, E., Mulya, S.P., Fuadina, L.N., 2018. Developing regional sustainability index as a new approach for evaluating sustainability performance in Indonesia. Environment and Ecology Research 6, 157–168.

Radovanović, M., Filipović, S., Golušin, V., 2018. Geo-economic approach to energy security measurement—principal component analysis. Renewable and Sustainable Energy Reviews 82, 1691–1700.

Ramadass, K., Smith, E., Palanisami, T., Mathieson, G., Srivastava, P., Megharaj, M., Naidu, R., 2015. Evaluation of constraints in bioremediation of weathered hydrocarbon-contaminated arid soils through microcosm biopile study. International Journal of Environmental Science and Technology 12 (11), 3597–3612.

Rassafi, A.A., Vaziri, M., 2005. Sustainable transport indicators: definition and integration. International Journal of Environmental Science & Technology 2 (1), 83–96.

Romero, J.C., Linares, P., López, X., 2018. The policy implications of energy poverty indicators. Energy Policy 115, 98–108.

Saleem, H., Jiandong, W., Mohammed Aldakhil, A., Nassani, A.A., Moinuddin Qazi Abro, M., Aqeel Khan, K.Z., Bin Hassan, Z., Mohd Rameli, M.R., 2019. Socio-economic and environmental factors influenced the united nations healthcare sustainable agenda: evidence from a panel of selected Asian and African countries. Environmental Science and Pollution Research 26 (14), 14435–14460.

Samuel-Johnson, K., Esty, D.C., 2000. Pilot environmental sustainability index report. In: Davos (Switzerland): World Economic Forum: Annual Meeting.

Shah, S.A.A., Zhou, P., Walasai, G.D., Mohsin, M., 2019. Energy security and environmental sustainability index of South Asian countries: a composite index approach. Ecological Indicators 106, 105507.

Singh, A.K., Issac, J., Narayanan, K.G.S., 2019. Measurement of environmental sustainability index and its association with socio-economic indicators in selected Asian economies: an empirical investigation. International Journal of Environment and Sustainable Development 18 (1), 57–100.

Sun, H., Mohsin, M., Alharthi, M., Abbas, Q., 2020. Measuring environmental sustainability performance of South Asia. Journal of Cleaner Production 251, 119519.

Sutton, P., 2004. A perspective on environmental sustainability. Paper on the Victorian Commissioner for Environmental Sustainability 1–32.

Sutton, P.C., 2003. An empirical environmental sustainability index derived solely from nighttime satellite imagery and ecosystem service valuation. Population and Environment 24 (4), 293–311.

Tabachnick, B.G., Fidell, L.S., Ullman, J.B., 2007. Using Multivariate Statistics, vol. 5. Pearson, Boston, MA.

Tambouratzis, T., 2016. Analysing the construction of the environmental sustainability index 2005. International Journal of Environmental Science and Technology 13 (12), 2817–2836.

UN Environment, 2019. Global Environment Outlook GEO 6: Healthy Planet, Healthy People. Nairobi, Kenya, vol. 745. University Printing House, Cambridge, United Kingdom.

Wackernagel, M., Rees, W., 1998. Our Ecological Footprint: Reducing Human Impact on the Earth, vol. 9. New society publishers.

Wang, J., Hu, Q., Wang, X., Li, X., Yang, X.J., 2016. Protecting China's soil by law. Science 354 (6312), 562–562.

Wendling, Z.A., Emerson, J.W., Esty, D.C., Levy, M.A., de Sherbinin, A., Emerson, J.W., 2018. Environmental Performance Index. Yale Center for Environmental Law & Policy, New Haven, CT, USA.

World Bank, 2019. Myanmar Country Environmental Analysis: A Road towards Sustainability, Peace, and Prosperity–Synthesis Report. World Bank.

World Bank, 2020. https://datahelpdesk.worldbank.org/knowledgebase/articles/906519-world-bank-country-and-lending-groups.

World Economic Forum - WEF - Global Leaders for Tomorrow Environment Task Force, Yale Center for Environmental Law and Policy - YCELP - Yale University, and Center for International Earth Science Information Network - CIESIN - Columbia University, 2002. Environmental Sustainability Index (ESI). Palisades, NY: NASA Socioeconomic Data and Applications Center (SEDAC). https://doi.org/10.7927/H4SB43P8 (Accessed 14 June 2020).

World Wildlife Fund (WWF), 1998. Living Planet Report 1998. World Wildlife Fund, Gland.

Yigitcanlar, T., Teriman, S., 2015. Rethinking sustainable urban development: towards an integrated planning and development process. International Journal of Environmental Science and Technology 12 (1), 341–352.

Yu, J., Wu, J., 2018. The sustainability of agricultural development in China: the agriculture—environment nexus. Sustainability 10 (6), 1776.

Zhang, D., Tu, Y., 2021. Green building, pro-environmental behavior and well-being: evidence from Singapore. Cities 108, 102980.

Zhou, N., He, G., Williams, C., Fridley, D., 2015. ELITE cities: a low-carbon eco-city evaluation tool for China. Ecological Indicators 48, 448—456.

Zhou, P., Ang, B.W., Poh, K.L., 2006. Comparing aggregating methods for constructing the composite environmental index: an objective measure. Ecological Economics 59 (3), 305—311.

CHAPTER

Pearls and perils of resources recovery and reuse technologies

15

Maksud Bekchanov[1,2]

[1]*Research Unit Sustainability and Global Change (FNU), Center for Earth System Research and Sustainability (CEN), University of Hamburg, Hamburg, Germany;* [2]*Center for Development Research (ZEF), University of Bonn, Bonn, Germany*

Chapter outline

1. Introduction

Enormous amount of waste and wastewater generated in both urban and rural areas is a major reason for air, soil, and water pollution, especially in developing countries (Lazarova et al., 2013). Disposal of untreated waste or release of untreated wastewater into fresh water sources is a serious threat which aggravates environmental pollution consequently leading to various water- and air-borne illnesses (Drechsel et al., 2010; Gebrezgabher et al., 2016). Given the increasing scope of environmental and health problems triggered by inadequate sanitation, the United Nations Sustainable Development Goals also underline the needs for improved sanitation measures in the developing countries (UNWATER, 2016). These measures particularly aim at better access to potable water supply and sewage systems in residential areas, reduction of open defecation, improved waste management, and increased recycling of waste and wastewater. With the increasing land scarcity and environmental

https://doi.org/10.1016/B978-0-12-824084-7.00007-2
Copyright © 2022 Elsevier Inc. All rights reserved.

control requirements, recycling the waste and reusing the recovered products for value creation will be more viable than the waste disposal (open dumping or land filling) (Tay and Show, 1997).

Under conditions of growing water scarcity induced by population growth, industrial development, and global warming effects, treated wastewater can be suitable complement to fresh water supply (Scheierling et al., 2011; Lazarova et al., 2013; Drechsel et al., 2015). Thus, treatment and reuse of wastewater not only improve sanitation and alleviate environmental concerns in the epoch of urbanization but also bear additional economic value added through recovering water, energy, and nutrients from waste and wastewater (Scheierling et al., 2011). Given the increasing costs of traditional ways of water supply augmentation (e.g., building reservoirs or interbasin water transfers) along with the rapid advancements in waste and wastewater treatment technologies, the costs of additional water supply through water treatment are expected to be competitive compared to the alternative options of water supply (Drechsel et al., 2015). Yet, distributing and matching water with varying quality for appropriate activities will be a challenge for water managers and policy makers (Drechsel et al., 2015; von Braun, 2016).

Depletion of phosphate mines and fossil fuel reservoirs leads to price shocks in food, energy, transportation, and fertilizer markets consequently threatening food and energy security (Ashlay et al., 2011; Cordell et al., 2011; Aleklett and Campbell, 2003; Höök and Tang, 2013). For counterbalancing such shocks and maintaining a stable and sustainable economic prosperity, transformation toward the increased uses of alternative and renewable sources of water, energy, and nutrients is gaining prominence. Recycling waste and wastewater can be a win-win option from both environmental and economic perspectives, consequently allowing not only for improving environmental habitats and increasing the value of ecosystem services but also for supplying food, energy, and water for production processes and direct consumption.

This study provides a review of various types of waste, respective treatment technologies, and available assets from waste recycling. Thus, first, the development stages of sanitation and waste management systems, and the waste availability and treatment levels across the regions of the world are presented. Next, the availability and reuse of waste and wastewater across the world are described before a brief discussion of the available options for waste and wastewater treatment. Then, poverty alleviation effects and environmental health risks related with Resources Recovery and Reuse (RRR) technologies are discussed. The last section summarizes the findings and provides final concluding remarks.

2. The development stages of waste and wastewater management and reuse

Problems of pollution and the need for sanitation especially in urban areas have been known over centuries, and the respective management practices have been evolved over time. As recently reported, four major epochs of the development of sanitation and waste management throughout the history are as follows (Ashley et al., 2011): (1) the use of night soil and sewage for farming purposes in period between 3000 BC and 1850; (2) the Era of sanitation awakening started from 1860 to 1960; (3) the period of wastewater reclamation and eutrophication monitoring continued between 1960 and 2000; and (4) the recent Era of ecological sanitation started from 2000s.

At the first stage, a waste from the residential areas, especially feces, sewage, and manure were either directly applied to croplands or recycled through composting before the applications. The use of night soil for improving soil quality was known and widely practiced in China as early as 3000 BC (Ashley et al., 2011; Marald, 1998). Human excreta was used as soil amendment in Japan since the 12th century and continued till the recent past (Matsui, 1997). Seeing a night soil man carrying buckets in the streets and collecting urine and feces was common in Singapore till mid-1980s. Following large-scale land degradation and consequent famines in the Middle Ages, sewage was also being applied for farming purposes in Germany and the United Kingdom. In the 19th century, England was importing large amounts of bones all across the European countries (Cordell et al., 2009) for applying it in agricultural lands. This technology was later improved for creating a liquid fertilizer through dissolving bones (Liu, 2005). During that period, night soil companies were functioning in New York city (Ashley et al., 2011).

In the second stage, health risks related with the use of fecal waste imposed the implementation of disease prevention and hygienic measures. Particularly, the cholera epidemic in Europe in 1850s increased the importance of sanitation measures (Ashley et al., 2011). Thus, the main focus of waste management in this period was disposing the waste outside of the living areas for preventing further illnesses and disease epidemics. Wide-scale construction of sewage systems and introduction of septic tanks and cesspits were specific characteristics of this period.

However, enormous amount of waste disposal into environmental systems increased environmental pollution problems. Increased environmental consciousness and the need for more sustainable management of wastes after 1960s started a new Era of environmental protection (Ashley et al., 2011). Waste and wastewater were required to be treated before discharging it into the rivers or lakes. Different methods of wastewater treatment such as physic-chemical and biological treatment methods were invented and applied. Wastewater was treated and widely used for irrigation purposes, for instance, in Israel. Organic waste was composted and used as fertilizer for crops.

Since 2000s, given the increased scarcity of fertilizer and energy resources, technologies of producing nutrients and energy such as biogas, electricity, fertilizer, and soil amendments have been developed and widely facilitated (Ashley et al., 2011). Particularly, these technologies aimed at separation of urine in the sewage system and its recycle for producing fertilizers, or composting fecal sludge or organic waste for further production of fertilizers and biogas (Tilley et al., 2014). The use of wastewater passed through advanced treatment process became more common in multiple sectors (agriculture, industry, and residential sites).

Indeed, these development tendencies in waste and wastewater treatment sector describes the changes in technological frontiers at global level. However, advancement level of waste and wastewater treatment largely varies across the countries. The developed countries of the world tend more toward "environmental-friendly" waste and wastewater treatment and reuse which offer multiple environmental and economic benefits through recycling (Table 15.1). Despite multiple benefits of waste and wastewater treatment and reuse, "pollution-inducing" practices of disposing waste and wastewater without adequate treatment are still common in developing countries of Africa, Latin America, and South Asia.

Pearce (2015) differentiated four types of mental models (concepts) of waste management across the world. These concepts consider different levels of technological advancement and roughly match with the technological progress level observed across the four epochs of waste management and sanitation discussed above: (1) nonrecognizant; (2) sanitation-oriented; (3) treatment-oriented; and (4)

Table 15.1 The comparison of "pollution-inducing" and "environmental-friendly" waste management.

	"Pollution-inducing" waste disposal	"Environmental-friendly" waste reuse
Collection	– Lack of latrines and septic tanks – Lack of waste collection – Open defecation	– Flush toilets, septic tanks, and latrines – Waste collection stations
Transportation	– Lack of organized transportation of waste – Discharge to drainage system	– On-site of centralized sewage system – Special trucks to transport waste
Treatment	– Lack of treatment or minimal treatment	– Screening plastic waste – Removal of pollutants
Disposal/reuse	– Disposal into dumping sites or discharge waste into water system	– Disposal to dumping site after proper treatment – Recycling soil conditioners, energy commodities, proteins, and effluents
Environmental effects	– Water and air pollution – Groundwater contamination – Land erosion and degradation – Reduced biodiversity	– Improved sanitation – Reduced water and air pollution – Reduced health risks
Economic effects	– Reduced environmental system and recreation benefits – Reduced agricultural yields	Recovery of nutrients, energy, and effluents

recovery-oriented. A noncognizant model does not consider a proper management of waste or sanitation, and appropriate infrastructure for waste collection or public facilities for sanitation does not exist. This model may characterize the conditions in urban slums across South America and Africa. A sanitation-oriented model prioritizes waste management for protecting health and avoiding human contact with waste. This approach may be more dominant in fast-growing second-tier cities across China and India. A treatment-oriented model aims at environmental protection in addition to health protection and thus considers the prevention of pollutants from leaking into environmental system. This model is more common in most cities of the developed world. A recovery-focused model considers waste and wastewater not only from sanitation and environmental protection perspective but also treats as an economic resource which can be recycled and returned to the production circle. This model is less common in practice compared to the other three mental models and shared only in few places across the world. Yet, as implied from the "Kuznetz curve," with the improved income levels and reduced technology costs, "environmental-friendly" waste and wastewater management systems should gradually replace the less advanced alternatives.

3. Waste and wastewater generation across the world

Large amount of waste and wastewater especially in urban areas is a potential resource valuable for recycled economic assets. Globally, total volume of wastewater is estimated to be between 0.68 and 0.96 km^3 per day or 250–350 km^3 per annum (GWI, 2009; FAO, 2010). It is almost 10%–15% of

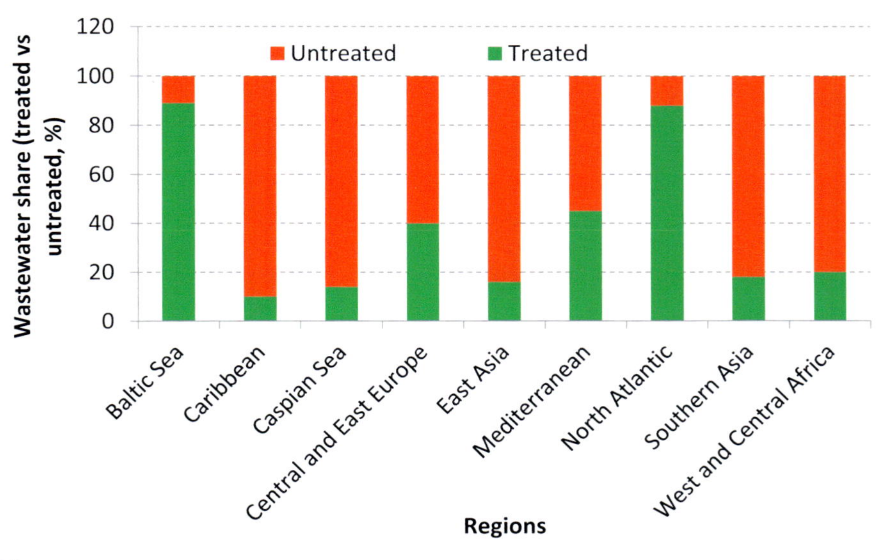

FIGURE 15.1

The ratio of wastewater treatment across the word.

Source: Author's presentation based on Corcoran et al. (2010).

annual agricultural water withdrawals (2504 km^3; Siebert and Döll, 2007). Yet, only 4% (32 million m^3 per day) of these wastewater passes through advanced treatment (GWI, 2009), while the remaining 96% is disposed in lakes or river stream with very limited or without treatment. Although a large share of wastewater is treated in West European and North American countries, wastewater treatment rates are very low in developing countries located in South and Southeast Asia (Fig. 15.1).

Release of untreated wastewater into fresh water aquifers not only reduces downstream water availability due to heavy pollution but also may have adverse effects on ecology of these water systems by launching eutrophication problems and degrading living habitats for aquatic organisms (Scheierling et al., 2011; Cai et al., 2013). Thus, adequate sanitation and appropriate treatment of wastewater are essential for both environmental and human health protection (Harada et al., n.d.). Moreover, wastewater treatment can be also turned into beneficial business, thus allowing for recovery of useful economic assets. As estimated, each 1 US$ investment in improved sanitation and wastewater treatment may yield returns worth of 3 to 34 US$ (Hutton and Haller, 2004). Reuse of wastewater resources can be also a potential option for considerably reducing water deficit in developing countries where irrigation water availability is a key challenge for sustainable agricultural production because of high population growth and temperature raise.

Massive quantity of municipal solid waste is another environmental threat that can be transformed into useful source of recycled energy and soil amendments. At present, daily 3.5 million tons (as of 2012) of municipal solid waste is generated across the world and is expected to increase over six million tons coming to 2025 (WEC, 2016, Fig. 15.2). While almost half of this waste is generated in the Organisation for Economic Co-operation and Development countries, rapid increase of waste

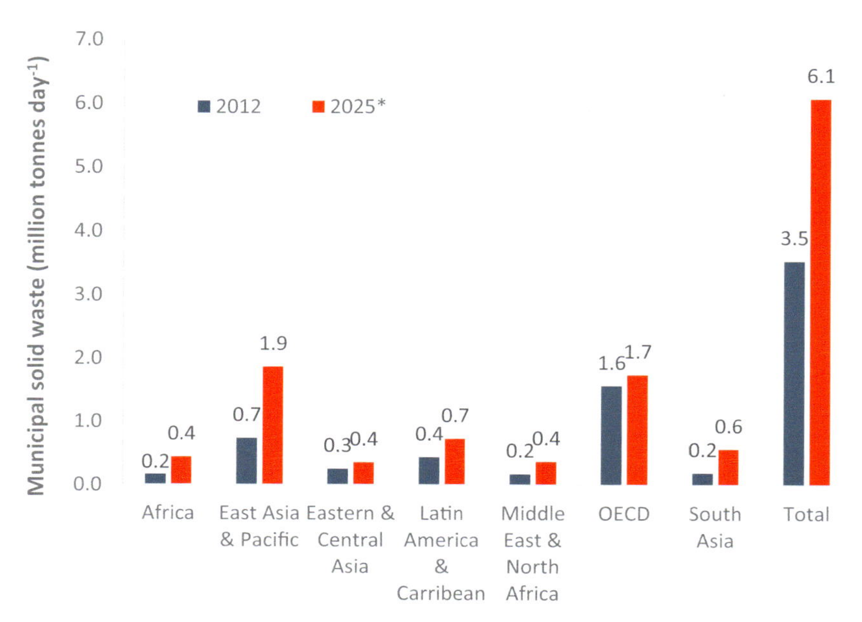

FIGURE 15.2

Daily municipal solid waste generation across the world regions.

Source: Author's presentation based on Hoornweg and Bhada-Tata (2012).

generation is expected in East Asian and Pacific countries till 2025. Almost half of this municipal solid waste is organic waste, which can be further composted or recycled to produce fertilizer or energy commodities (Fig. 15.3). In addition to wastewater and municipal solid waste, livestock manure and crop residues can be useful as soil amendments or biofuel.

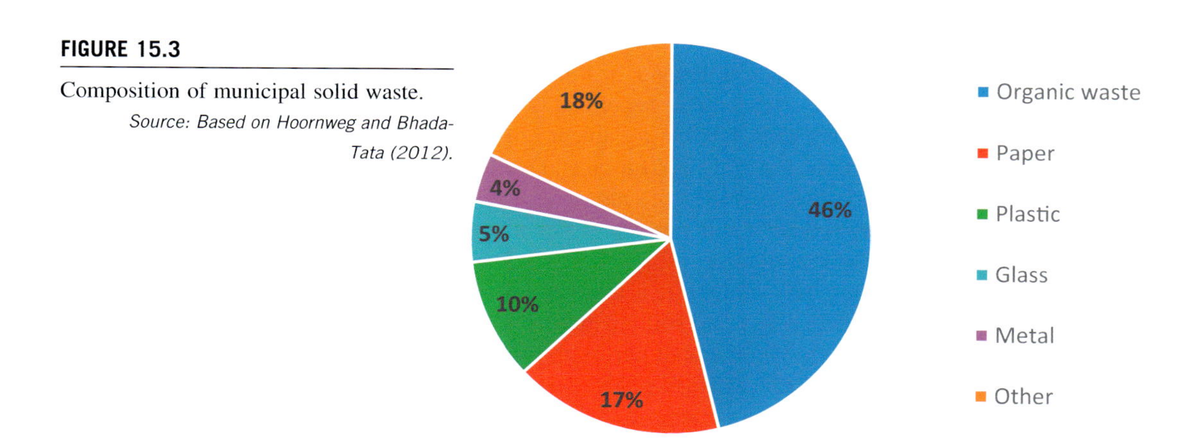

FIGURE 15.3

Composition of municipal solid waste.

Source: Based on Hoornweg and Bhada-Tata (2012).

4. Available resources recovery and reuse options

4.1 General description

Wastes such as municipal organic waste, sewage water, and fecal sludge can be recycled and reused in multiple ways of recovering valuable assets such as effluents (treated water), nutrients (phosphates, nitrogen, and protein), and energy (biogas, liquid fuel, and electricity) (Table 15.2). Sewage and drainage waters can be reused for irrigation or aquaculture after treatment and thus considerably

Table 15.2 Options of resource recovery and reuse (RRR).

		Intermediate products												
		Raw sludge	Bio-solids	Biogas	Dewatered sludge	Steam	Compost	Sludge centrate	Sludge ash	Syngas	Oil	Liquid	Treated water	Soldier fly
WASTE STREAM														
Waste-water	Sewage	x	x	x	x	x	x	x	x	x	x	x	x	
	Fecal sludge			x										x
	Urine											x	x	
	Drainage												x	
	Algae			x			x				x			
Organic waste	Food waste			x			x							
	Waste from food processing			x		x	x							
	Manure			x		x	x							x
	Crop residues					x	x							
FINAL OUTPUTS														
Effluents (Treated waste-water)	For irrigation												x	
	For aquifer recharge												x	
	For fish pond												x	
Soil nutrients	Fertilizer	x	x					x	x			x		
	Soil amendments		x		x		x							x
	Struvites							x						
	Cover crop	x					x							
Energy	Gas			x										
	Electricity			x						x				
	Heat					x								
	Fuel										x	x		
	Protein													x
Crop protection	Pesticides									x				
Building materials		x			x				x					

Source: Based on Pearce (2015) and Tan and Lagerkvist (2011).

improve water availability for agriculture, especially in dry regions. Organic food waste and animal manure can be also recycled (composted) and reused for cultivating crops as soil amendments or for cooking as biofuel. Some of these RRR technologies may allow for recovering multiple assets (e.g., not only water or fertilizer but also both or even energy in addition) from waste. Next subsections provide a detailed description of various options of recovering water, nutrients, and energy from the recycled waste and wastewater. For clarity, we separately describe recovery of particular asset (effluent, fertilizer, or energy) in each subsection, but it does not mean that a certain technology produces only a single type of asset.

4.2 Treated wastewater as an economic asset: current status and potential options

Wastewater treatment first of all aims at safe disposal of wastewater after treatment (sanitation benefits) and thus protection of environmental resources. Yet, effluents and nutrients embedded in wastewater may bear additional economic benefits through enhancing biomass production and energy recovery. Water treatment options vary depending on the purpose of the treatment, the complexity of the process and investment, and operating costs. In general, four steps of wastewater treatment can be differentiated: (1) primary treatment, (2) secondary treatment, (3) sludge treatment, and (4) advanced treatment (Razzak et al., 2013, Fig. 15.4). Primary treatment considers capturing large objects such as plastics and rag, removal of scum and grits, and separation of liquid and solid waste sequentially. In secondary treatment, water passed through primary treatment can be released to aeration or filtration

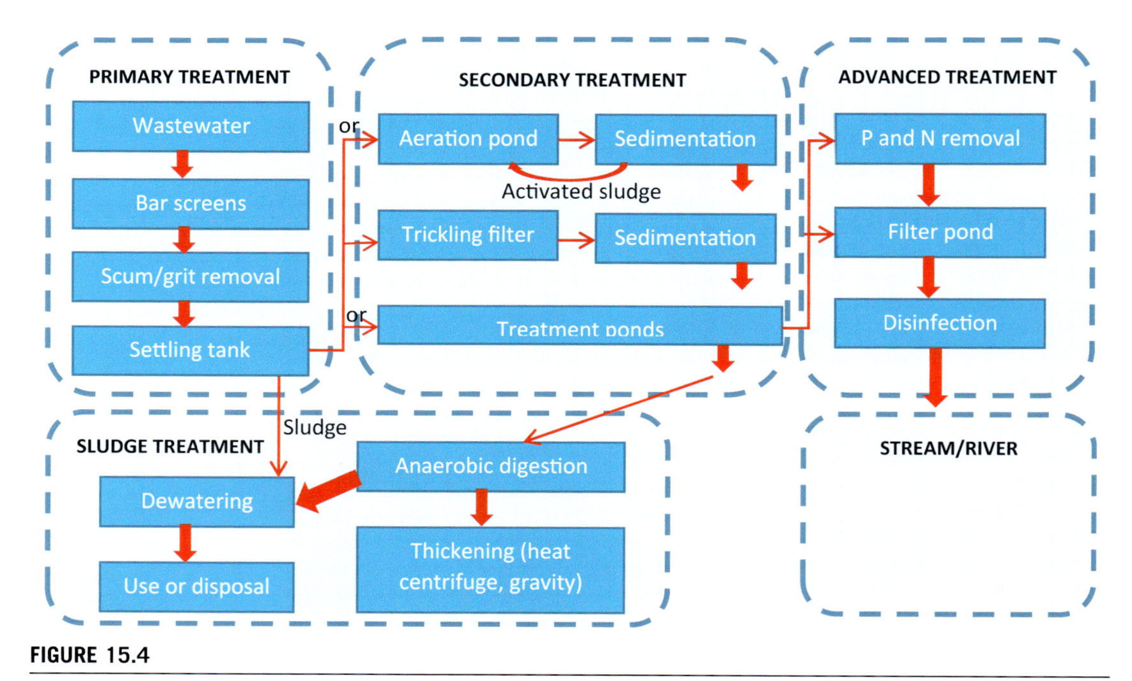

FIGURE 15.4

Wastewater treatment system.

Source: Adapted from Razzak et al. (2013).

Table 15.3 Wastewater reuses for different purposes in the world.

	Sectors	Share in total water reuse (%)
1	Irrigation	32
2	Landscape irrigation	20
3	Industrial activities	19
4	Environmental flow	8
5	Nonpotable residential use	8
6	Recreation	7
7	Recharging aquifers	2
8	Other	4

Source: Based on GWI (2009).

ponds or lagoons where soild waste will be sedimented. Sedimented waste from primary and secondary treatment will be further recycled in sludge treatment stage, while filtrated water from the lagoons will be further transferred for advanced treatment. In sludge treatment process, the solid sludge can be dewatered and disposed to dumping site or can be further recycled through incineration and thickening process to produce energy, compost, or nutrients. Meanwhile, the filtrated water may pass through advanced phosphate and nitrogen removal and clarification process before a release into water system or before a reuse for irrigation or landscape reclamation.

While wastewater reuses for agricultural and landscape irrigation are common practices, treated or untreated wastewater can be reused also for maintenance of river ecosystems, potable and nonpotable uses, recreation, fish production and aquifer recharge (World Bank, 2010, Scheierling et al., 2011, Lazarova et al., 2013, Hettiarachchi and Ardakian, 2016). At present, treated wastewater from different economic sectors is mostly reused for agricultural production (32%) (Table 15.3) because of its rich nutrient content (Fig. 15.5). Wastewater uses for irrigation are particularly common in areas near urban settlements (Scheierling et al., 2011). Except for agriculture, large portions of wastewater are also used for landscape irrigation (20%) and industrial activities (19%, Table 15.3).

Wastewater was estimated to be used over 6–20 million ha of croplands in total (World Bank, 2010; Drechsel et al., 2015). In China alone, wastewater is applied over 4.2 million ha irrigated lands, which represents 5.7% of country's total irrigated lands (Xie, 2009). According to some estimates, wastewater allows for producing about 10% of total crop production outputs from irrigation globally (Drechsel et al., 2010, Scheierling et al., 2011). It is used for irrigation of cultivating both food and fodder crops (Lautze et al., 2014). While some level of wastewater treatment is required to apply wastewater for irrigation purposes, there are also cases that wastewater is directly applied for irrigation in some countries of South and East Asia and Africa. Wastewater was properly managed and formally used for irrigation purposes in the developed countries such as Israel, Australia, and the United States of America; however, informal (or unplanned) use of wastewater is common both in China and India (Drechsel et al., 2015). Considerably lower costs of wastewater reuse compared to deep groundwater extraction or water transfer from the neighboring basins, for example, also add to its financial viability (Fig. 15.6). Yet, advanced treatment of wastewater through the removal of undesired vegetation or

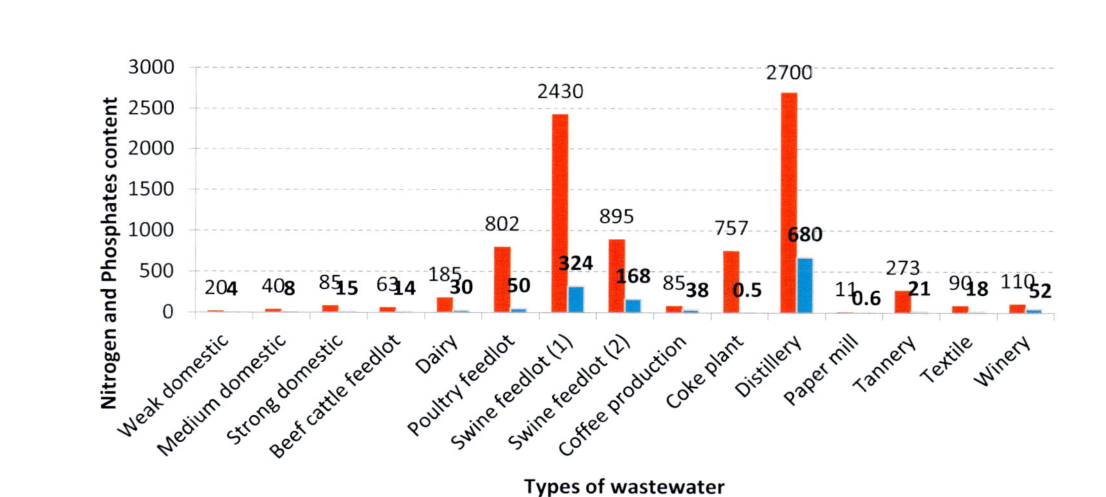

FIGURE 15.5

Nitrogen and phosphorus content of different types of wastewater.

Source: Based on Christenson and Sims (2011).

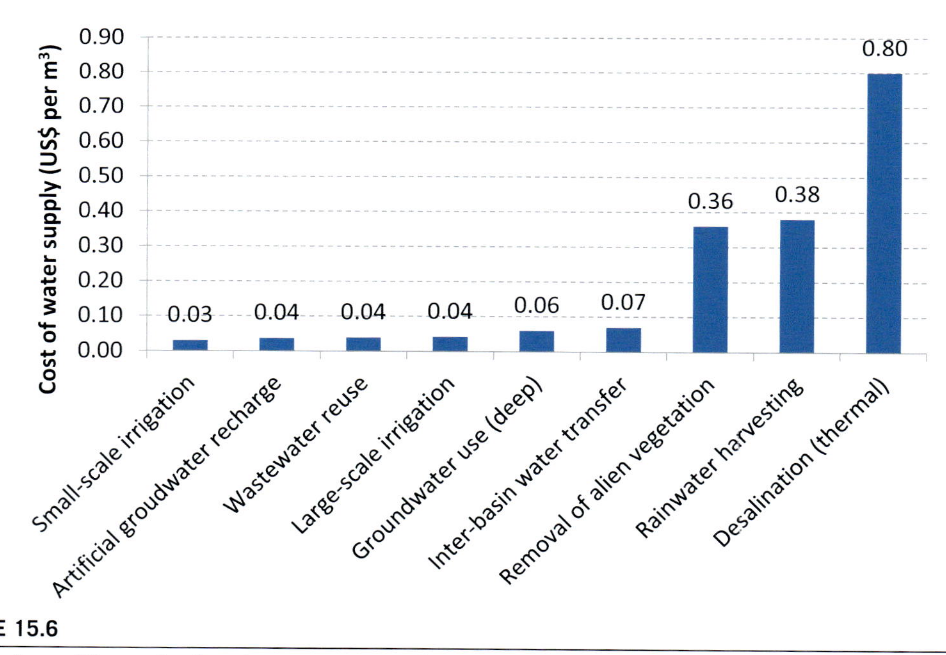

FIGURE 15.6

Comparison of the costs of wastewater reuse to alternative water supply options (a case study of India).

Source: Adapted from McKinsey and Company (2009).

desalinization for generating good quality water suitable for reuse in irrigation and nonirrigation activities (e.g., industry, maintenance of landscapes in municipal areas, and drinking) may come at much higher costs since it may demand large amounts of energy use and capital investments.

Wastewater is reused also for aquaculture (Drechsel et al., 2015; Tilley et al., 2014). Increased productivity of fish production was reported when reclaimed water was applied in fish ponds (Lautze et al., 2014). In ideal conditions, fish production may reach as high as 10 ton per hectare in wastewater ponds (Tilley et al., 2014). Effluents can be applied in fish ponds to maintain water supply. Sludge can be discharged to the pond to enrich the nutrient content of water and increase biomass of algae which is consumed by fish. Though this system cannot fully eliminate toxic elements in water, at least this system substantially reduces mechanical wastewater treatment costs (Tilley et al., 2014).

Cultivation of fodder crops, plants, and macrophytes in wastewater stabilization ponds or drying beds may also considerably improve feed stocks and provide construction materials for local village communities (Harada et al. n.d.; Tilley et al., 2014). Alternatively, nutrients in wastewater can be removed by cultivating microalgae in heavily polluted water systems (ponds, canals, etc.), and the biomass from this aquacrop later can be used as fish feed or bioenergy source (Drechsel et al., 2015). Removal of phosphorus, nitrogen, and toxic metals from wastewater also prevents unwanted phytoplankton blooms in aquatic systems (Cai et al., 2013). Some algal species (out of over 36,000 various species) are characterized by accumulation of oil and lipids in their cells and thus can be further used for producing not only animal feed and bioenergy but also soil amendments, pharmaceutical materials, and dyes (Razzak et al., 2013). *Chlorella vulgaris* and *Phormidium laminosum* are two main species with high protein and lipid content and widely investigated for their potential of removing phosphorus and nitrogen content from the wastewaters (Razzak et al., 2013). Microalgae can be grown in all types of wastewaters from municipal (Li et al., 2011; Chi et al., 2011), agricultural (Mulbry et al., 2008, 2009), and industrial sectors (Chinnasamy et al., 2010; Markou and Georgakakis, 2011).

Use of algal species for biofuel production may partially replace demand for biofuel crops and thus reduce land use requirements for cultivating biofuel crops (Singh et al., 2011; Pittman et al., 2011). It may in turn lead to availability of more land for food crops and lower food prices. According to some estimations, biofuel productivity in lagoons culturing microalgae is 12–14,000 L/ha per annum, which is twice as high as productivity of palm oil fields (5600 L/ha per annum, Cai et al., 2013). In addition to wastewater treatment and bioenergy production benefits, algae can also contribute to carbon fixation since its cultivation requires large amount of CO_2 consumption (Razzak et al., 2013). Yet, harvesting microalgae both through mechanical and chemical methods substantially increases the costs of bioenergy production from microalgae and reduces its competitiveness with other energy resources such as petroleum (Razzak et al., 2013). It is also reported that most of the studies on algae cultivation in polluted environments are conducted at laboratory scale, yet the results of some pilot projects on microalgae cultivation at larger scale showed inconsistent purification of wastewater and unstable biomass outputs (Cai et al., 2013). Thus, lack of reliable and cost-effective methods of harvesting and producing algae biomass at large scale may constrain the biofuel generation based on algae feedstock (Christenson and Sims, 2011).

In industry, fully or partially treated wastewater can be circularly reused in most sectors such as commercial laundries, car washing stations, textile industry, meat processing, beverage production, and power plants (Jimenez and Asano, 2008). Wastewater can be also used for cooling plants or heating the buildings. Moreover, wastewater can be applied for recharging aquifers through infiltration basins or injection wells (Lazarova et al., 2013). Wastewater use for refilling the depleted gas mines, for instance, may further prevent potential earthquake risks.

4.3 Nutrients from waste: current status and potential options

The importance of fertilizer for agricultural production and global food security is unquestionable though the criticality of phosphorus availability for meeting future food demands was not commonly recognized as of water and energy (Cordell et al., 2009). Global demand for phosphate is estimated to increase from 42.7 to 46.7 Mt by 2025 due to population growth and related increase in food demand (FAO, 2015). Given the higher birth rates and currently underdeveloped levels of agriculture, the highest share (more than 30%) of this additional fertilizer demand growth is expected to occur in South Asia (Fig. 15.7). Substantial increase of fertilizer demand is also expected in Latin American-Caribbean and East Asian regions (with shares of 26% and 19%, respectively).

At present, agriculture is not only the dominant user of water resources but also of fertilizer, consuming about 90% of phosphate resources (Smil, 2000; Rosemarin, 2004; Mayer et al., 2016). Although opinions on the time of full depletion of phosphate rocks vary, the estimated amount of phosphate rocks from the currently known mines may suffice only till 2100s even under very optimistic scenario, unless new supplies are found (Steen, 1998, Gunther, 1997; Cordell et al., 2009). Declining quality of the reserves and increasing costs of extraction and transportation has been commonly admitted by the fertilizer industries (Runge-Metzger, 1995; Smil, 2000; IFA, 2008; Cordell et al., 2009). Rapid depletion of phosphate deposits and their availability only in countable number of countries such as Morocco, China, the United States of America, Jordan, and South Africa would lead higher fertilizer prices and lower crop yields consequently threatening food security (Cordell et al., 2009; Jasinski, 2010, Cieslik and Konieczka, 2017). This would in turn increase poverty and hunger, especially for the poor in developing countries.

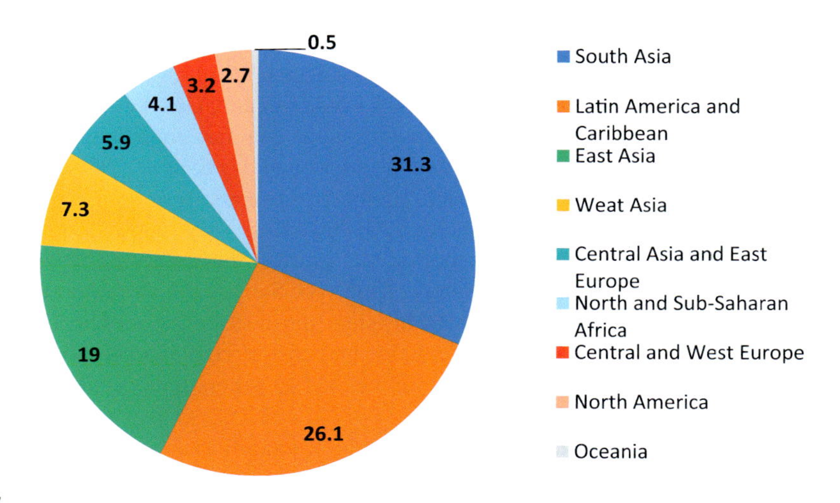

FIGURE 15.7

Estimations on increasing fertilizer demand share by regions.

Source: Adapted from FAO (2015).

Given the depletion of phosphate rocks in the near future, maintaining present and expected levels food security may require the dramatic transformation in phosphate production sources (Cordell et al., 2011). At present, phosphorus rocks and manure application contribute largely for overall phosphorus supply, though crop residues are also applied to nutrition plants. About 15 million tons (Mt) of mined phosphates are estimated to be used for fertilizer production per annum globally (Rittmann et al., 2011). However, its large portion (6–8 Mt) is being disposed to environmental systems through soil erosion and runoff losses, 5–7 Mt through animal waste, and 2–3 Mt through sewage waste (Rittmann et al., 2011). Thus, two main opportunities of increasing the life expectancy of world's phosphate deposits and counterbalancing the expected higher fertilizer and food prices and increasing national phosphorus security are more efficient use of fertilizer in agriculture and recycling waste (especially manure) and wastewater (Cordell et al., 2009, 2011).

Particularly, phosphate recovery from fecal sludge, urine, manure, crop residues, food waste, and other organic wastes (bone meal, ash, algae, and seaweed) may gain prominence in the long run (Karak and Bhattacharyya, 2011; Ashley et al., 2011; Cordell et al., 2009). According to modeling estimations, the recovery of phosphates from urine and feces, for instance, may potentially yield about 20% of phosphates supply after 2050s (Mihelcic et al., 2011). Rich nutrition content of human and organic waste, especially bone meal, allows for production of fertilizer and soil amendments for agriculture from these wastes (Fig. 15.8). As estimated, the production of compost or soil amendments from fecal sludge may yield also net benefits worth of US$ 10 per ton in contrast to its disposal which may cost about US$ 42 per ton (Strauss et al., 2003).

As estimated earlier, large potential of recovering phosphates from fecal sludge exists in South and East Asian countries such as India and China (Mihelcic et al., 2011). Given the reliance of Indian agriculture on phosphates imports, phosphate recovery from feces can be particularly important in this country. Especially under conditions of hot climate, the efficiency of waste treatment technologies

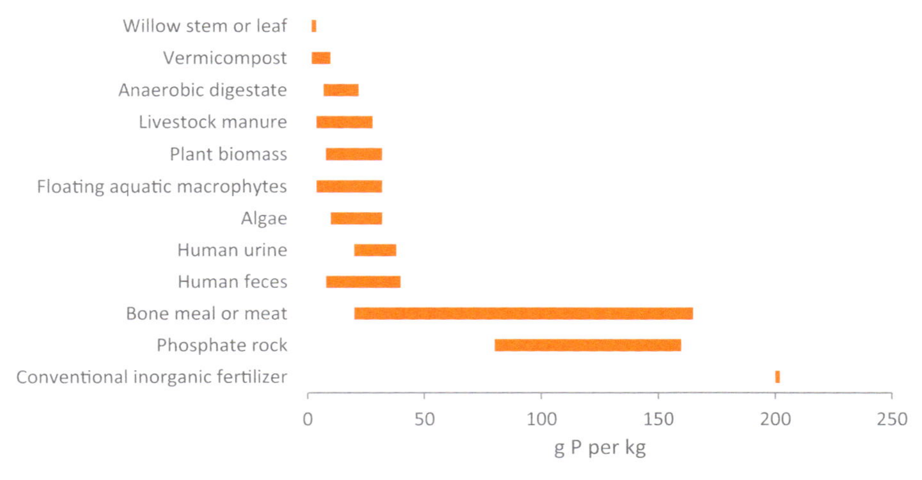

FIGURE 15.8

Phosphorus content of different types of waste.

Source: Modified after Roy (2016).

based on anaerobic digestion will be higher (Drechsel et al., 2015), thus increasing the feasibility of waste treatment technologies in India.

Multiple technologies exist for producing valuable fertilizers from waste. Morse et al. (1998) classified these technologies as follows: chemical precipitation, biological removal, crystallization, tertiary filtration, absorbent application, and sludge treatment. Although these technologies allow for safer application of recovered nutrients rather than direct application of fecal sludge for crop cultivation, their investment and operation costs, especially in developing countries may limit upscaling and wider impact (Cieslik and Konieczka, 2017). Cheaper options may include cultivation of cover crops and retention of crop residues to improve soil quality. Tan and Lagerkvist (2011) described various methods of recovering phosphorus and other nutrients (nitrogen, carbon, potassium, magnesium, etc.) from biomass such as rice and wheat straws, rice husks, pine wood, peach stones, sugarcane bagasse, sunflower shells, sewage sludge, and paper sludge ash and found out high phosphorus content of peach stone ash and sewage sludge ash especially.

Systematic analysis of using human excreta (feces and urine) for producing fertilizer was carried out at EAWAG (Tilley et al., 2014). EAWAG researchers classified four key stages of waste stream within the supply chain of the treatment system and reviewed various technologies for each stage. These four key stages are (i) collection of the waste (e.g., from latrines and septic tanks), (ii) its transportation from the residential site to treatment site (e.g., composting or other advanced methods of treatment), and (iv) final use for production purposes as a fertilizer.

Open defecation is a common practice in developing countries such as the ones in South Asia (Gupta et al., 2014). However, the collection of human waste requires changing the behavior of the people, building public and individual toilets, and constructing sewage systems for more effective sanitation. Installing ventilated improved pit and septic tanks may reduce the costs of sorting the waste in the later stages of waste treatment. Emptying septic tanks may be either done manually or using motorized machines. In small communities of Africa and South Asia, even using bikes for carrying urine containers was reported (Tilley et al., 2014). In advanced settlements, transportation can be done through sewage networks yet at higher capital costs. Waste treatment technologies vary depending on the purpose of recycling (e.g., fertilizer or biogas production) and availability of funds to establish them. Anaerobic baffled reactors and filters can be used to separate water from solid waste, consequently composting the solid waste for fertilizer production and releasing treated and disinfected water into environmental system. At cheaper costs, wastewater can be also treated in specially designed wetlands, stabilization reservoirs or lagoons that purify wastewater sequentially before reuse (Drechsel et al., 2015), and sedimented solid waste can be used for fodder or biomass production in these ponds. If wastewater is not directly used in the water treatment pond, treated wastewater can be diverted for irrigation purposes, for leaching fields or for recharging groundwater aquifers. Compost directly or after co-composting with additional nutrients can be applied in crop fields. Composting stations can also be additionally equipped with biogas reactors to produce biogas or electricity and thus increase the benefits from recycling.

Application of compost and direct use of effluents or fecal sludge after even minimal treatment may have considerable impact on crop biomass and yields. Given the fact that urine has higher phosphorus content rather than feces (Rose et al., 2015), Karak and Bhattacharyya (2011) reported the effects of urine application for the cultivation of various crops such as wheat, rice, corn, ryegrass, banana, cabbage, carrot, tomato, and spinach across several countries of the world and found out improved crop yields when urine was applied. In a similar review study, Singh and Agrawal (2008) also underlined the

Table 15.4 Effect of sewage sludge application on crop biomass and yields.

Crops	Sewage sludge amendment application rate	Effects on crop biomass and yields
Fescue	5.6 ton per ha	Yield increased by 30%
Corn	50–200 kg nitrogen per ha	Higher yield
Barley	10 ton per ha over 17 years	Increased dry matter and yield
Cotton	2:1 and 10:1—soil:sewage sludge ratio (together with tap water irrigation)	Increased seed production and fiber output
Maize	0–50 ton per ha	Increase in germination
Bahia grass	90–180 kg nitrogen per ha	50% increase of forage and improved spring crude protein
Sunflower	0–320 ton per ha	Increase in dry weight
Blue grama and tobosa grass	0–90 ton per ha	Increase in leaf area
Poplar tree	5:1 and 10:1—soil:sewage sludge ratio (together with tap water irrigation)	Increase in height and diameter
Apple tree	0–75 ton per ha over 2 years	Higher fruit yield

Source: Adapted from Singh and Agrawal (2008).

positive impact of applying sewage sludge on the yields of crops such as corn, barley, cotton, maize, sunflower, and different types of tress (Table 15.4).

Despite its yield and soil content improvement and soil humidity enhancement benefits, compost has lower comparative advantage over other fertilizers at present. The cost of compost that is adjusted considering its phosphate content can be considerably higher than the similarly adjusted prices for fertilizers with phosphates content (Fig. 15.9). Although adjusted price for *diammonium phosphates* (about 40% of compost price) is more expensive than other fertilizers and closer to compost price, it is because of additional nitrogen nutrients embedded in this fertilizer. Once the costs of transportation and application of compost are considered in comparative advantage analysis, willingness to buy and apply the compost by farmers may be further decreased given its bulky mass. Nevertheless, when the compost station is close to the farm and transportation of fertilizer increases due to bad road conditions, some level of compost application can be unavoidable. The comparative advantage of compost increases also due to its additional, positive external benefits such as the organic natural content of compost, sanitation benefits, and environment-friendly nature (preventing soil erosion, reduced phosphate contamination of return waters, and groundwater aquifers).

4.4 Energy from waste: current status and potential options

Energy security is crucial in many developing countries of the world since about 2.8 billion people will not have adequate access to modern energy facilities even coming to 2030 (IEA, 2010). Especially, about 550 million people in India and about 400 million in China live without electricity. These people mostly use solid fuels such as wood, crop residue, charcoal, and dung for cooking and heating despite enormous health risks of these cooking practices (Gebrezgabher et al., 2016). Generation of heat,

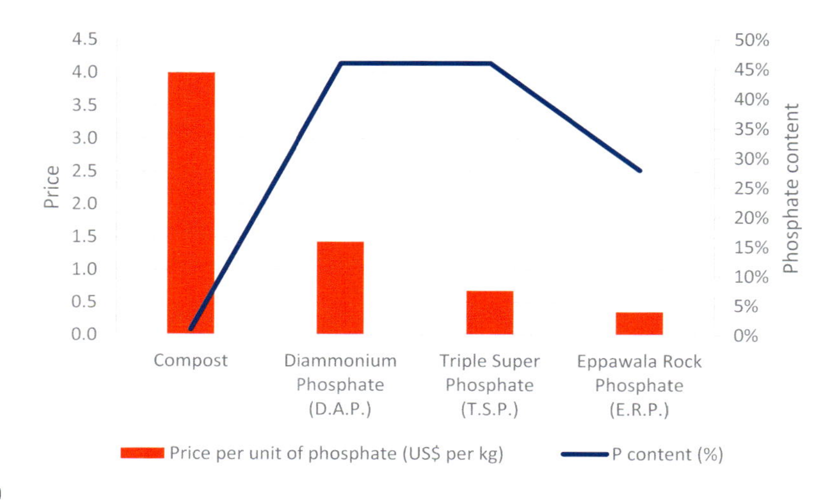

FIGURE 15.9

Prices per unit of phosphate content in different types of fertilizers.

Source: Calculated using data from Ceylon Fertilizer Company Ltd. (2016).

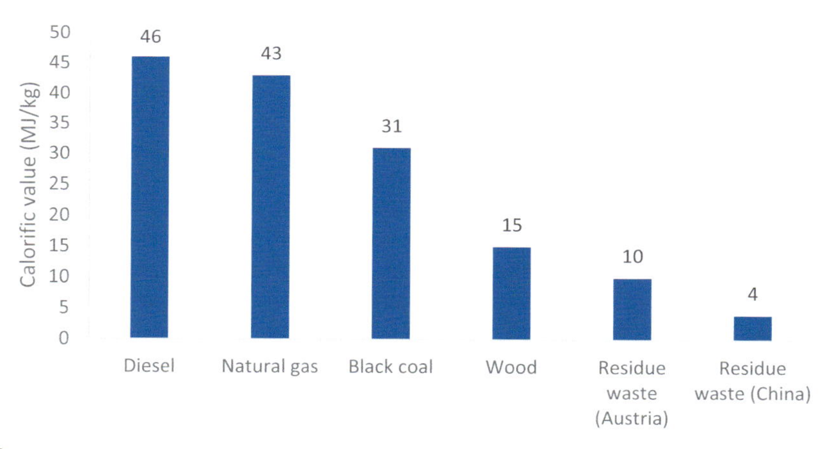

FIGURE 15.10

Calorific value of different energy sources.

Source: Based on WEC (2016).

electricity, biogas, and biofuel from waste can not only reduce environmental degradation effects of waste disposal but also supply additional energy resources though the calorific value per unit of waste is much smaller than alternative energy sources such as diesel, gas, coal, or wood (Fig. 15.10).

Waste and wastewater recycling requires enormous amount of energy (WEC, 2016). Thus, improving energy use efficiency in the sector not only allows for saving substantial volumes of energy at low cost but also for reducing carbon emissions largely. For instance, the use of wastewater from towers of cooling power plants can be effectively used for heating purposes while reducing energy consumption and heating costs. In addition to large amount of energy savings through improved technologies, waste from municipal and rural residential areas can be recycled to produce various energy commodities such as biogas, electricity, and liquid fuel (biodiesel).

Main approaches of recycling waste for energy production are (i) thermochemical treatment, (ii) biochemical treatment, and (iii) chemical treatment (Table 15.5). At present, 90% of processes aiming at recovering energy from waste (REW) are based on thermochemical treatment (WEC, 2016). Thermochemical treatment aims at burning waste at higher temperatures and thus using the heat

Table 15.5 Technologies of recovering energy from waste.

Treatment method	Treatment technology	Details of the technology	Output
Thermochemical treatment	Incineration	Mass burning at temperature higher than 1000°C	Heat and power
		Co-combustion with coal or biomass	
		Using pretreated waste fractions with higher energy contents	
	Thermal gasification	Conventional at temperature of 750°C	Hydrogen, methane, and syngas
		Passing waste into a kin at 4000–7000C	
	Pyrolysis	High pressure, no oxygen, and at temperature of 300–800°C	Char, gases, aerosol, and syngas
Biochemical treatment	Fermentation	Treating waste with bacteria in the absence of light (dark fermentation)	Ethanol, hydrogen, and biodiesel
		Treating waste with bacteria in the presence of light (photofermentation)	
	Anaerobic digestion	Waste treatment with microorganisms in the absence of oxygen	Methane
	Gas capture in dumping site	Extraction from dumping sites	Methane
	Microbial fuel cell	Conversion of the chemical energy of organic matter by catalytic reaction of microorganisms and bacteria	Power
Chemical treatment	Esterification	Reaction of an acid and an alcohol for creating an ester	Ethanol and biodiesel

Source: Adapted from WEC, 2016.

energy or producing biogas. Biochemical treatment considers composting the organic waste and treating it with microorganisms and bacteria which consequently allows for biogas and power generation. Chemical treatment of waste considers reaction of waste with acids and consequently producing ethanol or biodiesel.

About 130 million tonnes of municipal solid waste are recycled annually in over 600 plants of REW (Themelis, 2003). Global energy output from municipal solid waste recycling thus valued at US$25.32 billion annually (in 2013) (WEC, 2016). REW plants are located mainly in 35 countries and are built to deliver steam and electricity for heating and recover metals for reusing (Themelis, 2003). The largest market for REW commodities is the European Union, which accounts almost half of global market revenue in this sector (WEC, 2016). In Asia, Japan is a leader in REW, reusing almost 60% of its solid waste through incineration. REW facilities are relatively newly established in China where seven plants recycle over 1.6 Mt wastes per annum (Themelis, 2003). Yet, REW is very rapidly growing sector in this country and more than doubled during very short period of time between 2011 and 2015 (WEC, 2016).

Despite the availability of multiple options of advanced treatment of waste for recovering energy, especially in the developed countries of Europe and Asia, the share of energy produced from municipal solid waste is only 0.02% (0.7% × 3%) of global energy output (Fig. 15.11). Given much higher costs of producing energy using waste compared to other alternative options of energy production (Fig. 15.12), the magnitude of waste recycling for energy production purposes is limited currently. Perhaps with the improvement of REW technologies and consequent cost reductions, REW can be a more attractive option compared to the alternative ways of energy production in developing countries. Large plants of REW in urban areas may also reduce the production costs of electricity from waste due to scale effect and thus may improve the feasibility of energy production from waste.

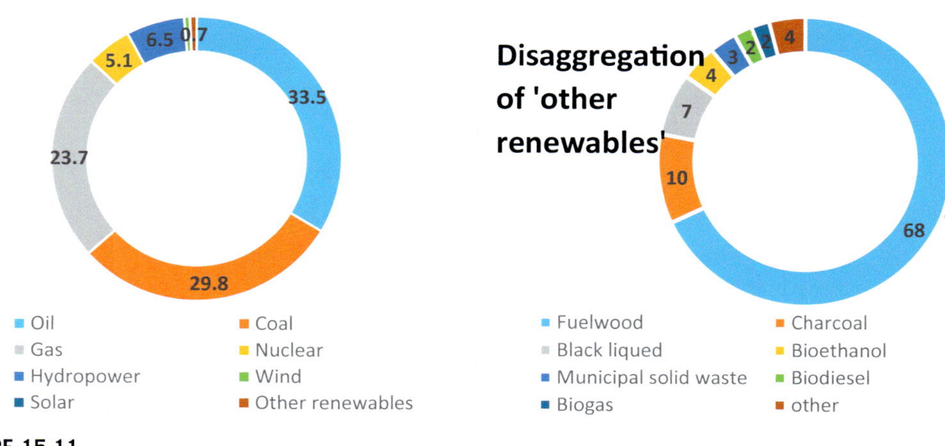

FIGURE 15.11

Main sources of energy supply globally (in 2016).

Source: Based on WEC, 2016.

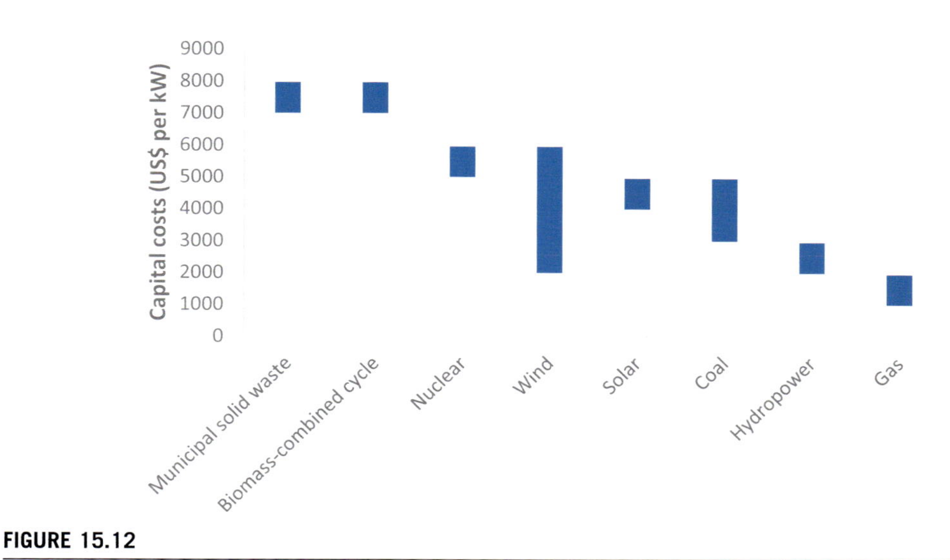

FIGURE 15.12

Costs of various energy production technologies in the United States.

Source: Based on WEC, 2016.

4.5 Construction materials and protein from waste

Nontraditional approaches of using waste such as fecal sludge from wastewater for producing construction and building materials or protein were also earlier reported. Fecal sludge from wastewater can be converted to inert and odorless ash through incineration process, and this ash can be mixed with clay for brickmaking to produce lightweight bricks (Tay and Show, 1997). Addition up to 20% of dry fecal sludge by weight was found not to considerably change brick's functional characteristics (Liew et al., 2004). Combusting sludge within bricks allows for creating small cavities which reduce vulnerability to freeze−thaw expansion (Alleman et al., 1990). Pelletized fecal sludge ash can be also used for producing masonry cement and lightweight concrete with moderate strength (Tay and Show, 1997). Burned fecal sludge through incineration can be also easily handled and disposed to land filling. Despite beneficial use of sludge ash for producing building materials, it is not always positively perceived by the producers of construction materials, especially in areas with abundant supply of conventional raw materials (Diener et al., 2014).

Fecal sludge is also used for rearing insect larva—black soldier fly (*Hermetia illucens*)—which is subsequently used as protein addition to animal feed (Diener et al., 2009; Nguyen, 2010). South African company, Agriprotein, uses this technology for producing feed for chicken and fish (van Huis et al., 2013). Many other studies also reported rearing insect larvae through using organic waste for producing feed not only for fish and chicken farms but also for farming frogs (Calvertet al., 1969; Hem et al., 2008; Ocio and Vinaras, 1979; Ogunji et al., 2007; St-Hilaireet al., 2007; van Huis et al., 2013). Productivity of larva mass can be higher especially when fecal sludge is applied together with municipal solid waste (Diener et al., 2009).

As experimented, one ton of fecal sludge with 40% dry solid content can yield 20 kg of dry animal meal from insect larvae with 35% protein content (Nguyen, 2010). Considering that fishmeal with 70%–80% protein content costs 0.7–1.2 US$ per kg (Diener et al., 2014), it can be estimated that protein or fishmeal obtained through the treatment of one ton of fecal sludge with insect larvae may worth of 7–12 US$. Under increasing prices for fish feed (tripled during the period between 2005 and 2013) owing to increasing aquaculture production, economic feasibility of animal feed production using insect larvae is likely to be improved (Naylor et al., 2009). Processed solids remained after insect larvae treatment can be used as soil amendment, thus further improving economic gains of insect larvae treatment (Diener et al., 2009).

5. Poverty alleviation and disease prevention effects of RRR

Recycling waste and wastewater resources for recovering effluents, nutrients, and energy not only provides additional economically valuable assets but also further improves water, food, and energy security in developing regions where these security improvements are highly demanded. Additional supply of water through wastewater treatment gain importance under increased frequency of droughts due to global warming (Meehl et al., 2007). When relevant reuses of water with low quality (return flows) reduce demand for freshwater uses (i.e., in agricultural irrigation). Improved water supply in turn would improve food and biomass outputs, consequently counterbalancing potential hunger and malnutrition risks expected due to temperature rise.

Improved water access together with increased availability of nutrients and energy resources is also essential for improved health of population and reduced incidents of illnesses among children. Safe access to water for drinking and sanitation and access to food supply at affordable prices are important as disease-preventive measures. Improved access to energy through waste and wastewater recycling, especially in winter months, may counterbalance frequent energy supply cuts in this period and thus indirectly add for disease prevention.

The establishment of well-organized waste recycling and wastewater treatment creates new job opportunities to poor people residing in developing regions (IGES, 2010). Thus, the wide-scale implementation of resource recycling and recovery technologies may have tremendous poverty alleviation effect. Yet, improving the working conditions and mechanization of waste and wastewater collection and treatment system can be essential to improve the status of the employees in this sector (Zhu et al., 2008).

6. Health and environmental risks related with RRR technologies

Despite multiple benefits available from the recycling and reusing waste and wastewater, their reuse does not come without environmental and health risks. Although urine application may boost crop yields, increased soil salinity and groundwater contamination can be a challenging issue especially when the urine application rates are too high (Karak and Bhattacharyya, 2011). Similarly, untreated application of feces, sludge, or sewage water may cause the accumulation of toxic content in the soil, higher carbon emissions, the spread of microbial organisms in the soil, and consequent contamination of both surface and groundwater resources. Indeed, proper treatment of waste and wastewater before any reuse may reduce these environmental risks. Especially, removal of phosphates from waste and wastewater may reduce environmental pollution and prevent or at least reduce eutrophication in water systems (Cordell et al., 2009). Yet, except high investment costs of advanced treatment technologies, their energy consumption and carbon footprint analysis should be additionally analyzed.

Direct and unplanned implementation of urine, feces, and sludge for crop cultivation also increases the health risks for plants, farmers, and consumers. High salinization or pollution of soils with toxic matter squeezes crop growth and reduce crop biomass and yields (Scheierling et al., 2011). Bacteria and viruses contained in waste, sludge, or wastewater can be transmitted to the farmers during the application process and trigger endemic and epidemic diseases. Chemical pollutants such as cadmium and mercury in sewage water and pharmaceuticals and antibiotics in waste also increase the risks of soil and groundwater contamination and consequent public health issues. Farm workers and consumers of vegetables and salads grown using feces, urine, and wastewater face an increased exposure to helminthic diseases such as hookworm and ascariasis and bacterio-viral diseases such as typhoid, diarrhea, and cholera (Scheierling et al., 2011). Especially in periods right before harvesting food crops, untreated use of wastewater, urine, and fecal sludge for irrigation may boost the incidents of these illnesses (WHO, 2004). Direct use of wastewater, urine, and fecal sludge for irrigation thus raises the issues of safeguarding farmers and public health in the developing countries. Given the possible contamination of urine after excretion, it should be prevented to be directly applied for crops during the last months of the preharvest season (Karak and Bhattacharyya, 2011). Moreover, it seems safer using feces and urine for biofuel, timber, and fodder crops rather than for food crops. Furthermore, adequate treatment of wastewater can be required before any irrigation reuse or discharge into water bodies for minimizing health or environmental risks. Different water quality standards apply for the reuse of effluent across the countries (Table 15.6).

Table 15.6 Effluent water quality standards for different reuse choices in selected countries.

	Country					
	India		Sri Lanka	Thailand	Costa Rica	Jordan
	DISF	UI	DISF	UI	DISF, LI	DISF, IVCC
pH	5.5–9.0	5.5–9.0	6.0–8.5	6.5–8.5	5.5–9.0	6.0–9.0
EC (mS/cm)				2000		
Turbidity (NTU)						10
TSS (mg/L)	100	200	50	30		50
O&G	10	10	10	5	30	8
COD (mg/L)	250		250			100
BOD (mg/L)	30	100	30	20	40	30
NH4–N (mg/L)	50	50	50			
TN (mg/L)						45
TP (mg/L)						30

Notes: DISF, *discharge into surface water;* IVCC, *irrigation of vegetables consumed cooked;* LI, *landscape irrigation;* UI, *unrestricted irrigation.*
Source: Adapted from Morel and Diener, 2006.

7. Conclusions

A brief overview of RRR technologies provided an insight on pros and cons of recovering effluents, nutrients, and energy from waste and wastewater. In general, potential of effluents for irrigation and environmental reuse is much higher and more economically viable than recovering nutrients and energy from waste and wastewater. The availability of large amount of wastewater from the sewage and drainage system can considerably enhance water supply under water scarce conditions and given the increasing costs of dam building and interbasin water transfers. Moreover, low quality water can be applied instead of freshwater in some relevant cases, consequently reducing the treatment and water supply costs. In developing countries with low-income level and abundance of lands, especially primary water treatment options such as filtration ponds can be economically and technically viable yet may require educational and extensional measures to improve the safety of effluents application. Advanced treatment options at higher costs perhaps can be limited only in remote areas where value of potable or industrial water is sufficiently high (for instance, in remote mining sites/towns).

Fertilizer from urine and fecal sludge is only the third best option among fertilizer augmentation measures, being feasible after the exhaustion of measures such as improving phosphates application efficiency and manure application. Improved phosphates application efficiency and livestock manure use are two best options preferable over fecal sludge compost use and urine application in terms of phosphates recovery potential (magnitude or availability) and implementation costs. Nevertheless, the potential of recovering phosphates from fecal sludge and urine may still allow for considerable recovery of phosphates and can be introduced once the other two better options reach their upscaling limits. Especially, reuse of fecal sludge and urine with minimal treatment can be recommendable in remote rural areas, which are disconnected from fertilizer markets or depend on heavy importing costs. Given the health concerns when applied for growing staple crops, using partially treated fecal sludge is more advisable for cultivating fodder (clover and sorghum), timber (trees), and fiber (cotton) crops. Advanced treatment of human waste for pelletized compost and soil amendments may come at higher costs than its direct application; therefore, their implementation can be limited to the production of some very economically valuable crops such as flowers or trees.

Energy recovery from waste through the use of advanced technologies can be much costly than effluents and fertilizer production from waste and wastewater. This option is characterized by lower economic and financial viability compared to other options of generating renewable energy such as solar power or wind power generation technologies. Thus, energy recovery from fecal waste and wastewater has very limited potential to generate energy at least in the nearest future. Nevertheless, using waste (manure, dung, crop residues, feces, and urine) for cultivating biofuel crops that are further used for household cooking or heating can be viable in remote rural areas without connection to the common energy grid. Summing up, economic relevance of particular RRR option is very case-specific and depends on environmental, geographic, demographic, socioeconomic, and institutional conditions of the implementation site.

Acknowledgment

This study was funded by Federal Ministry for Economic Cooperation and Development of Germany (BMZ) through a joint-project of International Water Management Institute (IWMI) and Center for development Research (ZEF) titled "Research and capacity building for inter-sectorial private sector involvement for soil rehabilitation" (PN10200176). The author was also supported by the Cluster of Excellence CLICCS at Universität Hamburg. Special thanks go to Dr. Nicolas Gerber for reviewing the manuscript and providing constructive feedback.

References

Aleklett, K., Campbell, C.J., 2003. The peak and decline of world oil and gas production. Minerals and Energy—Raw Materials Report 18 (1), 5–20.

Alleman, J.E., Bryan, E.H., Stumm, T.A., Marlow, W.W., Hocevar, R.C., 1990. Sludge-amended brick production: applicability for metal-laden residues. Water Science and Technology 22 (12), 309–317.

Ashley, K., Cordell, D., Mavinic, D., 2011. A brief history of phosphorus: from the philosopher's stone to nutrient recovery and reuse. Chemosphere 84, 737–746.

Cai, T., Park, S.Y., Li, Y., 2013. Nutrient recovery from wastewater streams by microalgae: status and prospects. Renewable and Sustainable Energy Reviews 19, 360–369.

Calvert, C.C., Martin, R.D., Morgan, N.O., 1969. House fly pupae as food for poultry. Journal of Economic Entomology 62, 938–939.

Ceylon Fertilizer Company Ltd, 2016. Data on Fertilizer Prices. Available online at. www.lakpohora.lk. (Accessed 4 April 2017).

Chi, Z., Zheng, Y., Jiang, A., Chen, S., 2011. Lipid production by culturing oleaginous yeast and algae with food waste and municipal wastewater in an integrated process. Applied Biochemistry and Biotechnology 165, 442–453.

Chinnasamy, S., Bhatnagar, A., Hunt, R.W., Das, K.C., 2010. Microalgae cultivation in a wastewater dominated by carpet mill effluents for biofuel applications. Bioresource Technology 101, 97–105.

Christenson, L., Sims, R., 2011. Production and harvesting of microalgae for wastewater treatment, biofuels, and bioproducts. Biotechnology Advances 29, 686–702.

Cieslik, B., Konieczka, P., 2017. A review of phosphorus recovery methods at various steps of wastewater treatment and sewage sludge management. The concept of "no solid waste generation" and analytical methods. Journal of Cleaner Production 142 (4), 1728–1740.

Corcoran, E., Nellemann, C., Baker, E., Bos, R., Osborn, D., Savelli, H. (Eds.), 2010. Sick Water? the Central Role of Wastewater Management in Sustainable development.A Rapid Response Assessment. United Nations Environment Programme, UN-HABITAT, GRID-Arendal. www.grida.no.

Cordell, D., Drangert, J.-O., White, S., 2009. The Story of phosphorus: global food security and food for thought. Global Environmental Change 19, 292–305.

Cordell, D., Rosemarin, A., Schröder, J.J., Smit, A.L., 2011. Towards global phosphorus security: a systems framework for phosphorus recovery and reuse options. Chemosphere 84, 747–758.

Diener, S., Zurbrügg, C., Tockner, K., 2009. Conversion of organic material by black soldier fly larvae: establishing optimal feeding rates. Waste Management & Research 27 (6), 603–610.

Diener, S., Semiyag, S., Niwagaba, C.B., Muspratt, A.M., Gning, J.B., Mbéguéré, M., Ennin, J.E., Zurbrugg, C., Strande, L., 2014. A value proposition: resource recovery from faecal sludge—can it be the driver for improved sanitation? Resources, Conservation and Recycling 88, 32–38.

Earthscan; Ottawa, Canada: International Development Research Centre (IDRC); Colombo, Sri Lanka. In: Drechsel, P., Scott, C.A., Raschid-Sally, L., Redwood, M., Bahri, A. (Eds.), 2010. Wastewater Irrigation and Health: Assessing and Mitigating Risk in Low-Income Countries. International Water Management Institute (IWMI), London, UK:, p. 404.

Drechsel, P., Wichelns, D., Qadir, M., 2015. Wastewater: Economic Asset in an Urbanizing World. Springer.

(Food and Agriculture Organization of the United Nations), 2010. AQUASTAT Database. Available at: http://www.fao.org/nr/water/aquastat/main/index.stm. (Accessed 13 April 2014).

FAO, 2015. World Fertilizer Trends and Outlook to 2018.

Gebrezgabher, S., Amewu, S., Taron, A., Otoo, M., 2016. Energy Recovery from Domestic and Agro-Waste Streams in Uganda: A Socioeconomic Assessment. International Water Management Institute (IWMI).

CGIAR Research Program on Water, Land and Ecosystems (WLE), Colombo, Sri Lanka, p. 52. https://doi.org/10.5337/2016.207 (Resource Recovery and Reuse Series 9).
Gunther, F., 1997. Hampered effluent accumulation process: phosphorus management and societal structure. Ecological Economics 21, 159–174.
Gupta, A., Spears, D., Coffey, D., Khurana, N., Srivastav, N., Hathi, P., Vyas, S., 2014. Revealed preference for open defecation: evidence from a new survey in rural North India. Economic and Political Weekly 49 (38).
GWI (Global Water Intelligence), 2009. Municipal Water Reuse Markets 2010. Media Analytics Ltd, Oxford, UK.
Harada, H., Strande, L., Fujii, S., n.d. Challenges and Opportunities of Faecal Sludge Management for Global Sanitation. Internet Source: www.eawag.ch.
Hem, S., Toure, S., Sagbla, C., Legendre, M., 2008. Bioconversion of palm kernel meal for aquaculture: experiences from the forest region (Republic of Guinea). Africal Journal of Biotechnology 7 (8), 1192–1198.
Hettiarachchi, H., Ardakian, R., 2016. Safe Use and Wastewater in Agriculture: Good Practice Examples. UNU.
Höök, M., Tang, X., 2013. Depletion of fossil fuels and anthropogenic climate change—a review. Energy Policy 52, 797–809.
Hoornweg, D., Bhada-Tata, P., 2012. What a Waste – a Review of Solid Waste Management. World Bank Urban development series. No. 15 Knowledge Papers. http://siteresources.worldbank.org/INTURBANDEVELOPMENT/Resources/336387-1334852610766/What_a_Waste2012_Final.pdf.
Hutton, G., Haller, L., 2004. Evaluation of the Costs and Benefits of Water and Sanitation Improvements at the Global Level. World Health Organization, pp. 1–87. Available at: http://www.who.int/water_sanitation_health/wsh0404summary/en/#.
International Energy Association (IEA), 2010. World Energy Outlook 2010. International Energy Agency (IEA) of the Organisation of Economic Co-Operation and Development (OECD), Paris.
IFA, 2008. Feeding the Earth: Fertilizers and Global Food Security, Market Drivers and Fertilizer Economics. International Fertilizer Industry Association, Paris.
Institute for Global Environmental Strategies (IGES), 2010. The 3Rs and Poverty Reduction in Developing Countries: Lessons from Implementation of Ecological Solid Waste Management in the Philippines (Philippines.
Jasinski, S.M., 2010. Phosphate Rock, Mineral Commodity Summaries. US Geological Survey. January.
Jiménez, B., Asano, T. (Eds.), 2008. Water Reuse: An International Survey of Current Practice, Issues and Needs. IWA Publishing, London, UK.
Karak, T., Bhattacharyya, P., 2011. Human urine as a source of alternative natural fertilizer in agriculture: a flight of fancy or an achievable reality. Resources, Conservation and Recycling 55, 400–408.
Lautze, J., Stander, E., Drechsel, P., da Silva, A.K., Keraita, B., 2014. Global Experiences in Water Reuse. International Water Management Institute (IWMI). CGIAR Research Program on Water, Land and Ecosystems (WLE), Colombo, Sri Lanka, p. 31. https://doi.org/10.5337/2014.209 (Resource Recovery and Reuse Series 4).
Lazarova, V., Asano, T., Bahri, A., Anderson, J., 2013. Milestones in Water Reuse. The Best Success Stories. IWA Publishing, London, UK.
Li, Y., Chen, Y., Chen, P., Min, M., Zhou, W., Martinez, B., Zhu, J., Ruan, R., 2011. Characterization of a microalga Chlorella sp. well adapted to highly concentrated municipal wastewater for nutrient removal and biodiesel production. Bioresource Technology 102 (8), 5138–5144.
Liew, A.G., Idris, A., Wong, C.H.K., Samad, A.A., Noor, M.J.M.M., Baki, A.M., 2004. Incorporation of sewage sludge in lay brick and its characterization. Waste Management and Research 22 (4), 226–233.
Liu, Y., 2005. Phosphorus Flows in China: Physical Profiles and Environmental Regulation. PhD-Thesis. Wageningen University, Wageningen, ISBN 90-8504-196-1.
Marald, E., 1998. I Motet Mellan Jordbruk Och Kemi: Agrikulturkemins Framvaxt Pa Lantbruksakademiens Experimentalfalt 1850–1907. Institutionen for idehistoria, Univ Umea.

Markou, G., Georgakakis, D., 2011. Cultivation of filamentous cyanobacteria (bluegreen algae) in agro-industrial wastes and wastewaters: a review. Applied Energy 88 (3), 389–401.

Matsui, S., 1997. Nightsoil collection and treatment in Japan. In: Drangert, J.-O., Bew, J., Winblad, U. (Eds.), Ecological Alternatives in Sanitation. Publications on Water Resources: No 9. Sida, Stockholm.

Mayer, B.K., Baker, L.A., Boyer, T.H., Drechsel, P., Gifford, M., Hanjra, M.A., Parameswaran, P., Stoltzfus, J., Westerhoff, P., Rittmann, B.E., 2016. Total value of phosphorus recovery. Environmental Science and Tehcnology 50 (13), 6606–6620.

McKinsey&Company, 2009. Charting Our Water Future. Available online at: www.mckinsey.com. (Accessed 4 May 2017).

Meehl, G.A., Stocker, T.F., Collins, W.D., Friedlingstein, P., Gaye, T., Gregory, J.M., Kitoh, A., Knutti, R., Murphy, J.M., Noda, A., Raper, S.C.B., Watterson, I.G., Weaver, A.J., Zhao, Z.C., 2007. Global climate projections. In: Solomon, S., et al. (Eds.), Climate Change 2007: The Physical Science Basis. Contribution of Working Group I to the Fourth Assessment Report of the Intergovernmental Panel on Climate Change (IPCC). Cambridge University Press, Cambridge.

Mihelcic, J.P., Lauren, M.F., Shaw, R., 2011. Global potential of phosphorus recovery from urine and feces. Chemosphere 84, 832–839.

Morel, A., Diener, S., 2006. Greywater management in low and middle-income countries, review of different treatment systems for households or neighbourhoods - Sandec Report No. 14/06. Sandec (Water and Sanitation in Developing Countries) at Eawag (Swiss Federal Institute of Aquatic Science and Technology), Dübendorf, Switzerland.

Morse, G.K., Brett, S.W., Guy, J.A., Lester, J.N., 1998. Review: phosphorus removal and recovery technologies. The Science of the Total Environment 212, 69–81.

Mulbry, W., Kondrad, S., Pizarro, C., Kebede-Westhead, E., 2008. Treatment of dairy manure effluent using freshwater algae: algal productivity and recovery of manure nutrients using pilot-scale algal turf scrubbers. Bioresource Technology 99, 8137–8142.

Mulbry, W., Kondrad, S., Buyer, J., Luthria, D., 2009. Optimization of an oil extraction process for algae from the treatment of manure effluent. Journal of the American Oil Chemists' Society 86, 909–915.

Naylor, R.L., Hardy, R.W., Bureau, D.P., Chiu, A., Elliott, M., Farrell, A.P., Foster, I., Gatlin, D.M., Goldburg, R.J., Hua, K., Nichols, P.D., 2009. Feeding aquaculture in an era of finite resources. Proceedings of the National Academy of Sciences (PNAS) 106 (36), 15103–15110.

Nguyen, H.D., 2010. Decomposition of Organic Wastes and Fecal Sludge by Black Soldier Fly Larvae. Asian Institute of Technology, Thailand.

Ocio, E., Vinaras, R., 1979. House fly larvae meal grown on municipal organic waste as a source of protein in poultry diets. Animal Feed Sciences and Technology 4 (3), 227–231.

Ogunji, J.O., Nimptsch, J., Wiegand, C., Schulz, C., 2007. Evaluation of the influence of housefly maggot meal (magmeal) diets on catalase, glutathione S-transferase and glycogen concentration in the liver of *Oreochromis niloticus* fingerling. Comparative Biochemistry and Physiology - A: Molecular Integrative Physiology 47 (4), 942–947.

Pearce, B.J., 2015. Phosphorus Recovery Transition Tool (PRTT): a transdisciplinary framework for implementing a regenerative urban phosphorus cycle. Journal of Cleaner Production 109, 203–215.

Pittman, J.K., Dean, A.P., Osundeko, O., 2011. The potential of sustainable algal biofuel production using wastewater resources. Bioresource Technology 102, 17–25.

Razzak, S.A., Hossain, M.M., Lucky, R.A., Bassi, A.S., de Lasa, H., 2013. Integrated CO_2 capture, wastewater treatment and biofuel production by microalgae culturing—a review. Renewable and Sustainable Energy Reviews 27, 622–653.

Rittmann, B.E., Mayer, B., Westerhoff, P., Edwards, M., 2011. Capturing the lost phosphorus. Chemosphere 84, 846–853.

Rose, C., Parker, A., Jefferson, B., Cartmell, E., 2015. The characterization of feces and urine: a review of the literature to inform advanced treatment technology. Critical Reviews in Environmental Science and Technology 45 (17), 1827–1879.

Rosemarin, A., 2004. The precarious geopolitics of phosphorous. Down to Earth: Science and Environment Fortnightly 27–31.

Roy, E.D., 2016. Phosphorus recovery and recycling with ecological engineering: a review. Ecological Engineering (in press).

Runge-Metzger, A., 1995. Closing the cycle: obstacles to efficient P management for improved global security. In: Tiessen, H. (Ed.), Phosphorus in the Global Environment. John Wiley and Sons Ltd, Chichester, UK, pp. 27–42.

Scheierling, S., Bartone, C.R., Mara, D.D., Drechsel, P., 2011. Towards an agenda for improving wastewater use in agriculture. Water International 36 (4), 420–440.

Siebert, S., Döll, P., 2007. Irrigation water use – a global perspective. In: Lozan, J.l., Grassl, H., Hupfer, P., Menzel, L., Schönwiese, C.-D. (Eds.), Global Change: Enough Water for All? Universitaet Hamburg/GEO, pp. 104–107.

Singh, R.P., Agrawal, M., 2008. Potential benefits and risks of land application of sewage sludge. Waste Management 28 (2), 347–358.

Singh, A., Nigam, P.S., Murphy, J.D., 2011. Renewable fuels from algae: an answer to debatable land based fuels. Bioresource Technology 102, 10–16.

Smil, V., 2000. Phosphorus in the environment: natural flows and human interferences. Annual Review of Energy and the Environment 25, 53–88.

St-Hilaire, S., Sheppard, D.C., Tomberlin, J.K., Irving, S., Newton, G.L., McGuire, M.A., Mosley, E.E., Hardy, R.W., Sealey, W., 2007. Fly prepupae as a feedstuff for rainbow trout, *Oncorhynchus mykiss*. Journal of World Aquaculture Sociology 38 (1), 59–67.

Steen, I., 1998. Phosphorus availability in the 21st Century: management of a nonrenewable resource. Phosphorus and Potassium 217, 25–31.

Strauss, M., Drescher, S., Zurbrügg, C., Montangero, A., Cofie, O., Drechsel, P., 2003. Co-composting of Faecal Sludge and Municipal Organic Waste: A Literature and State-Of-Knowledge Review. SANDEC/EAWAG – IWMI.

Tan, Z., Lagerkvist, A., 2011. Phosphorus recovery from the biomass ash: a review. Renewable and Sustainable Energy Reviews 15, 3588–3602.

Tay, J.-H., Show, K.-W., 1997. Resource recovery of sludge as a building and construction material- a future trend in sludge management. Water Science and Technology 36 (11), 259–266.

Themelis, N.J., 2003. An overview of the global waste-to-energy industry. In: Waste Management World (Review Issue) 2003-2004, pp. 40–47.

Tilley, E., Ulrich, L., Lüthi, C., Reymond, P., Schertenleib, R., Zurbrügg, C., 2014. Compendium of Sanitation Systems and Technologies, 2nd Revised Edition. Swiss Federal Institute of Aquatic Science and Technology (Eawag). Dübendorf, Switzerland.

UNWATER, 2016. Water and Sanitation Interlinkages across the 2030 Agenda for Sustainable Development (Geneva.

van Huis, A., Van Itterbeeck, J., Klunder, H., Mertens, E., Halloran, A., Muir, G., et al., 2013. Edible Insects: Future Prospects for Food and Feed Security. FAO, Rome, p. 201.

Von Braun, J., 2016. Expanding water modeling to serve real policy needs. Water Economics and Policy 2 (4), 1671004.

WEC, 2016. World Energy Resources. World Economic Council (WEC).

WHO, 2004. The Global Burden of Disease: 2004 Update. World Health Organization (WHO), Geneva, Switzerland. Available at: http://www.who.int/healthinfo/global_burden_dise.

World Bank, 2010. Improving Wastewater Use in Agriculture: An Emerging Priority. Energy Transport and Water Department Water Anchor (ETWWA).

Xie, J., 2009. Addressing China's Water Scarcity: Recommendations for Selected Water Resources Management Issues. The World Bank, Washington DC.

Zhu, D., Asnani, P.U., Zurbrügg, C., Anapolsky, S., Mani, S., 2008. Improving Municipal Solid Waste Management in India: A Sourcebook for Policy Makers and Practitioners. World Bank, Washington DC.

CHAPTER

The macroeconomic impact of climate change

16

Guller Sahin
Kütahya Health Sciences University, Evliya Çelebi Campus, Kütahya, Turkey

Chapter outline

1. Introduction

The historical course of greenhouse gas (GHG) levels shows that global average temperatures can be above 2.6–4.8°C and sea levels will be 0.45–0.82 m higher by the end of the 21st century (Chalmers, 2014). Depending on the increase in the amount of GHG, it is estimated that the global average temperatures will rise and that extreme weather events will intensify in climate models. Therefore, it is expected that problems, such as a decrease in biodiversity in the ecosystem, a decrease in agricultural productivity, deterioration of water and soil quality, water scarcity, an increase in pests and diseases, a shrinkage and extinction of ecological habitats, and an increase in sustainable food security will occur. Accordingly, if the amount of GHG cannot be reduced immediately, it is predicted that the global climate will change radically as never before. In this direction, it is assumed that the economic costs arising from climate change will increase cumulatively.

The economic aspect of climate change includes a reduction of carbon dioxide (CO_2) emissions, which create a GHG effect, at the rates stipulated in the Kyoto Protocol. However, reducing emission rates constitutes an important cost element for countries. Therefore, flexibility mechanisms have been developed in the Kyoto Protocol in order to achieve emission reductions at the lowest possible cost.

Handbook of Energy and Environmental Security. https://doi.org/10.1016/B978-0-12-824084-7.00003-5
Copyright © 2022 Elsevier Inc. All rights reserved.

Among these, the most discussed and remarkably flexible mechanism is emissions trading (Peker and Demirci, 2008). This is because emissions trading is accepted as the beginning of climate change policies and explains how to regulate emission reductions on an international scale (Damro and Mendez, 2003; Matthews and Paterson, 2005). Within this context, nature-based carbon offsetting solutions should be supported by incentives, as well as carbon pricing policies to mitigate and prevent climate change. Combining affordable carbon pricing with fossil fuel subsidies can prevent more emissions-intensive infrastructure from locking up by facilitating the transition to a cleaner development path. Sustainable economic growth may require less economic measures, such as raising emissions standards.

Detailed examination of the relationship between climate and economic outcomes is a crucial issue, as policymakers need reliable information to predict the impact of climate change (Olper et al., 2021). Climate shock arising from climate change incurs large-scale and integrated economic costs. Rising temperatures reduce not only output but also growth rates and political stability, thereby reducing industrial and agricultural production. Floods, storms, hurricanes, and earthquakes destroy habitats and assets such as roads and facilities, leading to economic production losses in future years. Determining the value of these production losses is an important component in assessing the welfare impact of climate change. Within this context, the economic value of an asset is the net present value of its expected future production, and the loss of output (income) caused by a disaster is equal to the value of the assets lost. The sum of output and asset losses, therefore, produces double the costs (Dell et al., 2012; Hallegatte and Vogt-Schilb, 2016).

According to the literature evidence, lower latitude tropical and subtropical regions, especially poor countries, are at serious macroeconomic risk (Mendelsohn and Massetti, 2017). In fact, according to the Global Climate Risk Index, all of the high-risk countries most affected by climate change are in Africa (Eckstein et al., 2020). The main underlying factors are that poor countries often have climates close to physical thresholds, limited access to adequate financial instruments for climate adaptation, and a lack of capital and technology. In these countries, mostly outdoor, exposure to heat or natural resource-intensive jobs are worked. However, it seems that climate change may provide certain advantages for a number of countries on the horizon by 2050. It is among the country-specific positive effects of climate change that Canada's agricultural productivity may increase, and water stress or drought events may not occur in northern hemisphere countries, such as Sweden and Finland (Bretschge and Valente, 2011; Woetzel et al., 2020). Although the short-term effects of climate change on the economy may be positive, in the long run, the negative effects will be wider and more dominant than the positive effects. Therefore, livable areas, drought-related food systems, agricultural lands, and the availability of natural resources will contain significant risks (Tol, 2018).

Climate change will also affect rich countries as well as poorer countries. Persistent increases in temperatures, extreme fluctuations in precipitation patterns, and more volatile weather events will have long-term macroeconomic effects by negatively impacting labor productivity, slowing investment and harming human health (Kahn et al., 2019). There are many tangible examples that prove the magnitude of the macroeconomic effects of climate change. For example, the economic damage caused by Hurricane Katrina in the United States of America in 2005 was 100 billion US$. Again in the United States of America, plants and animals are damaged by air pollution to the tune of 500 million US$ per year; it has been calculated that an expenditure of 40 US$ per person is required to reduce the damage caused by air pollution by 60%. In Russia, the per capita cost is 38 US$ due to increased health expenditure, and decreased productivity due to air pollution. When losses in the agricultural sector are included in this figure, the figure rises to 135 US$ per person (Keleş et al., 2015). It is clear that, regardless of the development stages of countries, the neglect of the issue of climate change continues to affect economies throughout the world.

While the source of GHG emissions is predominantly rich countries, the negative effects of climate change are generally seen in poor countries. This argument contains two policy implications. First, the rationale for drastic reductions in emissions is a call to consider the plight of poor countries and the impact imposed by rich countries. The other is whether promoting economic output is a better way to reduce emissions, given that poverty is one of the root causes of vulnerability to climate change (Tol, 2009).

The purpose of this study is to reveal the macroeconomic effects of global climate change. Within the scope of the purpose, firstly, a conceptual framework is presented for the costs of the effects of climate change. After this, an interaction process between the impact channels of climate change on the macroeconomy is evaluated in terms of economic growth, labor markets, labor productivity and sectors, and is explained using quantitative data. The study, evaluation, and conclusion are completed with policy recommendations.

2. Macroeconomic impact of climate change

There are several channels through which climate change can affect economic outcomes, from agriculture to political instability and health. The net effect of climate shock on economic performance is the sum of the direct and indirect effects. However, this is difficult to examine in detail because the range of mechanisms by which these effects can affect macroeconomic phenomena, positively or negatively, is quite wide. Identifying macroeconomic impact poses a fundamental problem of complexity; quantification poses challenges. Calculations are incomplete unless they take into account ecological impact, such as human mortality and biodiversity loss, which are difficult to measure economically and from which economic consequences are only gradually emerging. For example, it is difficult to quantify how much of the cost of a long hurricane season or worse-than-normal flooding can be explained by climate change, as opposed to natural variation. At the same time, even if the impact of climate change on economic mechanisms is known individually, the problem of how these mechanisms interact to shape macroeconomic outcomes is not clearly known (Bergholt and Lujala, 2010; Dell et al., 2008; Tol, 2018). However, the relationship between climate and the economy is a feedback process. Climate change is modeled as stochastic shocks that strike the productivity of labor, energy efficiency, capital stock, and inventories of firms. In this direction, technical changes are experienced in both the energy and manufacturing sectors. Innovation determines the cost of energy produced by dirty and clean technologies; this affects the mix of production in energy technology and the amount of CO_2 emissions. Therefore, structural change in the economy is closely linked to climate dynamics. At the same time, climate shock changes business cycles, technical change trajectories, and economic growth (Lamperti et al., 2018, 2020). As a result, a cyclical interaction is observed in the climate—economy relationship.

The effects of climate change emerge with increasing temperatures and fluctuations in precipitation regimes (Fig. 16.1). This situation increases the frequency and severity of climate-related natural disasters and causes large amounts of economic losses. For example, the average cost of global flood losses in 2005 was approximately US$ 6 billion per year. This figure rises to US$ 52 billion by 2050,

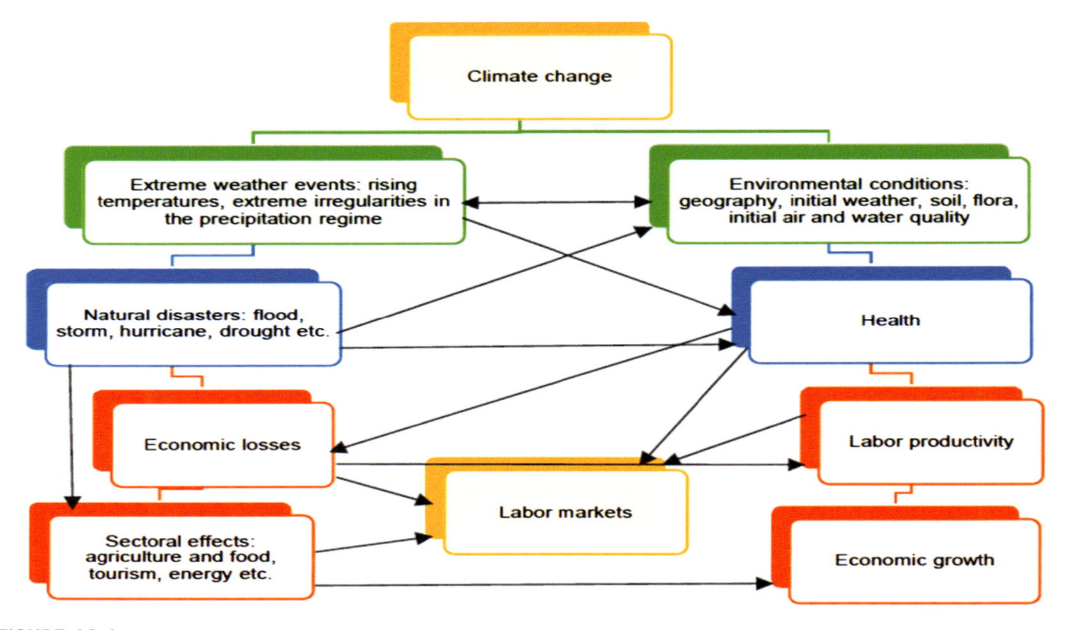

FIGURE 16.1

Macroeconomic effects of climate change.

Source: Figure was adapted and redraw from reference from Başoğlu, 2014.(Başoğlu, 2014).

when only socioeconomic change is foreseen. The figure, together with other damage caused by climate change, is estimated to be US$ 1 trillion or more per year (Hallegatte et al., 2013). In the period 1999–2018, more than 12,000 extreme weather events killed approximately 495,000 people globally, and created an economic burden of around US$3.54 trillion (purchasing power parity) (Eckstein et al., 2020).

There are two main risks in determining the cost of natural disasters in climate models. First, it underestimates the effects of asset losses from disasters on economic output flows in the production function. The other risk is that the capital stock in the production function underestimates the output impact of natural disasters and underestimates the importance of rebuilding capacity as a critical determinant of welfare losses (Hallegatte and Vogt-Schilb, 2016). However, environmental and economic values are often viewed as contradictory to each other in any discussion on climate change. It is under consideration that a choice must be made between protecting nature and ensuring growth, and less emissions only mean more costs (Doganova and Karnoe, 2015). In the climate literature, it is argued that deterioration (changes) is caused by market failure; while in the growth literature, climate change is seen as the biggest source of externality and is characterized as market failure (Dean and McMullen, 2007; Pretis, 2021). In this context, improvement of market failure may increase the effective supply of natural capital. Quantitative increases and quality improvements in natural capital supply, which is an important component of the production function, will increase output and accompany economic value (Hallegatte et al., 2011). Consequently, efforts to reconcile climate change and economic output are gaining more and more importance.

The macroeconomic impact of climate change is mostly measured with a static approach as the extent to which the climate in a given period affects welfare for that period. The dynamic approach, on the other hand, focuses on saving and capital accumulation (Fankhauser and Tol, 2005). In this context, macroeconomic effects are seen mainly on economic growth, labor markets, labor productivity, and related sectors.

2.1 Effects on economic growth

The asymmetrical impact of climate change between poor and rich countries is particularly serious when it concerns not only income levels but also economic growth, because then the impact and scope increase over time (Collier et al., 2008). Conceptually, climate shock affects economic growth through productivity and output channels (Dell et., 2012). While in the case of the productivity channel, climate change reduces productive capacity; in the case of the output channel, it reduces efficiency (Adom and Amoani, 2021). In the empirical literature on climate economics, information is given that climate change negatively affects economic performance. See the following studies: Abidoye and Odusola (2015); Alagidede et al. (2015); Baarsch et al. (2020); Bowen et al. (2012); Dell et al. (2008); Dell et al. (2012); Fankhauser and Tol (2005); Kahn et al. (2019); Kompas et al. (2018); Raddatz (2007, 2009); Stern (2007); and Tol (2018). However, it is stated in the literature that efforts to mitigate the effects of climate change may possibly accompany economic growth and produce widespread positive effects, including improved output and productivity. This argument is supported by the fact that since natural capital is a production function input, improvements in environmental quality can lead to an increase in the amount of input and raise the level of output (Adom and Amoani, 2021; Tol, 2009). Adom and Amoani (2021), Dell et al. (2009), Ferreira et al. (2020), Hallegette et al. (2011), Tol and Dowlatabadi (2001), and Tol and Yohe (2006) show that adapting to climate change can reduce negative effects on growth.

Climate shock can affect economic growth through physical capital and new investment. Since physical capital has a valid useful life under certain temperature and environmental conditions, climate shock may cause physical capital to depreciate before its expected period and increase the depreciation margin (Bretschger and Valente, 2011). At the same time, persistent climate shock may lead to more frequent capital investment (Fankhauser and Tol, 2005). This effect can lead to significant capital losses, particularly in high-income countries that invest around 20% of their income annually in fixed capital (Stern, 2007). When savings rates are constant, the negative effects of climate shock on output can produce a proportionate reduction in total investment. This may reduce future production and consumption due to the "capital accumulation effect." If saving rates are endogenous, economic agents may change their saving behavior to adapt to the effects of future climate shock. This, in turn, may suppress absolute and per capita growth expectations due to the "savings effect." In endogenous growth models, the effects of capital accumulation and savings may worsen through changes in the rate of technical progress and labor productivity. The capital accumulation effect is particularly important when technological change is endogenous and may be greater than the direct effects of climate shock (Fankhauser and Tol, 2005). When saving rates are not stable, the negative effects of climate shock, a loss of primary production resources, and a decrease in their efficiency can reduce both savings, due to low-income levels and investments and due to a decrease in capital demand by reducing output. The increase in the difference between savings and investments can change the external debt stock and foreign trade balance (Eboli et al., 2010).

Although an increased risk of climate shock may reduce the expected rate of return on physical capital, the risk can also serve to increase the relative return on human capital. In this case, a substitute for human capital investment can be created even though physical capital investment falls. Changing the composition of capital stock can provide a driving force for the adoption of new technologies, leading to improvements in total factor productivity (Skidmore and Toya, 2007).

Climate change may reduce natural capital endowment by causing the loss of productively used land (Guo et al., 2021; World Bank, 2010). For example, Strobl (2008) states that soil losses from storms reduce growth in the coastal states of the United States by an average of 0.8%. Therefore, a decrease in land and ecosystem services can slow economic growth by lowering the total capital stock. If natural capital losses can be covered by physical capital increases, growth may not be affected by natural capital losses. However, resources needed for necessary investment may create additional tax burdens and savings increases. This process may cause welfare losses by reducing the amount of consumption (Hallegatte, 2012).

Climate shock can also affect institutional structures and create obstacles to stable growth. The need for additional expenditure to protect against the effects of shock may not be considered appropriate by certain sections of society, and failure to make other necessary investment may lead to social conflict (Hallegatte, 2012). In this context, a social environment that includes conflict may disrupt the social and political stability necessary for a good economic structure, resulting in lower economic performance than countries with better institutional structures. Baarsch et al. (2020), Adom and Amoani (2021), and Noy (2009) emphasize the importance of institutional quality in the macroeconomic output costs of climate shock.

Due to ineffective emissions reduction policies, it is predicted that economic losses due to climate change will increase cumulatively. If mitigation measures are not taken, the costs of climate change will be equivalent to losing at least 5% of global annual gross domestic product (GDP). If the risk is broadened and the spectrum of impact is taken into account, damage estimates can rise to 20% or more of GDP (Stern, 2007). For example, low agricultural productivity and rising sea levels are projected to reduce GDP in South and Southeast Asia by 5.5% and global GDP by 1.5% in 2060. These estimates do not include increased health costs and productivity losses due to local pollution in most countries (OECD, 2014). A rise in temperatures above 3°C would shrink the world economy by 18% over the next 30 years; if the Paris Agreement targets are achieved, it is assumed that there will be a 4% economic loss. However, in four different temperature increase scenarios, it is seen that the climate effects will affect the economies of Middle Eastern and African countries the most (Table 16.1). In a scenario in which rising temperatures will be most severe, China is at risk of losing about 24% of GDP, while approximately 10% of the US, Canadian, and UK economies and about 11% of Europe's GDP are at risk (Guo et al., 2021).

2.2 Impact on labor markets

Climate change affects labor markets both directly and indirectly (Fig. 16.2). The direct effects are medium and long-term effects as a result of possible changes in extreme weather events and climatic events. These affect certain sectors, such as agriculture and tourism, more than other sectors, and it takes a long time for markets to adapt to these effects. Examples of direct impact are more investment in coastal retaining walls to prevent damage from rising sea levels, and insurance companies raising their premiums to reflect increased claims associated with climate change claims (Medhurst, 2009;

Table 16.1 Temperature rise scenario by mid-century.

Temperatures / Regions	Well below 2°C increase	2.0°C increase	2.6°C increase	3.2°C increase
	Paris target	The likely range of global temperature gains		Severe case
Simulating for economic loss impacts from rising temperatures in %GDP, relative to a world without climate change (0°C)				
World	-4.2%	-11.0%	-13.9%	-18.1
OECD	-3.1%	-7.6%	-8.1%	-10.6%
North America	-3.1%	-6.9%	-7.4%	-9.5%
South America	-4.1%	-10.8%	-13.0%	-17.0%
Europe	-2.8%	-7.7%	-8.0%	-10.5%
Middle East and Africa	-4.7%	-14.0%	-21.5%	-27.6%
Asia	-5.5%	-14.9%	-20.4%	-26.5%
Advanced Asia	-3.3%	-9.5%	-11.7%	-15.4%
ASEAN	-4.2%	-17.0%	-29.0%	-37.4%
Oceania	-4.3%	-11.2%	-12.3%	-16.3%

Source: Guo et al. (2021).

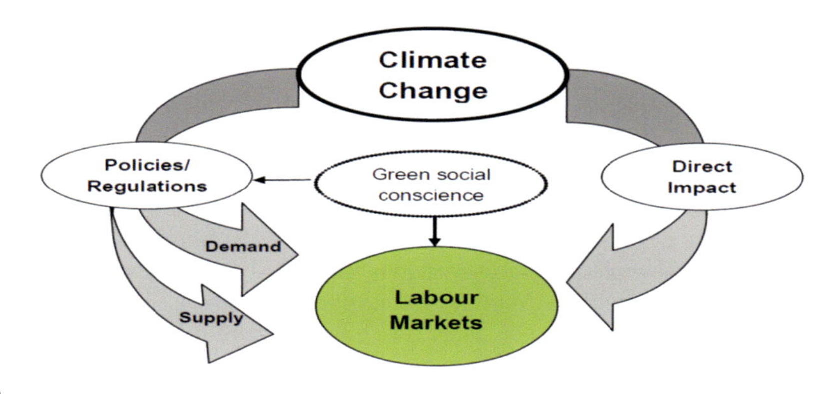

FIGURE 16.2

Impact of climate change on labor markets.

Source: Martinez-Fernandez et al. (2010).

OECD, 2012). Indirect effects occur through policies and regulations. This impact is the identification and analysis of policy options that authorities will use to maximize economic opportunities and minimize any negative impact on employment and the economy, especially during the transition to low-carbon technologies. The impact of policies and regulations affects labor and the supply of, and demand for, products and services. For example, though emissions trading schemes and carbon taxes increase the costs associated with carbon-intensive activities, thanks to the price mechanism, businesses and consumers are informed to replace carbon-intensive activities with low or zero carbon-intensive activities. Regulations and green public procurement is determined to transition to low-carbon activities, including green technologies, certification for green building codes, and waste recycling (Château, 2011; Medhurst, 2009; OECD, 2012).

Labor markets are indirectly affected by differences in consumer preferences. Global awareness of climate change results in increased societal awareness, independent of policy and regulation. For that reason, consumer demand for goods and services with green qualities is increasing. This causes a number of companies to prefer green initiatives more as a part of their relationship with customers, in line with their corporate responsibility approach. As a result, the growing "green social conscience" (consumer preferences, policies, and regulations) is a factor driving industries and labor markets (Martinez-Fernandez et al., 2010; OECD, 2012).

With the effects of climate shock on labor markets, especially as economies transition to greener growth, to what extent and how new jobs will be created, regulated or destroyed will gain importance. The possible effects of green jobs created by a green economy on labor markets are as follows (Martinez-Fernandez et al., 2010; OECD, 2012; Scott, 2014; UNEP et al., 2008):

- In certain cases, the transition to green growth may result in the creation of new jobs as newer products and technologies emerge. New job opportunities can arise, for example, through investment in adaptation measures to make ecosystems more resilient, strengthen infrastructure, water management, shoreline protection, and low-carbon technologies.
- A number of employment areas may be relocated, such as switching to ecological resources instead of using fossil fuels, recycling instead of landfilling and waste incineration.
- Certain jobs may disappear without replacement, such as the discontinuation of production of plastic materials due to prohibitions.
- Many existing jobs, especially plumbing, electrical, metal, and civil works, can be repurposed with a variety of skills, working methods, and new green ideas.

Although the transition to a low-carbon and resource-efficient green economy is seen as a necessity, there is no consensus on the effects of the green economy on labor markets. On the one hand, there is the opinion that green jobs are a pillar to solve problems in the labor market, while on the other hand, it is considered that the replacement of green jobs with old-style manufacturing jobs may create a new type of unemployment problem (Martinez-Fernandez et al., 2010).

There are concerns that measures to be taken for climate change mitigation and adaptation will cause a decrease in demand in the labor market, as technology may replace jobs, especially in local economies and among businesses that support these economies. Expectations that the measures will change job profiles in certain sectors and require new skills increase these concerns. However, when these opportunities are properly exploited, the local economy will be able to lead the transition to a greener labor market and more sustainable economic growth (OECD, 2012). Therefore, the net effect of climate change on employment may develop in a positive way. Within this context, employment gains will be possible with net increases of 0.5%–2% worldwide. Therefore, by 2030, a strong business potential will have been created by creating potentially additional jobs, especially in the agriculture, forestry, energy, recycling, construction, and transportation sectors. The creation of new markets for low-carbon energy technologies and other low-carbon goods and services is assumed to be another area that will create significant business opportunities (ILO, 2017; Stern, 2007). It is estimated that the cumulative investment amount to be made in low-carbon energy generation technologies by 2050 will exceed $13 trillion and that the global market will be above $500 billion annually (Hirst,

2006). Springboard (2006) explains that major shifts toward low-carbon energy technologies will accompany changes in employment patterns, and in line with the scale of investment, current employment in these sectors will increase from 1.7 million people worldwide to over 25 million by 2050. Arias (2009), Babiker and Eckaus (2007), on the other hand, state that measures regarding emissions restrictions cannot create a net employment increase.

2.3 Effects on labor productivity

In terms of impact channels, health problems, heat stress, relative humidity distribution, and other extreme weather events, arising from the intensity of climate shock and increasing temperatures, affect workforce capacity. Problems, such as water scarcity, malnutrition, premature deaths, pest infestations, and various diseases as a result of climate change, include a number of factors, such as physical and cognitive labor productivity, conflict and health, all of which have integrated economic consequences. Reducing the workload of employees to prevent heat stroke, increasing the frequency of short breaks, and limiting their efforts to prevent excessive bodily exertion all reduce labor productivity (Abidoye and Odusola 2015; Kahn et al., 2019; Kjellstrom et al., 2009; Lecocq and Shazili, 2007; Woetzel et al., 2020; Scott, 2014). Exposure to higher temperatures before serious health effects occur includes reduced mental ability, reduced working capacity, and an increased risk of accidents. All of these effects result in lower labor productivity. Gosling et al. (2019), Graff Zivin and Neidell (2012), Hsiang (2010), and Sullivan (2014) explain that climate change reduces labor productivity in their studies.

Climate shock impairs the health and productivity of millions of workers, thereby increasing labor costs. At the same time, occupational health management interventions are becoming a cost-increasing factor. For example, considering the population growth and employment structure in China, high temperature subsidies for people working on extremely hot days are estimated to have been 38.6 billion yuan per year for the 1979–2005 period, with this figure expected to increase to 250 billion yuan per year by 2030 (Kjellstrom et al., 2009; Zhao et al., 2016). Jobs exposed to heat in 2017 equaled approximately half of GDP, accounted for around 30% of growth and accounted for about 75% of the workforce. Effective outdoor daylight hours lost annually due to decreased average labor productivity are estimated to have increased by about 15% by 2030 compared to today. However, the effects on growth can affect both the supply and demand sides of the economy. The supply side outcomes are that they can have direct effects on labor and capital productivity, destroying the capital stock and reducing the capital services derived from that stock. The demand side effects, on the other hand, may occur in a chain manner. For example, negative prospects for climate change can lower property prices, and falling property prices may reduce government tax revenues (Woetzel et al., 2020).

2.4 Sectoral impact

Although there is no consensus in theory or practice on the mechanism explaining changes that climate shock may cause on the economy, leading indicators play an important role in determining the magnitude of the effects. The impact may have different types of disasters and different degrees of risk on various sectors (Fig. 16.3). Some of these are the share of climate-sensitive sectors in the economy, and the indirect effects of climate change on nonclimate-sensitive sectors. The main sectors that are highly sensitive to the direct effects of climate shock, due to changing resource and land structure and extreme weather events, are agriculture, tourism, energy, finance-insurance, transportation,

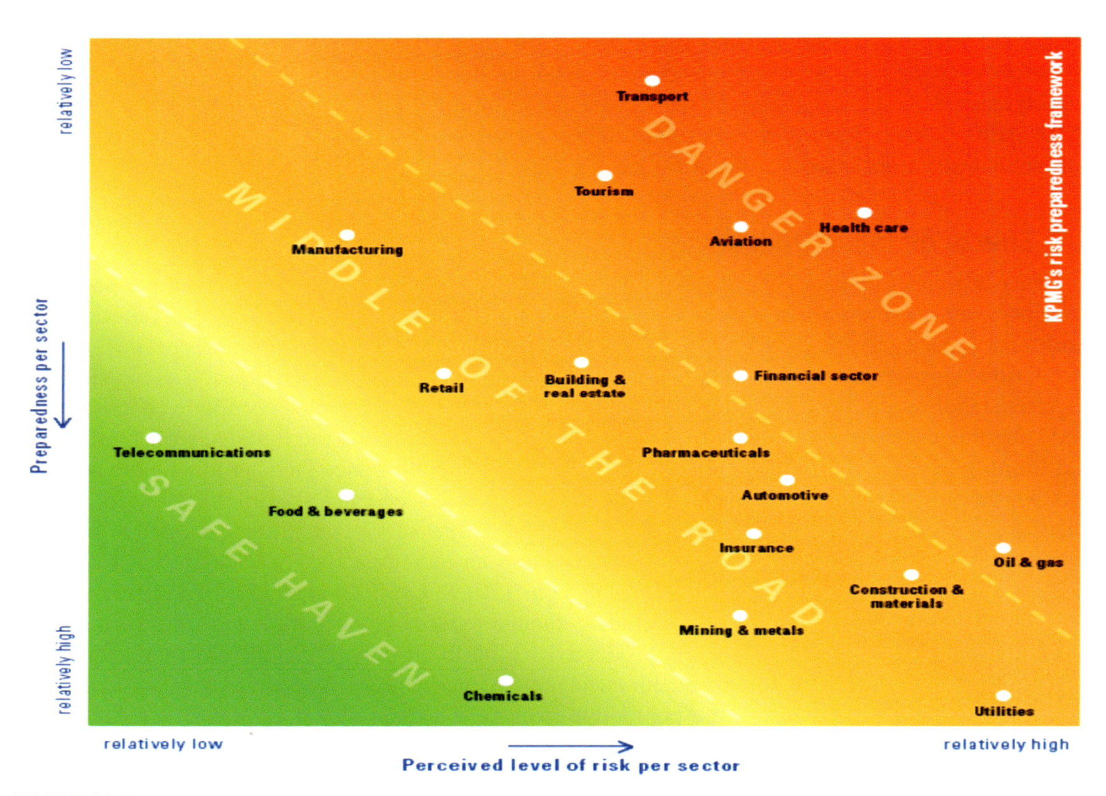

FIGURE 16.3

Identification of sectors at risk from climate change.

Source: Medhurst (2009).

infrastructure, and health (Lecocq and Shazili, 2007; Loayza et al., 2012; Martinez-Fernandez et al., 2010; Medhurst, 2009). For example, unless clean coal technologies develop rapidly, a number of sectors, such as coal mining, are likely to be adversely affected by climate events. In other sectors, such as ecological energy, an increase in demand is forecast. In other industries, such as the automotive industry, firms may have to transform products to remain competitive. The impact on different industry groups depends on the extent of exposure to higher energy costs, international competition, and conversion coverage. In certain cases, the driver of change may be the need to reduce costs, while in other cases, it may be the need to respond to different market preferences. In either case, significant changes to processes and products may be required (Medhurst, 2009).

2.4.1 Impact on agriculture and food sectors

As a result of climate shock, temperature, amount of precipitation, and sunlight duration, which all affect agricultural productivity, may change; depending on the rise in sea levels, certain regions are at risk of flooding and changes in the salinity of groundwater. As a result of these effects, crop yield, food production and food quality will be adversely affected. While increased CO_2 intensity is predicted to benefit crop productivity at lower temperature rises, product quality is expected to decline. In order to increase product quality, conditions, such as the amount of water in the soil and nutrient levels, must also be met. The provision of these conditions is becoming more and more difficult and is dependent on the severity and frequency of extreme weather events. These difficult conditions also pose a food safety risk. In general, climate shock makes it difficult to grow the same crops under the same locations (IPCC, 2019). Therefore, climate change reduces agricultural productivity and crop production, causing an increase in food prices, a food security risk, an increase in imports, and a decrease in exports. These global challenges also involve a complex and integrated agenda that brings with it issues of agroecosystem resilience, livelihoods, eco-efficiency, and sustainability (Battisti and Naylor, 2009). Studies by Alene (2010), Batten (2018), Cline (2008), Keane (2009), Lobell and Burke (2008), and Smith and Myers (2018) are examples in the literature of the negative effects of climate shock on the agricultural sector.

The effects of the observed evidence of climate shock on agriculture and the food sector are still evident in various parts of the world today, as the negative effects of climate trends are more far-reaching than the positive effects. Within this context, climatic effects reduce the yield of basic crops, such as wheat, corn, cotton, sugar beet, barley, and rice in certain regions at higher or lower latitudes. At the same time, they increase the price volatility of agricultural products and decrease the quality of food. The capacity of farmers to adapt to climate change is limited, and it is estimated that if temperatures increase by 3°C or more, the adaptation capacity will be exceeded, especially in tropical and subtropical regions close to the equator (Cameron, 2014; IPCC, 2019; Millner and Dietz, 2015).

A more than 30% increase in the global per capita food supply since 1961 has been accompanied by increases in nitrogen fertilizers and water supplies. The food system is under pressure from non-climatic stressors, such as demand for animal-derived products, population and income growth, and climate change. These factors influence the dimensions of availability, access, and the use and stability of the food system. Food security may be increasingly affected by future climate shock. Within this context, due to this climate shock, a 1%–29% increase is predicted in grain prices by 2050. Therefore, consumers will be affected by higher food prices, with these effects varying regionally. At the same time, production in many regions will be adversely affected by the change and expansion of the distribution of pests and diseases. Given the integrated mechanism of increasing extreme weather events, the risks of food system disruption will also increase (IPCC, 2019).

As a result of the drought in Southern Africa in 2015, agricultural output decreased by 15%. It is estimated that the probability of a yield decrease of more than 10% in wheat, corn, soybean, and rice will increase from 6% to 18% in any year by 2050. Although yield gains are expected in certain regions, as in Russia and Europe in general, climate shock on a global scale will work against the agricultural sector (Woetzel et al., 2020). With the world population estimated to be 9.2 billion by 2050, it is expected that total food production will increase by 60%–70% in order to adequately feed people (Battisti and Naylor, 2009). A number of economies dependent on the agricultural sector may lose more than 50% of their total agricultural output by 2080, even when carbon fertilization effects are included (Keane et al., 2009). Gross added value from agriculture to the global economy was US$ 2.7 trillion in 2016, up from US$1.9 trillion in 2013. In 2013, it was stated that the annual cost of malnutrition was 3.5 trillion US$ (FAO, 2013).

2.4.2 Impact on the tourism sector

For the effects of climate change on the tourism sector, the following predictions have been made: rising sea levels and acidification of the oceans will threaten infrastructure and natural attractions in coastal tourism; rising temperatures will shorten winter sports seasons, shift demand to higher latitudes, and destroy a number of ski resorts; natural disasters will cause changes in biodiversity affecting ecotourism; and changes in precipitation will affect the amount of water available (Nicholls, 2014; UNWTO, 2003). At the same time, it is stated that the species of coral reefs will be endangered, the shrinking of coastal land areas will threaten island states which depend on tourism, and that drought, epidemic diseases, and heat waves will affect tourism and keep tourists away from these destinations (Woetzel et al., 2020). Therefore, all these factors will damage the tourism sector and reduce revenues from the sector.

The impact of climate shock on the tourism sector may change the economic structure on a large scale. A decrease in tourism activities can lead to sectoral unemployment by reducing the demand for labor. It may also result in reductions in investment, such as in construction, the transportation network, and infrastructure facilities (UNWTO, 2003). Tourism activities have important economic consequences by playing a role in the redistribution of income from urban areas to rural areas, from developed countries to underdeveloped countries, and from north to south. Tourism revenues have a large share in balance of payments, especially in island states and in underdeveloped countries. Therefore, the tourism sector offers poorer countries significant employment opportunities and poverty prevention potential (UNWTO, 2007).

2.4.3 Impact on the energy sector

In the future, it is expected that the energy sector will be affected by policy responses to climate change at different levels. The increasing severity and frequency of climate shock presents increasing challenges for energy supply and demand, energy sources, facilities, and distribution. Encouraging investment in low-carbon technologies poses significant challenges for governments in achieving carbon reduction targets. Limiting global warming to 1.5°C, in order to mitigate the effects of climate shock predicted for the second half of the 21st century and the goal of transitioning to low-carbon energy by 2040, means completely transforming global energy markets (IPCC, 2014a). Cayan et al. (2006), Hanson et al. (2007), and Vanrheeen et al. (2004) provide examples of the negative effects of climate change on the energy sector.

Climate shock can change the energy sector, sometimes individually and sometimes in an integrated way. Within this context, the possible effects of climate shock on the energy sector are as follows (Griffiths et al., 2009; IPCC, 2014b; Zamuda et al., 2013):

- Thermoelectric power generation facilities are at risk due to the following: decreases in the amount of water available; increases in air and water temperatures, which reduce cooling efficiency; increases in the risk of water pollution, endangering the local ecology; and increases in the possibility of exceeding water thermal intake or waste water limits.
- The energy infrastructure in coastal areas is at risk due to rising sea levels, increased intensity, and frequency of storms and hurricanes, as well as flooding.
- Climate shock potentially disrupts oil and gas production, refining and distribution, as well as electricity generation and distribution.

- Oil and gas production is sensitive to reduced water volumes given the volumes of water required for enhanced oil recovery, fracturing, and refining.
- Ecological energy sources, especially bioenergy, solar energy, and hydropower, can be affected by changing precipitation patterns and the intensity of rising temperatures.
- Electrical transmission and distribution systems may lose their efficiency by carrying less current due to increased temperatures, and may suffer increased physical damage from extreme weather events. This situation may adversely affect economic activities by causing loss of time due to large-scale and long-term power cuts.
- An increase in temperature can increase the demand for electricity for cooling and reduce the demand for fuel and natural gas for heating.
- Fuel transport by barge and rail is susceptible to increased disruption and delays during more frequent periods of drought and flooding that affect water levels in rivers and ports.
- Because of the large quantities of water used for cooling in nuclear power plants, water scarcity can interrupt the effective use and power production of nuclear power plants.

Some of the impacts mentioned above (such as increased temperatures of ambient water used for cooling) are expected to occur in all regions. Other impacts may vary more by region and differentiate the vulnerabilities faced by various stakeholders. However, regional differences do not mean regional isolation as energy systems become more and more integrated. Integrated factors can create additional challenges. For example, combinations of extreme weather events can create short-term peaks in demand, reducing the system's flexibility and supply, thereby limiting its ability to respond to demand (Zamuda et al., 2013).

3. Evaluation and conclusion

Climate change is an undeniable reality that has gained a global identity. Scientific evidence shows that although temperature increases have positive short-term effects in certain countries at high latitudes, the global net impact of climate change is negative. In particular, as a result of extreme fluctuations in precipitation regimes and increasing temperatures, climate change creates serious costs on economies. Although the changing climatic conditions create large-scale macroeconomic effects, especially in poorer countries and in tropical and subtropical regions close to the equator, the range of influence is expected to expand in the process. Therefore, the biggest long-term threat facing the global economy is climate change. Within this context, it is expected that the macroeconomic reflections of climate shock on economic growth, labor markets, labor productivity, and various sectors will become much more dominant.

In recent years, an increasing body of econometric evidence has been collected at local, regional, and global levels, which link the effects of climate change to economic consequences. The broad consensus on the role of climate shock in economic growth is that rising temperatures, especially in low-income countries, have significant negative effects. Climate shock has the following effects: it decreases growth; reduces the quantity of factors of production and lowers their productivity; reduces the expected rate of return of physical and natural capital and increases depreciation; changes productive capacity and increases the trend of production; and increases capital investment and worsens institutional quality. However, the decrease in physical capital investment can create substitutes for human capital investment. This change in the composition of the capital stock may trigger the adoption of new technologies, resulting in an increase in total factor productivity. At the same time, it can produce effects in productivity and output situations to adapt to climate change, reduce the negative impact of climate shock on production, and improve production capacity. The effects of climate change on labor markets show a more optimistic picture, especially due to the emergence of new employment areas due to the creation of green jobs within the green economy. In the case of labor productivity, which is another channel where climate change affects macroeconomics, physical and cognitive labor productivity decreases, the risk of accidents increases, health problems occur, working capacity decreases, and labor costs increase. The effects of climate shock on economic sectors are more significant and involve greater risks in the short term, especially in the agriculture, food, tourism, and energy sectors. The long-term perspective is that climate change will seriously affect all sectors. Within this context, an increase in the frequency and severity of the effects will play an important role in shaping sectors in the future. As a result, serious changes will occur in the global economy.

Consequently, the debate on how to choose between preventing climate change and promoting growth within the historical path has lost its validity today. Within this context, Hubert Reeves emphasizes the importance of the subject saying: "We are at war with nature. If we win, we will lose." Scientific evidence for the current effects of climate change and climate—economy simulations for its possible effects reveal the urgency of the issue. The fact that developed countries are less affected by the short-term effects of climate shock, or that certain countries gain advantage in the short term, does not change this fact. This is because the long-term effects will work against all countries. The negative picture between climate change and the economy requires effective and urgent measures to mitigate and adapt to climate change. However, scientific evidence shows that the measures taken at the local, regional, and global dimensions in the fight against climate change are not yet at the desired level.

4. Policy recommendations

In order to reduce the macroeconomic costs of climate change, first of all, integrated solution proposals for many policy areas, such as efficient and effective use of resources, waste management, competition and growth, should be implemented with a cross-border approach. Global cooperation in emissions reduction policies is increasingly needed to sustain growth, as economic costs are projected to be particularly significant over the next few decades. However, there is a consensus in the literature on the necessity of turning to market-based economic and financial instruments. In particular, climate change adaptation strategies are expected to decrease economic costs at local, regional, and global levels in the long term, even if they increase in the short term. For climate change mitigation and adaptation, ensuring maximum efficiency and effectiveness in resource management by reusing and renewing existing products, so reducing both resource consumption and reducing the pressures on the environment with less energy are among the effective policy recommendations.

Adaptation to climate change in employment policies can be made in a growing market, especially with the combination of strong government policies on ecological energy. The weight of the green economy and sustainable investment in emissions reduction policies should be increased. Long-term investment integration can improve the risk sharing of hard-to-predict climate damage, unless climate shock is strongly correlated across regions. Biodiversity should be protected in the agricultural sector, more food should be produced with less input, food diversity should be developed, sustainability in agricultural ecosystems should be ensured, and plant breeding should be emphasized in crop improvement strategies. Other policy recommendations include reducing food waste and changing consumption patterns with less GHG-dense foods (for example, replacing animal products with plant-based foods). More durable infrastructure investment can be made in places at risk in the tourism sector; winter sports providers can be directed to the use of artificial snowmakers and move to higher altitudes. Thanks to structural changes in the economy and developments in energy technologies, fossil fuels can be reduced by replacing them with biofuels.

References

Abidoye, B.O., Odusola, A.F., 2015. Climate change and economic growth in Africa: an econometric analysis. Journal of African Economies 24 (2), 277–301.

Adom, P.K., Amoani, S., 2021. The role of climate adaptation readiness in economic growth and climate change relationship: an analysis of the output/income and productivity/institution channels. Journal of Environmental Management 293, 1–12.

Alagidede, P., Adu, G., Frimpong, P.B., 2015. The effect of climate change on economic growth: evidence from sub-Saharan Africa. Environmental Economics and Policy Studies 18 (3), 417–436.

Alene, A.D., 2010. Productivity growth and the effects of R&D in African agriculture. Agricultural Economics 41 (3–4), 223–238.

Arias, C., 2009. Going green to make green hiring and looking for sustainable jobs at colleges and corporations. Sustainability 2 (3), 152–156.

Baarsch, F., Granadillos, J.R., Hare, W., Knaus, M., Krapp, M., Schaeffer, M., et al., 2020. The impact of climate change on incomes and convergence in Africa. World Development 126, 1–13.

Babiker, M.H., Eckaus, R.S., 2007. Unemployment effects of climate policy. Environmental Science and Policy 10, 600–609.

Başoğlu, A., 2014. Küresel iklim değişikliğinin ekonomik etkileri. KTÜ Sosyal Bilimler Enstitüsü Dergisi 7, 175–196.

Batten, S., 2018. Climate Change and the Macro-Economy: A Critical Review. Bank of England Working Paper No. 706, pp. 1–48.

Battisti, D.S., Naylor, R.L., 2009. Historical warnings of future food insecurity with unprecedented seasonal heat. Science 323, 240–244.

Bergholt, D., Lujala, P., 2010. Climate-related natural disasters, economic growth, and armed civil conflict. Journal of Peace Research 49, 147–162.

Bowen, A., Cochrane, S., Fankhauser, S., 2012. Climate change, adaptation and economic growth. Climatic Change 113, 95–106.

Bretschger, L., Valente, S., 2011. Climate change and uneven development. The Scandinavian Journal of Economics 113 (4), 825–845.

Cameron, E., 2014. Climate Change: Implications for Agriculture. Key Findings from the Intergovernmental Panel on Climate Change Fifth Assessment Report. Available from: https://climate.gov.ph/files/Agriculture_Briefing_Web_EN.pdf (accessed 29.07.21).

Cayan, D., Luers, A.L., Hanemann, M., Franco, G., 2006. Scenarios of Climate Change in California: An Overview. A Report from: California Climate Change Center, pp. 1–47.

Chalmers, P., 2014. Climate Change: Implications for Buildings. Key Findings from the Intergovernmental Panel on Climate Change Fifth Assessment Report. Available from: https://climate.gov.ph/files/Buildings_Briefing_EN_Web.pdf (accessed 29.07.21).

Château, J., Saint-Martin, A., Manfredi, T., 2011. Employment impacts of climate change mitigation policies in OECD: a general-equilibrium perspective. OECD Environment Working Papers No 32, 1–31.

Cline, W.R., 2008. Global warming and agriculture. Finance and Development 23–37.
Collier, R., Conway, G., Venables, T., 2008. Climate change and africa. Oxford Review of Economic Policy 24 (2), 337–353.
Damro, C., Mendez, P.L., 2003. Emissions trading at Kyoto: from EU resistance to union innovation. Environmental Poltics 12 (2), 71–94.
Dean, T.J., McMullen, J.S., 2007. Toward a theory of sustainable entrepreneurship: reducing environmental degradation through entrepreneurial action. Journal of Business Venturing 22, 50–76.
Dell, M., Jones, B.F., Olken, B.A., 2008. Climate change and economic growth: evidence from the last half century. NBER Working Papers Series No 14132, 1–46.
Dell, M., Jones, B.F., Olken, B.A., 2009. Temperature and income: reconciling new crosssectional and panel estimates. American Economic Review 99 (2), 198–204.
Dell, M., Jones, B.F., Olken, B.A., 2012. Temperature shocks and economic growth: evidence from the last half century. American Economic Journal: Macroeconomics 4 (3), 66–95.
Doganova, L., Karnoe, P., 2015. Clean and profitable: entangling valuations in environmental entrepreneurship. In: Antal, A.B., Hutter, M., Stark, D. (Eds.), Moments of Valuation: Exploring Sites of Dissonance. Oxford University Press, UK, pp. 5–21.
Eboli, F., Parrado, R., Roson, R., 2010. Climate-change feedback on economic growth: explorations with a dynamic general equilibrium model. Environment and Development Economics 15 (5), 515–533.
Eckstein, D., Künzel, V., Schäfer, L., Winges, M., 2020. Global Climate Risk Index 2020. Germanwatch e.V, Berlin.
Fankhauser, S., Tol, R.S.J., 2005. On climate change and economic growth. Resource and Energy Economics 27, 1–17.
FAO (Food and Agriculture Organization of the United Nations), 2013. Food Outlook: Biannual Report on Global Food Markets. FAO Publications, pp. 1–133.
Ferreira, J.J.M., Fernandes, C.I., Ferreira, F.A.F., 2020. Technological transfer, climate change mitigation, and environmental patent impact on sustainability and economic growth: a comparison of European countries. Technological Forecasting and Social Change 150, 1–8.
Gosling, S., Zaherpour, J., Szewczyk, W., 2019. Assessment of Global Climate Change Impacts on Labour Productivity. Available from: https://www.gtap.agecon.purdue.edu/resources/download/9260.pdf (accessed 27.07.21).
Graff Zivin, J.S., Neidell, M.J., 2012. The impact of pollution on worker productivity. American Economic Review 102 (2), 3652–3673.
Griffiths, J., Zabey, E., Boffi, A.-L., 2009. Water, Energy and Climate Change: A Contribution from the Business Community, pp. 1–17. World Business Council for Sustainable Development North America Office.
Guo, J., Kubli, D., Saner, P., 2021. The Economics of Climate Change: No Action Not an Option. Swiss Re Institute Publishing, Zurich, Switzerland.
Hallegatte, S., Heal, G., Fay, M., Treguer, D., 2011. From Growth to Green Growth: A Framework, pp. 1–37. World Bank Policy Research Working Paper No. 5872.
Hallegatte, S., 2012. A framework to investigate the economic growth impact of sea level rise. Environmental Research Letters 7, 1–7.
Hallegatte, S., Colin, G., Nicholls, R.J., Corfee-Morlot, J., 2013. Future flood losses in major coastal cities. Nature Climate Change 3 (9), 802–806.
Hallegatte, S., Vogt-Schilb, A., 2016. Are Losses from Natural Disasters More than Just Asset Losses? the Role of Capital Aggregation, Sector Interactions and Investment Behaviors, pp. 1–26. World Bank Policy Research Working Paper No. 7885.
Hanson, C.E., Palutikof, J.P., Livermore, M.T.J., Barring, L., Bindi, M., Corte-Real, J., et al., 2007. Modelling the impact of climate extremes: an overview of the MICE project. Climatic Change 81, 163–177.

Hirst, N., 2006. Energy Technology Perspectives, Scenarios and Strategies to 2050. OECD/IEA, pp. 73–91.
Hsiang, S.M., 2010. Temperatures and cyclones strongly associated with economic production in the caribbean and Central America. Proceedings of the National Academy of Sciences 107 (35), 15367–15372.
ILO (International Labour Office), 2017. Work in a Changing Climate: The Green Initiative. International Labour Office, Geneva, Switzerland.
IPCC (Intergovernmental Panel on Climate Change), 2014. In: Pachauri, R.K., Meyer, L.A. (Eds.), Climate Change 2014: Synthesis Report. Contribution of Working Groups I, II and III to the Fifth Assessment Report of the Intergovernmental Panel on Climate Change [Core Writing Team. IPCC, Geneva, Switzerland.
IPCC (Intergovernmental Panel on Climate Change), 2014. Climate Change: Implications for the Energy Sector. Available from: www.worldenergy.org/assets/images/imported/2014/06/Climate-Change-Implications-for-the-Energy-Sector-Summary-from-IPCC-AR5-2014-Full-report.pdf (accessed 25.07.21).
IPCC (Intergovernmental Panel on Climate Change), 2019. Chapter 5: Food Security (accessed 25.07.21). https://www.ipcc.ch/site/assets/uploads/2019/08/2f.-Chapter-5_FINAL.pdf.
Kahn, M.E., Mohaddesby, K., Ng, R.N.C., Pesaran, M.H., Raissi, M., Yang, J.-C., 2019. Long-term Macroeconomic Effects of Climate Change: A Cross-Country Analysis. IMF Working Paper No. WP/19/215, pp. 1–58.
Kean, J., 2009. Climate Change Mitigation in the Agricultural Sector: The Role of Aid for Trade. Overseas Development Institute, pp. 1–14.
Kean, J., Page, S., Kergna, A., Kennan, J., 2009. Climate Change and Developing Country Agriculture: An Overview of Expected Impacts, Adaptation and Mitigation Challenges, and Funding Requirements. Internetional Centre for Trade and Sustainable Development – International Food & Agricultural Trade Policy Council, pp. 1–49.
Keleş, R., Hamamcı, C., Çoban, A., 2015. Çevre Politikası. İmge Kitabevi, Ankara.
Kjellstrom, T., Kovats, R.S., Lloyd, S.J., Holt, T., Tol, R.S.J., 2009. The direct impact of climate change on regional labor productivity. Archives of Environmental and Occupational Health 64 (4), 217–227.
Kompas, T., Pham, V.H., Che, T.N., 2018. The effects of climate change on GDP by country and the global economic gains from complying with the Paris climate Accord. Earth's Future 6 (8), 1153–1173.
Lamperti, F., Dosi, G., Napoletano, M., Roventini, A., Sapio, A., 2018. Faraway, so close: coupled climate and economic dynamics in an agentbased integrated assessment model. Ecological Economics 150, 315–399.
Lamperti, F., Dosi, G., Napoletano, M., Roventini, A., Sapio, A., 2020. Climate change and green transitions in an agent-based integrated assessment model. Technological Forecasting and Social Change 153, 1–22.
Lecocq, F., Shazili, Z., 2007. How Might Climate Change Affect Economic Growth in Developing Countries?: A Review of the Growth Literature with a Climate Lens, pp. 1–52. World Bank Policy Research Working Paper No. 4315.
Loayza, N.V., Olaberria, E., Rigolini, J., Christiaensen, L., 2012. Natural disasters and growth: going beyond the averages. World Development 40 (7), 1317–1336.
Lobell, D.B., Burke, M.B., 2008. Why are agricultural impacts of climate change so uncertain? The importance of temperature relative to precipitation. Environmental Research Letters 3, 1–8.
Martinez-Fernandez, C., Hinojosa, C., Miranda, G., 2010. Greening Jobs and Skills: Labour Market Implications of Addressing Climate Change, pp. 1–69. OECD Local Economic and Employment Development Working Paper Series.
Matthews, K., Paterson, M., 2005. Boom or bust? The economic engine behind the drive for climate change policy. Global Change, Peace and Security 7 (1), 59–75.
Medhurst, J., 2009. The Impacts of Climate Change on European Employment and Skills in the Short to Medium-Term: A Review of the Literature. GHK, pp. 1–47.
Mendelsohn, R.O., Massetti, E., 2017. The use of cross-sectional analysis to measure climate impacts on agriculture: theory and evidence. Review of Environmental Economics and Policy 11 (2), 280–298.

Millner, A., Dietz, S., 2015. Adaptation to climate change and economic growth in developing countries. Environment Development Economics 20 (3), 380–406.
Nicholls, M., 2014. Climate Change: Implications for Tourism. Key Findings from the Intergovernmental Panel on Climate Change Fifth Assessment Report. Available from: https://climate.gov.ph/files/Tourism_Briefing_Web_EN.pdf (accessed 21.07.21).
Noy, I., 2009. The macroeconomic consequences of disasters. Journal of Development Economics 88 (2), 221–231.
OECD (Organisation for Economic Co-operation and Development), 2012. Enabling Local Green Growth: Addressing Climate Change Effects on Employment and Local Development. OECD Local Economic and Employment Development Programme Report, pp. 1–114.
OECD (Organisation for Economic Co-operation and Development), 2014. Shifting gear: policy challenges for the next 50 Years. OECD Economics Department Policy Notes No 24, 1–13.
Olper, A., Maugeri, M., Manara, V., Raimondi, V., 2021. Weather, climate and economic outcomes: evidence from Italy. Ecological Economics 189, 1–18.
Peker, O., Demirci, M., 2008. İklim değişikliğinin bilim ve ekonomi perspektifinden analizi. Süleyman Demirel Üniversitesi İktisadi Ve İdari Bilimler Fakültesi Dergisi 13 (1), 239–251.
Pretis, F., 2021. Exogeneity in climate econometrics. Energy Economics 96, 13.
Raddatz, C., 2007. Are external shocks responsible for the instability of output in low-income countries? Journal of Development Economics 85 (1), 155–187.
Raddatz, C., 2009. The Wrath of God : Macroeconomic Costs of Natural Disasters, 5039. World bank policy research working paper No. WPS, pp. 1–35.
Scott, M., 2014. Climate Change: Implications for Employment. Key Findings from the Intergovernmental Panel on Climate Change Fifth Assessment Report. Available from: www.cisl.cam.ac.uk/system/files/documents/ipcc-ar5-employment-briefing-print-en.pdf (accessed 01.08.21).
Skidmore, M., Toya, H., 2007. Do natural disasters promote long-run growth? Economic Inquiry 40 (4), 664–687.
Smith, M.R., Myers, S.S., 2018. Impact of anthropogenic CO_2 emissions on global human nutrition. Nature Climate Change 8, 834–839.
Springboard, S., 2006. The Business Opportunities for SMEs in tackling the causes of climate change. Vivideconomics 1–28.
Stern, N., 2007. The Economics of Climate Change: The Stern Review. Cambridge University Press, Cambridge, UK.
Strobl, E., 2008. The Economic Growth Impact of Hurricanes Evidence from Us Coastal Conties. Institute for the Study of Labor Discussion Paper Series No. 3619, pp. 1–41.
Sullivan, R., 2014. Climate Change: Implications for Investors and Financial Institutions. Key Findings from the Intergovernmental Panel on Climate Change Fifth Assessment Report. Available from: https://www.pnas.org/content/pnas/107/35/15367.full.pdf (accessed 01.08.21).
Tol, R.S.J., Dowlatabadi, H., 2001. Vector-borne diseases, development & climate change. Integrated Assessment 2, 173–181.
Tol, R.S.J., Yohe, G.W., 2006. A review of the Stern review. World Economics 7 (4), 233–250.
Tol, R.S.J., 2009. The economic effects of climate change. Journal of Economic Perpectives 23 (2), 29–51.
Tol, R.S.J., 2018. The economic impact of climate change. Review of Environmental Economics and Policy 12 (1), 4–25.
UNEP, ILO, IOE, ITUC (United Nations Environment Programme, International Labour Organization, International Organisation of Employers, International Trade Union Confederation), 2008. Green Jobs: Towards Decent Work in a Sustainable, Low-Carbon World (accessed 05.08.21). http://www.unep.org/labour_environment/PDFs/Greenjobs/UNEP-Green-JobsReport.pdf.

UNWTO (World Tourism Organization), 2003. Climate Change and Tourism. 1st International Conference on Climate Change and Tourism, Tunisia (accessed 07.08.21). www.unwto.org/archive/global/event/1st-conference-climate-change-and-tourism.

UNWTO (World Tourism Organization), 2007. Tourism & Climate Change Confronting the Common Challenges (accessed 07.08.21). https://docplayer.net/10741092-Tourism-climate-change-confronting-the-common-challenges.html.

Vanrheenen, N.T., Wood, A.W., Palmer, R.N., Lettenmaier, D.P., 2004. Potential implications of PCM climate change scenarios for Sacramento-San Joaquin river basin hydrology and water resources. Climatic Change 62, 257−281.

Woetzel, J., Pinner, D., Samandari, H., Engel, H., Krishnan, M., Boland, B., et al., 2020. Climate Risk and Response: Physical Hazards and Socioeconomic Impacts. McKinsey Global Institute Report, pp. 1−150.

World Bank, 2010. World Development Report 2010: Development and Climate Change. World Bank Publications, Washington.

Zamuda, C., Mignone, B., Bilello, D., Hallett, K.C., Lee, C., Macknick, J., et al., 2013. U.S. Energy Sector Vulnerabilities to Climate Change and Extreme Weather. U.S. Department of Energy's Office of Policy and International Affairs, National Renewable Energy Laboratory (accessed 08.08.21). http://energy.gov/sites/prod/files/2013/07/f2/20130710-Energy-Sector-Vulnerabilities-Report.pdf.

Zhao, Y., Sultan, B., Vautard, R., Braconnot, P., Wang, H.J., Ducharne, A., 2016. Potential escalation of heat-related working costs with climate and socioeconomic changes in China. PNAS 113 (17), 4640−4645.

CHAPTER

Sustainable development: a case for urban leftover spaces

17

Jasim Azhar

Architecture Department, King Fahd University of Petroleum and Minerals, Dhahran, Saudi Arabia

Chapter outline

List of abbreviations

TOADS Temporarily Obsolete Abandoned Derelict Sites
VPS Visual Preference Study

1. Introduction

In dense cities, an outward urbanized expansion (Burgess, 2000) has been widely recognized as disastrous for economic, environmental, and social terms (Chen et al., 2008). The urban land expansion has surpassed the growth of urban population (Seto et al., 2011) that generates a demand for more expansion and development, which makes cities to depend on their surrounding hinterlands and beyond, both for food and diverse services (Billen et al., 2012). A shift in edges between urban and rural makes the boundaries obscured while resulting in the loss of up to 40% of the species in some of the most biologically diverse areas around the world (Pimm and Raven, 2000). Over the recent century, urban hinterland has been broadened to become global. In this situation, cities face enormous challenges to develop sustainable, healthy, and livable urban environments and will have to compromise with problems like limited urban space, resources scarcity of food, environmental pollution, and loss of biodiversity. Scholars (Scannell and Gifford, 2010; Bolund and Hunhammar, 1999; Barnett, 2009) have asserted that there is a need to preserve a sense of place, promote ecosystem services, and provide

Handbook of Energy and Environmental Security. https://doi.org/10.1016/B978-0-12-824084-7.00022-9
Copyright © 2022 Elsevier Inc. All rights reserved.

safety to people from environmental dangers generated by anthropogenic operations in cities. The role of sustainable development can be emphasized from within the cities, where leftover spaces have no purpose can cause a threat to public safety. Such spaces could be used not only to mitigate the negative effects of climate change but also can enhance the quality of life in cities. To address this impasse, there is a probable for redesigning urban leftover spaces as a part of sustainable development approaches.

This study concentrates on stressing the potential of urban leftover spaces to be transformed in order to satisfy the needs of the future and playing a significant role in urban revitalization. For this, it is necessary to recognize the experiences of people with urban leftover spaces and their perceptions of such spaces. Peoples' experiences of urban leftover spaces are altered as if the aesthetic quality of these spaces changes. This research assesses the interrelationships between how people perceive and what they prefer about the qualities of urban leftover spaces. Photographs were presented to individuals to see as how they evaluate the spaces considering their surroundings. This was accomplished by examining people's preferences for leftover spaces. Prior research has ignored to find out and infer what and how people visually prefer such spaces to be like. Therefore, this inquiry proposes to highlight the potential of feasible design solutions for urban leftover spaces.

2. Urban leftover spaces

According to Jacob (1961), where urban centers create problems, they present solutions like creation of jobs and enhancing the quality of life for people. Williams (n.d.) suggested how recent years have laid more stress on various attributes of a city such as its layout, density, composition, shape, construction forms, and volume can affect its sustainability. Leftover spaces result from uncoordinated development. They may be in between stages of formal development waiting to be used in future or they may just remain idle with, recognized as "out of place." There is, however, an opportunity to make use of such spaces, use them properly (Cresswell, 1996). Research also recommends such spaces may show abandoned property by owner or may be regarded as a hazard to public safety. Such leftover spaces may exist as small spaces between buildings or vacant lots here and there, attracting debris and trash (National Vacant Properties Campaign, n.d.). Wilkinson (2011) seconds this by adding such spaces become a threat to economic opportunities, environment, housing, and neighborhood development. Aruninta (2004) points out how recent remodeling of economies across the global has contributed to a rapid increase in the emergence of leftover spaces. Trancik (1986) set up an analytical investigation on leftover spaces 30 years ago. In his view, these spaces have no positive impact on the surroundings, he referred to them as "lost spaces." These lost spaces failed to connect urban elements reasonably, did not have measurable boundaries, and were ill-defined. Le Goix et al. (2019) adds, as prices of real estate have increased since the 1990s, the value of lost spaces with no specific shape and size has increased, notably in valuable cities. Endensor (2005) explains the planning category of leftover spaces as scattered areas left behind during planning stage. Creating zones and land-use policies helps draw boundaries between landscape, forms division between uses, and separates the private and public realm (Trancik, 1986). Nipesh (2012) describes the functional category of leftover spaces as those spaces within a city that have lost utilization for time. Geographical leftover spaces are the ones created by geographical features, for example, rivers and hills. Gallagher (2010) suggests leftover spaces can emerge from declining population, urban decay, natural catastrophe, or conflict. Sola-Morales (1995) emphasizes on the untapped potential of such void spaces, as they create a sense of freedom and possibilities. According to Greenberg et al. (1990), various scholars define such spaces as "Temporarily Obsolete Abandoned Derelict Sites (TOADS)." Different terminologies by various

Table 17.1 Thoughts and terms used by researchers to describe leftover spaces.

1967	Turner	Liminal space	Threshold spaces that are neither here nor there; they are betwixt and between the positions assigned and arrayed by law, custom, convention, and ceremony (Turner, 1967).
1971	Northam	Vacant parcel	Five conditions: Land small in size, often irregular in shape; parcels with physical limitations, such as steep slope or flood hazard; corporate reserve parcels held for future expansion or relocation; parcels held for speculation; institutional reserve parcels set aside by public or quasi-public entities for future development (Mhatre, 2013).
1989	Burroughs	Interzone	Spaces that become the point of maximum visibility of coagulum that cuts the space and becomes the concentration of experiences like in front of buildings which are related to time and movement (Burroughs 1989).
1991	Lefebvre	Thirdspace	"Firstspace" perspective is focused on the "real" material world and sociality (the "secondspace" perspective that interprets the "imagined" representations of the world) through the insertion of a "thirdspace": that of spatiality. Thereby a trialectic is created with the thirdspace being a space of extraordinary openness, a place of critical exchange (Lefebvre, 1991).
1995	Auge	Nonspace	Spaces of institutions formed in relation to certain ends like transport, transit, commerce, and leisure. These spaces are never totally completed, but nonspaces are the real measure of our time (Auge, 1995).
1996	Loukaitou-sideris	Cracks in the city	Cracks are the "in-between" spaces—residual, underutilized, and often deteriorating—that frequently divide physical and social worlds (Loukaitou-Sideris, 1996).
1999	Delgado	Interstitial space	Transitional spaces that may well be situated in city centers, to be crossed and circulated, as opposed to fixed places (Huffschmid, 2012).
2000	Doron	Dead space	Spaces that cannot be zoned or delimited and that present suspension, solitude, and silence within the bustling city (Doron, 2000).
2005	Groth and Corjin	Indeterminate space	New, transitional reappropriations that are assumed by civil or "informal" actors coming from outside the official, institutionalized domain of urban planning and urban politics (Groth and Corjin, 2005).
2007	Worpol and Knox	Slack space	Places where the participants don't want to be seen or heard by others. These spaces are dependent on the level of tolerance from society (Worpole and Knox, 2007).
2007	Franck and Stevens	Loose space Tight space	A dynamic space that allows people to carry out their desired actions while recognizing the presence and rights of others, whereas tight space is static in action as it is dependent on action or regulation (Franck and Steven, 2007).
2010	Jarnang	Undefined space	Interspaces that have lost their former functions and have been invaded by new groups and uses.

Based on Azhar et al. (2018).

researchers are vacant, loose, liminal, derelict, transitional, neglected, and indeterminate. Table 17.1 summarizes the terminology adopted by many researchers while relating to leftover spaces and how such spaces were described (Azhar et al. 2018).

Urban space classifications present a model for planning urban forms (Azhar and Gjerde, 2016). According to Sandalack and Uribe (2010), the functional attributes and aesthetics considerations create various built forms. The interpretations of Sola-Morales (1995) and Trancik (1986) are based on cause-and-effect criteria, which contribute to the regeneration of such spaces, being neglected. De Girolamo (2013) stresses that urban planning and design should give importance to the aspect of temporariness, and society like Gap Filler focuses on temporary spaces in order to revive cities to benefit the people by promoting engagement and knowledge sharing among the public. Access land is viewed as a valuable resource because of space scarcity; hence, leftover spaces may have a great potential to offer economic benefit.

Urban form's relentless development often neglects the qualities and probable of urban leftover spaces. According to Doron (2006), such spaces are often underutilized, not possessing an official usage, informal parking lots, and can be found anywhere in open/closed, public/private, and interior/exterior zones. Azhar and Gjerde (2016) suggest how rooftops can be categorized as leftover spaces and other spaces are in front of, between, at the back of, and at the side of buildings on microlevel (see Fig. 17.1). A leftover territory may occur in between objects, usually provoking a social or physical crisis. In urban neighborhood, there are many criteria including current activity, ownership, formation, management, accessibility, climate and sun orientation, shape, surrounding, background, and scale. Hajer and Reijndorp (2001) argue how size and accessibility will not determine the nature or value of a space but by how the space contributes to the ambience of environment.

3. Visual preferences

The study adopts the model of environmental aesthetics from Nasar (1998), Gjerde (2015), and Redies (2015) (see Fig. 17.2). All approaches addressed the aesthetic responses of people, both formal and symbolic, which are recognized through the visual form and include aesthetic parameters. The initial stage involves the interaction of a user with the built environment that creates a stimulus. Second stage is concerned with the external information of the context that is visually perceived and processed. The information about the built environment is then divided into the three parameters of perception of beauty (the affective state), cognition (cultural experience), and associations or meanings that arouse feelings. These three parameters are both interrelated and work independently (Gjerde, 2015; Redies, 2015). The third stage establishes the knowledge of stage two and makes an individual preference assessment of an object, whether they like or dislike it. Last, stage four creates an aesthetic experience through an internal neural mechanism.

Lynch (1960) refers to city design as an art piece that has multiple associations and functions attached to it. How people create an image of a city is related to their experiences and how spaces function. Nasar (1994) claims visual study approach is effective in explaining how people perceive different environments. According to Rapoport (1977), how people feel about the surrounding environment reflects their intuitions and judgements. Kaplan and Kaplan (1982) contradicted this by claiming perceptions can be interpreted through preferences. Feelings or emotions, appearance and impression may affect judgements. However, accumulation of individual feelings about the

FIGURE 17.1

Six different types of leftover space in Wellington city (A) In front of a building, (B) Backyard of a building, (C) Building rooftop, (D-E) Enclosed by two and three sides (F) Underneath a building.

environment make preferences. Bell (1999) explained perception as using different senses to gain and comprehend information. All senses are used as a group, yet some senses may be more significant than the other. All information gathered by senses is sent to the brain of interpretation to ensure understanding of the environment and an experience of it (Bell, 1999). Kaplan (1985) argues people respond not only to things but also to the surrounding settings; hence, they respond differently to specific spaces

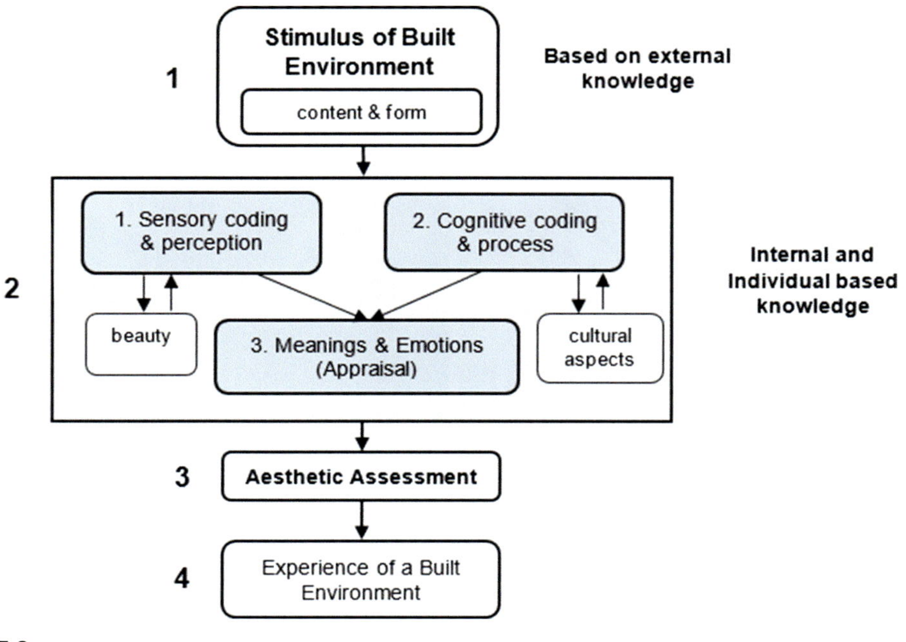

FIGURE 17.2

Diagram of analytical framework for experiencing built environment.

Adapted from Nasar (1998); Gjerde (2015) and Redies (2015).

depending on their knowledge, context, and interpretation. Nasar (1994) suggests use of pictures would give similar responses as being there in the scene in person. Much prior research has used pictures or slides for taking our people's preferences (Hartig and Staats, 2006; Herzog, 1989; Kaplan, 1973). In addition, Nelessen (1979) confirmed this approach and referred to Image-Based Survey/ Visual Preference Study (VPS), especially while engaging with nontechnical people. Besides, Kaplan (1985) confirms the usage of photographs for understanding perceptions and preferences of people in place of actual spaces. All these reasons aided the decision to use photos for the VPS of leftover spaces.

4. The method

The aim of this VPS was to examine the potential of public–private urban leftover spaces dispersed around Wellington City. To help ensure that any visual improvements to these spaces will enhance the appreciation and uses of such spaces, it is necessary to first find out how the public reacts to the visual quality of a leftover space and then design them according to their preference. Two different studies were conducted. In the first study, participants were asked to identify areas they liked or disliked, describe the physical features on which their evaluations were based on and their design preferences for such space. The second study was built upon the first study, in which the participants were to do a semantic differential rating for visualized designed space that they prefer most. The participants were shown three alternative design solutions. In the first study, 94 people took participation, which took 23 min (on average) to complete the study, whereas, in the second study, 97 participants answered the survey questions (see Table 17.2). Both studies involved different participants and six months were taken to complete it.

The study was based on a photo-based questionnaire in which six type leftover spaces with natural and built structures were shown while the emphasis of the pictorial frame was made on the spaces themselves. None of the photos contained people since they have been found to be problematic (Herzog, 1989). In Study 1, three pictures were shown for the same type of leftover space and participants had to give rating and select a descriptive attribute that best fitted why they gave a particular score to each picture. To understand the reasons behind the scoring, a 5-point Likert scale was simplified (Benson, 1971), and a new 3-point Likert scale (1 (-1 and -2), = dislike, 2 (0) neutral, and 3 ($+1$ and $+ 2$) = like) was created by merging the relevant averages. The Likert scale measured the preferences for each space. A list of unique attributes was mentioned as a reason for their selection of leftover space (see Fig. 17.3). The photos were carefully selected for each type of leftover space. A content analysis was used to examine the frequency of words and important phrases used by the participants in this section.

The photomontage visuals in Study 2 were produced based on the results of Study 1, and three alternative design modifications for each type of leftover spaces were produced (see Fig. 17.4). According to Stamps (1990, 1992), reaction to a photomontage gives similar responses to those experienced in the real world, and stamps found a correlation of preferences for places represented through photographs with on-site experience. All one-point perspective photos were treated in Photoshop to reconstruct the spaces after each change, with an emphasis on one specific attribute in each leftover space, noting that these attributes changed with the spaces. All the leftover spaces were designed without changing the current usage of the site. Study 2 had two sections. The first presented three design modifications for each of the six leftover spaces, with a 3-point Likert scale for rating the

Table 17.2 Distribution of participants in study 1 and study 2.

Demographic distribution of study 1 (N = 94) *Frequency in numbers with Percentage*		
Gender	**Frequency**	**Percentage (%)**
Male	41	44
Female	53	56
Total	94	100
Demographic distribution of study 2 (N = 99) *Frequency in numbers with Percentage*		
Gender	**Frequency**	**Percentage (%)**
Male	42	43.2
Female	55	56.7
Total	97	100
Prefer not to answer	02	

A

B

C

Please score each image from the scale -2 (Dislike) to 2 (Like)

	Dislike (-2)	Somewhat Dislike (-1)	Neither Like or Dislike (0)	Somewhat Like (1)	Like (2)
A	○	○	○	○	○
B	○	○	○	○	○
C	○	○	○	○	○

From the list below tick up to three things which you feel need to be **removed or added to improve the space**

- ☐ *Provide more vegetation*
- ☐ *Remove car parking space*
- ☐ *Create seating space*
- ☐ *Change surface colors / materials*
- ☐ *Insert change in level*
- ☐ *Improve entrance to building*
- ☐ *Improve lighting*
- ☐ *Remove garbage bin*
- ☐ *Other*

for all three options

FIGURE 17.3

A Sample From the Questionnaire: where participants had to choose one photo and select at least three options for its improvement.

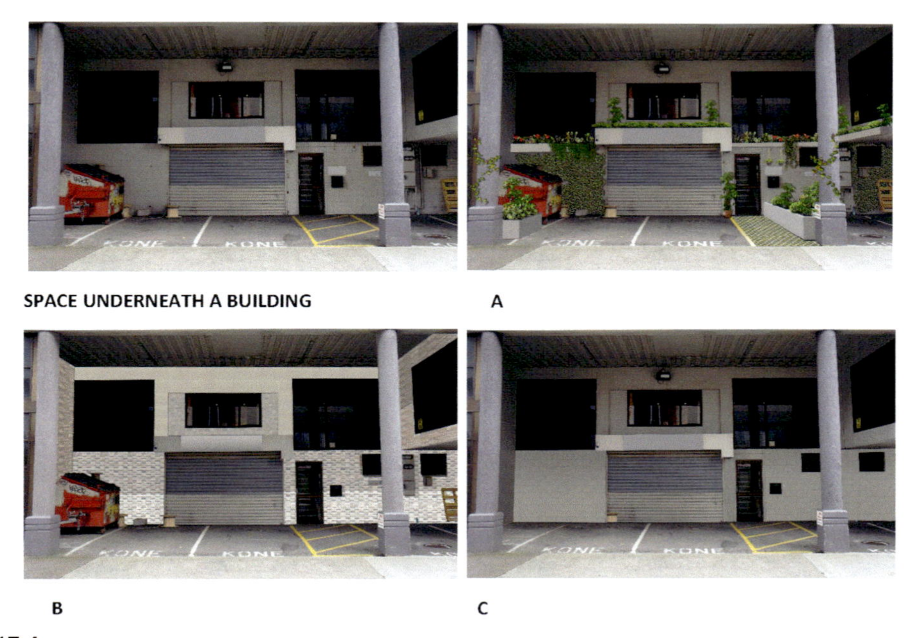

FIGURE 17.4

Options presented to participants in Study 2.

modifications ("Dislike = 1," "Neither like nor dislike = 2," "Like = 3"). Second section was related to the semantic differential measures which sought each participant's reaction to the redesigned space through a series of stimulus concepts (see Fig. 17.5). The concepts (adjectives) were evaluated through a 5-point bipolar rating scale. Stem and Noazin (1985) concluded that a 5-point scale had maximum reliability and validity for bipolar adjective scales. This section investigated reactions to the concepts of attractiveness (ugly-beautiful), satisfaction (annoying-pleasing), buildable (impossible-realizable), usability (boring-interesting), and mood (constrained-energetic). Likert scale reveals how many people agree or disagree with a particular statement, whereas the semantic differential scale decides how much of a trait or quality the item has when rated using a bipolar scale defined by adjectives (Osgood and Snider, 1969).

5. The findings

Study 1 investigated the likings on a Likert scale (1–3) for three photos of various leftover spaces with unique attributes. It became obvious from the results that the likings of males and females were similar for all spaces, as both genders agreed that providing more cleanliness and maintenance by removing garbage bins could transform the spaces. Females favored providing more vegetation, creating clear pathways, creating seating space, and allowing graffiti on walls more than males, while males preferred removing the car parks, changing surface materials/colors, providing more shade, improving the maintenance and cleanliness, and installing wind turbines and solar panels more than

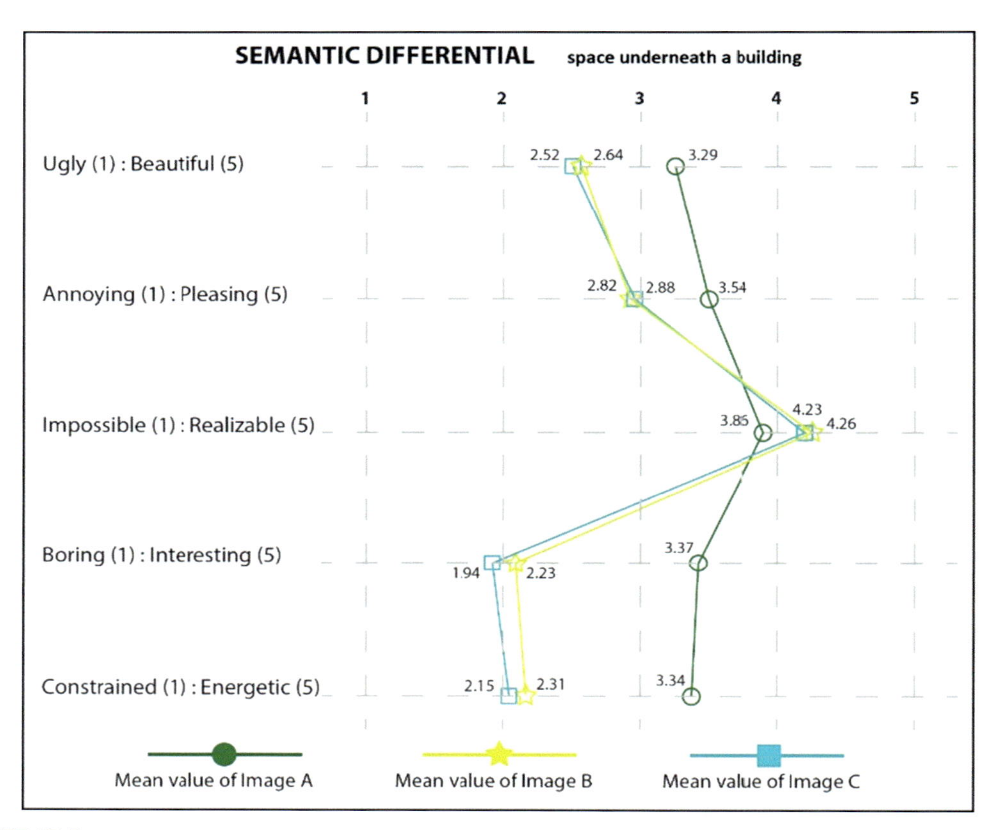

FIGURE 17.5

A sample of participant's attitude for three options for space underneath a build ng.

females (see Fig. 17.6). Other less favored solutions ascribed to the removal of the boundary walls for spaces in front of a building.

The most favored design solution among genders of Study 2 ascribed to include vegetation in all leftover spaces (Fig. 17.7). The space in front of a building had a particular preference ranking. The first preference was given to the removal of boundary walls, whereas the second most preferred design was providing more vegetation. A correlation test for all designed spaces with vegetation revealed that the most preferred image had a strong, positive association and correlated with all affective appraisals (semantic differentials) except for the bipolar category of "impossible-realizable." This suggested category was perhaps independent of the association and was not influenced by the image's likability. The results deduced that designing a space with vegetation is something the public would appreciate, but this is not recognized as a practical solution. The suggestions for improving the spaces were about providing amenities for the public, such as providing community gardens, designing a recreation space, and providing a food market.

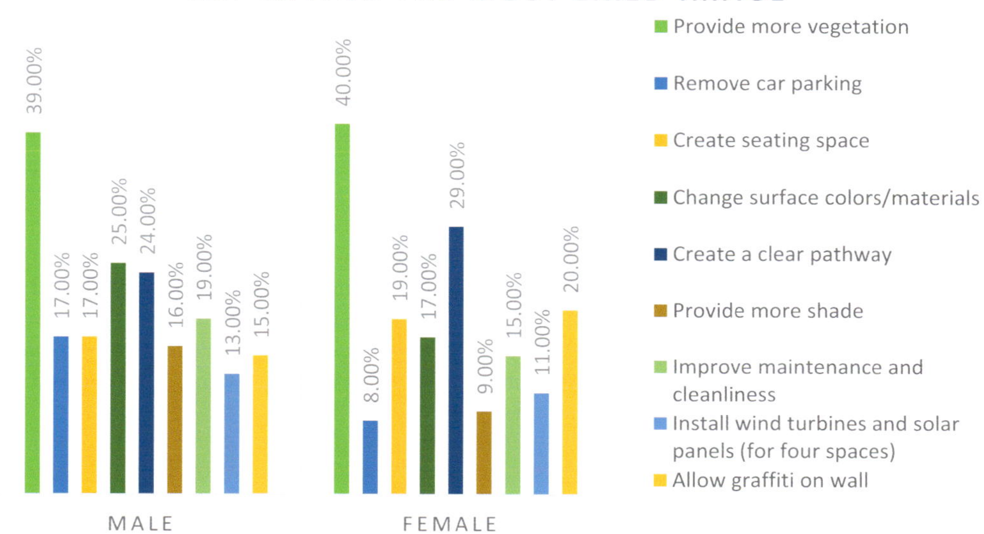

FIGURE 17.6

Overall modifications suggested by male and female participants in study 1.

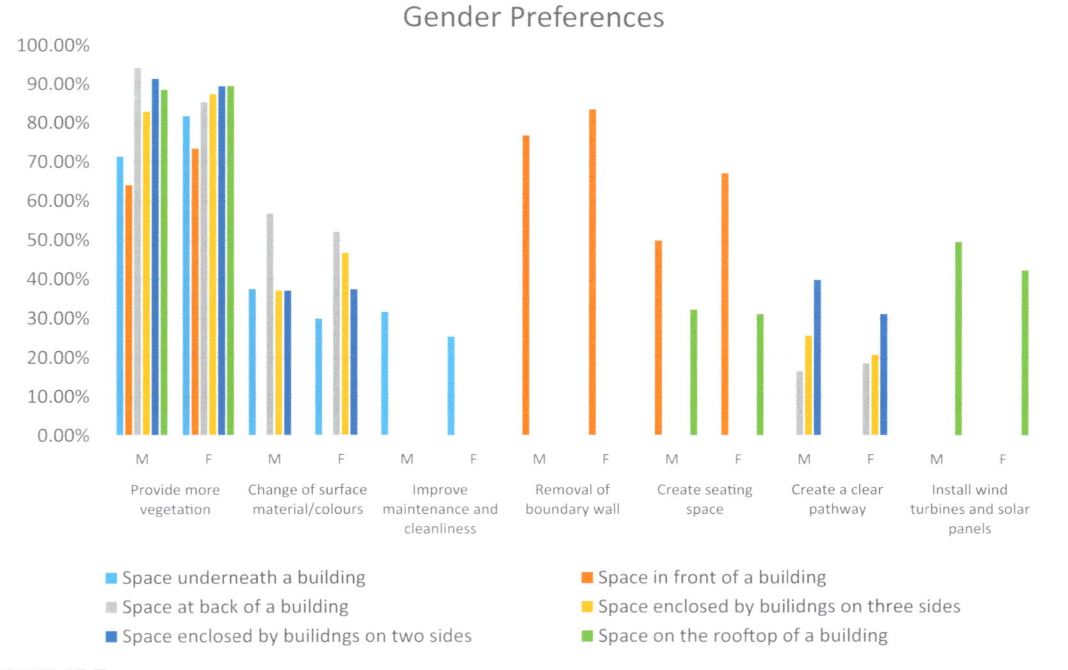

FIGURE 17.7

Preferences by gender group in study 2.

6. Conclusion

According to Kaplan and Kaplan (1982) and Nasar (1990), the cognitive process comprises the perception of a visual attribute based on which an individual assesses the environment. This paper has focused on identifying those attributes that could enhance the aesthetics and usability of space from the public's perspective. Both studies revealed that people have a common preference for providing more vegetation. This attribute was the most recurring design solution for each type of leftover space. Studies 1 and 2 affirmed that installing various types of plants such as trees, shrubs, climbers, and ground cover was liked by all participants. The solutions of providing more vegetation were seen as attractive. However, the maintenance of public—private spaces needs to be acknowledged, which depends on the owner's goodwill. Providing more vegetation also could help in mitigating adverse climate change effects and enhancing urban life by providing environmental and social services. Ulrich (1986) found that vegetation alters the subjective and perceptual function of the brain. Both study suggests natural regeneration and human intervention are key contributors for improving leftover spaces. There might seem complexity in the varying results of usefulness of a leftover space, but they do direct toward thoughtful design schemes, keeping in mind the aesthetic quality of a space.

Creative solutions are required to curb the negative effects of environment in cities, which lead to social cohesion and quality of life. Urban areas can be managed, developed, and used efficiently by implementing sustainable principles. The efficiency in cities is realized through the densification of settlements or by vacant land use. A pivotal aspect in establishing urban environmental security could be attained through an appropriate ratio between investment in spaces and bioactive spaces. The green spaces are a notable resource for any real estate that encourages a healthy lifestyle and influences the aesthetics quality. Besides, local government authorities should encourage pro-environmental practices, aid local initiatives, and provide essential information to the community. Small-scale initiatives such as urban farming, living streets, pocket parks, etc., could play a role in minimizing some of the environmental threats. In addition, this study provides a starting point for developing sustainable cities by upgrading the quality of well-being by effective usage of leftover spaces and designing them with people's preferences.

References

Aruninta, A., 2004. Controversies in Public Land Management Decision-Makings: Case Study of Land Utilization in Bangkok, Thailand. Chulalongkorn University, Bangkok.

Auge, M., 1995. "Non-Places: Introduction to an Anthropology of Supermodernity", Social Science and Urban. Verso, London.

Azhar, J., Gjerde, M., 2016. Rethinking the role of urban in-between spaces. Journal of Architectural Science Review, ASA Conference Proceedings, Taylor & Francis, Adelaide.

Azhar, J., Gjerde, M., Vale, B., 2018. Urban Leftover Spaces: Transformation from 'Within'. Journal ARCH-DESIGN 18/V. International Architecture Design Conference Proceedings 129—140, 18. In this issue.

Barnett, J., 2009. Environmental security. In: Kitchin, R., Thrift, N. (Eds.), International Encyclopaedia of Human Geography. Elsevier, pp. 553—557.

Bell, S., 1999. Landscape, pattern, perception and process. In: Bell, S. (Ed.), E&FN Spon, New York.

Benson, P.H., 1971. How many scales and how many categories shall we use in consumer research? A comment. Journal of Marketing 35, 59—61.

Billen, G., Garnier, J., Barles, S., 2012. History of the urban environmental imprint: introduction to a multidisciplinary approach to the long-term relationships between Western cities and their hinterland. Regional Enviornmental Change 12 (2), 249–253.

Bolund, P., Hunhammar, S., 1999. Analysis: ecosystem services of urban areas. Ecological Economics 29, 293–310.

Burgess, R., 2000. The compact city debate: a global perspective. In: Jenks, M., Burgess, R. (Eds.), Compact Cities: Sustainable Urban Forms for Developing Countries, E & FN Spon, pp. 9–24.

Burroughs, W., 1989. Interzone. Viking Press, New York.

Chen, H., Jia, B., Lau, S.S.Y., 2008. Sustainable urban form for Chinese compact cities: challenges of a rapid urbanized econo Jarnang, M. 2010. "The city undefined room". Master of Landscape. Faculty of Landscape Planning, my. Habitat International 32 (1), 28–40. https://doi.org/10.1016/J.HABITATINT.2007.06.005.

Cresswell, T., 1996. In Place/out of Place: Geography, Ideology, and Transgression. University of Minnesota Press, Minneapolis; London.

De Girolamo, F., 2013. Living landscapes -landscapes for living paesaggi abitati time and regeneration: temporary reuse in lost spaces. Planum. The Journal of Urbanism.

Doron, G., 2000. The dead zone and architecture of transgression. CITY Analysis of Urban Trends, Culture, Theory, Policy, Action 4, 247–263.

Doron, G., 2006. The derelict land and the Elephant. The Field Journal 1, 10–23.

Edensor, T., 2005. Industrial Ruins. Berg Publishers, New York.

Franck, K., Steven, Q., 2007. Loose Space: Possibility and Diversity in Urban Life. Routledge, London.

Gallagher, J., 2010. Reimagining Detroit: Opportunities for Redefining an American City. Wayne State University Press, Detroit, MI.

Greenberg, M., Popper, F., et al., 1990. The TOADS: a new American urban epidemic. Urban Affairs Quarterly 25 (3), 435–454.

Gjerde, M., 2015. Street Perceptions: A Study of Visual Preferences for New Zealand Streetscapes, Doctoral Thesis. Victoria University of Wellington, Wellington, New Zealand.

Groth, J., Corijn, E., 2005. Reclaiming urbanity: indeterminate spaces, informal actors and urban agenda setting. Journal of Urban Studies 503–526.

Hartig, T., Staats, H., 2006. The need for psychological restoration as a determinant of environmental preferences. Journal of Environmental Psychology 26 (3) (2006).

Hajer, M., Reijndorp, A., 2001. In Search of New Public Domain: Analysis and Strategy. NAI Publishers, Rotterdam.

Herzog, T., 1989. A cognitive analysis of preference for urban-nature. Journal of Environmental Psychology 9 (1989), 27–43.

Huffschmid, A., 2012. From the City to Lo Urbano: Exploring Cultural Production of Public Space in Latin America. Ibero-American University, pp. 119–136.

Jacobs, J., 1961. The Death and the Life of the Great American Cities. Vintage press.

Kaplan, R., 1973. Predictors of environmental preference—designers and 'clients'. In: Preiser, W.F.E. (Ed.), Environmental Design Research, Dowden, Hutchinson & Ross, Stroudsburg, PA.

Kaplan, S., Kaplan, R., 1982. Cognition and Environment—Functioning in an Uncertain World. Praeger, New York, NY (1982).

Kaplan, 1985. The analysis of perception via preference - a strategy for studying how the environment is experienced. Landscape Planning 12 (2), 161–176.

Le Goix, R., Giraud, T., Cura, R., Le Corre, T., Migozzi, J., 2019. Who sells to whom in the suburbs? Home price inflation and the dynamics of sellers and buyers in the metropolitan region of Paris, 1996-2012. PloS One 14 (3), e0213169. https://doi.org/10.1371/journal.pone.0213169.

Lefebvre, H., 1991. the Production of Space, Translated by Donald Nicholson-Smith. Blackwell, Oxford.

Loukaitou-Sideris, A., 1996. Cracks in the City: Addressing the Constraints and Potentials of Urban Design. Journel of Urban Design, University of California, Los Angeles.
Lynch, K., 1960. The Image of the City. MIT press, Cambridge.
Mhatre, P. (2013). Vacant and Abandoned Lands: A Theory Paper, Neighborhood Revitalization. Accessed on 13.04.2018, Retrieved from: https://pratikmhatre99.files.wordpress.com/2013/11/vacant-and-abandoned-lands.pdf
Nasar, J., 1990. The evaluative image of the city. Journal of the American Planning Association 56 (1), 41–53. https://doi.org/10.1080/01944369008975742.
Nasar, J., 1994. Urban design aesthetics: the evaluating qualities of building exteriors. Environment and Behaviour 26. Sage Publication.
Nasar, J., 1998. The Evaluative Image of the City. Sage Publication, Thousand Oaks, California.
National Vacant Properties Campaign. (n.d.). National Vacant Properties Campaign. Retrieved from http://www.vacantproperties.org.
Nelessen, A., 1979. "History of the Visual Preference Survey in Visions for a New American Dream: Process, Principles and an Ordinance to Plan and Design Small Communities, American Planning Association (eds.1994.), Chicago.
Nipesh, 2012. "Urban Voids & Shared Spaces", Deep within an Exploration. Retrieved from. https://nipppo.wordpress.com/2012/05/07/urban-voids/.
Osgood, C.E., Snider, J., 1969. Semantic Differential Technique: A Sourcebook. Aldine Publications, Chicago.
Pimm, S., Raven, P., 2000. Biodiversity. Extinction by numbers. Nature 403, 843–845.
Rapoport, A., 1977. Human Aspects of Urban Form: Toward a Man-Environment Approach to Urban Form and Design. Pergamon Press, New York: Oxford.
Redies, C., 2015. Combining universal beauty and cultural context in a unifying model of visual aesthetic experience. Frontiers in Human Neuroscience 9 (218), 1–20.
Sandalack, B.A., Uribe, G.A., 2010. Open Space Typology as Framework for Design of the Public Realm. Retrieved from: https://www.scribd.com/doc/313567532/231997201-Typology-of-Public-Space-Sandalack-Uribe-pdf.
Scannell, L., Gifford, R., 2010. Defining place attachment: a tripartite organizing framework. Journal of Environmental Psychology 30, 1–10.
Sola-Morales, D., 1995. Terrain vague. In: Davidson, C. (Ed.), Anyplace. MIT Press, Cambridge, MA, pp. 118–123.
Seto, K.C., Fragkias, M., Guneralp, B., Reilly, M.K., 2011. A meta-analysis of global urban land expansion. PLoS ONE 6.
Stamps, A., 1990. Use of photographs to stimulate environments. A meta-analysis. Perceptual and Motor Skills 71, 907–913.
Stamps, A., 1992. Perceptual and preferential effects of photomontage simulations of environments. Perceptual and Motor Skills 74, 675–688.
Stem, D., Noazin, S., 1985. The effects of a number of objects and scale positions on graphic position scale reliability. In: Lusch, R.E., et al. (Eds.), AMA Educators' Proceedings. Marketing Association, Chicago, pp. 370–373.
Trancik, R., 1986. "Finding Lost Space", Theories of Urban Design, first ed. John Wiley & Sons, New York.
Turner, V., 1967. Betwixt and between: the liminal period *in* rites de Passage. In: The Forest of Symbols: Aspects of Ndembu Ritual. Cornell University Press, Ithaca, New York.
Ulrich, S., 1986. Human responses to vegetation and landscapes. Journal of Landscape and Urban Planning 13, 29–44.
Williams, K., n.d. Can Urban Intensification Contribute to Sustainable Cities?. An International Perspective, Oxford Centre for Sustainable Development Oxford Brookes University, Routledge, London

Wilkinson, L., 2011. Vacant Property : Strategies for Redevelopment in the Contemporary City. Retrieved 17 Jan 2016 From: https://smartech.gatech.edu/bitstream/handle/1853/40778/LukeWilkinson_Vacant%20Property.pdf.

Worpol, K., Knox, K., 2007. The Social Value of Public Spaces. Joseph Rowntree foundation, New York.

CHAPTER

Evaluating climate change towards sustainable development 18

Vikniswari Vija Kumaran[1], Nazatul Faizah Haron[2], Abdul Rahim Ridzuan[3], Mohd Shahidan Shaari[4], Nur Surayya Saudi[5] and Noraina Mazuin Sapuan[6]

[1]*Universiti Tunku Abdul Rahman, Kampar Campus, Perak, Malaysia;* [2]*Universiti Sultan Zainal Abidin, Kuala Terengganu, Terengganu, Malaysia;* [3]*Universiti Teknologi Mara, Cawangan Melaka, Melaka, Malaysia;* [4]*Universiti Malaysia Perlis, Perlis, Malaysia;* [5]*Universiti Pertahanan Nasional Malaysia, Kuala Lumpur, Malaysia;* [6]*Universiti Malaysia Pahang, Pahang, Malaysia*

Chapter outline

1. Introduction

There is a growing awareness that our planet Earth is rapidly changing, and scientists have related those occasions to climate change. The Earth's surface temperature has been gradually rising since the late 19th century. The change is the result of a huge scale release of carbon dioxide (CO_2)-related products into the air. Since 2001, CO_2 emissions have been more rapid and seen in an upward trend.

Handbook of Energy and Environmental Security. https://doi.org/10.1016/B978-0-12-824084-7.00014-X

Copyright © 2022 Elsevier Inc. All rights reserved.

Banett (2017) indicated that about 97% of peer-reviewed climate studies conclude that the sources of higher CO_2 release that causes climate change are derived from burning fossil fuels, industrial activities, and deforestation. Based on Fig. 18.1, the upward trend can be seen from 1960 to 1979, and a downward trend was observed until 1981. The rest trend showcased an upward trend while reducing trend between 2012 and 2016.

The overall outcomes of an increasing trend have resulted in rising sea temperatures, coral reef corruption, and harming the ocean life. The loss of man mass, floods, and other outcomes of rising sea levels includes human passings and property losses worth loads of billions of dollars. Global warming has other disastrous outcomes, such as mortality from heat waves, starvation, storms, and fires, to name a couple. The overall misfortune from these occurrences amounts to trillions of dollars per year, increasing year after year.

Vulnerable nations are among the hardest hit from climatic changes that mitigate the needs for immediate action to prevent climate change. The Intergovernmental Panel on Climate Change (IPCC) in 2001 stated that relying on their economic, social, environmental, and geographic location, numerous nations are more vulnerable to climate change than others. For instance, small island governments, like the Maldives and nations with low-lying seaside regions, are particularly vulnerable to rising sea levels, whereas high-elevation inland nations are not. Last mentioned, the IPCC (2014) concluded that disparities in vulnerability are caused by each climatic and nonclimatic cause, with the latter containing multidimensional inequalities frequently caused by uneven development processes.

Moreover, the vulnerability of climate change through the rising temperatures might have the greatest effect on the African region to the Southeast Asian region over the next 30 years, as stated by Verisk Maplecroft (2018). Climate change is now having a severe impact on agriculture, such as in India and Myanmar. Furthermore, unlike most industrialized countries, these countries have a lack of funding for mechanized agriculture.

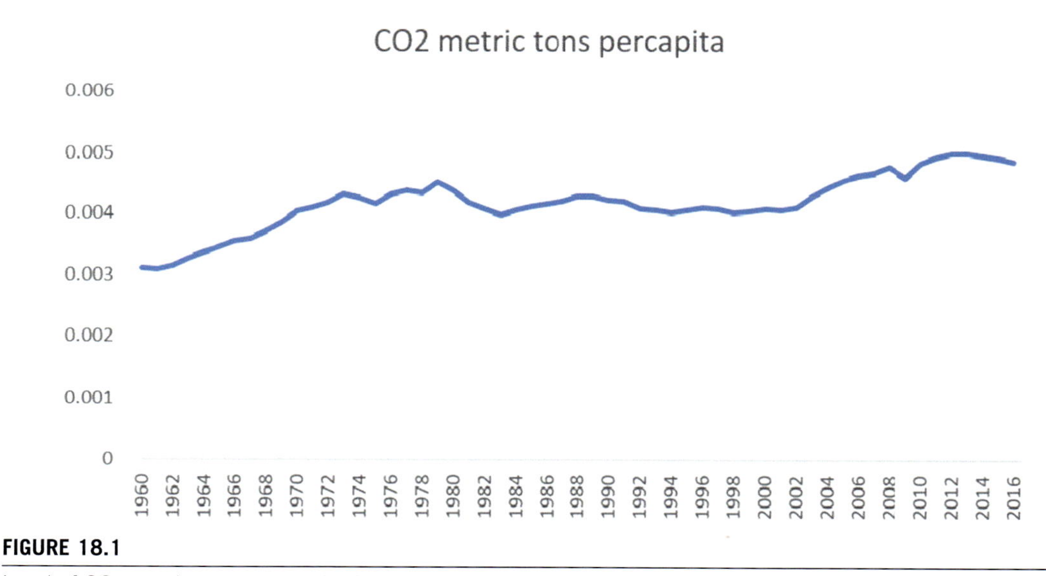

FIGURE 18.1

Level of CO_2 metric tons per capita for the world between 1960 and 2016.

Through successive assessment reports, the IPCC's concept of vulnerability has remained largely unchanged, and it has driven toward a couple of discoveries through scholastic composition. Eakin and Luers (2006) have made a comprehensive social–biological framework worldview for tending to human and environmental sources of vulnerability and developing proper measurements. Beroya-Eitner (2016), on the other hand, has laid out the challenges with an emphasis on the ecological factor. Meanwhile, the vulnerability index developed by the University of Notre Dame (2015) fits within a comprehensive paradigm because it includes both social and ecological components in its sectoral indices. Bohringer and Jochem (2007) consolidated different vulnerability indicators, examined 11 composite indices, including the Human Development Index (HDI), and inferred that each had structural flaws. According to Klugman et al. (2011), the HDI is the most broadly utilized composite index, and it has gotten the most flak. The mathematical nature of the model that supports the composite index has been the focus of most criticism. The mathematical design of a composite index can be depicted as far as factor standardization, collection weighting, and the aggregator capacity's practical structure as expressed by Greco et al. (2019). Most composite environmental index has restrictive mathematical constructions, which is normally acknowledged.

The following part of this chapter will depict further the causes, effects of climate change toward ASEAN countries and vulnerable nations, integration between climate change and sustainable development, and relevant policies and practical framework for action.

2. Causes of climate change

In this globalization and industrial revolution era, numerous countries, including developed and developing countries, are moving efficiently and effectively by utilizing the available resources, engaging in large-scale production of manufactured goods, and declining the cost of production of goods and services to achieve rapid growth in the economy (Hassan, 2009). However, everything has its opportunity costs. The industrial revolution leads to environmental degradation. Not only that, to improve a country's economy, it required a large amount of energy consumption, mostly achieved by consumption of fossil fuels, which caused environment pollution such as CO_2 emissions, and eventually led to global warming and climate change. Extreme competition among countries, enhancement in developing countries' growth, market liberalization, and globalization accelerated these developments. The most observable effect of the industrial revolution is climate change, in which the frequency of occurrence has been increased year by year.

Climate change is the greatest environmental warning to humans, caused by greenhouse gases (GHGs) formation and increment. GHGs have the property that trapped and absorbed infrared radiation, which is net heat energy emitted from the Earth's surface and reradiating it back to the Earth's surface. CO_2 is defined as a gas that has no color, no odor, and is nonpoisonous formed through the burning of carbon and through respiration of living things (Udara et al., 2019). In 1990, CO_2 emissions were recorded at 22.15 Gt, and it had increased dramatically to 36.14 Gt in 2014. The global temperature will increase as well as the CO_2 emissions increase. Fig. 18.2 shows the percentage of global greenhouse emissions by the economic sector. From the pie chart, we can conclude that electricity and heat production together with agriculture, forestry, and land use produce more GHGs than other sectors.

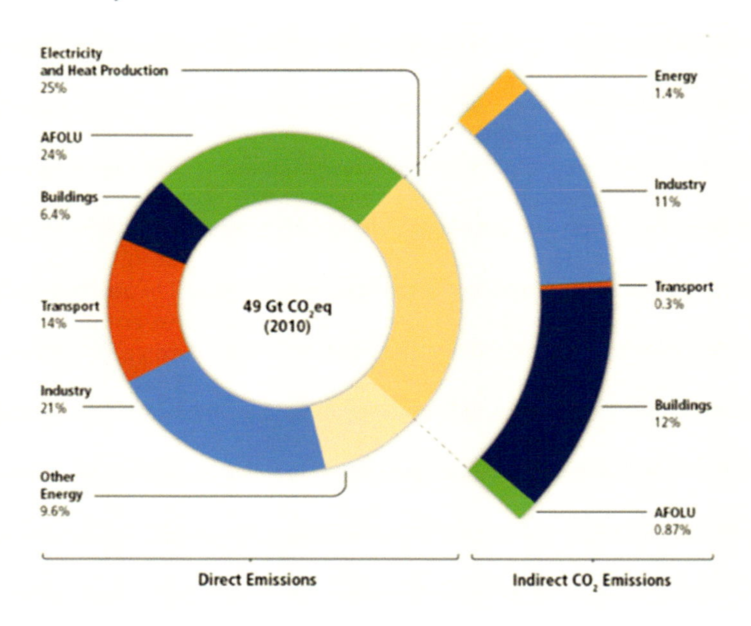

FIGURE 18.2

Global greenhouse emission by economic sector.

Source: IPCC (2014).

This is aligned with the industrial revolutions, where the atmospheric concentration of CO_2 has been increasing rapidly due to human activities, and it has now reached a threatening level. Human activities such as the burning of oil, coal, and gas for vehicular transportation and power generation and deforestation were the primary cause of CO_2 concentration increased in the atmosphere on the Earth. Climate change happens due to GHG emissions, and it causes numerous issues to the world such as global warming, ecosystem imbalance, economic issues, technological problems, and societal problems.

GHGs include CO_2, methane, and water vapor. The major sources of GHGs are based on the combustion of fossil fuels and the devastation of areas that contain a large volume of carbon, such as rainforests. Global warming is expected to raise the temperature to 1.5°C in 20 to 30 years if the temperature consistently increases. Thus, it is crucial to meet the reality of climate change in most countries and develop an efficient way to mitigate the problem.

Based on Fig. 18.3, CO_2 emissions based on fossil fuel and industrial processes are the highest gases compared to others. GHGs have been increased steadily in atmospheric concentrations since industrial revolutions. Most countries attach their industrial production, such as the United States, to burning fossil fuels to generate electricity, heat, and transportation. Over the past three decades, GHG and CO_2 emissions have risen almost 1.6% every year because of the big consumption of fossil fuels,

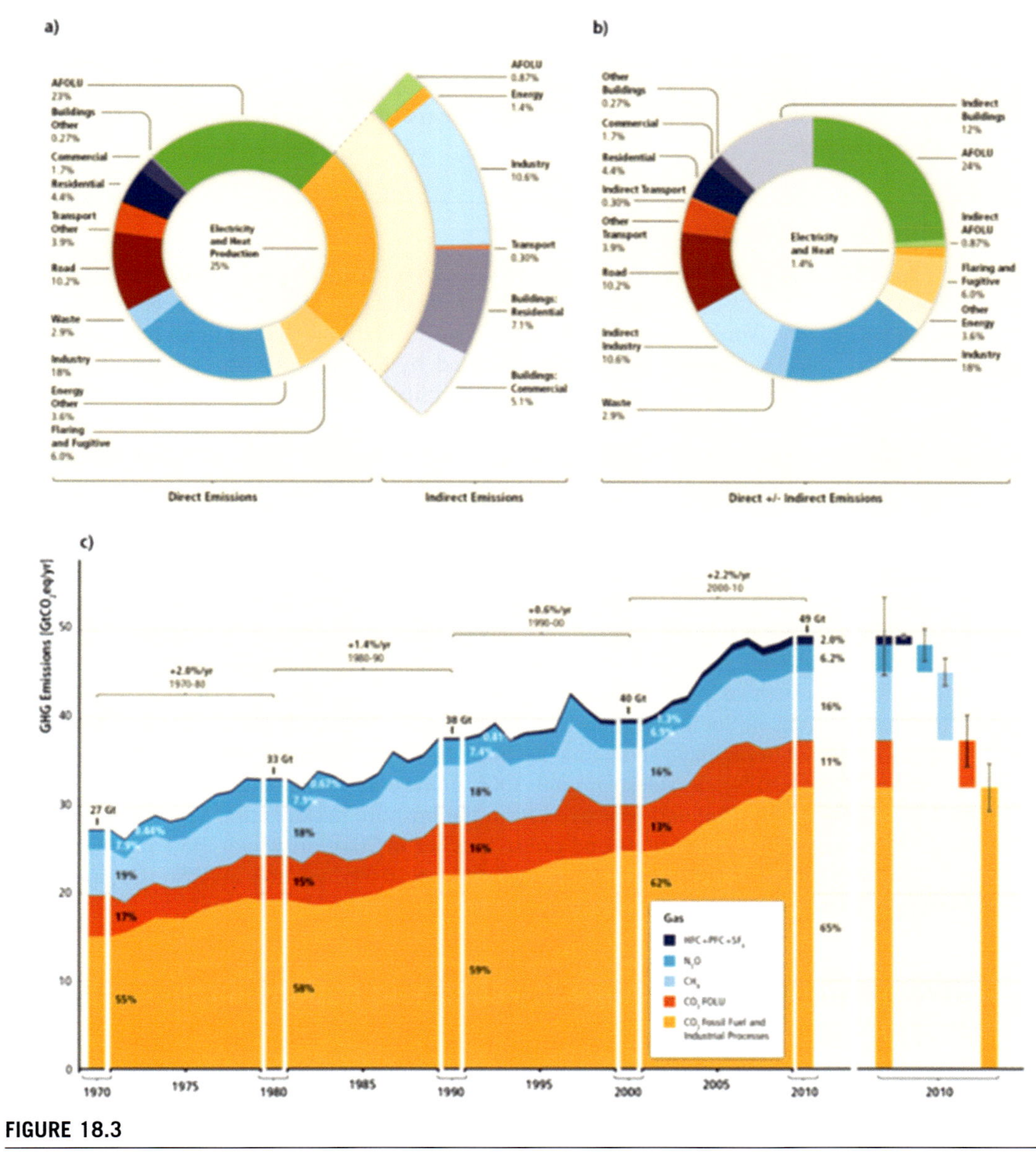

FIGURE 18.3

Global emissions by gas.

Source: IPCC (2014).

which have increased 1.9% annually. Moreover, human activities are anticipated to cause a high temperature of about 1.0°C of global warming, beyond the industry level, 0.8°C−1.2°C (IPCC, 2018).

In addition, the CO_2 levels in the atmosphere keep increasing from 1750 to 2020. The high intensity of CO_2 also causes a greenhouse effect due to the absorption of the heat energy from the sun and traps the heat close to the Earth's surface (NOAA, 2021). Thus, the heat energy will cause a rise in the temperature of the Earth's atmosphere and further contribute to the rising sea level, which is considered as global warming. This will overall affect the weather and wildlife population, and habitats. When extreme weather appears, it will increase the risk of wildfires, crop production reduced, and clean water shortage.

3. The impact of climate change

According to Watson et al. (2018) mentioned that the Amazon has the highest biodiversity globally, housing approximately 30% of the world's known species and around 390 billion trees living in the Amazon forest. Due to this, Amazon is recognized as the world's largest tropical rainforest. Moreover, around 20% of the world's oxygen turnover is produced by the Amazon forest area. Therefore, Amazon is a significant character in reducing CO_2 levels released into the atmosphere and fighting against climate change. Sullivan (2019) found that the Amazon was burned as an initiative to develop the agriculture sector, a major contributor to Brazil. By destroying the Amazon forest area, Brazil could have more places for agricultural development.

Moreover, Nangoy (2019) highlighted that Indonesia was recognized as the fifth-largest contributor of GHG because Indonesia burns its forest frequently for agriculture purposes. Indonesia had burned around 857,756 ha and 529,267 ha of forest in 2019 and 2018, respectively. Jong (2019) mentioned that the CO_2 released by Indonesia in 2019 is the double emissions from the fires in Amazon in which Indonesia had released approximately 708 million tons of GHG in 2019.

Moreover, toxic gases were released from deforestation activities in Indonesia in 2019. Thus, it had caused serious impacts such as possessing the worst haze level among the year and affecting the residents' health conditions in Southeast Asia such as Singapore, Thailand, Malaysia, and others. Last but not least, the habitat of orangutans and tigers had been destroyed from these deforestation activities (Indonesian Fires Burnt, 2019). Besides, around 3000 homes, 25,000 koalas, 30% of the koalas' habitat, and 33 people were killed and destroyed from the bushfire in Australia. The climate change caused Australia to experience the hottest and driest weather in 2019, in which the temperature had reached 41.9°C on December 18, 2019. This worsened and spread the bushfires in Australia and caused approximately 25.5 million acres of the area to be burned.

4. Understanding climate change in vulnerable nations through four dimensions

According to the United Nations, vulnerable nations are characterized as Least Developed Countries (LDCs), Landlocked Developing Countries (LLDCs), and Small Island Developing States (SIDS). These nations are distinguished as the least contributor of climate change; however, they face the disproportionality influenced by the adverse consequences of climate change and are regularly least ready to adapt due to their underlying requirements, a geological drawback. Simultaneously, these nations are among the most grounded advocate for more robust action on climate change as they face genuine difficulties that influence all parts of the day by day lives of their populaces. Consequently, this part presents the synergisms of the impacts of climate change within the context of vulnerable nations in social, economic, cultural, and environmental dimensions, respectively. The comparative analyses have been taken by the International Institute for Sustainable Development (IISD) to start the initiative to provide the general baseline of national adaptation policy and practice in developing countries. The review involves 15 African and Asian countries, which suggests the highest levels of vulnerability and the lowest levels of readiness to adapt are Bangladesh, Burkina Faso, Kenya, Mali, Ethiopia, and Uganda.

4.1 Social

Climate can be defined as an outcome of a complex geo-atmospheric ecological system and complex system that can be surprising by behaving in unanticipated ways. Furthermore, climate change is only one of many interacting phenomena of global environmental change caused or affected by human activity. Greater climate changes for a nation such as floods, drought, heat stress, species loss, and ecological change can be experienced directly toward conceptualization as connected phenomena with common causes are climate science. Therefore, when it comes to climate change and society, climate science plays a basic part.

Social factors related to convictions, interpersonal organizations, and ethnicity that influence the thoughts, decisions, and behaviors of people, groups, or communities. Convictions can be affected by getting the data and sorting the data. Channel of data conveyance and roads for financial and social help are the informal communities that an individual or family makes and has a place. Economic status and structures of livelihood influence the sort of interpersonal organization that one may have a place in.

According to Blaikie et al. (1994), accessing data is significant in influencing vulnerability. Lack of dispersal of data to the public will decrease resilience in the face of environmental risk exposure which generally increments vulnerability (Vincent, 2007). Besides, Adger (2000) expressed that developing countries dominate where their regular assets subordinate rustic occupations. It is anything but an indication of versatility and limited adapting in the rustic populace. Heavy reliance on natural resources, for example, woodland products, increments vulnerability of climate change. An increased vulnerability of households and communities that depend heavily on natural resources will lead to climate change.

The structure and composition of the populace that incline communities, groups, and gatherings of people can influence climate change and fluctuation. The structure and composition of populations are identified with elements like age, family size, reliance proportion, and extent of the proficient and economically active populace. In connection to family estimates, huge families will generally have a high number of socially and monetarily dormant wards. Thus, it makes them more defenseless against environmental change. Families straightforwardly extent their assets toward the government assistance of the family part. In this way, bigger families with bigger assets contribution will expand the environmental change.

Other than that, education plays a considerable role in gaining access to data, resources, and nonenvironment delicate occupations, which will minimize vulnerability. However, the vulnerability to climate change is increment because of the most untaught part in the general public, and ignorant people will, in general, have the least employable expertise in nonenvironment delicate areas and rely more upon environment touchy types of occupation. Brooks et al. (2005) indicated that illiteracy serves as a barrier to understanding the mind-boggling nature of dangers and reacts to the appropriate measure.

4.1.1 *Few cases of the social impact on climate change*

Extreme climate or changes in vocalization can meddle with the capacity of people or families in vulnerable nations to earn enough to make a living. Severe weather can destroy various assets that occupants rely on to support their occupations, including natural resources, well-being, infrastructure, and financial capital (International Institute for Sustainable Development, 2003). The effect of a solitary tempest or hefty downpour can limit economic activity for weeks or months (Douglas et al., 2008). As observed before, climate change could affect the industries that provide employment opportunities. This is particularly significant for the travel industry communities in Mombasa (Kenya), Rio de Janeiro (Brazil), Tallinn (Estonia), and Venice (Italy) (Awuor et al., 2008; Amelung et al., 2007; Sgobbi and Carraro, 2008). Livelihood impacts are particularly genuine for the helpless society, directly affecting food purchase, bill payments, and local interaction. When income sources are intruded on, helpless families regularly need to forfeit nourishment, children's education, and remaining abundance of meeting their necessities (Satterthwaite et al., 2007; Ruth and Ibarraran, 2009).

4.2 Economic

Economic factors are all around recognized in assume part of adding to the weakness, fluctuation, and flexibility to environmental change. Cannon (1994) expressed that a strong economy goes about as a safety net in environmental risk and hazard exposure. People that have great admittance to the resources arguably have a security net in the case of environmental danger and exposure. It allows them to attract different resources to keep up with their livelihoods to minimize the resultant effect of such antagonistic occasions. However, people with restricted economic privilege have a more serious level of reliance and are ostensibly less strong on account of shocks coming about because of progress and changeability. A few markers are utilized to evaluate the economic status like occupation, financial exercises, material abundance, and income.

Source diversity of income provides some form of welfare protection against shocks. In circumstances where people are more limited to a solitary type of revenue, they are less inclined to work on their ability in a crisis period. People limited to one sector for occupation are bound to neglect to get their necessities in a crisis period since they cannot attract different assets to limit the resultant effect of unfavourable occasions. Climate sensitive occupations like agriculture rely upon climate change for production and are vulnerable to climate change. Changing environmental conditions upsets economic production of climate-sensitive occupation, which decreases the wage for support of livelihood.

4.2.1 *Economy under pressure, especially for the industrial*

The high temperature will prompt a steep increase in economic losses (Pachuari et al., 2014). As indicated by a previous study by Jonathan et al. (2019), assessments of future misfortunes represent 2%–10% or more of global gross domestic product (GDP) per year. Extreme occasions and slow-onset changes will annihilate the infrastructure and disrupt the services needed for activities, including electricity, water, and sewage supplies. Meanwhile, catastrophic events are getting more continuous and pulverizing. Centre for Research on the Epidemiology of Disaster (2020) has discovered 396 occasions over the previous decade's annual average, influencing 95 million individuals around the world, bringing about an economic deficiency of $103 billion. These damages are practically unimaginable for families living in vulnerable nations to overcome.

According to a study distributed recently in the Proceedings of the National Academy of Sciences, from 1961 to 2010, the rising temperature cut the per-individual GDP of the world's vulnerable nation by 17%–31%. There was a wide gap in economic output between vulnerable nations and developed countries by 25% (Marshall et al., 2019). As per a new report from the Swiss Re Institute (2021), nations worldwide could be altogether lost about 10% of the total economic worth from climate change if the temperature keeps increasing in the current direction by 2050. A worldwide temperature alteration's belongings are far and wide, and range in seriousness, and no country is insusceptible to the harm that changing climate examples could cause.

For these cases, the vulnerable nations have endured harsher impacts to some degree since they are concentrated in already hotter parts of the world, like Africa, South Asia, and Central America. In such places, a tick up in temperatures can immediately cut work usefulness and farming yields while expanding levels of viciousness, wrongdoing, suicides, sickness, and mortality. The International Labor Organization (2019) has distributed a report entitled "Chipping away at a Hotter Planet: The Impact of Heat Stress on Labor Productivity and Decent Work," which brought up that regardless of whether the high-temperature increment will be restricted to 1.5° continuously by 2030, the climate change will cause a loss of productivity that will reach 2.2% of the multitude of working hours every year. Particularly for individuals who just rely upon open air work like in farming or may depend on natural resources, both of which are vulnerable against an evolving climate.

4.3 Cultural

Culture has a role in humanity's response to climate change. Society reaction to each dimension changes the global climate that is intervened by culture. Climate change undermines cultural dimensions of life and livelihoods that incorporate the material and lived parts of culture, identity, community attachment, and sense of place. There is an important cultural dimension to how society reacts and adjust to climate-related dangers. Culture intercedes changes in the climate and changes in societies.

Understanding mitigation and adaption to climate change can be comprehended by way of life. Culture additionally has its influence in outlining climate changes as a marvel of worry to society. Culture is implanted in the predominant production methods, utilization, ways of life, and social association that emanate ozone-depleting substances. The consequences of the emission impact climate changes and are given meaning through cultural interpretation of sciences and risk. Physical sway on the climate changes like outrageous climate occasion, sea fermentation, coral bleaching, dry season, and less overall rain coupled with heavier downpour occasion will modify on a very basic level the way that individuals live by changing the climate they live in.

How individuals react to and sway by climate changes can be coordinated by regional cultural variants in expectation, conviction, ways of knowing, and values. Other than that, the general effect of climate changes will shift, and it relies upon human endeavors to relieve and adjust. Expanded exertion can limit the drawn-out impacts, while the inability to act could incredibly fuel the risks. Social standards and assumptions will impact the activities that a populace will take. Therefore, climate changes and culture have a mutual influence on each other.

4.3.1 *Cultural impacts on climate change cases*

Based on the previous study by Craeynest (2010), stated that the potential global losses and damage caused by climate change between 2000 and 2200 were US$275 trillion, but it excluded noneconomic losses and damage. Noneconomic or market value, losses and damages include cultural and social impact (Serdeczny et al., 2016). In addition to losses, job shifts brought about by climate change and other environmental impacts may likewise progressively change or potentially adapt different parts of community life or relationship, leading to cultural changes (Morrissey and Smith, 2013; Snyder et al., 2003). According to Morrissey and Smith (2013), elements of cultural and social roles develop in response to interactions with specific physical landscapes. Changes in the landscape or its sustainable interaction will change not only traditional knowledge but also identity.

According to the previous study by Marra et al. (2012), due to the unique vulnerability of Pacific atoll countries to climate change, the lives of Tuvaluans people have been affected by climate change. For instance, the need for adaptions to move inland or raise the building houses has been consistently carried out as the shores have dissolved and flowing flooding has expended recurrence and seriousness. In the drought in 2011, essentially every part of Tuvaluans' daily life was affected by the need for exacting water apportioning, and the long-term drought could impact crops, and other natural resources continue to be seen and felt.

4.4 Environmental

There is developing acknowledgment in the research community to observe why the countries are in more danger of climate change. Thus, this section will review the causes in the scope of the environment. As we probably are aware, an Earth-wide temperature boost can bring about numerous changes to the climate. An Earth-wide temperature boost additionally can cause an ascent in the ocean level, expanded danger of droughts and floods, prompting the deficiency of seaside land, a change of precipitation patterns and dangers to biodiversity. The United Nations Framework Convention on Environmental Change (UNFCCC) revealed that the low-lying and small island developing countries to be especially defenseless against the unfriendly impacts of the environmental change and recognizes the special needs of LDCs. These high-hazard locales are fundamentally located in sub-Saharan African and, to lesser broaden South Asia with vulnerabilities related to increased seaside immersion, increased water pressure, increased exposure to irresistible infection, changes in river hydrology, and adjustments to the greatness and recurrence of outrageous occasions (IPCC, 2017; World Bank, 2020; Ford et al., 2015).

The LDCs, LLDCs, and SIDS are disproportionately influenced by the adverse consequences of climate change because of their underlying limitations and topographical inconveniences. However, simultaneously, they contribute least to climate change. The International Organization for Migration reported that LDCs are among the most vulnerable to climate change because of a lack of capacity and development. In 2007, a major socioeconomic effects happened in LDCs, where 23 million individuals were affected (UNCTAD, 2017) due to floods particularly in Haiti, Angola, Malawi, Bangladesh, Nepal, Myanmar, Madagascar, Mozambique, Bangladesh and Vanuatu (Joined Countries General Get together, 2018). These environmental calamities are not only huge in human and financial misfortune withing the LDCs but also risk the accomplishment in the Reasonable Advancement Objectives (Sustainable Development Goals [SDGs]).

5. Case of ASEAN countries

CO_2 emissions are the cause of environmental degradation, and it is always a serious issue that is a concern at all times. The burning of fossil fuels and other human-induced behaviors contributed to the increase in CO_2 emissions. However, among the energy, the selected ASEAN-5 countries, Indonesia's GDP and CO_2 emissions were recorded at the highest level. Cox and Cox (2020) found that energy input is important in stimulating a country's economy. Besides, good practice in adopting renewable energy in the region is vital as the energy demand in the ASEAN region keeps increasing that leads the government to be aware due to obstinate decline in quality of the environment from the combustion that causes negative externalities to the economy and gives impact to human health and reduces the overall productivity in the long-run. If the growth of the economy of one country increases, then this could lead to an increase in CO_2 emissions. As a result, GDP and CO_2 emissions will record the same moving pattern. For example, the GDP of the United States increased from $18.2 trillion in 2015 to $19.5 trillion in 2017; at the same time, CO_2 emissions in the United States increased from 400 ppm in 2015 to 405 ppm in 2017.

Fig. 18.4 shows the annual CO_2 the ASEAN countries had released from 1990 to 2018. Overall, within the period, the annual CO_2 released by Indonesia was maintained at the highest level, while the annual CO_2 released by the Philippines was maintained at the lowest level, and others were fluctuated. Vehicles and various industrial activities had released GHGs, which is CO_2 that would lead to an increase in the global temperature. In 2018, the global temperature was recorded at 1.42°F, which is beyond the 20th century average and classified as the fourth hottest year in NOAA's climate record. The high temperature resulted in global warming and thus caused the United States to face 14 weather

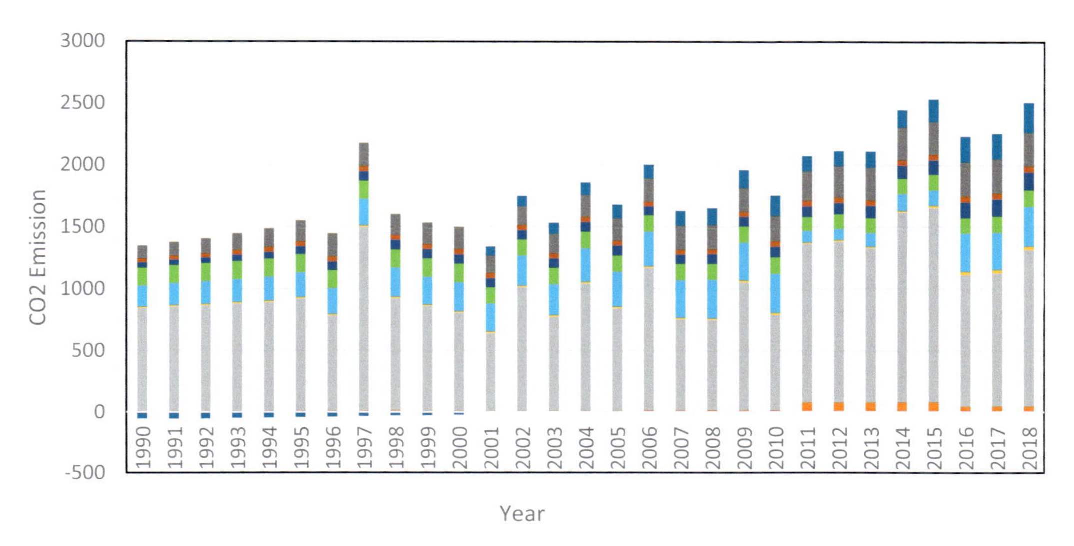

FIGURE 18.4

Annual CO_2 emissions in ASEAN countries from 1990 to 2018.

Source: World bank (2021).

and climate disasters. This indicates that high temperatures would lead to global warming, resulting in climate change and thus affect environmental sustainability.

Southeast Asia is the most vulnerable region to be affected by climate change. According to the Global Climate Risk Index 2019, among the 10 countries in ASEAN, Vietnam, Indonesia, the Philippines, and Thailand are among the top 10 countries most affected by climate change globally, which shows the seriousness of climate change in the ASEAN region. One factor contributing to climate change in ASEAN is deforestation, especially in Malaysia, Indonesia, and Cambodia. These countries have the world's largest forestlands, and many trees have been cut down for agricultural purposes. Not only deforestation, rapid economic growth and urbanization also contribute to climate change in ASEAN countries. Many big cities in ASEAN countries are vulnerable to floods due to misplanning in the development and construction. According to the Global Climate Risk Index 2019, among the ten countries in ASEAN namely Vietnam, Indonesia, the Philippines, and Thailand are among the top 10 countries most affected by climate change globally, which shows the seriousness of climate change in the ASEAN region. One factor contributing to climate change in ASEAN is deforestation, especially in Malaysia, Indonesia, and Cambodia. These countries have the world's largest forestlands, and many trees have been cut down for agricultural purposes. Not only deforestation, rapid economic growth and urbanisation also contribute to climate change in ASEAN countries. Many big cities in ASEAN countries are vulnerable to floods due to misplanning in the development and construction. Tree cover losses defined as the removal of trees due to human and natural causes. This could give long-term and often irreversible impacts on carbon storage and biodiversity. Examples of activities are deforestation and fire in the Amazon. Besides, the amount of carbon released through the biologically diverse ecosystems is 2.64 billion tons, equivalent to the annual emissions of 570 million cars.

Furthermore, based on Global Climate Risk Index (2019) from 1998 to 2017, at least 526,000 people passed away worldwide as an impact of more than 11,500 severe weather conditions. The effects of climate change have already been felt. Polar ice shields are melting, temperatures are increasing, sea levels are rising, and glaciers are shrinking. These are the direct consequences of climate change that all the residents in the world can witness.

Based on Eckstein et al. (2021), Puerto Rico, Myanmar, and Haiti are classified as the most impacted countries from 2000 to 2019. Then followed by an ASEAN country, the Philippines. In these two decades, these are the most impacted countries, as shown through average weighted ranking (CRI score) and the specific results relating to the four indicators analyzed. Based on the report, countries with slow growth and low per capita GDP have a high probability of getting impacts from climate change, seen in some ASEAN countries such as Indonesia, Laos, Malaysia, Myanmar, Philippines, Thailand, and Vietnam. In comparison, developed countries such as Australia, Japan, New Zealand, South Korea, and China could have lower impacts of climate change due to the technology to control the carbon emission and net agricultural benefit with expected increasing crop yields.

Besides, the COVID-19 pandemic has triggered most countries with movement restrictions and lockdown. This had reduced economic activity and thus resulted in a reduction in CO_2 emissions. As evidence, China, which is considered a country that records high economic activities, had successfully reduced its CO_2 emissions by 25% due to the coronavirus lockdown (World Meteorological Organization, 2020).

Other factors such as foreign direct investment (FDI), trade, financial development after the crisis, and technology advancement would cause the economy's high growth, and these are also the reasons behind the high carbon emission in most ASEAN countries. Since FDI inflows are increasing in ASEAN countries, it was expected to cause the CO_2 emissions because FDI inflows will boost the economic growth. This will encourage more business activities such as production, export, and import. However, the combustion of fossil fuel from the production factory will result in emissions of GHG such as CO_2, especially power plants and an industry that uses its electricity and heat production. In 2017, the industry contributed over half of the CO_2 emissions. However, it is undeniable that FDI inflows might also positively impact the environment as more modern and environmentally friendly technology is transferred from one country to another through a connection of FDI (OECD, 2002). Thus, control of carbon emission is crucial to have stable growth and clean environment.

Besides, financial development after the crisis is causing the economic growth of countries in Asia. For example, Malaysia, Thailand, the Philippines, and Vietnam had grown about 20% in their debt ratio from 2010 to 2017, while Indonesia did not significantly increase their debt ratio. Thus, this indicates that the private sectors in the countries above have increased their financial leverage to finance their businesses or expand their businesses. As the business has more funds available, it will most likely increase its production line or invest in a project like having more manufacturing plants (OECD, 2018). On the other hand, the households in Thailand and Malaysia also have a higher demand for more credit as they want to invest in real estate and consumer goods. When the spending level among households increases, it is believed to contribute to economic growth because businesses earn more profit and therefore produce more CO_2 (OECD, 2018). Lastly, by living in the 21st century, technological advancement has made every individual's life more convenient and improved their lifestyle.

However, technological advancement could also negatively affect the environment. The GHG emissions have increased by approximately 8% because of digital technology. In addition, the CO_2 emissions contributed by OECD countries in digital technology have shown an upward trend over time. The report increased about 450 million tons starting from 2013 (The Shift Project, 2019). Using technology without limit, especially in the industrial area, would cause high pollution. Therefore, companies or factories are recommended to use effective technology to reduce and control their CO_2 emissions level.

6. Relevant policies to mitigate climate change

By having a clear view regarding the factors contributing to CO_2 emissions, policymakers and industry could generate more specific solutions to reduce the impact generated by those factors and ease the implementation process to reduce CO_2 emissions. According to Cox and Cox (2020), CO_2 emissions and GDP was heading in the same direction. Therefore, to minimize the impact generated by GDP, the government should apply restrictions such as carbon tax to restrict a country's economic activity. As a result, to avoid paying a high amount of carbon tax, a factory might reduce their operation hour or production volume, leading to a decline in economic activity. Cap and trade is a type of policy in which government will set a specific amount of CO_2 that a factory can release, and a penalty will be given to those factories that fail to meet the restricted level. If a factory can maintain or expand its production level and with a much lower CO_2 emissions level than the restricted level set by the government, they can sell the additional restricted amount of CO_2 to other factories. As a result, the extra income would motivate the factory to increase their production efficiency and thus result in lower CO_2 emissions (Center on Budget and Policy Priorities, 2015).

On the other hand, in this 21st century, most factories and companies have applied and worked on Industry 4.0. It helped them to increase their cost efficiency and enabled them to generate a higher amount of profit. Therefore, companies or factories are recommended to use the profit generated to invest in green technology that would thus reduce and control their CO_2 emissions level. For example, Kwiatkowski et al. (2019) found that Industrial Emissions Control Technology is an effective way for a company or factory to maintain and control its CO_2 emissions level. Hence, using renewable energy technology at the industry level would result in lower CO_2 emissions levels because it is a technique that transforms renewable resources like solar, wind, and hydro and others to the electricity needed for production. Therefore, companies or factories can apply Industrial Emissions Control Technology or Renewable Energy Technology in the production process to reduce CO_2 emissions. Additionally, companies or factories should also implement a proper way to handle their production waste and replace harmful chemical material needed in production with environmentally friendly material. For example, the World Trade Organization (2005) is negotiating to reduce widespread import tariffs to zero. The green products in the range make it more cost-effective and easier to implement for customers and users around the world. Meanwhile, the UNFCCC Green Climate Fund (2020) supports projects, plans, policies, and other activities in developing countries. Currently, the focus of the discussion is to provide additional financing as the main challenge faced by many of the poorest countries in the world.

Promoting innovation is one of the post-2015 sustainable development objectives and supports the acknowledgment of other reasonable SDGs, for example, eradicating poverty and hunger, ensuring access to energy and health, and promoting sustainable economic growth. Policies should support the development of an enabling environment for active technology, innovation-focused government, and market policies and regulations, and play a more active and participatory role for the wider business community. The role of the innovation developments toward climate change on a worldwide scale is to the turn of events and diffusion of a broad variety of new clean technologies in both developed and developing countries. United Nations (1993) express that clean technology innovation is the key to tending to worldwide climate change difficulties, which has become a broad consensus. Mitigation and adaption (UNFCCC, 2006), technological development and technological diffusion help private sector participants develop and implement many cleantech solutions (Lybecker and Lohse, 2005). Technology can be transferred and shared through trade associations, collaborative innovation frameworks, and bilateral or regional government to government or private sector technology and innovation partnerships.

Furthermore, to provide the optimal innovation and technological environment, the developed and developing countries could attract all foreign innovations and encourage investment for innovations domestically and across national borders. Policies must be defined and implemented to facilitate innovation and collaborative technology partnerships, and it might help companies, consumers, and the whole economy move up the innovation value chain. Technology innovation also enables measuring the competitive tax rates, capital depreciation, tax incentives for the R&D and investment, tax holiday, and tax-free geographic zones. As for the governments, it can help to encourage innovation through investments in infrastructure such as roads, pipelines, or reliable access to electricity, and a policy to stimulate and allow for FDI and also a robust global market mechanism as a co-financing tool that will assist in integrating a country into global supply chains. Moreover, high technology innovation might require a consistent investment in education, especially for advanced research institutions. The market mechanism is proving its value in facilitating private capital access to private technology in India. These mechanisms should continue as essential options in a new global approach architecture that can combat climate change effectively (ICC Commission on Environment and Energy, 2010).

Environmental quality has become an important factor in promoting a better life. This is because environmental issues like natural disaster and climate change are happening globally, which further affects the whole ecosystem, including human life. Thus, the whole community needed to take part in reducing the impact on environmental quality. Environmental awareness should be created for students as they are still in a stage of learning. The schools or universities could organize an event that exhibits the green technology invented by students or researchers to get the attention of businesspeople interested in investing in green technology. Gray (2017) found that researchers at the University of Surrey have discovered a supercapacitor that could be used for electric cars. The supercapacitor is believed to let the car travel for six to eight hours without stopping for recharging purposes, and the charging time is expected to be the same as the time spent on filling the car with petrol. Therefore, the electric car can be used to replace the car that used petrol to reduce CO_2 emissions since the burning of petrol emits CO_2 into the atmosphere.

Thus, all the policies with a specific focus on regulation, renewable energy, innovation, and community could help a country control carbon emissions and reduce the impact of climate change. Besides, the focus on those elements will also increase economic growth and living quality of the community, which could overall help a country achieve sustainable development.

7. Integration between climate change and sustainable development

Sustainable development has become the national agenda in line with SDGs, and climate change is one of the total 17 goals. Sustainable development is closely connected to climate change, after the two major agreements in 2015, the 2030 Agenda for Sustainable Development and the Paris Agreement. Climate change affects natural and human living conditions, as well as socioeconomic progress. From the perspective of society, high development could cause greenhouse emissions and sensitivity to climate change (Kumaran et al., 2021). This is because the rise of GDP in a country tends to produce more output with high technology. However, advanced countries have shown that the rapid progress in economic growth coupled with effective policy and management system to control greenhouse emissions can reduce the impact of climate change.

For developed countries, GDP will keep increasing when the economy is in a boom period where people with more income will increase dependency on electric appliances and transportations, which contributes to higher consumption of energy and high pollution. On the other hand, for developing countries and LDCs, GDP is fluctuating and might face recession during financial and health crises. During the recession, people will focus more on basic life necessities consumption and reduce unnecessary consumption. In this situation, the energy-intensive sectors may be affected, and hence the energy consumption and CO_2 emissions will be reduced. However, in later stages during the expansion period, the GDP exceeds the threshold parameter and cause the CO_2 emission to increase again, and the unavailability of renewable energy would cause a high carbon emission level.

The government should strictly impose effective climate change policies and relevant laws and regulations to penalize companies that violate the rules to improve environmental quality. It is essential to increase environmental awareness among the citizens in every country because it is quite difficult to control energy usage as the population grows. By increasing citizens' awareness, the government can easily impose regulations to counter environmental degradation issues and save the planet from the risk. With more adaptation and mitigation, the plant could be saved. For example, using resources that can save energy such as solar panels, household items with solar function, and energy-saving could overall control carbon emissions and lead to a country's sustainable development.

7.1 "Sustainomics"—A practical framework for action

The main concern of policymakers is the growth of the country, poverty, inequality, food security, and environmental quality. Therefore, the integration between climate change adaptation and mitigation aligns with our national sustainable development policies. "Sustainomics" has incorporated the three fundamental components: economy, social, and environment.

To begin, sustainable development is described as extending the development that allows individuals and communities to achieve their needs while maintaining the resilience of the economy, society, and environment. Making development more sustainable (MDMS) is an efficient method that could allow us to solve the significant issue without interruption since many untenable behaviors are simpler to detect and remove.

Second, the three pillars of the triangle of sustainable development must be controlled consistently. Social, economic, and environmental components are among these to achieve sustainable development. Social component will be more on focusing on the empowerment and values of society. In contrast, the economy and environment will be more initiative to grow and keep them sustainable.

Third, the thinking on extending beyond traditional boundaries. This is how society, investors, and more disciplines effectively handle space and time. Climate change is a worldwide topic that has been debated for centuries and impacts every human being on the planet Fig. 18.5 shows a typical graph

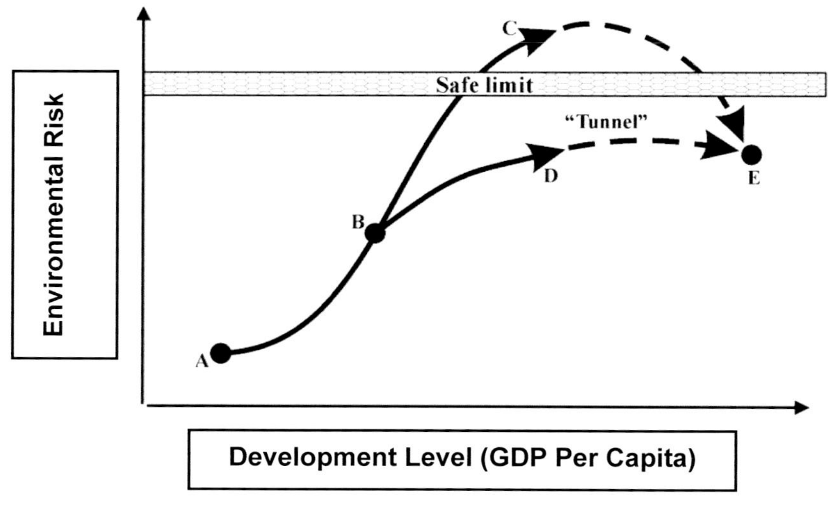

FIGURE 18.5

Environmental risk versus development level.

Adapted from Munasinghe (1995a).

linking environmental risk (measured in GHG emissions per capita) to a country's level of development (measured in GDP per capita). A typical developing nation may be on curve AB which shows the growing economy, whereas an industrialized or developed country may be on curve C. MDMS lays the groundwork for the future. Industrialized countries are with levels of "hazardous" chemicals that exceed safe limits. They may mitigate climate change by changing their development patterns and pursuing the CE future growth path to disentangle the relationship between carbon emissions and economic growth.

To withstand the consequences of climate change, the poorest and most vulnerable nations require an adaptation safety net. Meanwhile, middle-income nations may devise effective methods to "tunnel" through (along with BDE). The growth from the industrial revolution would turn the tunnel below the safe limit. Therefore, in this situation, technology and financial assistance will be needed to expand and become more sustainable by adopting a less carbon-intensive method that could reduce the climate vulnerability.

The sustainomics framework also offers policymakers few practical tools, including new and old techniques that may be applied creatively. This helps in finding and executing the most desirable climate policies focusing on three main goals: economic, society, and environment and the tradeoffs between the diverging goals. The very peaked route ABCE depicted in Fig. 18.5 is the result of economic crisis and environmental externalities. Effective policies could help reduce such distortions while also permitting transit through the BDE's long-term sustainability tunnel and overall achieving sustainable development.

8. Summary

Economy development is important for a country to enhance its reputation and strengthen its home currency and competitiveness. However, the high volume of CO_2 emissions has brought few negative impacts such as worsening the environmental quality, affecting society's health condition, global warming, and others. Apart from that, climate change and sustainable development are two issues that are intertwined and constitute a severe threat to human lives and livelihood. Although the difficulties are complicated, both problems can be overcome if we start right away. We now know enough to take the first steps toward more sustainable growth, which would change the uncertain business environment into a more secure future.

Anticipating the future affected by the environmental change is perhaps the main challenge within recent times. The effect of climate change is more exceptional toward the vulnerable nation because of lack of capacity and capability. The issue is significantly more troublesome when the catastrophic event occurs and brings about temperature changes, like sea dissemination and terrestrial biological system reactions, where it will be modified as the climate changes. Conclusion: A few suggestions can be close as to reaction with these phenomena to reduce the impacts, for instance, agriculture sharing where farming was distinguished as a priority sector for adaption to climate change. Other than that, the improvement in the monitoring and evaluation systems is needed to assess the effectiveness of current investments in adaptation action, and more attention should be given at the subnational level in adaptation planning. These proposals are lined up with those developed under the Task Force on Displacement under the United Nations Framework Convention on Climate Change (TFD, 2018).

References

Adger, W.N., 2000. Social and ecological resilience: are they related? Progress in Human Geography 24 (3), 347–364.

Amelung, B., Nicholls, S., Viner, D., 2007. Implications of global climate change for tourism flows and seasonality. Journal of Travel Research 45, 285–296.

Awuor, C.B., Orindi, V.A., Adwera, A.O., 2008. Climate change and coastal cities: the case of Mombasa, Kenya. Environ Urban 20, 231–242.

Banett, L., 2017. Deforestation and climate change. A publication of climate institute (1400).

Beroya-Eitner, M.A., 2016. Ecological vulnerability indicators. Ecological indicators 60, 329–334.

Blaikie, P., Cannon, T., Davis, I., Wisner, B., 1994. At Risk: Natural Hazards, People, Vulnerability, and Disasters. Routledge, London, UK.

Brooks, N., Adger, W., Kelly, M., 2005. The determinants of vulnerability and adaptive capacity at the national level and the implications for adaptation. Global Environmental Change Part A 151–162. https://doi.org/10.1016/j.gloenvcha.2004.12.006.

Böhringer, C., Jochem, P.E., 2007. Measuring the immeasurable—A survey of sustainability indices. Ecological economics 63 (1), 1–8.

Cannon, T., 1994. Vulnerability analysis and the explanation of natural'disasters. Disasters, development and environment 1, 13–30.

Center on Budget and Policy Priorities, 2015. Policy Basics: Policies to Reduce Greenhouse Gas Emissions. Retrieved from: https://www.cbpp.org/research/policy-basics-policies-to-reduce-greenhouse-gas-emissions.

Centre for Research on the Epidemiology of Disaster, August 20, 2020. Disaster Year in Review 2019. Retrieved from: https://www.preventionweb.net/publications/view/71642.

Climate change 2014: synthesis report., Contribution of Working Groups I, II and III to the fifth assessment report of the Intergovernmental Panel on Climate Change, 2014. Pachauri, R.K., Allen, M.R., Barros, V.R., Broome, J., Cramer, W., Christ, R., Dubash, N.K., p 151

Cox, P.G., Cox, S., 2020. The Recent History of GDP Growth, Co2 Emissions, and Climate Policy Paralysis, All in One Table-Runner. Resilience. Retrieved from: https://www.resilience.org/stories/2020-04-30/the-recent-history-of-gdp-growth-co2-emissions-and-climate-policy-paralysis-all-in-one-table-runner/.

Craeynest, L., 2010. Loss and Damage from Climate Change: The Cost for Poor People in Developing Countries. Action Aid. Available at: http://www.actionaid.org/publications/loss-and-damage-climate-change-cost-poor-people-developing-countries.

Douglas, I., Alam, K., Maghenda, A., McDonnell, Y., McLean, L., Campbell, J., 2008. Unjust waters: climate change, flooding and the urban poor in Africa. Environ Urban 20, 187–205.

Eakin, H., Luers, A.L., 2006. Assessing the vulnerability of social-environmental systems. Annual Review of Environment and Resources 31, 365–394.

Eckstein, D., Künzel, V., Schäfer, L., 2021. Global Climate Risk Index 2021, Germanwatch. Available online: https://reliefweb.int/sites/reliefweb.int/files/resources/GlobalClimateRiskIndex2021_1_0.pdf (accessed on 8 October 2021).

Ford, B.Q., Dmitrieva, J.O., Heller, D., Chentsova-Dutton, Y., Grossmann, I., Tamir, M., ... Mauss, I.B., 2015. Culture shapes whether the pursuit of happiness predicts higher or lower well-being. Journal of Experimental Psychology: General 144 (6), 1053.

Gray, A., January 9, 2017. 5 Tech Innovations that Could Save Us from Climate Change. World Economic Forum. Retrieved from: https://www.weforum.org/agenda/2017/01/tech-innovations-save-us-from-climate-change/.

Greco, S., Ishizaka, A., Tasiou, M., Torrisi, G., 2019. On the methodological framework of composite indices: A review of the issues of weighting, aggregation, and robustness. Social indicators research 141 (1), 61–94.

Harris, J.M., Roach, B., Codur, A.M., 2019. The Economic of Global Climate Change. Global Development and Environment Institute Tifts University. Retrieved from: https://www.bu.edu/eci/files/2019/06/The_Economics_of_Global_Climate_Change.pdf.

Hassan, A., 2009. Risk management practices of Islamic banks of Brunei Darussalam. The Journal of Risk Finance 10 (1), 25–37.

ICC Commission on Environment and Energy, 2010. Market Mechanisms in the Post-2012 GHG Regime: Role and Shape of Future Greenhouse Gas and Carbon Markets.

Indonesian fires burnt 1.6 million hectares of land this year: researchers, 2019. CNA. Retrieved from: https://www.channelnewsasia.com/news/asia/indonesia-fires-burnt-1-6-million-hectares-of-land-haze-12144822.

International Institute for Sustainable Development, 2003. Livelihoods and Climate Change. IISD, Manitoba, Canada.

International Labour Organization, 2019. Working on a Warmer Planet: The Impact of Heat Stress on Labour Productivity and Decent Work. Retrieved from: https://www.ilo.org/wcmsp5/groups/public/—-dgreports/—-dcomm/—-publ/documents/publication/wcms_711919.pdf.

IPCC, 2014. Climate Change 2014: Synthesis Report. Contribution of Working Groups I, II and III to the Fifth Assessment Report. Intergovernmental Panel on Climate Change, Geneva, Switzerland, p. 151.

IPCC, 2017. IPCC Special Report on Climate Change, Desertification, Land Degradation, Sustainable Land Management, Food Security, and Greenhouse gas fluxes in Terrestrial Ecosystems. Summary for Policymakers. Intergovernmental Panel on Climate Change.

IPCC, 2018. Global warming of 1.5°C. An IPCC Special Report on the impacts of global warming of 1.5°C above pre-industrial levels and related global greenhouse gas emission pathways, in the context of strengthening the global response to the threat of climate change, sustainable development, and efforts to eradicate poverty. V. Masson-Delmotte, P. Zhai, H. O. Pörtner, D. Roberts, J. Skea, P.R. Shukla, A. Pirani, W. Moufouma-Okia, C. Péan, R. Pidcock, S. Connors, J. B. R. Matthews, Y. Chen, X. Zhou, M. I. Gomis, E. Lonnoy, T. Maycock, M. Tignor, T. Waterfield (eds.).

Klugman, J., Rodríguez, F., Choi, H.J., 2011. The HDI 2010: new controversies, old critiques. The Journal of Economic Inequality 9 (2), 249–288.

Kumaran, V.V., Munawwarah, S.N., Ismail, M.K., 2021. Sustainability in ASEAN: the roles of financial development towards climate change. Asian Journal of Economics and Empirical Research 8 (1), 1–9.

Kwiatkowski, S., Polat, M., Yu, W., Johnson, M.S., 2019. Industrial Emissions Control Technologies: Introduction. In: Encyclopedia of Sustainability Science and Technology. Springer Nature.

Lybecker, K.M., Lohse, S., 2015. Innovation and Diffusion of Green Technologies: The Role of Intellectual Property and Other Enabling Factors. Global Challenges Report.

Marra, J.J., Kenner, M.L., Finucane, D., Spooner, M.H., Smith, M.H., 2012. Climate Change and Pacific Islands: Indicators and Impacts. Report for the 2012 Pacific Island Regional Climate Assessment (PIRCA).

Marshall, B., Noah, D., Josie, G., 2019. Climate Change Has Worsened Global Economic Inequality. School of Earth, Energy & Environmental Science. Retrieved from: https://earth.stanford.edu/news/climate-change-has-worsened-global-economic-inequality#gs.5j97x5.

Morrissey, J., Smith, A.O., 2013. Perspectives on Non-economic Loss and Damage: Understanding Values at Risk from Climate Change. CDKN, ICCCAD, German Watch, MCII. United Nations University UNU-EHS.

Munasinghe, M., 1995. Making Growth More Sustainable, Ecological Economics, vol. 15, pp. 121–124.

Nangoy, F., 2019. Area Burned in 2019 Forest Fires in Indonesia Exceeds 2018 – Official. Reuters. Retrieved from: https://www.reuters.com/article/us-southeast-asia-haze/area-burned-in-2019-forest-fires-in-indonesia-exceeds-2018-official-idUSKBN1X00VU.

NOAA, 2021. National Centers for Environmental Information. State of the Climate. Global Climate Report. https://www.ncdc.noaa.gov/sotc/global/202113.. (Accessed 5 April 2022).

OECD, 2002. Reading for Change – Performance and Engagement across Countries. OECD, Paris.

OECD, 2018. Economic Outlook for Southeast Asia, China and India 2018: Fostering Growth through Digitalisation. OECD Publishing, Paris. 9789264286184-en.

Pachauri, R.K., Allen, M.R., Barros, V.R., Broome, J., Cramer, W., Christ, R., Dubash, N.K., 2014. Climate Change 2014: Synthesis Report. Contribution of Working Groups I, II and III to the Fifth Assessment Report of the Intergovernmental Panel on Climate Change. IPCC, p. 151. Retrieved 10 April, 2020 from: https://epic.awi.de/id/eprint/37530/1/IPCC_AR5_SYR_Final.pdf.

Ruth, M., Ibarraran, M.E., 2009. Introduction: distributional effects of climate change—social and economic implications. In: Ruth, M., Ibarraran, M.E. (Eds.), Distributional Impacts of Climate Change and Disasters. Edward Elgar Publishing, Cheltenham, UK.

Satterthwaite, D., Huq, S., Pelling, M., Reid, H., Lankao, P., 2007. Adapting to Climate Change in Urban Areas: The Possibilities and Constraints in Low- and Middle-Income Nations. Human Settlements Discussion Paper Series. IIED, London.

Serdeczny, O., Waters, E., Chan, S., 2016. Non-economic loss and damage in the context of climate change: Understanding the challenges. Discussion Paper/Deutsches Institut für Entwicklungspolitik, ISBN 978-3-88985-682-1.

Sgobbi, A., Carraro, C., 2008. Climate Change Impacts and Adaptation Strategies in Italy: An Economic Assessment. Fondazione Eni Enrico Mattei. Berkeley Electronic Press. Working Paper 170. http://www.bepress.com/feem/paper170 (last accessed 2 September 2010).

Snyder, R., Williams, D., Peterson, G., 2003. Culture loss and sense of place in resource valuation: economics, anthropology, and indigenous cultures. In: Lentoft, S., Minde, H., Nilsen, R. (Eds.), Indigenous Peoples: Resource Management and Global Rights (107−123). Eburon, Delft, The Netherlands.

Sullivan, Z., 2019, 26 August. The Real Reason the Amazon Is on Fire. TIME. Retrieved from: https://time.com/5661162/why-the-amazon-is-on-fire/.

Swiss Re Institute, 2021. Major Economies Could Lose Roughly 10% of GDP in 30 Years. Retrieved from: https://www.swissre.com/media/news-releases/nr-20210422-economics-of-climate-change-risks.html.

TFD, 2018. https://environmentalmigration.iom.int/sites/default/files/2018_TFD_report_16_Sep_FINAL-unedited.pdf.

The Shift Project, 2019. Lean ICT − towards Digital Sobriety. Retrieved from: https://theshiftproject.org/wp-content/uploads/2019/03/Lean-ICT-Report_The-Shift-Project_2019.pdf.

Udara, A., Wadu, M., Samarasinghalage, T., 2019. Global research on carbon emissions: a scientometric review. Sustainability 11 (14), 3972.

UN, 1993. Agenda 21, Chapter 34, United Nations Document A/CONF.151/26/Rev. 1, vol. I (Annex II.

UNFCCC, 2006. Technologies for Adaptation to Climate Change. Climate Change Secretariat (UNFCCC), Bonn.

Vincent, K., 2007. Uncertainty in adaptive capacity and the importance of scale. Global Environmental Change 17 (1), 12−24.

Watson, J.E.M., Evans, T., Venter, O. et al., 2018. The exceptional value of intact forest ecosystems. Nat Ecol Evol 2, 599−610. https://doi.org/10.1038/s41559-018-0490-x.

World Bank, 2020. GDP (Current US$). Retrieved from: https://data.worldbank.org/indicator/NY.GDP.MKTP.CD.

World Meteorological Organization, 2020. Economic Slowdown as a Result of Covid Is No Substitute for Climate Action. Retrieved from: https://public.wmo.int/en/media/news/economic-slowdown-result-of-covid-no-substitute-climate-action.

World Trade Organization, 2005. Trading into the Future. Retrieved from: https://www.wto.org/english/thewto_e/whatis_e/tif_e/understanding_text_e.pdf.

Further reading

Adger, W.N., Huq, S., Brown, K., Conway, D., Hulme, M., 2003. Adaptation to climate change in the developing world. Progress in Development Studies 3 (3), 179–195. https://doi.org/10.1191/1464993403ps060oa.

CO_2 Human Emissions, 2017. Main Sources of Carbon Dioxide Emissions. Retrieved from:https://www.che-project.eu/news/main-sources-carbon-dioxide-emissions.

Connolly-Boutin, L., Smit, B., 2016. Climate change, food security, and livelihoods in sub-Saharan Africa. Regional Environmental Change 16, 385–399. https://doi.org/10.1007/s10113-015-0761-x.

Corlew, L.K., O'Donnell, C., Charlene, B., Ashley, M., Xu, Y., Bruce, H., 2012. The Cultural Impacts of Climate Change: Sense of Place and Sense of Community in Tuvalu, a Country Threatened by Sea Level Rise.

Dumenu, W.K., Obeng, E.A., 2016. Climate change and rural communities in Ghana: social vulnerability, impacts, adaptations and policy implications. Environmental Science & Policy 55, 208–217. https://doi.org/10.1016/j.envsci.2015.10.010.

European Environment Agency, 2020. The European Environment: State and Outlook 2020. European Communities. Retrieved from: https://www.eea.europa.eu/publications/soer-2020.

Gasper, R., Blohm, A., Ruth, M., 2011. Social and economic impacts of climate change on the urban environment. Current Opinion in Environmental Sustainability 3 (3), 150–157. https://doi.org/10.1016/j.cosust.2010.12.009.

Joe, M.C., February 20, 2020. Why Climate Change and Poverty Are Inextricably Linked. Global Citizen. Retrieved from. https://www.globalcitizen.org/en/content/climate-change-is-connected-to-poverty/?template=next.

Jong, H.N., 2019. Indonesia forest-clearing ban is made permanent, but labeled "propaganda." Mongabay (Accessed 15 September 2019).

Justine Marrion, M., 2020. Climate Change, Culture and Cultural Rights. Retrieved from: https://www.ohchr.org/Documents/Issues/CulturalRights/Call_ClimateChange/JMassey.pdf.

Katherine, K., Ware, J., 2019. Counting the Cost 2019: A Year of Climate Breakdown. Christian Aid. Retrieved from: http://caid.org.uk/climate-breakdown.

Mohamed Shaffril, H.A., Ahmad, N., Hamdan, M.E., Samah, A.A., Samsuddin, S.F., 2020. Systematic literature review on adaptation towards climate change impacts among indigenous people in the Asia Pacific regions. Journal of Cleaner Production 120595. https://doi.org/10.1016/j.jclepro.2020.120595.

Nunez, C., 2019. What Is Global Warming? National Geographic. Retrieved from: https://www.nationalgeographic.com/environment/global-warming/global-warming-overview/.

Piers, B., Terry, C., Ian, D., Ben, W., 1994. At Risk: Natural Hazards, People Vulnerability and Disaster 1st Edition. https://doi.org/10.4324/9780203428764.

PONS IP, 2019. Companies Owning Intellectual Property Rights Earn 20% More Revenue Per Employee than Other Companies in EU. Retrieved from: https://www.ponsip.com/en/blog/companies-owning-intellectual-property-rights-earn-20-more-revenue-employee-other-companies-eu.

Renee, C., June 20, 2019. How Climate Change Impacts the Economy. Columbia Climate School. Retrieved from: https://news.climate.columbia.edu/2019/06/20/climate-change-economy-impacts/.

Rienne, W., Latka, C., Britz, W., 2021. Who Is Most Vulnerable to Climate Change Induced Yield Changes? A Dynamic Long Run Household Analysis in Lower Income Countries. Climate Risk Management. https://doi.org/10.1016/j.crm.2021.100330.

Roncancio, D.J., Cutter, S.L., Nardocci, A.C., 2020. Social vulnerability in Colombia. International Journal of Disaster Risk Reduction 50, 101872. https://doi.org/10.1016/j.ijdrr.2020.101872.

Tiffany, S., Lui, S., Burfitt, B., 2018. Effects of climate change on society, culture and gender relevant to the Pacific Islands. Pacific Marine Climate Change Report Card 201–210.

UN, 2019. World Population Prospects 2019 Highlights [Internet]. Available from: https://population.un.org (Accessed: 19 September 2019).

UNCTAD, 2017. So Much Done, So Much to Do (UNCTAD/DOM/2018) (Accessed 5 June 2018).

UNFCC, 2020. Report of the Green Climate Fund to the Conference of the Parties. Retrieved from: https://unfccc.int/sites/default/files/resource/cp2020_05E.pdf.
World Bank, 2021. CO_2 emissions (metric tons per capita).
Xu, X., Wang, L., Sun, M., Fu, C., Bai, Y., Li, C., Zhang, L., 2020. Climate change vulnerability assessment for smallholder farmers in China: an extended framework. Journal of Environmental Management 276, 111315. https://doi.org/10.1016/j.jenvman.2020.111315.

CHAPTER

19 Response to energy and environmental challenges: drivers and barriers

Liliana N. Proskuryakova

National Research University Higher School of Economics, Moscow, Russia

Chapter outline

1. Introduction: The main approaches to identification of grand challenges

Grand challenges tend to persist over years, and sometimes, decades. These issues are hard to address and to overcome. However, they may change or seize to exist; new issues may arise. Researcher points to the lack of methods for holistic framing of complex sociotechnical challenges (Sinfield et al., 2020). The most common approach to watch for grand challenges is trend monitoring (Cagnin et al., 2013), analysis of weak signals (Bredikhin 2020), brainstorming (such as "virtual idea blitz") (Bacq et al., 2020), and other methods applied in future studies.

Future studies and future-oriented analysis (FTA) are essential for the development of capacities to anticipate and address evolutionary and disruptive changes, including grand challenges. Regular research and analytical exercises that imply trend and weak signal monitoring make organizations more adaptive and prepared to coming changes through the introduction of new solutions—technologies and business models. In addition to FTA activities, identification of grand challenges is easier with networking and cooperation within and across organizations, stakeholders, and societies (Cagnin et al., 2013).

A variety of methods associated with imagination may also be instrumental in the identification of grand challenges, namely gaming, science fiction, vision-building, and scenario analysis (Bina et al., 2017; Ingeborgrud, 2018). As it is the case in foresight and some other participatory studies, the

Handbook of Energy and Environmental Security. https://doi.org/10.1016/B978-0-12-824084-7.00020-5
Copyright © 2022 Elsevier Inc. All rights reserved.

process of identifying the grand challenges could be more important that the outcome. In the course of the exercise, its participants have to agree on the vision of desired future, strategic goals, and action plans on its achievement. In this setting, the grand challenges would represent barriers for making the future vision come true.

Pressure-State-Response (PSR) framework is the approach offered by Rapport and Friend (1979) and adopted by the Organization for Economic Cooperation and Development (OECD) to conceptualize environmental, sustainability, and climate-related problems and solutions, as well as outline measurement (performance) approaches (OECD 1993). The PSR framework (Fig. 19.1) reflects causal relations between human activities (economic activity that harms the environment and policy measures aimed at environment protection) and the changes in the environment and state/quality of natural resources. This framework is particularly helpful for formulating responses to grand energy and environmental challenges as it underlines the cause—effect relationships and shows the interconnection of environmental, economic, societal, and other issues (Smith et al., 2014). This general model was revisited by OECD to incorporate the development and classification of environmental indicators, thus making it widely known and applicable across the world. The PSR model was later advanced and

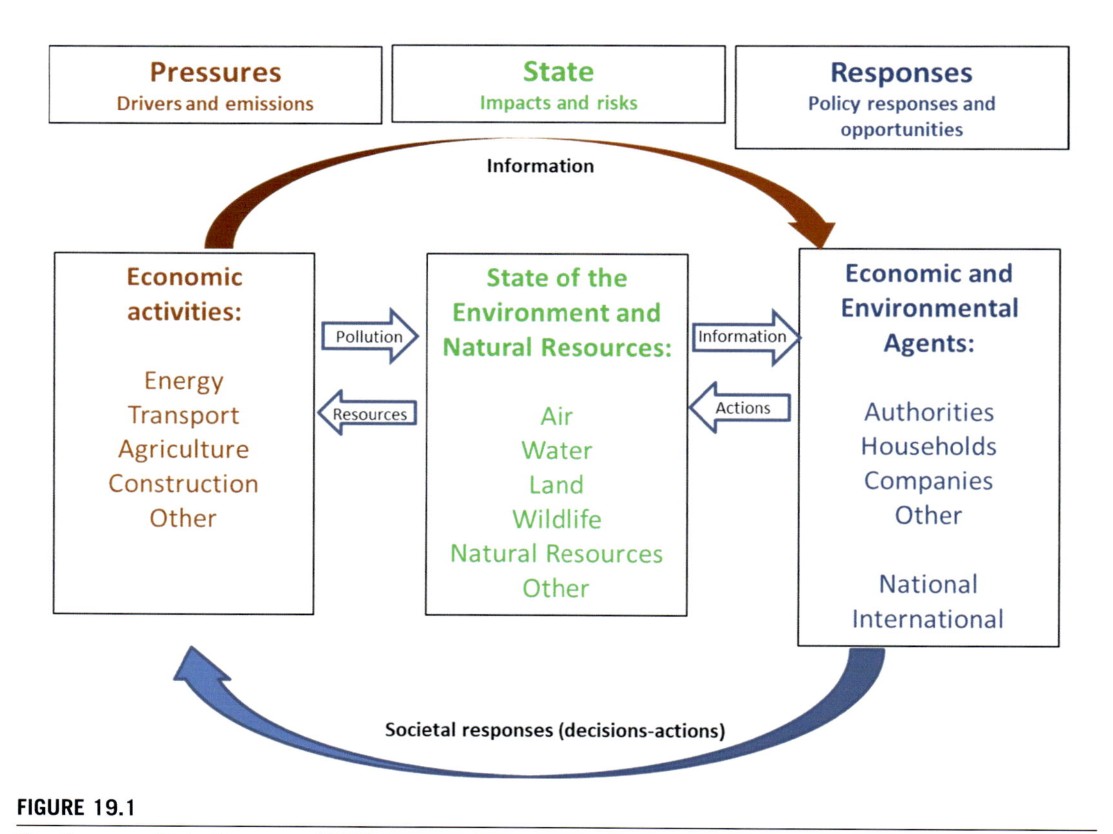

FIGURE 19.1

The Pressure-State-Response (PSR) framework.

Adapted from OECD (1993).

applied by various international organizations for policy- and action-oriented responses to grand challenges: the US Environmental Protection Agency (EPA 1994), UNEP (1994) and other. In 2021 International Energy Agency (IEA) and other international organizations employed this tool to address the new challenge requiring measurement and policy responses—climate change (OECD, 2021).

The PSR framework was further extended by the European Environment Agency (EEA) and transformed into a more comprehensive DPSIR (drivers, pressures, state, impact, and response model of intervention) model. This causal framework is similarly used for describing the bidirectional interactions between society and the environment, and the identification of correct and effective responses (EEA, 1999). This model has been further advanced to take into account more complex interactions between the model components, track progress in addressing the problems (challenges), and tackle multiple problems (challenges) (Smith et al., 2014).

DPSIR provides a means of selecting and organizing data/indicators in a way useful for decision-makers and the public. By highlighting the relationships between the environment and economic dimensions of sustainable development, it also helps policy-makers design policies that address problems at the appropriate level. There is now a further need to extend the PSR framework to cover the environmental/social interface of sustainable development in order to better track the course toward a sustainable future.

Response to grand energy and environmental challenges requires the concerted efforts of international organizations, national governments, business, civil society organizations (CSOs), and society at large. Media plays a crucial role in addressing grand challenges through awareness rising and forming people's opinions and attitudes, as well as communicating research outcomes to nonexpert audience. The possibility to address grand challenges through science has increasingly attracted the attention of researchers over the last decade (Efstathiou, 2016).

Key stakeholders involved in the identification and formulating response to grand challenges are international organizations. They are best suited to mobilize the necessary resources and actors, assure wide coverage and legitimacy. One example is the Urban Climate and Resiliency-Science Working Group that was established in 2018 for scientific analysis of grand challenges related to climate resilience of cities (González et al., 2021). Other examples include the Millennium Ecosystem Assessment, Intergovernmental Panel on Climate Change, and Intergovernmental Platform on Biodiversity and Ecosystem Services (Suni et al., 2016).

International organizations, especially the intergovernmental ones (IGOs), have the capacity to make a major input in system transformations to respond to grand challenges. By creating regional and global norms and regulations, these organizations could provide the environmental conditions for the necessary solutions to be created and applied. IGOs could offer new vision, instruments, and practices, as well as harmonize their applications across countries (Nilsson, 2017).

Other important actors are governments. They could identify and monitor grand challenges, as well as set visions through national foresight exercises. To enforce the desired solutions, national agencies need to integrate foresight outcomes in national strategic planning (Proskuryakova, 2019) and enforce them through instruments like sustainable procurement (Pot, 2021). As grand challenges are of transnational (transboundary) nature, they could not be tackled by any government alone, even a member of G7. Therefore, international cooperation efforts have to be reflected in foreign and development assistance policies.

Companies, especially multinationals, also play a role in the identification and response to grand challenges. This may be done in many different ways, including responsible research and innovation aimed at addressing sustainability challenges (Stahl et al., 2019), ethical organizational innovation, and new extended reporting that goes beyond existing Environmental Social and Governance indexes to measure companies' harmful impacts. Additional steps may be taken to involve private sector organizations in addressing grand challenges in addition to the already existing international regulatory frameworks for global business (such as 2011 UN Guiding Principles on Business and Human Rights), as well as voluntary business initiatives (such as Global Compact) (Giuliani, 2018).

There are stakeholders in the third sector as well. These include CSOs, social enterprises, and grassroot initiatives that are rarely involved in the identification of grand challenges and the search for responses. Nevertheless, their role is important in implementing sustainable solutions (including eco-innovations), formulating green and inclusive approaches to economic activities, particularly the transition to circular economy (Colombo et al., 2019). The involvement of CSOs may be enhanced through the use of latest technologies, communicating with local communities through awareness rising, and organization of community action (Mohan et al., 2021).

2. Contemporary energy and environmental challenges and their implications for security

There are several energy and environmental grand challenges that are typically addressed by policymakers and researchers. Most commonly these issues correspond to the Sustainable Development Goals and other major international agreements (i.e., the Paris Agreement). These are discussed in Chapter 28 "Energy and Environment: Sustainable Development Goals and Global Policy Landscape." Among the frequently addressed issues are: making energy sources clean and affordable, limiting environmental degradation and loss of biodiversity, addressing the depletion of natural resources, preventing and mitigating climate change, reusing and recycling of waste. These challenges and their implications for security are reviewed below and summarised in Table 19.1.

Clean and affordable energy is in the focus of the seventh Sustainable Development Goal (SDG7). SDG7 is integrated in national strategies and plans, and serves as land mark for nongovernmental actors. Researchers from different fields and backgrounds contribute to addressing this challenge through new technologies and applications (Lee et al., 2015). Clean energy is associated with decreasing energy intensity and energy saving, diversifying energy mix by adding renewable energy sources that should partially or fully substitute fossil fuels. Both diversification of energy mix and lowering the dependence on fossil fuels (with price fluctuations, projected depletion, and export barriers) are associated with higher energy security. Moreover, harvesting and use of clean energy sources generate very little or zero emissions, and, consequently, have very low or zero impact on human health and environment. This levels up the quality of life and increases social and environmental security.

There is a close interlink between availability and quantity of natural resources, economic globalization, and consumption of different energy resources. The analysis undertaken in fossil fuel-rich economies shows that economic globalization and renewable energy consumption mitigate emissions, while economic growth, urbanization, and nonrenewable energy consumption significantly worsen environmental quality (Majeed et al., 2021).

Climate change has arguably become the key challenge of the 21st century due to the magnitude of expected negative consequences. If the most devastating scenarios come true, climate change will have

Table 19.1 Summary of energy and environmental grand challenges and the implications of related responses for security.

Grand challenge	Implications for energy security		Implications for environmental security		Other security implications (human, food security, climate change, etc.)	
	Positive	Negative	Positive	Negative	Positive	Negative
Clean and affordable energy	Diversification of energy mix Lowering dependence on hydrocarbon imports Lowering the negative impact of energy price fluctuations	Variable nature of renewables, their dependence on weather Lower efficiency of renewables compared to traditional sources Renewables require rare earth elements	Lower or zero emissions related to clean energy extraction (harvesting), transportation, and use	Emissions from clean energy equipment manufacturing and disposal Some impact on wildlife	Lower or zero negative impact of clean energy on climate Lower or zero of clean energy impact on human health	Unemployment in traditional energy sectors Lack of access to clean energy technologies for poor countries/ communities
Environmental degradation and loss of biodiversity	The search for new clean and efficient energy sources, technologies, and materials to compensate for natural resource depletion and take over new markets	Additional costs for energy companies to mitigate their negative environmental impacts	Environment conservation, lowering the negative impacts on environment and biodiversity	Limited resources cause the need to prioritize environmental threats: only a few may be fully addressed	Higher quality of life, better ecosystem services Positive impact on food security	Additional costs for energy companies to mitigate their negative social impacts
Depletion of natural resources	New technologies for exploration and extraction of unconventional reserves The shift to	Lower natural resource base Increasing prices of natural resources	Lower use of natural resources leads to lower emissions and waste, lower impact on wildlife	Extraction of unconventional resources may lead to higher negative impact on environment and biodiversity	Switch to cleaner alternatives with lower impact on environment, climate, and human health	Lower access or less affordable natural resources or related products Extraction of unconventional resources may

Continued

Table 19.1 Summary of energy and environmental grand challenges and the implications of related responses for security.—cont'd

Grand challenge	Implications for energy security		Implications for environmental security		Other security implications (human, food security, climate change, etc.)	
	Positive	**Negative**	**Positive**	**Negative**	**Positive**	**Negative**
	clean energy sources					lead to negative social impacts (due to contamination of soil and water)
Increasing volume and rate of waste generation	Scaling up waste-to-energy solutions More efficient use of waste heat and other resources; lean production	Higher fines and penalties for energy companies that may be included in final energy prices	Waste threats lead to advancing environment conservation and protection	Higher negative impact on environment and biodiversity	More public awareness of waste problems; recycling and reuse of materials, resources, and goods	Environment pollution leads to lower food and human security, as well as negative impact on wildlife

Source: Author's own compilation based on Le et al., 2015; Reinhart et al., 2016; Hussain et al., 2020; Scherer et al., 2020; Huang et al., 2020; Stringer et al., 2021; Abeldaño Zuñiga et al., 2021; Majeed et al., 2021; Proskuryakova, 2021; Proskuryakova and Loginova, 2021; Proskuryakova 2022.

severe repercussions on water security, food security, environmental, and social security (Huang et al., 2020; Stringer et al., 2021; Abeldaño Zuñiga et al., 2021). This global challenge has close interconnections with the other global challenges, as the impact of climate change has adverse effects for the environment, biodiversity, agriculture, habitat, and more.

Environmental degradation and loss of biodiversity occur primarily due to anthropogenic activity and threaten ecosystems and societies across the world. The associated security threats include pollinator loss, nutrient depletion, soil compaction, ocean acidification, and inappropriate land and sea use or overuse (that lead to habitat degradation). In order to minimize the negative and avoid catastrophic impacts, strategic regulations and interventions are required. In order to prioritize security threats that have to be addressed through actions and limited resources, three criteria may be applied—importance, neglect, and tractability. Research and policy priorities have to consistently assess and address these priority areas together in order to effectively respond to global environmental challenges (Scherer et al., 2020).

Addressing the depletion of natural resources is not only a factor of economic development (energy use, in particular). Natural resource depletion positively and significantly influences CO_2 emissions: a 1% increase in natural resource depletion could increase CO_2 emissions by 0.0286%. Other factors that positively influence CO_2 emissions are urbanization, economic growth, and trade openness (Hussain et al., 2020). Studies that address the problem of depleted domestic fossil fuels through imported hydrocarbons show higher environmental burden and varied impacts for different industries: regional, national, or even global (for exporters). For some time, it would be possible to extend the natural resource use, for instance, by introducing a cap for different consumers (Kumar et al., 2019). However, in the long run, the solutions should focus on restructuring the economy and switching to alternative (renewable and environment friendly) resources.

Growing volumes and increasing rate of waste generation, including municipal solid waste, present significant threats to personal (health) and environment (land, water, and air) security. The adverse effects at all stages of waste handling (collection, processing, and disposal) coupled with increasing resource mining and urbanization magnify these impacts. The largest share of waste is generated by extractive industries, and this fact adds one more argument to rethinking of resource use in search of potential solutions to this grand challenge (Reinhart et al., 2016). The waste management reform should start with high level legislative and regulatory acts, that shoud further be translated into strategies and action plans, and implemented through national initiatives (projects). Special attention should be paid to awareness rising, separate waste collection, more efficient resource use, and various waste recycling solutions (Proskuryakova, 2021).

3. Key drivers for resolution of energy and environmental challenges

The main drivers that help addressing energy and environmental challenges include the existing international agreements (regional and universal), cities' and regions' contributions (such as C40 Cities Climate Leadership Group or zero emission targets set by California and Scotland), business initiatives (such as RE100 and Climate Smart Agriculture LCTPi), and civil society campaigns (such as Earth Hour). The role of international law is highly important: the international agreements are discussed in Chapter 28 "Energy and Environment: Sustainable Development Goals and Global Policy Landscape."

However, it is important to go beyond the international regulations and national agencies, and involve subnational actors. The summary of drivers discussed below is presented on Fig. 19.2.

The subnational actors range from cities and regions to companies and research teams. Their potential for climate mitigation is tentatively assessed as substantial—up to 20 Gt CO_2 equivalent, which is much higher than existing Nationally Determined Contributions (NDCs[1]) (Hsu et al., 2020). To unleash this potential, governmental actors should offer information support and enabling environment to support the full implementation of subnational projects, adding up to national and international efforts. Bottom-up initiatives at grassroot, local, and regional level are rarely counted, but have the potential to make a substantial contribution to addressing energy and environmental challenges, such as climate change.

Global energy and environmental governance previews participation of new subnational actors and their interaction with traditional ones. The new energy and environmental challenges also imply new issues that have to be addressed by governance institutions, including new communications and processes. Many issues in this area still have to be addressed by researchers: categorization of actors,

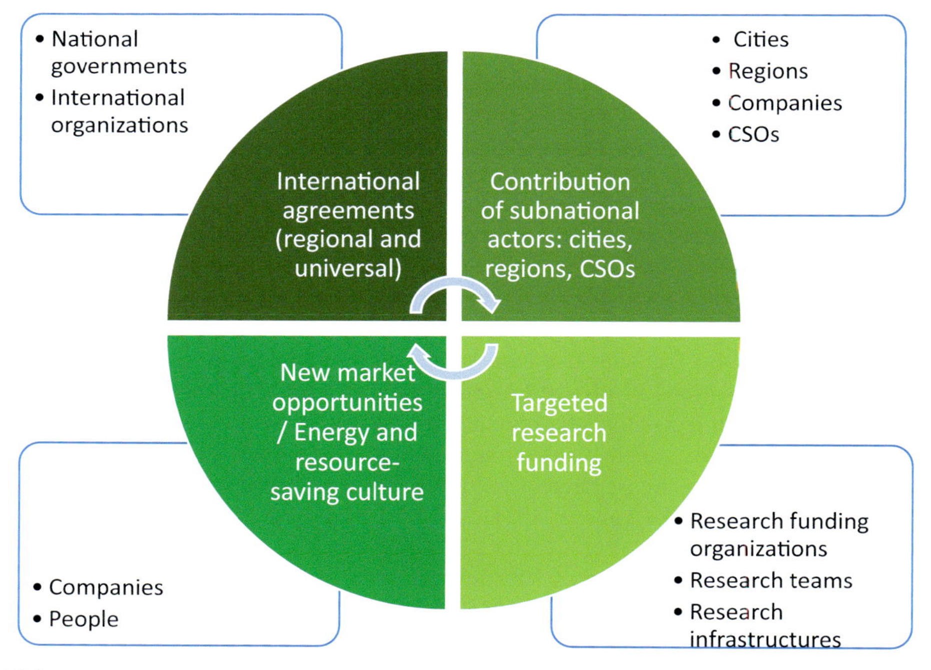

FIGURE 19.2

Key drivers that support the responses to grand challenges and related actors.

Source: Author's own analysis based on Proskuryakova, 2018; Hsu et al., 2020; Overland and Sovacool, 2020; Gu et al., 2020; Nchanji and Lutomia, 2021.

[1]The Paris Agreement requires each country to set and report their post-2020 climate actions, referred to as their NDCs.

analysis of their roles and impact, and assessment of effectiveness of institutions and the entire governance ecosystem in addressing the thematic and governance challenges (Hsu et al., 2020).

An important driver is the allocation of financial resources by governments and private donors to support energy and environment-related research and breakthrough solutions. Although funding priorities usually cover key energy and environmental challenges, there are a number of biases that hinder the effect of this driver. First, the small number of technology priorities receives the bulk of research budgets. On the one hand, it is good, as it allows concentrating funds to quickly achieve a breakthrough. On the other hand, many technology solutions are left with very small or no funding. For example, in renewable energy sector, the funding is focused on solar and wind, while other renewable power technologies receive much smaller support.

Part of these solutions to new challenges has to be identified through social sciences that are typically left out when energy and environmental research funding is distributed. The analysis of donor funding distribution shows that social sciences often receive less than 0.2% of all funding. More equitable and mindful allocation of funding would drive the reposes to energy and environmental challenges. It implies allocation of resources proportionally to the magnitude of challenge, aligning finance priorities with the main trends, more funding transparency and coordination, and earmarked support for social sciences research (Overland and Sovacool, 2020).

Advancing the energy and resource-saving culture changes people's behavior and attitudes that are key to more sustainable practices. However, the positive interlink between resource scarcity and resource-saving and proenvironmental behavior does not automatically appear in any country or social group. This link has emerged only in countries with future-oriented culture and for individuals with future-oriented traits (Gu et al., 2020). Other factors that are associated with energy and resource-saving behavior of individuals are environmental responsibility and concerns about environmental issues and health. Different resource-saving behaviors of the analysis revealed that energy and resource-saving behaviors are influenced by different goals and their cost was different. These conclusions offer direct implications for policymakers who wish to promote citizens' energy and resource-saving behavior (Liobikienė and Minelgaitė, 2021).

New market opportunities are a powerful driver to address grand challenges through new technologies, goods, and services. For instance, Schirrmeister and Warnke (2013) outline "waste-based (open) innovation" and "city-level open innovation platform" as examples of such opportunities that are suitable for aligning technological and social innovations in order to execute a structural transformation. The green innovations diffusion is assisted by technology transfer, development assistance, knowledge (R&D) spillovers, and learning-by-doing (Verdolini and Galeotti, 2011; Popp, 2019). Eventually, grand challenges may change the production and consumption patterns by adding the sustainability approach (Nchanji and Lutomia, 2021).

4. Key barriers for addressing energy and environmental challenges

The first barrier to addressing energy and environmental issuesis linked with defining grand challenges and framing them. The expert methods for holistic framing of complex sociotechnical challenges could be biased. Therefore, researchers are looking for new approaches, such as big data analysis (Gokhberg et al., 2020) and comprehensive success factor analysis (Sinfield et al., 2020). Grand challenges go beyond borders, across industries and societies that makes it difficult to design and implement policy responses. Moreover, several grand challenges could be intertwined, as it is the case with energy and environmental challenges (Wang et al., 2019). Other barriers and possible approaches to overcome them are discussed below and summarized in Fig. 19.3.

An important barrier to finding responses to grand challenges is proper assessment of the challenges, their sources, and magnitude. The concept of footprint applicable to the environment, water, and carbon helps in overcoming this barrier. It tells that the problem may develop in nonobvious locations, in places of consumption (not extraction). The complexity also lies in the "fundamental interdisciplinary and integrative nature" of footprint studies, which makes them difficult to plan, execute, and translate into policy and action (Hogeboom, 2020).

Other key barriers to addressing energy and environmental challenges are industrial path dependency and orientation on economic (GDP) growth as the most important national goal. Path dependency arises from inertia in traditional industries with long investment cycles, like energy. It is also grounded in urgent decisions that have to be made under pressure and is applicable to corporate and national policy-making. Under pressure, decision-makers are likely to adhere to well-known solutions and are not looking for long studies and evaluations. This is particularly a barrier for mission-oriented innovation policy that should be put in place to address grand challenges (Reale, 2021).

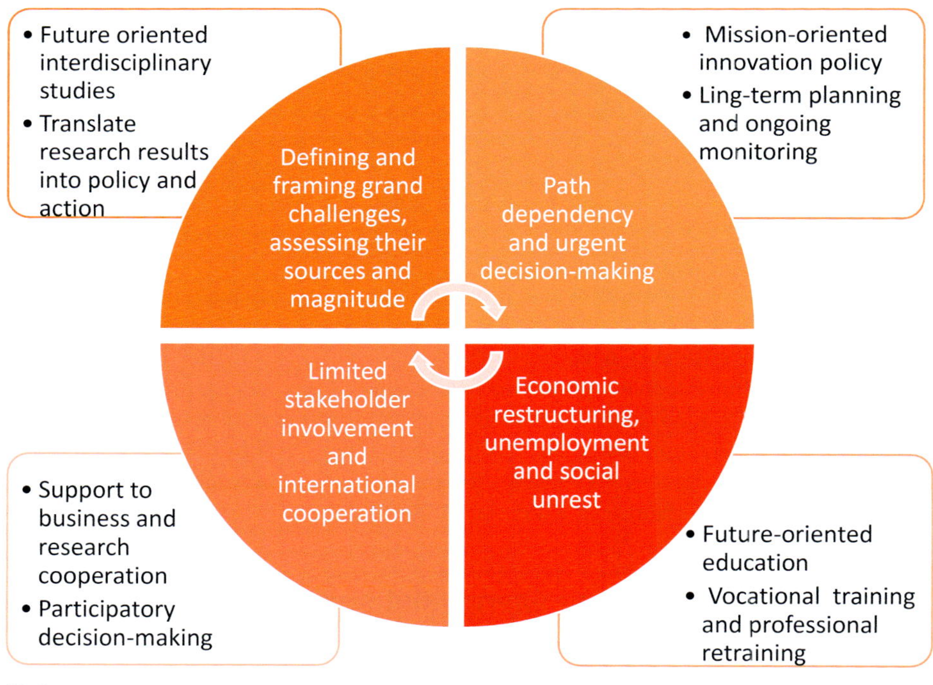

FIGURE 19.3

Key barriers that support the responses to grand challenges and related responses.

Source: Author's own analysis based on de Montluc, 2010; Giannopoulos, 2017; Hake and Proskuryakova, 2018; Smol et al., 2018; Proskuryakova et al., 2018; Wang et al., 2019; Hogeboom, 2020; Reale, 2021.

Little public awareness about energy and environmental challenges prevents putting in place the solutions. Fist, consumers are less inclined to change behavior patterns, adhere to resource-saving and resource efficient actions that may gradually establish new cultural practices. Second, with low awareness, people cannot offer new solutions that nourish user innovation and grassroot initiatives. No doubt that public awareness is one of the major driving forces in energy transition and the shift to the circular economy (Smol et al., 2018).

In addition to international agreements, comprehensive responses to grand challenges require international cooperation. Lack or cross-border cooperation at all levels could be a barrier. Intergovernmental cooperation and state support to business cooperation, particularly in monopolistic industries, is a prerequisite for large-scale international projects. This high-level cooperation is prone to many macro-level risks associated with international politicy and economic relations (de Montluc, 2010).

Researcher responses to grand challenges are also best addressed by international research consortia, and research isolationism is a barrier. National efforts are no longer enough to tackle complex research problems. The benefits of international research teams include multinational and interdisciplinary perspectives, more substantial resources (financial, research infrastructure, and knowledge platforms), and other. To overcome this barrier, national governments have to join international research programs or make other arrangements to make international research cooperation available to scientists (Giannopoulos, 2017).

Green transition previews structural changes in the economy. While new green sectors are created, some sectors are restructured or disappear. Unemployment and related social unrest could be a significant barrier on this way. A number of new educational programs will be developed, while only a share of employees could afford professional retraining or their employers will pay for it. At the same time, state support is required for professional and vocational training of low-income individuals so that they could also benefit from economic development opportunities of the low-carbon transition (Wang et al., 2019).

5. Designing responses to grand challenges

Responses to grand challenges require transitions in the energy and other sectors and domains with a cross-sectoral approach and comprehensive policy planning. The nature of grand challenges related to energy and environment requires cooperation and coordination of sectoral policies and governance levels. Evidence-based policy measures should be designed based on the outcomes of future studies and common visions developed through foresight exercises. These exercises typically employ a range of methods, including trend monitoring as the main method to identify grand challenges. Finally, decision-makers have to be prepared to accept recommendations that may suggest fundamental changes, e.g., related to a radical redistribution of resources or authority.

Responses to grand challenges have to be divided into several comprehensive and complex tasks that could be addressed simultaneously and/or in parallel. Some tasks will be directed toward pulling together international efforts through international agreements and programs. Other tasks should be directed toward involving a multitude of different actors, including government agencies, business, regions and cities, and CSOs. The collective input of all these actors in addressing grand challenges could be significant, but requires enabling environment and a new governance framework. Researchers could support policy-makers with evidence base, but research priorities have to be identified and research funding has to be allocated in a more equitable (across disciplines and technology groups) and transparent way, imply more coordination and future trend orientation.

Resource-saving culture is grounded on every-day behavioral practices that are influenced by people's beliefs and goals: orientation toward future, environmental responsibility, and concerns about environmental issues and health. Resource-saving culture, values, and practices represent an important driver for addressing grand challenges. Other drivers include international agreements, cities' and regions' contributions, business initiatives and civil society campaigns, allocation of funding for researching grand challenges, advancing resource-saving culture, and grasping new market opportunities.

Industrial path dependency is particularly noticeable in traditional industries like energy and metallurgy that are among the largest pollutants. It is a barrier not only for environmental security but also for energy security, as conservative thinking prevents companies from innovative behavior under the conditions of natural resource depletion. The paramount importance of climate change and its consequences has shifted attention from economic to sustainable development and green growth. Other barriers to address global challenges include lack of stakeholders' involvement and cross-border cooperation.

Drivers that facilitate and barriers that hinder the responses to grand challenges are numerous and multilevel. They have to be identified and systematized for each challenge that is being tackled. Here, again, future studies are instrumental to map all issues and factors; identify and involve all stakeholders; support the consensus on future vision; identify research, technology, social, and business activities, and policy action.

6. Conclusions

There are multiple energy and environmental challenges faced by the humankind in the 21st century. However, grand challenges are few and include assuring clean and affordable energy, environmental degradation and loss of biodiversity, depletion of natural resources, and increasing volume and rate of waste generation. Approaches to addressing these challenges should be well thought and comprehensive. They may be identified with future studies or existing tools for complex systems analysis. The implementation of these approaches is assisted by a set of drivers and hindered by a number of barriers.

Grand energy and environmental challenges should be addressed by cumulative efforts of a variety of stakeholders—international organizations, national governments, business, CSOs, and society at large. Both top-down and bottom-up initiatives are instrumental for tackling global problems. There is no single solution that may be effectively implemented by a one actor or even few actors. Most grand challenges require international regulations and cooperation, as they go beyond national borders. Other important drivers are the actions of subnational actors, new market opportunities, and new cultural patterns.

The barriers to addressing a grand challenge start with its very definition and framing, identifying their sources and magnitude. The effects and implications of grand challenges are numerous, and it is hard to grasp them all and analyze in a systematic manner. When it comes to implementation of proposed solutions, designed actions confront with path dependency and urgent decision-making. It is also difficult to address grand challenges in unstable economic situation or social setting, with lack of stakeholder involvement, and limited international cooperation.

Possible responses to grand challenges require structural changes in the energy and other domains with a cross-sectoral perspective and integrated policy planning. Policy-making should rely on evidence from future studies and common visions developed by stakeholders. Decision-makers need to refrain from short-sighted approach and be prepared to take difficult decisions. People and companies should be open to changes and be ready to change their every-day resource consumption, mobility, and other habits and practices.

Acknowledgments

The chapter is based on the study funded by the Basic Research Program of the National Research University Higher School of Economics.

References

Abeldaño Zuñiga, R.A., Lima, G.N., González Villoria, A.M., 2021. Impact of slow-onset events related to Climate Change on food security in Latin America and the Caribbean. Current Opinion in Environmental Sustainability 50, 215–224.

Bacq, S., et al., 2020. The COVID-19 Virtual Idea Blitz: Marshaling social entrepreneurship to rapidly respond to urgent grand challenges. Business Horizons 63 (6), 705–723.

Bina, O., et al., 2017. The future imagined: exploring fiction as a means of reflecting on today's Grand Societal Challenges and tomorrow's options. Futures 86, 166–184.

Bredikhin, S.V., 2020. Approaches to disruptive change: the contribution of complexity science to futures studies. Futures 124, 102624.

Cagnin, C., Havas, A., Saritas, O., 2013. Future-oriented technology analysis: its potential to address disruptive transformations. Technological Forecasting and Social Change 80 (3), 379–385.

Colombo, L.A., Pansera, M., Owen, R., 2019. The discourse of eco-innovation in the European Union: an analysis of the eco-innovation action plan and horizon 2020. Journal of Cleaner Production 214, 653–665.

de Montluc, B., 2010. Russia's resurgence. Prospects for Space Policy and International Cooperation 26 (1), 15–24.

EEA, 1999. State and Pressures of the Marine and Coastal Mediterranean Environment. EEA, Copenhagen, p. 44 pp. FAO. https://www.eea.europa.eu/publications/medsea/download.

Efstathiou, S., 2016. Is it possible to give scientific solutions to Grand Challenges? On the idea of grand challenges for life science research. Studies in History and Philosophy of Science C: Studies in History and Philosophy of Biological and Biomedical Sciences 56, 48–61.

EPA, 1994. A conceptual framework to support the development and use of environmental information. In: Environmental Statistics and Information Division. Office of Policy, Planning and Evaluation. EPA 230-R-94-012. USEPA, Washington DC.

Giannopoulos, G.A., 2017. Strategic management and promotion issues in international research cooperation. Case Studies on Transport Policy 5 (1), 9–21.

Giuliani, E., 2018. Regulating global capitalism amid rampant corporate wrongdoing—reply to "Three frames for innovation policy". Research Policy 47 (9), 1577–1582.

Gokhberg, L., Kuzminov, I., Khabirova, E., Thurner, T., 2020. Advanced text-mining for trend analysis of Russia's extractive industries. Futures 115, 102476.

González, J.E., et al., 2021. Urban climate and resiliency: a synthesis report of state of the art and future research direction. Urban Climate 38, 100858.
Gu, D., et al., 2020. Concern for the future and saving the earth: when does ecological resource scarcity promote pro-environmental behavior? Journal of Environmental Psychology 72, 101501.
Hake, J., Proskuryakova, L., 2018. New Energy Sources, Technologies, and Systems: The Priority of Social, Climate, and Environmental Issues. Foresight and STI Governance 12 (4), 6–9.
Hogeboom, R.J., 2020. The water footprint concept and water's grand environmental challenges. One Earth 2 (3), 218–222.
Hsu, A., et al., 2020. Beyond states: harnessing sub-national actors for the deep decarbonisation of cities, regions, and businesses. Energy Research and Social Science 70, 101738.
Huang, J., et al., 2020. Declines in global ecological security under climate change. Ecological Indicators 117, 106651.
Hussain, J., Khan, A., Zhou, K., 2020. The impact of natural resource depletion on energy use and CO2 emission in Belt & Road Initiative countries: a cross-country analysis. Energy 199, 117409.
Ingeborgrud, L., 2018. Visions as trading zones: national and local approaches to improving urban sustainability. Futures 96, 57–67.
Kumar, V.V., Hoadley, A., Shastri, Y., 2019. Dynamic impact assessment of resource depletion: a case study of natural gas in New Zealand. Sustainable Production and Consumption 18, 165–178.
Lee, D.-J., et al., 2015. Clean, efficient, affordable and reliable energy for a sustainable future. Energy Conversion and Management 102, 1–3.
Liobikienė, G., Minelgaitė, A., 2021. Energy and resource-saving behaviours in European Union countries: the Campbell paradigm and goal framing theory approaches. Science of the Total Environment 750, 141745.
Majeed, A., et al., 2021. Modeling the dynamic links among natural resources, economic globalization, disaggregated energy consumption, and environmental quality: fresh evidence from GCC economies. Resources Policy 73, 102204.
Mohan, M., et al., 2021. Afforestation, reforestation and new challenges from COVID-19: thirty-three recommendations to support civil society organizations (CSOs). Journal of Environmental Management 287, 112277.
Nchanji, B.E., Lutomia, C.K., 2021. COVID-19 challenges to sustainable food production and consumption: future lessons for food systems in eastern and southern Africa from a gender lens. Sustainable Production and Consumption 27, 2208–2220.
Nilsson, A., 2017. Making norms to tackle global challenges: the role of Intergovernmental Organisation. Research Policy 46 (1), 171–181.
OECD, 1993. OECD core set of indicators for environmental performance reviews. In: A Synthesis Report by the Group on the State of the Environment. OECD, Paris.
OECD, 2021. International Program for Action on Climate. https://www.oecd.org/climate-change/ipac/.
Overland, I., Sovacool, B.K., 2020. The misallocation of climate research funding. Energy Research and Social Science 62, 101349.
Popp, D., 2019. Environmental Policy and Innovation: A Decade of Research. National Bureau of Economic Research. Working Paper Series, No. 25631.
Pot, W.D., 2021. The governance challenge of implementing long-term sustainability objectives with present-day investment decisions. Journal of Cleaner Production 280, 124475. P 2.
Proskuryakova, L., 2018. Updating energy security and environmental policy: energy security theories revisited. Journal of Environmental Management 223, 203–214. https://doi.org/10.1016/j.jenvman.2018.06.016.
Proskuryakova, L., 2019. Foresight for the 'energy' priority of the Russian science and technology strategy. Energy Strategy Reviews 26, 100378.

Proskuryakova, L.N., 2021. Policy and governance for waste management in Russia. In: Singh, P., Milshina, Y., Tian, K., Borthakur, A., Verma, P., Kumar, A. (Eds.), Waste Management Policies and Practices in BRICS Nations. CRC Press. Ch. 13.

Proskuryakova, L., 2022. The interaction of environmental systems and human development in a time of wild cards. A big data enhanced foresight study. Journal of Environmental Management. In preparation.

Proskuryakova, L., Loginova, I., 2021. Energy and Environment: Sustainable Development Goals and Global Policy Landscape. In: Singh, Pardeep, Milshina, Yulia, Tian, Kangming, Borthakur, Anwesha, Verma, Pramit, et al. (Eds.), *Energy and Environmental Security in Developing Countries*. Springer Nature, pp. 355–374.

Proskuryakova, L., Saritas, O., Sivaev, S., 2018. Global water trends and future scenarios for sustainable development: The case of Russia. Journal of Cleaner Production 170, 867–879.

Rapport, D., Friend, A., 1979. Towards a Comprehensive Framework for Environmental Statistics: A Stress -response Approach. Statistics Canada Catalogue 11-510. Minister of Supply and Services Canada, Ottawa.

Reale, F., 2021. Mission-oriented innovation policy and the challenge of urgency: lessons from Covid-19 and beyond. Technovation 107, 102306.

Reinhart, D., Bolyard, S.C., Berge, N., 2016. Grand Challenges – management of municipal solid waste. Waste Management 49, 1–2.

Scherer, L., et al., 2020. Global priorities of environmental issues to combat food insecurity and biodiversity loss. Science of the Total Environment 730, 139096.

Schirrmeister, E., Warnke, P., 2013. Envisioning structural transformation — lessons from a foresight project on the future of innovation. Technological Forecasting and Social Change 80 (3), 453–466.

Sinfield, J.V., Sheth, A., Kotian, R.R., 2020. Framing the intractable: comprehensive success factor Analysis for grand challenge. Sustainable Futures 2, 100037.

Smith, C., et al., 2014. Conceptual Models for the Effects of Marine Pressures on Biodiversity. Technical Report. Devotes Projects. http://www.devotes-project.eu/wp-content/uploads/2014/06/DEVOTES-D1-1-Conceptual Models.pdf.

Smol, M., et al., 2018. Public awareness of circular economy in southern Poland: case of the Malopolska region. Journal of Cleaner Production 197, 1035–1045. P 1.

Stahl, B.C., et al., 2019. Ethics in corporate research and development: can responsible research and innovation approaches aid sustainability? Journal of Cleaner Production 239, 118044.

Stringer, L.C., et al., 2021. Climate change impacts on water security in global drylands. One Earth 4 (6), 851–864.

Suni, T., et al., 2016. National Future Earth platforms as boundary organizations contributing to solutions-oriented global change research. Current Opinion in Environmental Sustainability 23, 63–68.

UNEP, 1994. World Environment Outlook: Brainstorming Session. ENEP/EAMR, Nairobi, 94-5. Environmental Assessment Programme.

Verdolini, E., Galeotti, M., 2011. At home and abroad: an empirical analysis of innovation and diffusion in energy technologies. Journal of Environmental Economics and Management 61 (2), 119–134.

Wang, C., Guan, D., Cai, W., 2019. Grand challenges cannot Be treated in isolation. One Earth 1 (1), 24–26.

CHAPTER

Engagement and relational governance

20

Richard R. Reibstein

Environmental Law and Policy, Department of Earth and Environment, Boston University, Boston, Massachusetts, United States

Chapter outline

1. Transformation of the set of systems for living

In order to achieve energy and environmental security, government must operate to address the needs of humanity as a whole, in the context of the living biosphere. In order for that to happen, the current set of systems by which we produce the means and manner of living must change. We must cease overuse of materials and the overproduction of wastes, we must transform how we make energy so that we do not impact our climate and disperse toxic materials, and we must change how we provide food and shelter and how we conduct transportation and entertainment in ways that don't cause extinctions, degradation of ecosystems, and pollution of water, land, and air. We must reorganize so that we don't

Handbook of Energy and Environmental Security. https://doi.org/10.1016/B978-0-12-824084-7.00011-4

Copyright © 2022 Elsevier Inc. All rights reserved.

poison ourselves and our homes, and we must find ways to arrest the overcrowding of the earth. When this list of necessary actions is presented, it is tempting to throw up one's hands and say, it's all too overwhelming. It is useful to recognize that as concerns overpopulation, we know the answer, and it is within our means. There is a clear historical lesson that serves to show us that we can change the world as it can be changed: the empowerment of women, education, and increased prosperity has led to reductions in population. Developed countries where these have occurred have also seen dramatic reductions in population growth. As population can be changed by lifting people out of poverty and ignorance, and empowering them, all the other changes that need to be made can be achieved in the same way. The way forward is a better way than continuing business as usual.

2. The necessity of citizen engagement[1]

But how do we move forward? The most important key to our success is engagement. The word "citizen" means someone who belongs to a society. All people are citizens of the planet, people who belong here, and on recognition of that fact, a foundation can be laid upon which sensible societies can be built. If we recognize that we are all part of the problem and we are all part of the solution, then systemic change can occur. If everyone is involved in governance, then governance will be more likely to reflect their needs. If citizens of the world society of human species leave governance to others, then the fate of the world will depend on whether the authorities can be trusted. The concept of trust in governing authorities requires much development. Attention to this matter must be paid if the tools of governance are to be used to bring about the transformation of our systems that must occur for the rate of species extinction to be reduced, for climate change to be slowed, for the dispersion of toxics to be ceased, and for the health of all living things to be protected. People must regard themselves as citizens and concern themselves with how they are governed, and recognize that their responsibility is to act to ensure governance benefits the whole, and is not captured for the limited interests of a particular group.

2.1 Insufficiency of engagement—the governance of electricity distribution

There needs to be sufficient engagement by citizens with all levels of government so that it effectively represents common interests. To discuss particular examples, consider how the provision of electricity is governed in the United States. Even though the United States is a democracy, not many citizens have any idea. They know that they receive a bill from the electric company, but they likely do not know who or what process oversees the company's operations. In the United States, the early development of the electrical system was heavily influenced by Samuel Insull,[2] who argued that an electric company represented a "natural monopoly." This made sense in that you don't want 10 different companies stringing electrical wires through town, making a dense mess of wires. You want one company doing it. But a monopoly is in a position to control prices, and without competition can raise the price as high as it will go, and this violates the public interest. Therefore, in the United States, electric companies were brought under public utility commissions or departments of public utilities, who have had the power to approve rates, and thus the companies were not able to push the price to the highest point the

[1]This can be identified as the fifth D: Democracy.

[2]In 1892, Insull, previously Thomas Edison's private secretary, became president of Chicago's electrical utility, expanded its services, and created holding companies that bought and sold stock in electrical companies.

market could bear. The idea has been that electric companies, though owned by private investors, were allowed to make a reasonable profit, but no more than that. They were to be regulated in the public interest. During the New Deal, it became apparent that state public utility regulation was not effectively addressing interstate sales of electricity, and that the utilities were being used by investors as reliable sources of income that could be leveraged through holding companies to acquire massive amounts of debt to use in financial speculation. This was recognized as one of the causes of the 1929 stock market crash. The Federal Power Act[3] began decades of national regulation to ensure that our electrical energy system was built and operated in the public interest.

2.2 Investment incentives divergent from public interest

It is possible to look at the rise of nuclear power as an indication of how this simple principle was forgotten, because many utilities overinvested in excessively expensive systems, some of which were abandoned.[4] One cause of this phenomenon was the fact that a public utility commission would approve a reasonable rate of return from a utility's investment, and thus large capital projects brought stronger returns to investors. Another was the focus on growth in sales, which was an interest of the investor-owners and other businesses, and not necessarily that of the served community. We are right now in the process of a difficult transition, from a business model that has paid good returns to investors,[5] and which depended on a large generation station sending power out to customers (one-way), to a system in which people have solar panels on their roofs and can get paid for adding power to the system (two-way distributed generation). Because we now have the potential for energy storage with advanced batteries, we have the ability to create "microgrids," in which communities or discrete areas can have local control over the transmission of energy. Because we have intelligent systems, we can change pricing to motivate people to change their energy use, and reduce the "peak" loads—the highest amount of use at one time—in order to reduce the need for energy. These and many other changes present the opportunity for an efficient, clean system. But it makes investing in utilities a different deal. Not only will profits decrease as energy use decreases but also with much more community and user involvement and control, utilities will no longer be convenient "cash cows" for investor owners, but systems operated for the profit and good of all.

---

[3]The Federal Water Power Act of 1920 (16 USCS § 791a), pertaining to hydroelectric projects, was significantly amended in 1935 to regulate electric companies in interstate commerce. Reform targeting abuses of electrical company holding companies was accomplished the same year by the Public Utility Holding Company Act, which required registration with the Securities and Exchange Commission and disclosure of financial information. Licensing of such plants is carried out by the Federal Energy Regulatory Commission (FERC).

[4]For example, the German Economic Research Institute DIW in 2019 published a retrospective that concluded that none of the 674 nuclear plants that have been built have met competitive criteria, but had support for military or other reasons. In the United States, "Many reactors are subject to cost-plus regulation that guarantees their operators a fair financial return. The regulation's costs are grafted onto the price of electricity." https://www.diw.de/documents/publikationen/73/diw_01.c.670581.de/dwr-19-30-1.pdf. The principle of providing electricity in the public interest does not mean that public entities are reliable stewards, as for example, the ruinous investment in nuclear plants by the Washington Public Power Supply System (WPPS), a municipal corporation, which led to a default of more than $2 billion on bond payments in 1982. The need for citizen involvement to increase the chances that the public interest will in fact be served applies to public agencies as well.

[5]This history explains why getting the "utilities" card in the game Monopoly was a good strategic move.

2.3 A current example of divergence

The evolution of our electrical system from one-way centralized distribution to local control and an integrated, modernized, intelligent grid, is clearly desirable, but it is not happening quickly. In some places, utilities are actively opposing the growth of distributed energy.[6] How is it that utilities can resist the change that we so obviously need? One answer is that there is insufficient involvement by citizens. How many citizens know about public utility commissions? How many have ever attended a PUC meeting? How many can master the detailed technical information that is presented at these meetings? How many know how electricity is governed - that there are state commissions, but that many matters are handled at the federal level? For example, a group calling itself the New England Ratepayers Association but clearly representing the interests of utilities who don't want to see distributed generation come about, in April 2020 petitioned the Federal Energy Regulatory Commission (FERC) to assert authority over state "net metering" programs.[7] How many citizens know that net metering programs are the reason that people are putting solar panels on their roofs, as they govern how people will receive payment for the energy produced by those panels? For years, utilities have argued that people with solar panels should receive less money, on the basis that it is somehow unfair that they receive more than the utility receives (usually we are dealing with the difference between retail and wholesale rates). The additional argument has been made (and was made by the Ratepayers Association) that paying those who contribute energy from solar panels is unfair to the poor, who cannot afford to put panels on their roof. It is true that those wealthy enough to put panels on their roof end up in a better position than those who can't, but this is not their doing, it is a symptom of the system that now requires the bulk of the money comes from those who can't provide solar power. There is certainly inequity involved, but it is also certainly in the interest of society for much more of our energy to come from solar power, and to evolve a system in which customers also become providers. This involves changing from a system in which someone has the opportunity to extract cash, to one in which cash extraction is no longer the essential purpose of the system, but rather the provision of energy (We also need systems that ensure equitable pricing of electricity). The Ratepayers Association has asked the FERC to assert its authority so that it can mandate that all net metering programs pay no more to individual solar providers than the utilities get for providing electricity. A victory for them could destroy the economic incentive to put solar panels on roofs. That is not in the public interest. The overriding public interest, in order to save this world from drowning in rising seas and other climate change disasters, is to convert as quickly to clean energy as possible. Distributed systems are essential to our survival. But how many citizens have ever even heard of FERC?

---

[6]For example, Florida utility companies have "hindered potential rivals seeking to offer residential solar power. They have spent tens of millions of dollars on lobbying, ad campaigns, and political contributions. And when homeowners purchase solar equipment, the utilities have delayed connecting the systems for months." https://www.nytimes.com/2019/07/07/business/energy-environment/florida-solar-power.html. See also: Leah Stokes, *Short Circuiting Policy*, Oxford University Press, 2020, (Winner, Best Energy Book, 2020; American Energy Society).

[7]The petition was opposed by comments from 450 organizations supporting existing state net-metering programs that have created incentives for solar panel installations. On July 16, after receiving 57,000 opposing comments and 22 in support, FERC dismissed the petition. https://environmentamerica.org/news/ame/statement-federal-energy-regulatory-commission-acts-wisely-dismissing-new-england.

3. Sufficiency of engagement—some simple principles

To change the electrical system so that it is cleaner and not destructive of our vital environment, we need a citizenry aware enough to support experts who can represent their interests in Public Utility Commission hearings. There need to be enough people belonging to associations who will pay enough experts to counter the influence of the experts now being paid by the utilities who want to keep their current business models intact. To change the system to one that can serve the public interest, we need enough information to be generated and provided so that people will understand the importance of such things as FERC petitions that could harm us all, and how to submit comments to FERC officials, how to appear and speak at FERC hearings, and how to demand that FERC commissioners represent the public interest; and if they don't, to ensure that Congress takes action to change the statute so that the agencies regulate in the public interest.

3.1 Overcoming the barrier of complexity

The federal and state authorities are one aspect of electrical system governance. Another is the governance of the grid, an arcane and complex system that very few understand. How many people have ever heard of the meetings on capacity that the grid operators conduct? These are attended by experts, professionals in the field, and virtually unknown to everyone else. But they are critical. Understanding the simple principle that the grid needs to be modernized to allow for distributed generation—many sources such as solar panels feeding in, as well as power going out—is a touchstone for effective participation in these complex matters. For example, in many states, the utilities have denied "interconnection" to independent power sources, such as solar or wind or other generators that they don't own, because the grid does not have the capacity to handle the input of electricity from those sources. They also may charge very large interconnection fees. The utilities have been able to get legislatures to set caps on the amount of independent electricity that is fed into the system. By these means they have dramatically slowed the growth of distributed energy. It may in fact be the case that the grid cannot handle the input of energy in many places, and that the caps are necessary—as the grid exists at this time. But the grid can be modernized. The grid can be updated so that it can handle the input of distributed sources. The question then becomes whether this is happening and how we can make it happen if it is not. In Massachusetts, Democratic Governor Deval Patrick[8] had enormous success in stimulating growth in clean energy. Approximately 50,000 jobs were created, solar energy grew from 900 thousand MegaWatt hours in 2006 to over 4.7 million MegaWatt hours in 2015, and Patrick's energy officials instituted a requirement for all utilities to develop "Grid Modernization" plans. The next governor, Republican Charles Baker, has not continued this focus. Because the grid has not become more capable of receiving distributed sources, the caps on solar panels continue to slow its growth, and the potential for Massachusetts to quickly evolve a system of clean energy is stalled. At the monthly capacity meetings, the grid operators review the capacity of the system to handle new sources and limit new interconnections. How many citizens know about capacity meetings, caps, interconnections, and grid modernization planning? How many know that the DPU and FERC are supposed to govern the system in their interest and can work with the grid operators and all relevant parties to force it to evolve in the way that meets the public need? If citizens knew this, they might

[8]Governor of Massachusetts from 2007 to 2015.

press their agencies and chief executives to make the system work in the public interest. When governors are dependent on the support of powerful interests to win and stay in office, the interest of the citizen falls by the wayside. But this need not be the case, if many citizens understand the simple story of what is going on and organize to have a voice strong enough to ensure representation of their interests.

We can appeal directly to our representatives and press them to work on these issues. In the same way that we elect representatives to office in order to serve us, we can join organizations that raise money to pay experts to represent our interests in critical forums. We can become informed about which private companies are actually behaving in more responsible ways[9] and patronize them. For example, in many states, municipalities have been given the right under state law to implement Community Choice Aggregation, and the city or town can then choose clean energy providers as the default electricity generator for its residents. Citizens of municipalities without such programs can organize and work for their establishment. Citizens of states without such laws can work for their institution. States can help towns that don't have the staff to implement this. The federal government can act to make it easier for choice to be exercised.

4. Citizen engagement anywhere

In any country with democratic institutions, citizens can seek participation in government operations. The Economist Intelligence Unit (EIU), the research and analysis division of the consulting group associated with the Economist magazine, reported in 2018 that there are 20 full democracies in the world, 55 "flawed" democracies (including the United States), and 39 "hybrid" regimes. It may be assumed that in all of these countries, participation may be possible, although perhaps limited, curtailed, or ignored. But even in the 53 countries labeled "authoritarian," citizen engagement is often possible. For example, authoritarian China has seen powerful citizen movements to address environmental issues. Most striking is that the EIU's 2018 Democracy Index reported that although there was a distinct trend of decline in the functioning of democratic governance processes, that there was substantial increase in political participation throughout the world, even in countries experiencing widespread skepticism about the performance of government.

An important step was taken in the formalization of rights to engagement in governance with the UN's 1998 Aarhus Convention, which had the objective of "guaranteeing the rights of access to information, public participation in decision-making, and access to justice in environmental matters." The convention entered into force in 2001 and has 47 parties in Europe and Central Asia. The rights recognized under the convention are similar to US rights under the Administrative Procedure Act, the Freedom of Information Act, and of many other countries allowing suit by citizens against the government, transparency in governmental operations, and open meetings laws. Such suits are often limited by the concept of "standing," whether a plaintiff can stand before the court and ask for a remedy, and the Aarhus Convention enlarges this concept, providing standing to environmental organizations as well as to individuals.[10]

---

[9]For example, Green Mountain Power of Vermont, which is a "B-Corp" (socially responsible) and claims it is pursuing a new way of doing business, in order to provide clean energy. https://greenmountainpower.com/.
[10]https://ec.europa.eu/environment/aarhus/pdf/Commission_report_2019.pdf, p. 6.

4.1 Progress

Examples of suits that have made a difference are the 2019 victory by the Urgenda organization, compelling the Dutch government to accelerate its efforts to combat climate change,[11] the 1993 case by Philippine attorney Juan Antonio Oposa against the granting of licenses for excessive logging, which established the right to sue to protect ecosystems on behalf of future generations,[12] or the cases in Colombia, India, and New Zealand in which rivers (the Atrato, Ganges, and Whanganui rivers) were recognized as having natural rights, associated with the rights of plaintiffs dependent on those rivers maintaining ecological health.[13]

In a healthy democracy, citizen input is sought out—and interested parties have advance notice of regulatory determinations, so that they can provide their comments before decisions are made. Hearings are made convenient for participation, and citizens can prompt the development of new policies, as well as influence policy-making direction. In well-functioning systems, citizens have access to legislators as well as regulators, and their comments are attended to. These are fundamental components of public participation as well as access to courts, which is typically necessary only when other processes have broken down and controversies have arisen because people have not been involved or heard.

4.2 Obstacles

In many countries, bureaucracies that have been targets of corruption efforts may not have the capacity to consult or listen to citizens, because that would frustrate the designs of powerful parties seeking to bend the government agencies to their will. Agencies, chief executives, legislators, and judges may all perceive personal interest in pleasing powerful constituencies, ignoring the expressed desire or need of the people. In all of these situations, it may even be dangerous to speak out and countless examples show how deadly attempts to engage with entrenched governmental processes can be. For example, political opponents of Russian premiere Vladimir Putin have been killed or jailed with alarming frequency, the movement for democracy in Hong Kong is being slowly crushed, and the free press was falsely accused of fabricating news almost daily by President Trump, itself a fabrication. But sometimes it is the lack of government that causes engagement to be dangerous, as for example, the murders of indigenous leaders in the Amazon, by those who wish to conduct mining, cattle ranching, and other operations for commercial gain[14]; or the harassment of environmental advocates who stand in the way of development, either through litigation or destruction of personal reputation. An early example of this was the hiring of private detectives to dig up dirt on consumer advocate Ralph Nader when he exposed dangerous auto design by General Motors. His exposing of this underhanded effort turned the tide and helped win

[11]https://www.urgenda.nl/en/themas/climate-case/.
[12]https://www.elaw.org/content/philippines-oposa-et-al-v-fulgencio-s-factoran-jr-et-al-gr-no-101083.
[13]https://celdf.org/2017/05/press-release-colombia-constitutional-court-finds-atrato-river-possesses-rights/.
[14]https://news.mongabay.com/2019/12/murders-of-indigenous-leaders-in-brazil-amazon-hit-highest-level-in-two-decades/.

support for his cause, and the United States benefited from the establishment of consumer protection and environmental regulation as a result. But SLAPP suits—Strategic Litigation Against Public Participation—by major corporations have persisted, for example, by the companies producing lead products accusing lead researchers of ethical violations,[15] (unproven), or when the Daishowa paper company in Canada sued the Lubicon tribe for boycotting their products—court rulings allowed the boycott to go forward, but a 1998 ruling did not require the company to pay the tribe's costs of litigation, meaning that "you have to have the money to speak out."[16]

Despite these problems and the recent surge in authoritarian regimes worldwide, the larger trend is toward democracy. *Encyclopedia Britannica* summarizes recent history: "During the 20th century the number of countries possessing the basic political institutions of representative democracy increased significantly. At the beginning of the 21st century, independent observers agreed that more than one-third of the world's nominally independent countries possessed democratic institutions comparable to those of the English-speaking countries and the older democracies of continental Europe. In an additional one-sixth of the world's countries, these institutions, though somewhat defective, nevertheless provided historically high levels of democratic government. Altogether, these democratic and near-democratic countries contained nearly half the world's population." Citizen engagement may be difficult and dangerous in many societies, but the long-term historical trend argues for continued effort.

5. The power of basic principles

Each of the issues affecting our lives and the tasks of transforming our societies so that they are sustainable through time, without damage to the future of the biosphere and its living inhabitants, are enormously complex. But simple principles can become the foundation, the touchstone that guides us through the mazes of each topic, and the bewildering array of topics. The idea of citizen engagement is a power principle. Who has the right to use resources on which all life depends? Currently powerful interests may feel that others playing a role in decision-making is interference with their economic rights or long-standing prerogatives, but the simple, clear principles of openness, fairness, and the commonality of environmental health require that decision-making be brought out into the open. Maybe those with existing business plans will be allowed to continue—but there must be no deals behind doors that can harm others or that go against the needs of the people. These are simple principles that can be applied to complex matters. For example, electricity is a very complex matter, and involving yourself in issues about how to govern the process of providing electricity only adds to the complexity. But citizens can understand some simple points: the entire system should not serve the interests of investors, but investors should only be allowed to make a reasonable return on investments that serve the public interest. Citizens can understand that the business model right now is not serving the public interest. It is not anger but some increased sophistication that we need. We do not need to overthrow the system, but to make it work as originally intended. We can learn how it can change to better reflect everyone's common need for clean energy. Armed with this simple principle, it is possible to organize political action for the change

---

[15]Herbert Needleman, the researcher who proved the connection between lead poisoning and declines in intelligence; and historians Gerald Markowitz and David Rosner, authors of *Deadly Deceit* and other works exposing the prolead propaganda of the industry.

[16]The quote is from Kevin Thomas, spokesman for Friends of the Lubicon. The boycott was stopped for three years, and the Sierra Legal Defense Fund picked up the $400,000 cost of litigation. http://sisis.nativeweb.org/lubicon/nov1098.html.

we need. If enough people become involved, then people with expertise will likely emerge to respond to the questions asked. If enough people remain interested, then questions and proposals will be aired and reviewed. It is this public discussion, anchored by the simple vision of a just and humane system that allows reasonable returns, replacing one bound to the goal of financial gain above all, which is the key activity in which citizens can engage to change the world for the better.

6. How society can invest in its own future?

Related to this issue and to the current state of noninvolvement which allows our current dysfunction to persist, not that many know about the enormous subsidies for fossil fuels. These subsidies have been estimated by a 2019 International Monetary Fund study as $5.2 Trillion (6.5% of global Gross Domestic Product).[17] If we switched those subsidies to clean sources, we could rapidly accelerate progress in their development. Unfortunately, conservatives have managed to capture much of public opinion with the idea that government shouldn't "pick winners," but just let the marketplace decide who will prosper. But the marketplace is already greatly skewed in favor of fossil fuels by these massive subsidies. The claim of the fossil fuel supporters that government subsidy of clean energy is subverting the market is the height of hypocrisy. It is not in the public interest for fossil fuels to be used, but it is vital to our common interests that clean energy be developed. The essence of good government is that which serves the common needs. The idea that government should tie its own hands and let the current makeup of the marketplace in which certain players have attained dominant positions—rather than a rationalized marketplace managed for the benefit of the entire population—determine things is an abdication of the responsibilities we hold toward ourselves, each other, and future generations, as well as all other living things. It is a denial of the fact that we are related to each other as common inhabitants of one earthly home.

6.1 Government is our tool to serve the common interest

We must use government to protect the biosphere, and that means we must pick winners—not specific companies, but the category of cleaner sources—they must be the winners and it is legitimate to influence the development of the marketplace so that they win. We should not use government to determine who does what, but we can rely on the essentials of free enterprise, which allows anyone to begin and profit from their own energy and creativity. We should not "pick winners" to extent that governments control the outcome of marketplace activities, but to the extent that they favor the development of that which the public needs and discourage that which harms us all. This is a long-established and time-honored idea, going back to the granting of monopolies by kings and queens, and the corporate charter, and the massive subsidies for fossil fuels that were granted when they were seen as essential to defense and our economy. The argument that the government should not pick winners is a selective argument applied in order to stop the development of industries competing with current business operations. Although this too is a complex area, involving how subsidies are hidden in

[17]https://www.imf.org/en/Publications/WP/Issues/2019/05/02/Global-Fossil-Fuel-Subsidies-Remain-Large-An-Update-Based-on-Country-Level-Estimates-46509. "... coal and petroleum together account for 85% of global subsidies. Efficient fossil fuel pricing in 2015 would have lowered global carbon emissions by 28% and fossil fuel air pollution deaths by 46%, and increased government revenue by 3.8% of GDP."

tax and other detailed laws, armed with the simple principle that we the people have the right to have our government act to favor the development of clean energy, we can institute systems of competition in which the best proposals receive support. Although the development of new clean energy technologies such as batteries, better motors, and better heating and cooling involves many complex issues, the simple concept that government can and must act to promote them is key to our progress. In order to grasp and use this concept, we must disabuse ourselves of the idea that government should stand back and play no role in the marketplace, and the pernicious idea that for the government to influence the marketplace is socialism or communism, and destructive. Communist governments have indeed provided examples of how centralized economic planning can be destructive, of the environment, of prosperity, and of freedom. But there is much that government can and must do to promote and support the development of that which the public needs that is completely congruent with the concept of free enterprise.

6.2 Using all the tools of governance to transform all systems

Engaged environmental citizens should be concerned that all the available tools of government be used to bring about the transformations needed. It is not just energy that can be transformed, but the use of materials and the production of goods, which must become a circular economy, because we can no longer coast on a downhill slide with a throughput system that transforms materials into waste at a frightening rate, and disseminates that waste into the biosphere, filling our oceans with plastic, our air with pollutants, and our bodies with substances that don't belong in biological organisms. In contrast to the governance system that governs energy, which is opaque and complex and distant from citizen experience, the environmental laws intended to limit pollution are far better known. The problem here is that the public has been misled to think of them as damaging to the economy. The success of laws to reduce air and water pollution and to clean up contamination on land has been significant, but limited by the contraction of funding and the diminishment of political support that resulted from the success of opponents in spreading the notion that we cannot afford them. The truth is that pollution laws are difficult only for polluting industries, and that they inspire and lift up cleaner industries, and it is that transition that has been resisted by entrenched interests. It is possible to understand this phenomenon in terms of the interest any investor has in seeing the returns that were expected when money was dedicated to a purpose, but that investors have been able to learn how to work the political system is an issue for all concerned. The clean economy would continue to evolve with continued development of antipollution law. By the 1990s, environmental agencies were initiating new efforts to prevent pollution, and not just control it after it has been generated. This far more efficient approach has been stifled by the interest in continued business as usual, which has learned how to exert pressure in many ways that citizens don't even see or which sounds like rational policy.[18] In the same way, the

[18]Quiet lobbying of state legislatures and agencies dampened a 1990s movement to adopt toxics use reduction laws and discouraged programs that urged the purchase of substitutes for products that become hazardous household wastes; efforts to "modernize" the US Toxic Substances Control Act included preemption of state toxic laws; and work with trade agencies included countering international efforts to expand limits on toxic products. A 2014 article quotes a Dow Chemical spokesman referring to its more than $5 million expenditures in the first quarter of the year as "collaboration" on policy. https://publicintegrity.org/politics/chemical-industry-among-big-spenders-on-lobbying-this-quarter/. Lobbying strategy includes advocating for better science as a means of hindering regulatory action by questioning its scientific justification. https://www.theguardian.com/us-news/2019/may/22/internal-emails-reveal-how-the-chemical-lobby-fights-regulation, https://www.latimes.com/business/hiltzik/la-fi-mh-a-look-inside-the-chemical-industry-20150515-column.html.

transformation of our systems of transportation is technologically feasible, as are our forms of shelter, and our methods of raising and obtaining food. All the systems on which we are dependent can be transformed, but the existence of powerful interests benefiting from things as they are exerts a strong inertial force that is compounded by expertise in governmental operations that is not shared by a disorganized public. But while it is tempting to blame our problems on those who resist the changes we need, it is citizen complacency that is the central issue, and where change can relieve us of complicity in our own dilemma.

7. Sensible transition through joint envisioning instead of disruptive revolution

The key to transformation to sustainable societies, a matter essential to our survival, hangs on the ability of engaged, environmental citizens to recapture government for the people. This is the central task on which to focus so that we can enable the creative forces for energy reform, food justice, people-powered transportation, habitat protection, environmental justice, and the cessation of toxic impacts to blossom. Government of, by, and for the people is government that is used for realization of the people's concerns, and the foremost concern today is reorganization of the economy so that it does not pollute, does not cause extinctions, does not remove mountain-tops, does not despoil the ocean, does not overharvest, and does provide sufficient protection to everyone so that they do not have to raze their local environment or race to exploit what's left in order to survive. The fundamental issue before us is to see the world we can have—to jointly envision that new economy that provides well-being to all, and also allows for the maximum amount of shared freedom, for it is the idea of freedom that keeps us in the anarchic situation where we think government should be limited. This is a false idea of freedom, for it is freedom only for those who are winning the current economic competition. The history of human evolution amply establishes that it is cooperation that has provided us with our wealth and progress—and so we should be helped to be free to cooperate—to share equitably would release our instincts to give freely. This form of government action is the opposite of coercion, such as a communist dictatorship employs.

7.1 Transitioning from corporate competition to cooperative enterprise

To illustrate, note that the concept that corporations operating for the purpose of generating profit for a few owners benefits society because it provides incentive for productive activities has led to the law of corporations that allows for limited liabilities and various tax advantages. Society has decided to provide incentives on the idea that this benefits everyone, but it doesn't. Many are left without necessities, new businesses are shut out of dominated markets, and corporate managers now get hundreds of times the return that employees get (the difference between management and staff income was more than an order of magnitude less only decades ago).[19] But society can insist that its representatives in government change the system of incentives. They have the right, in a government that is intended to represent the people, to demand that laws favor cooperatives, or profit-sharing companies, or profit-

[19]A 2017 study by the Economic Policy Institute found that in 2016, the average compensation of Chief Executive Officers in America's largest firms was 271 times that of the typical worker. In 1965, CEO pay was 20 times workers'. https://files.epi.org/pdf/130354.pdf.

making companies that adhere to responsible behavior, and that they keep markets free of domination, and that taxes and liabilities and rules of operation provide incentives that favor organizations that produce for the common good. Such a system is not one that mandates what everyone does, nor removes their freedom, nor installs an inefficient centralized planning function. It can remain free and creative, in fact more free and open than the one we have now. All we need to do is recognize that we have the right to change the rules and the vision for how it should change. What we need is a sense of the many tools of governance we can use to motivate action.

With such a determination, engaged citizens can come together to develop the means by which governance can be used to prompt the transformations we need. Subsidies and grants can be provided to companies that take back products at the end of their useful life. Matched with fees on those who create wastes, a push—pull effect can prompt rapid evolution of more responsible production. The burden on waste generation can be lifted from municipalities and taxpayers.

Cities can be given funding for ensuring that vehicle traffic no longer dominate city roads, but can be parked at the city's edge, and ample means for getting around without burning fuels provided. Taxes on international transfers of funds can be imposed so that large investors can't so easily capture local industries, and so that opportunities are provided to local entities, and revenues used to protect natural assets. Much now spent on useless nuclear threats, a destructive impulse of the first and worst order, can be used instead for protection of ecosystems and the development of clean energy. Basic universal income, education, and women's rights will help reduce population growth. Regulations can prompt the continued mitigation of pollution, and purchasing preferences can promote the continued growth of cleaner substitutes. Government funding of research on safer alternatives will pay society back many times, and help us meet our obligation to future generations.

8. Recognizing how primitive our vision has been

The tools of government are not just the simple, familiar ones of punishment. Environmental law began with the idea of stopping pollution by fining polluters. By not evolving past that,[20] society became stuck, and the environmental agencies became tagged as police to be feared and avoided. But environmental agencies began to evolve the concept of partnership, of voluntary initiatives, and this was successful in many areas, but was misinterpreted as a capitulation to business opposition, because it happened at the same time that opposition to environmental laws was building (and in too many cases, it *was* capitulation). Reduced funding meant that in order to support efforts to work cooperatively with businesses to find ways to reduce pollution (as opposed to fining them), money had to be shifted from enforcement programs. Even when capitulation was not the intent of leaders, the effect was to reduce enforcement, and so the success of voluntary programs has not been appreciated.

But we can envision a new environmental movement that is more sophisticated than what has come before. It would not just demand that enforcement against polluters be reinstated—which must

[20]Agencies did evolve past that, establishing programs of partnership and assistance with pollution prevention, energy efficiency and compliance, but the public perception of that evolution has not caught up, partly because these efforts happened at the same time as resource cuts to enforcement, and were perceived as an unwarranted "softening" (the reduction in enforcement was, but the establishment of assistance was and is warranted). Environmental progress requires *both* helping and enforcing hands. There should have been no shifting of funding, but rather the addition of assistance as a complement to enforcement.

happen—but that successful programs of partnership with those of good will—they do exist in business as in every sphere—be expanded. The pollution prevention movement of the 1990s was extremely effective, but its success has been ignored. The recycling movement of the 1980s has been forgotten. The local food movement of today is not getting the attention it needs. The clean energy movement is understood to be successful, but people do not see how it is being squelched.

The environmental citizen of today cannot be expected to become expert in all the arcane regulatory hearings and legislative development that opponents concerned with keeping their expected returns can master (because they can pay for expertise to master it for them). But today's environmental citizen can recognize how important it is to be engaged in all areas of governance—that the government must be made to belong to the people and to reflect the common interest, and can organize and work with others to share responsibilities and knowledge. In order for this to happen, there must be a vision that transformation of the world is necessary and can occur. It is not an angry, disruptive revolution that we need, but an informed, open, and careful process of planning. It can occur with cooperative and engaged interactions inspired by hope and knowledge of what is possible. The transformed world is feasible and what prevents us from having it is our neglect of the fact that we can respectfully cooperate to bring it into being.

8.1 The fundamental nature of relationship

Government is a form of relationship. It is how we relate to occupy the earth as one species. This is not its original purpose, which was fragmented into many locations and shaped by historical accident. But it is the current, pressing reality. When we see that we are part of it, and that we can make it work by relating to each other, taking others' needs as important as our own, we create a relational system that has the purpose and vitality that a competitive system does not have. Competition does indeed produce incentives and excellence, but it becomes destructive if not conducted within a context of cooperation. We must use government to create that context. We must learn to relate to each other sufficiently to see our common interest, in order to make it manifest. The fate of the world depends on this mutual identification, with each other, as citizens of our towns, states, countries, and the world.

A government that stands back and lets those who currently dominate our economic system continue to exploit our common resources, to the ruin of the present and future, is a government that denies our relationship to each other and to all who will yet come to be. By recognizing mutuality, we can overcome the disease of unrestrained competition. By transforming that negative freedom into the positive freedom we can have by working together, we can begin the work of transforming all of our systems to a set that maintains, rather than degrades, the very ground we stand on, the air we breathe, the water we drink, and the life forms that depend on us to act with responsibility.

CHAPTER

21

Nationally determined contributions to foster water-energy-food-environmental security through transboundary cooperation in the Nile Basin

Muhammad Khalifa[1], Maksud Bekchanov[2] and Balgis Osman-Elasha[3]

[1]*Institute for Technology and Resources Management in the Tropics and Subtropics (ITT), Cologne University of Applied Sciences, Cologne, Germany;* [2]*Research Unit Sustainability and Global Change (FNU), Center for Earth System Research and Sustainability (CEN), University of Hamburg, Hamburg, Germany;* [3]*African Development Bank Group, Immeuble Zahrabed Avenue du Dollar, Tunis, Tunisia*

Chapter outline

1. Introduction

Water is a central resource to achieve energy and food securities. Globally, there are 310 river basins that are shared between two or more countries (McCracken and Wolf, 2019). Management of transboundary water is one of the greatest challenges that face future development (Tahiya, 2019). In such basins, using the shared water resources might be a major source of tension and conflict, since

Handbook of Energy and Environmental Security. https://doi.org/10.1016/B978-0-12-824084-7.00025-4
Copyright © 2022 Elsevier Inc. All rights reserved.

unilateral actions can harm other riparian countries, especially those located at downstream parts of the basins (Kasymov, 2011). For the majority of transboundary basins (around 60%), there are no treaties that govern water use and promote transboundary cooperation between their riparian countries (De Stefano et al., 2010). Lack of trust and cooperation between riparian countries can lead to serious disputes and conflicts about these countries regarding using the shared water resources and, can consequently, jeopardize energy and food securities. This is the case of many transboundary river basins (e.g., The Nile and Euphrates/Tigris river basins). The likelihood of conflict increases when a combination of two factors exists: (i) rapid changes in the basin's physical setting (e.g., building a dam) and (ii) inability of the existing institutional setups to absorb and manage this change (Wolf, 2009). According to Farinosi et al. (2018), future demographic and climatic conditions are expected to increase the hydropolitical risk in many of these conflicted basins.

Within the transboundary setting, crossing the complex physical, ecological, political, juridical, and social factors makes it difficult to allocate, manage, and use the shared water resources (Choudhury and Islam, 2015). This might indicate possible risks on water, food, and energy of the riparian countries under transboundary disputes. Sharing benefits between riparian countries and cooperative management of such transboundary basins beyond political boundaries can foster peace with the environment and people and have the potential to maximize the profit gains for the riparian countries compared to the unilateral management approach (Whittington et al., 2005). However, this shift from conflict to cooperation is controlled by many factors, including creating mutual interests and mobilizing financial resources to achieve these interests (Yoffe et al., 2003; Choudhury and Islam, 2015). We argue that holistic approaches such as the Water-Energy-Food (WEF) Nexus and actions to adapt and mitigate the impacts of climate change can create reasonable and compelling foundations to foster transboundary cooperation between countries of these basins, with an aim to ensure water, energy, and food securities in the region.

WEF Nexus is an emerging concept that promotes holistic management of the three sectors in a coordinated manner. It aims at identifying synergies between the three sectors in order to be increased and promoted and trade-offs to be decreased or eliminated (Hoff, 2011). This approach has received a wide acceptance from both academia and policy (Mahlknecht et al., 2020). Building a regional WEF Nexus framework for a transboundary river basin would help in framing mutual interests. On the other hand, climate change is a threat multiplier that has impacts on multiple sectors. This threat requires global actions and financial resources that are beyond the capacities of individual countries, with a huge financial gap in committed and mobilized financial resources (Zhang and Pan, 2016). The Paris Agreement is adopted by countries aiming at holding global temperature raise below 2°C and try to constrain this change even below 1.5°C (UNFCCC, 2015). Meeting the climate target requirement of the Paris Agreement necessitates a substantial reduction of the greenhouse gas emissions in various sectors and a strong commitment to the Nationally Determined Contributions (NDCs) provided by all participating countries. It is believed that joint adaptation and mitigation are indispensable to accelerate achieving the goals of keeping global warming below 2 or 1.5°C levels.

The NDCs of the Paris climate agreement together with the United Nations Sustainable Development Goals (SDGs) of the global 2030 agenda represent important windows into countries' plans and actual needs. The SDGs that aim at maintaining economic prosperity, inclusive growth, and environmental sustainability were formulated by world leaders at the same year when the Paris Agreement was adopted (UN, 2015). These two sets of commitments and plans display multiple synergies, and many countries share comparable ambitions and needs when it comes to climate action and development goals. Analyzing synergies between the NDCs and the SDGs between riparian countries would help in identifying entry points for transboundary cooperation. Transboundary cooperation is not only important to directly build synergies between such countries but it also creates an enabling environment to attract more investments and climate finance and would reduce the needed implementation cost compared to unilateral actions (UNECE, 2009). Identifying entry points for transboundary cooperation among riparian countries would be a key step to achieve this transformation. Despite the importance of this particular information, systematically identifying such entry points has received little attention.

Against the above background, the current research focuses on the Nile Basin, which is the largest in Africa, with 11 countries sharing its water. We aim at analyzing the NDC−SDG synergies related to improving energy-water-food-environmental security, with an ultimate goal to identify entry points for transboundary cooperation between these countries through focusing on the WEF Nexus thinking and climate action perspectives. To this end, key questions need to be answered are: What are the priority actions of the riparian countries as agreed upon in their NDCs? What are the existing overlaps in these plans and actions between these countries? What are the actions the most relevant to achieve the regional SDG goals but did not gain sufficient attention in the national commitments (NDCs)? Answering these questions would shed light on the potential synergies and would help in identifying entry points for transboundary cooperation. Differences between countries in their ability to attract climate finance would emphasize the need and usefulness of the transboundary cooperation approach. The input data needed for the current analysis were derived from the NDC−SDG connection platform (DIE and SEI, 2020). Data on the mobilized climate finance from multilateral development banks (MDBs) were retrieved from the joint report on MDBs' climate finance (MDBs, 2019). This analysis is structured as follows:

(i) Investigate the NDC−SDG interactions related to improving energy-water-food-environmental security across the 11 Nile countries;
(ii) Determine priority climate actions of these countries in correspondence to each SDG;
(iii) Highlight climate change vulnerability and impacts on the Nile Basin countries;
(iv) Identify climate measures/actions that are very relevant to the regional context but did not gain sufficient attention in the NDCs;
(v) Evaluate climate finance flow from MDBs, as an example of variation in climate finance received by the Nile countries.

2. Study area

The Nile River basin is drained by the Nile River and its tributaries and is the second-largest basin in Africa. The area of the basin is about 3.2 million km^2 and accounts for almost 10% area of the African continent (Swain, 2011). The Nile is the longest river in the world with a length of 6670 km (El Bastawesy et al., 2015). The Nile River waters are formed in the Ethiopian Highlands and the Equatorial Lakes and flow toward the north. The river reaches the Mediterranean Sea at the end after crossing extremely arid regions such as the Saharan Desert. The territory of the basin is shared by 11 countries, namely: Egypt, Ethiopia, Eritrea, Sudan, South Sudan, Uganda, Kenya, the Democratic Republic of the Congo, Rwanda, Burundi, and Tanzania (Fig. 21.1).

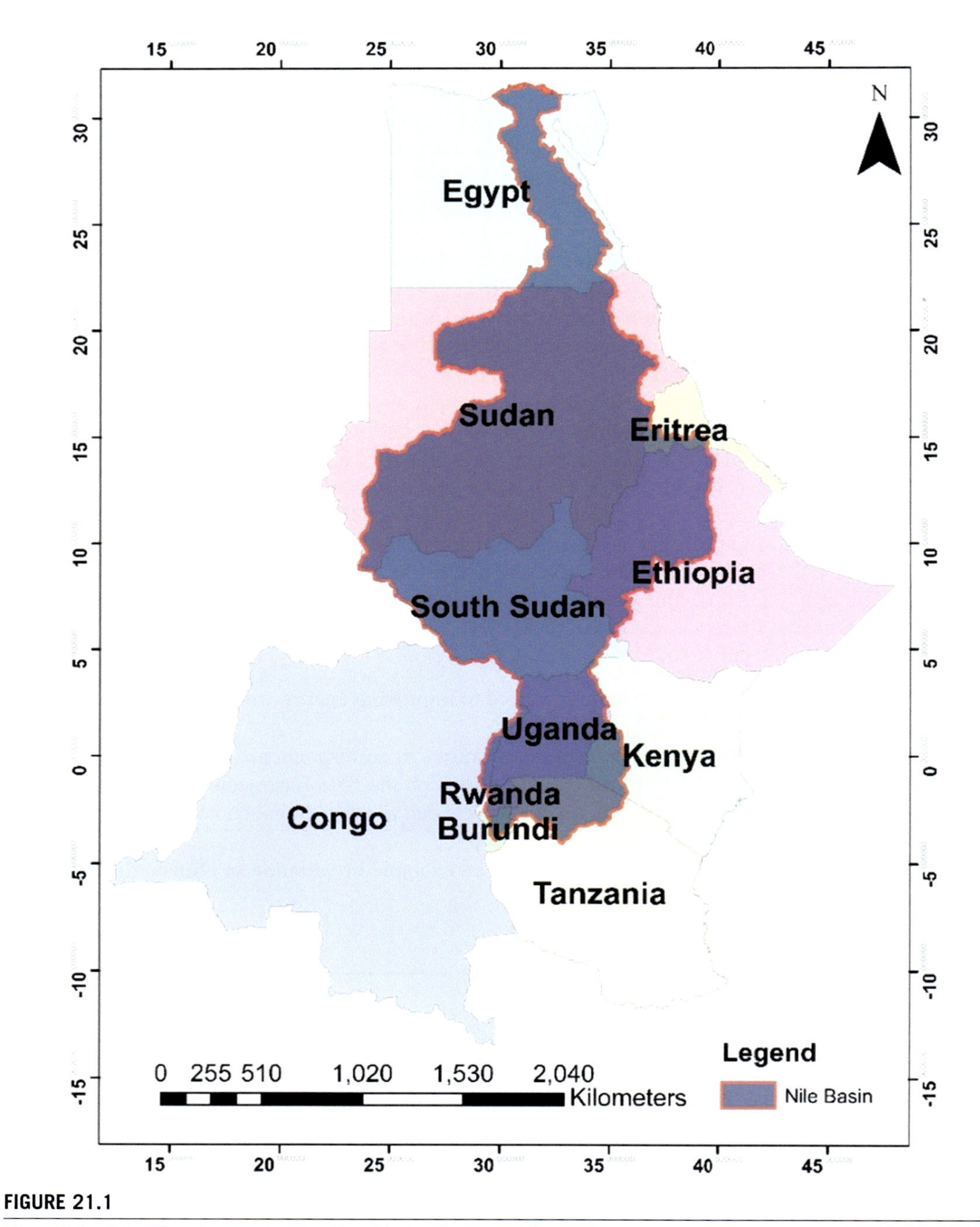

FIGURE 21.1

Map of the Nile River basin showing the 11 riparian countries.

Source: Authors' work.

The total population residing in the basin is about 437 million (as of 2012) (UNEP, 2013). More than 70% of the population lives in rural areas (Kimenyi and Mbaku, 2015). The river basin is the main supplier of freshwater to irrigation, residential areas, hydroelectricity supply, and fishery in the region. As the territory of the basin is shared by several countries, distribution of the scarce water resources is challenging. This basin is experiencing accelerating transboundary disputes among its riparian countries (Wheeler et al., 2020); especially after Ethiopia started to build the Grand Ethiopian Renaissance Dam (GERD), located only 20 km from the Ethiopian–Sudanese border. When completed, GERD will be the largest hydropower dam in Africa with a generation capacity of 6000 MW of power and will store around 74 billion cubic meters (Desso, 2019), almost equals the historical annual flow of the Nile that measured at Aswan Dam in Egypt (Sutcliffe and Parks, 1999). Based on the two factors suggested by Wolf (2009), such a major unilateral action from one riparian country coupled with the absence of effective national or regional institutions to absorb the expected large physical, environmental, and socioeconomic conditions in the basin can intensify the existing transboundary conflict and may endanger the water, energy, and food securities in the downstream countries, i.e., Sudan and Egypt. Climate change adaptation and mitigation options agreed in the NDCs of the individual riparian countries are therefore required—along with other adjustments—to be more holistic and need basin-wide coordination to be effective to sustain water, food energy production, and to ensure environmental security in the region.

Securing water, energy, and food for everyone seems to be a pressing challenge for the Nile countries. As evidenced from key indicators of water, energy, and food security shown in Table 21.1, the majority of the riparian countries of the Nile Basin are still beyond achieving security in these three important sectors. In fact, only Egypt exhibits good status in these indicators. Based on the data in Table 21.1, on average, 54.6% and 42.4% of the people in the basin do not have access to basic water sources and electricity, respectively. Moreover, more than 26% of the Nile countries' population suffer from moderate or severe food insecurity.

3. Climate change impacts, vulnerability, and adaptations in the Nile Basin

The 11 countries that share the Nile Basin will benefit from information regarding current and future climate projections as well as the potential socioeconomic scenarios to enable governments' response to climate change impacts. Since climate change will affect future water systems, sharing information on climate change trends will help the Nile Basin countries develop policies and action plans that will ensure the planning and management of the shared water resources in a transparent and fair manner. It will also help them avoid potential water use–related conflicts.

Following recent trends in the decline of per capita water availability in Africa, any degree of increase in global water temperatures will seriously affect the Nile Basin's water balance. Based on recent studies, the Nile Basin has shown a high level of sensitivity related to its natural flow to changing rainfall patterns in the Ethiopian Plateau. For example, a 20% increase in rainfall can lead to an 80% increase in the Nile's flow at Aswan. Similarly, a 20% decline in precipitation will also reduce the flow by 80%. An increase of over 2 C could also exhaust the water excess that Egypt currently enjoys (annual water supply of 63.7 km^3 against an estimated annual use of 61.7 km^3). On the other hand, a 20% increase in rainfall would result in 11% increase in Lake Victoria Basin outflows,

Table 21.1 Water supply, water withdrawals, energy availability, food access, and income levels across the riparian countries of the Nile River basin.

	Burundi	DR of Congo	Egypt	Eritrea	Ethiopia	Kenya	Rwanda	South Sudan	Sudan	Tanzania	Uganda
(1) Share in the territory of the basin (%)	0.4	0.7	9.5	0.8	11.5	1.6	0.7	19.5	44	3.7	7.6
(2) Population (million) (2020)	11.9	89.6	102.3	3.5	114.9	53.8	12.9	11.2	43.8	59.7	45.7
(3) GDP per capita (US$ per capita) (2019)	261.3	580.7	3019.2	642.5	855.8	1816.5	820.0	1119.7	441.5	1122.1	794.3
(4) Total renewable water resources per capita (m^3/inhabitant/year) (2013–2017)	1158	15,762	596.2	2143	1147	611.3	1110	4537	926.2	1761	1460
(5) Water withdrawal (km^3)	0.3	0.6	68.4	0.4	5.6	3.2	0.2	0.7	26.9	5.2	0.6
(6) % Population with access to basic water sources (2017)	60.4	43.2	99.1	51.6	29.6	58.9	57.7	40.7	60.3	56.7	42.0
(7) % Population with access to electricity (2019)	10.9	8.7	99.7	46.5	46.7	84.5	52.6	1.1	47.3	39.5	28.9
(8) Prevalence of severe food insecurity in the population (%) (2018)	No data	No data	7.8	No data	14.1	19.1	No data	63.7	16.4	36.9	20.6

Sources: (1, 4,5): AQUASTAT (2020); (2): UNDESA (2020); (3): World Bank (2020a); (6): WHO and UNICEF (2019); (7): IEA, 2020; (8): World Bank (2020b).

indicating a relatively low sensitivity (Sayed, 2004). While a 2 C increase in temperature might reduce the water flow in Lake Victoria by 50% on average (UNEP, 2013).

The IPCC (2007) projects that 75 to 250 million people in Africa will suffer increased water stress due to climate change by 2020. High climatic diversity and volatile seasonal distribution of the water resources are the main characteristics of the Nile Basin. For example, some areas in the Basin are witnessing long periods of intense rainfall and high levels of evaporation (e.g., equatorial region). Irrigated agriculture and livestock production are the key economic activities in this region. These activities are highly sensitive to climate change and extremes such as increasing fluctuations in rainfall, droughts, and floods, which can affect peoples' livelihoods and threaten food security situations (Tate et al., 2004).

The Upper Nile Basin which extends over Western Ethiopia, South Sudan, and Uganda is generally vulnerable to climate change and extremes, particularly since agriculture is highly dependent on rainfall, exposing it to the impact of fluctuating rainfall and temperature variations. The region has recently witnessed a drought incident that caused drastic reductions in crop yields and food insecurity. A study has revealed that in spite of projected increases in regional precipitation due to climate change, the frequency of hot and dry years is likely to also rise due to global warming. It also indicated that the sharp increases in extreme conditions will impact agriculture production and result in food shortages. Coupled with increased population, the food insecurity situation in the Upper Nile Basin will be further aggravated in the coming years (Coffel et al., 2019).

Climate change poses great risks to agriculture including fisheries threatening rural livelihoods and exacerbating cross-border and rural–urban migration problems. The climate-related losses in Kenya alone between 1997 and 2000 amounted to US$5 billion or more than 14% of GDP.

More pressures are caused by climate change on the regions that are already facing socioeconomic and development challenges. For example, the climate-induced sea-level rise and encroachment of saltwater threaten the Nile delta in Egypt with its dense population. A one-meter increase in sea-level rise could cause a loss of 4500 km^2 farmland, negatively impacting the supply and quality of freshwater and may lead to the displacement of over six million people. Moreover, the whole Nile Basin is at risk of increased temperature and evaporation, which are likely to disturb hydropower generation.

It is important that development options in the Nile Basin countries should take into consideration the climate change impacts. Some adaptation options identified in the National Adaptation Plans of some Nile Basin countries include strengthening and modernizing the climate observation systems to improve early warning and forecast; developing human and institutional capacity; incorporating local knowledge and involvement of the highly vulnerable communities in the adaptation efforts. Increased investments in renewable energy will contribute to the adaptation, improved livelihoods and resilience, as well as mitigation and low-carbon development. Integrated Water Resource Management has also proven to be an effective adaptation mechanism.

4. Analysis of the NDC–SDG synergies for the Nile Basin riparian countries

Based on the data derived from the NDC–SDG connection platform, the number of NDCs varies among the Nile's riparian countries. As shown in Fig. 21.2, it ranges between 13 (Democratic Republic "DR" of Congo) and 95 (Uganda). Analysis of these NDCs of the Nile countries revealed multiple

Country	No. of NDCs
Uganda	95
Sudan	72
South Sudan	66
Eritrea	62
Egypt	60
Tanzania	56
Burundi	49
Ethiopia	40
Rwanda	37
Kenya	18
DR Congo	13

FIGURE 21.2

Total number of NDCs of the Nile Basin countries.

Sustainable Development Goal (SDG)	Nile's average	Africa's average	Global average
SDG15: Life on land	15	14	13
SDG7: Affordable and clean energy	14	17	16
SDG2: No hunger	13	13	13
SDG6: Clean water ad sanitation	8	9	9
SDG17: Partnership for goals	8	8	86
SDG8: Decent work and economic growth	7	4	4
SDG11: Sustainable cities and communities	6	7	9
SDG13: Climate action	6	6	6
SDG9: Industry, innovation and infrastructure	5	5	7
SDG3: Good health and well-being	4	3	3
SDG14: Life below water	4	3	60
SDG5: Gender equality	3	1	1
SDG4: Quality education	3	3	3
SDG12: Responsible consumption and production	2	3	3
SDG10: Reduced inequalities	2	1	1
SDG1: No poverty	1	2	2
SDG16: Peace, justice and strong institutions	1	0	0.5

FIGURE 21.3

The connection between NDCs and SDGs of the Nile Basin as a regional average and compared with Africa and global averages.

linkages with SDGs. Most of the climate actions of the Nile countries as formulated in their NDCS were found to be related to five SDGs—in addition to Increased Climate Action (SDG13)—namely (i) life on land (SDG15), (ii) affordable clean energy (SDG7), (iii) no hunger (SDG2), (iv) clean water and sanitation (SDG6), and (v) partnership for goals (SDG17). This is consistent with Africa's averages, which show emphasis on the same five SDGs, but with some deviation from the global NDCs, which additionally put much emphasis on the goal of life below water (SDG14) (Fig. 21.3). The SDG ranking was calculated as a percentage of the sum of NDCs related to specific SDG for all countries to the total number of the Nile's NDCs.

5. Analysis of priority climate actions and relevant SDGs

5.1 Climate actions and "affordable and clean energy" (SDG7)

The two main categories of climate actions in NDCs of the Nile countries are related to (i) wider adoption of clean and renewable energy, and (ii) improvement of energy efficiency (Table 21.2). All of the Nile countries except Kenya (n = 10) included clean and renewable energy in their NDCs. Especially, the importance of bioenergy (n = 7), hydropower (n = 7), solar energy (n = 6), wind energy (n = 6), and geothermal energy (n = 6) are underlined in most of the NDCs. Less prioritized energy production options are waste-to-energy systems (n = 3), nuclear energy (n = 1), and tidal energy (n = 1). High risks related to nuclear energy and high costs of tidal energy and waste-to-energy systems explain the unpopularity of these options.

Most of these countries (n = 7) indicated their commitment to enhancing energy efficiency. Yet, only a few NDCs considered detailed plans of enhancing the energy efficiency either through improving the efficiency of lighting (n = 2), or more efficient electric appliances (n = 3), or cleaner cooking stoves (n = 3). Since the majority of the population of the basin lives in rural areas, the latter option can be an effective measure to improve energy and environmental security at a low cost. The majority of the actions related to energy security noted in the NDCs of the Nile Basin riparian countries focused on energy supply aspects. Energy access (n = 3), affordability (n = 1), and reliability (n = 1) were mentioned only in a few NDCs. Thus, these aspects should additionally be assessed in the future updates of the NDCs.

5.2 Climate actions and "clean water and sanitation" (SDG6)

The water system is the most vulnerable sector to climate change, and adaptation measures are essential to improve the resilience of this sector. Our results indicate that both water supply and water management measures received considerable attention from the Nile countries (n = 7 and n = 6, respectively) (Table 21.3). While the countries located in the arid zone (e.g., Sudan and Egypt) have obvious reasons to enhance their water supply due to scarcity, upstream countries such as Uganda and Tanzania may face issues related to water supply due to underdeveloped infrastructure.

From a water management perspective, the majority of the countries focused on rainwater harvesting (n = 6), integrated water resources management (n = 6), and efficient water uses (n = 7). Efficient water uses are especially relevant in the irrigated agriculture context (n = 6). Rainwater harvesting options are appropriate in areas where high precipitation occurs in short periods before the long duration of droughts. Groundwater management in contrast opted only in a few countries such as

Table 21.2 Strategies for enhancing energy security highlighted in the NDCs of the Nile Basin riparian countries.

Climate action	Burundi	DR Congo	Egypt	Eritrea	Ethiopia	Kenya	Rwanda	South Sudan	Sudan	Tanzania	Uganda	Count
Clean and renewable energy	x	x	x	x	x		x	x	x	x	x	10
Waste-to-energy systems	x							x	x			3
Bioenergy	x			x			x	x	x	x	x	7
Access to energy		x							x	x		3
Hydropower		x			x		x	x	x	x	x	7
Solar energy		x	x			x	x			x		5
Decentralization of energy supply		x					x		x	x		4
Energy efficiency			x	x	x		x	x	x		x	7
Transition from fossil fuel–based energy			x	x				x	x	x		5
Financial mechanism for energy			x								x	2
Pricing schemes			x									1
Nuclear power			x									1
Clean heating and air conditioning			x	x						x		3
Efficient lighting				x					x			2
Efficient appliances				x	x				x			3
Clean cooking solutions				x						x	x	3
Solar energy				x	x			x	x	x	x	6
Wind energy				x	x	x		x	x	x		6
Geothermal energy				x	x	x		x	x	x		6
Transition from coal to energy				x								1

Power plant construction				x	x		x	x	x			5
Tidal energy						x						1
Off-grid and smart-grid systems									x		x	2
Codes, labeling, and standard establishment									x			1
Affordable energy										x		1
Reliable energy										x		1

Source: Authors' calculation based on DIE and SEI (2020).

Table 21.3 Strategies of improving water security highlighted in the NDCs of the Nile Basin riparian countries.

Climate action	Burundi	DR Congo	Egypt	Eritrea	Ethiopia	Kenya	Rwanda	South Sudan	Sudan	Tanzania	Uganda	Count
Water access				x				x	x		x	4
Water security				x			x		x			3
Water supply			x	x	x			x	x	x	x	7
Water management	x			x			x	x	x	x		6
Rainwater harvesting			x		x		x	x	x	x		6
Groundwater management			x						x	x		3
Integrated water resources management	x						x	x	x	x	x	6
Hydrological knowledge development	x						x		x			3
Surface water management			x									1
Efficient use of water	x		x	x	x		x	x			x	7
Irrigation improvement	x		x	x	x		x					5
Wastewater management			x		x			x		x		4
Water treatment			x		x							2
Water rcuse			x							x		2
Water desalination			x	x								2
Water policy and regulations			x									1
Transboundary management			x									1
Protection of wetlands and lakes			x					x			x	3
Hydropower				x	x							2

Water quality improvement					x		x	x				3
Watersheds management							x	x		x		3
Access to sanitation										x		1

Source: Authors' calculation based on DIE and SEI (2020).

Egypt, Sudan, and Tanzania. This low level of emphasis on groundwater may be related to the abundant rainfall and availability of surface water in the upstream riparian countries compared to the arid downstream countries and because of high costs and energy requirements to develop this resource.

Despite its high relevance in the region, water quality aspects did not gain sufficient attention from the riparian countries. Only Ethiopia, Rwanda, and South Sudan have explicitly referred to improving water quality in their NDC plans, and only Egypt and Ethiopia referred to water treatment. Developing unconventional water resources also received low attention, strategies such as water reuse (n = 2) and desalination of seawater (n = 2) have received little attention.

Given the interdependence of all riparian countries for efficient uses of the natural resources and effectively cope with climate change impacts in the river basin, transboundary water management was only admitted in the NDC by Egypt. As the largest economy in the region and with high reliance on irrigated agriculture, the downstream location of the country makes it very vulnerable to upstream water system changes. Nevertheless, basin-wide cooperation is essential for all riparian countries when aiming at achieving common goals of coping with climate change, poverty, and food security requires.

5.3 Climate actions and "zero-hunger" (SDG2)

Increasing water scarcity in contrast to the growing population makes food security a challenging task in the Nile Basin. Therefore, it is not surprising that most of the countries mentioned climate-smart and sustainable agriculture as an important adaptation option in their NDCs (n = 9) (Table 21.4). Especially, climate-resistant seeds and increased crop varieties were underlined in the NDCs (n = 8). Agriculture productivity in all Nile countries, except Egypt, are rather too low. For example, the average yield of cereal crops in Sudan is only 0.5 tons per hectare compared to 7 tons per hectare in Egypt (FAOSTAT, 2020). Improving agricultural productivity, therefore, represents a high priority across the Nile River basin, as reflected in the respective NDCs. While emphasizing drought resilience, options of efficient fertilizer use and improved soil management practices were considered for eliminating hunger only by two countries, however.

The livestock sector is very sensitive to climate-induced drought and reduced vegetation. Yet, livestock ownership largely determines the wealth of the rural families in the region. Consequently, most of the NDCs emphasized the resilience of the livestock sector to effectively cope with climate-driven natural calamities (n = 8). Yet, the storage of the feed and fodder did not gain such importance in the NDCs (n = 3). Similarly, the storage of food was only mentioned by two countries as a potential option to enhance food security. Perhaps, the efficiency of the storage is quite low due to the risk of intense damage by insects and pests.

5.4 Climate actions and improved "life on land" (SDG15)

Climate actions related to supporting ecosystems (SDG15), as indicated in the NDCs of the riparian countries of the Nile Basin, focus on the forest sector. Particularly, all countries underlined afforestation as an effective mitigation option in their NDCs (n = 11). Most of the countries also reported the importance of conservation, restoration, and rehabilitation measures (n = 9) and forest management (n = 7). Carbon sequestration (n = 6) and fuelwood consumption reduction (n = 5) were also emphasized as a mitigation option by some countries (Table 21.5).

Table 21.4 Strategies to achieve food security are highlighted in the NDCs of the Nile Basin riparian countries.

Climate action	Burundi	DR Congo	Egypt	Eritrea	Ethiopia	Kenya	Rwanda	South Sudan	Sudan	Tanzania	Uganda	Count
Climate-smart and sustainable agriculture	X	x		X		x	X	x	x	x	x	9
Improved livestock resilience	X		x	X	X			x	x	x	x	8
Climate-resistant seeds and crops	X		x	X	X			x	x	x	x	8
Food storage					X						x	2
Maintained genetic resources and diversity	X		x						x		x	4
Diversification of crops	X		x		X				x		x	5
Sustainable irrigation			x				X	x			x	4
Livelihood and income diversification for farmers	X				X			x	x	x		5
Use of traditional knowledge					X					x		2
Food production	X	x	x	X	X		X	x	x			8
Increased agriculture productivity	X	x	x	X	X		X	x	x			8
Improved food security			x		X			x	x			4
Agroforestry development	X						X		x			3
	X				X				x			3

Continued

Table 21.4 Strategies to achieve food security are highlighted in the NDCs of the Nile Basin riparian countries.—cont'd

Climate action	Burundi	DR Congo	Egypt	Eritrea	Ethiopia	Kenya	Rwanda	South Sudan	Sudan	Tanzania	Uganda	Count
Storage and production of feed and fodder												
Improved soil management practices	X						X	x				3
Fertilizer use	X						X					2
Sustainable agriculture	X											1

Source: Authors' calculation based on DIE and SEI (2020).

Table 21.5 Strategies for improving life on land highlighted in the NDCs of the Nile Basin riparian countries.

Climate action	Burundi	DR Congo	Egypt	Eritrea	Ethiopia	Kenya	Rwanda	South Sudan	Sudan	Tanzania	Uganda	Count
Forest management	X			X			x	x	x	x	x	7
Afforestation	X	X	x	X	x	x	x	x	x	x	x	11
Conservation, restoration, and rehabilitation of ecosystems	X	X	x	X	x			x	x	x	x	9
Fuel wood consumption reduction	X						x	x	x	x		5
Ecosystem resilience to natural disasters	X				x							2
Species protection	X				x			x				3
Carbon sequestration		X	x		x			x		x	x	6
Biodiversity protection		X			x			x			x	4
Wetlands protection and restoration			x								x	2
National parks and other protected areas				x				x				2
Desertification prevention				x								1
Land erosion prevention					x							1
Genetic resources conservation							x					1
Deforestation prevention								x		x	x	3
Wildfires prevention								x		x		2
REDD + implementation										x		1
Ecosystem planning											x	1

Source: Authors' calculation based on DIE and SEI (2020).

Despite the high relevance of some strategies, only one or two countries mentioned such options as wetlands protection and restoration, the establishment of national parks, prevention of land erosion and desertification, and implementation of mechanisms such as the Reduced Emissions from Deforestation and Forest Degradation (REDD+).

5.5 Actions to support "partnership for goals" (SDG17)

Cooperation among the riparian countries of the Nile River basin is essential for the fulfillment of the commitments to improve energy, water and food security, and sustain the ecosystem described above. NDCs considered some aspects of the cooperation across and beyond the basin. Most of the countries pointed out financial resource's mobilization (n = 10), technology transfer (n = 10), and capacity building (n = 11) as important actions of promoting partnership (Table 21.6). Indeed, the role of finance, technology access, and developing capacities are key factors to implement the NDCs and achieve development goals. Yet, many other aspects of cooperation through the enhancement of climate-conscious trade (n = 1), research data generation and sharing (n = 1), and policy coordination (n = 1) did not gain sufficient attention despite their importance.

6. Mobilizing financial resources

Financial support and potential economic gains are important factors that provide incentives and vehicles for the transformation from conflict to cooperation in transboundary basins. It depends on many factors (Yoffe et al., 2003; Choudhury and Islam, 2015), and encompasses several effective instruments at different stages of the transformation process. There are several examples where mobilizing finance has played a key role in fostering cooperation over transboundary water (Motsert, 2005). Thus, mobilizing financial resources to foster cooperation between riparian countries in transboundary basins can play a crucial role in securing water, energy, food, and environment security and minimizing transboundary conflicts. Consequently, this can create an enabling environment for sustainable development that attracts investments in the main sectors. This transformation to cooperative water management can, in turn, minimize the required capital cost while accelerating the implementation of the NDCs and achieving the SDGs.

Mobilizing finance to achieve water, food, and energy securities and to adapt and mitigate climate change could be one of the most pressing challenges that most developing countries are facing. Many of the countries of the Nile River basin managed to attract substantial amounts of funds provided by the MDBs for maintaining their climate actions. Analysis of these flows of climate finance shows that the total amount received by these countries in 2019 represents 6.3% of the total commitments of the MDBs in that year (~US$61.6 billion). However, the amount of financial resources mobilized to the individual countries largely varies, Egypt attracting most of the climate funds and Eritrea the least. In addition to policy relevance, political stability remains the main factor for improving the investment climate.

During the period between 2015 and 2019, funds amounted to more than US$15 billion have been committed for climate action by MDBs in the Nile countries. Among the riparian countries, Egypt has received around US$6 billion, which represents the lion share (~39.6%) of the total climate finance mobilized to the Nile countries during this period. Ethiopia and Kenya have also received substantial financial resources amounting to US$2.9 and 2.6 billion, respectively. Countries such as Eritrea, South

Table 21.6 Strategies to build partnerships are highlighted in the NDCs of the Nile Basin riparian countries.

Climate action	Burundi	DR Congo	Egypt	Eritrea	Ethiopia	Kenya	Rwanda	South Sudan	Sudan	Tanzania	Uganda	Count
Financial resources mobilization		x	x	X	X	x	X	x	x	x	x	10
Need for technology transfer	X	x	x	X	X	x	X	x	x		x	10
Capacity building	X	x	x	X	X	x	x	x	x	x	x	11
Data generation and data processing	X											1
Promotion of climate-conscious international trade				X								1
Promotion of sustainable technologies				X				x			x	3
Investment in least developed countries				X	X							2
South–South cooperation				X				x				2
North–South cooperation				X				x				2
Policy coordination								x				1
Need for additional financial resources											x	1
Partnership											x	1
Public–private partnership											x	1

Source: Authors' calculation based on DIE and SEI (2020).

Sudan, Burundi, and Sudan have received negligible amounts of funding to support their national climate mitigation and adaptation measures compared to the other riparian countries (Fig. 21.4).

Total: 15.15 US$ billion (2015–2019)

Comparing the climate finance per capita across the countries changes the ranking of the countries largely. However, the per capita financing stays low in countries such as Sudan, DR Congo, South Sudan, Burundi, and Eritrea (Fig. 21.5). While mobilizing finance from local, unilateral, and

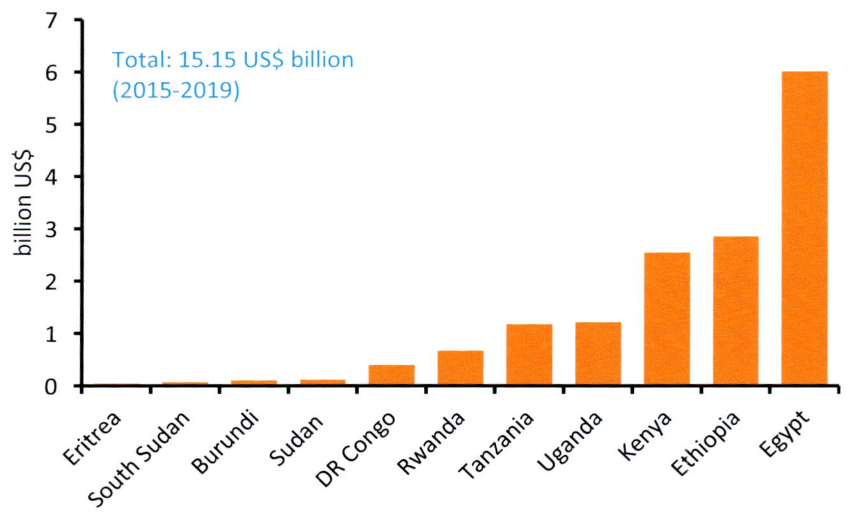

FIGURE 21.4

Climate finance recieved by the riparian countries of the Nile Basin for the period of 2015–2019.

Source: Authors' assessment based on MDBs (2019).

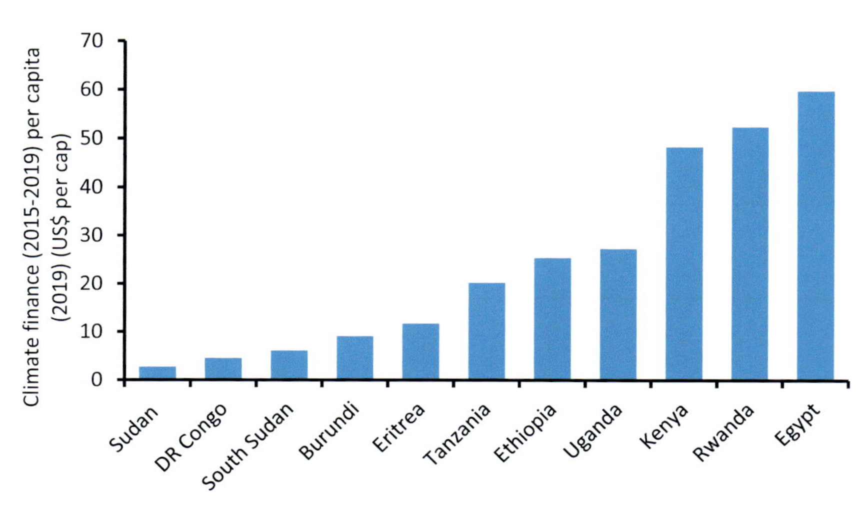

FIGURE 21.5

Climate finance (2015–2019) per capita (2019) across the riparian countries of the Nile Basin.

Source: Authors' assessment based on MDBs (2019).

multilateral bodies depends on the policies and priorities of these institutions, lack of adequate human and institutional capacities for developing financially attractive climate projects can be one of the important reasons for this variation in receiving climate finance among riparian countries.

7. Key findings and recommendations

Our analysis revealed that most of the climate actions, as formulated in the NDCs, of the riparian countries of the Nile Basin are related to SDGs 15, 7, 2, 6, and 17. This stresses the importance of these SDGs and indicates the relevance of following a nexus approach, especially in the sectors of water, energy, and agriculture (food security) in this region. We detected multiple overlaps in climate action exist between these countries, as documented in their NDCs, which highlights the potential for transboundary cooperation. The identified overlaps in the NDCs and SDGs of the Nile countries can be used as entry points to build action synergies and promote transboundary cooperation.

As exhibited by the actions related to SDG 17 (partnership for goals), all the Nile Basin countries have committed to cooperate with international organizations for the fulfillment of the NDCs. Especially, climate actions related to promoting renewable energies, water harvesting, efficient irrigation water uses, climate-smart agriculture, afforestation, and forest conservation are underlined by most of these countries. Despite high relevance to the region, enhancing the wider adoption of efficient cooking stoves and construction of the waste-to-energy systems were considered only in the NDCs of a few countries. Although hydropower was underlined as a relevant option for generating renewable energy by upstream countries, it went against the interests of the downstream regions that depend on stable upstream water supplies. Maintaining safe water sanitation, improving water quality, increasing the use of wastewater and desalinated seawater also did not receive much attention in the NDCs, despite their importance for public health and sustainable ecosystems. NDCs did not also sufficiently cover sustainable soil management and efficient fertilizer use measures that can greatly enhance food supplies.

According to previous studies, this region has huge potentials for solar energy (e.g., in Egypt and Sudan), and for hydropower generation (e.g., Ethiopia). Therefore, we think energy represents a key entry point that the Nile countries can cooperate on. Tapping these large potentials of renewable energy in these countries and exporting it to other countries within the Nile would have multiple benefits not only on sustaining socioeconomic development but also for ensuring environmental security and sustainability of natural resources base in these countries. For example, a shift toward renewable energy would reduce the dependency on wood fuel, which represent a high share of energy source in some of these countries. This consequently would create synergies with climate action that are connected to SDG15, which aims at afforestation, conserve and restore forest and important ecosystems (section 2.3.1). Additionally, in the case of hydropower, multipurpose dams can provide a reliable source of water, which can create synergies to sustain water availability to ensure water access for everyone (SDG6) (Section 21.3.4) and reduce hunger (SDG2), by providing water for irrigation (Section 21.3.3).

Availability of climate funds from MDBs and other local, unilateral, and multilateral sources, technology transfer and capacity building co-benefits incentivize them to commit to the NDCs. Although the Nile region is attracting substantial climate finance from the MDBs, for instance, there are large differences between the individual countries in the received amounts, which stresses the importance of regional actions and political stability. In addition to the partnership to get funds from international donors, a partnership among the riparian countries on developing human and institutional capacities is essential.

Effective coordination of the committed climate actions and achieving synergetic effects of maintaining water, energy, food, and environmental securities is central. The identified synergies and goals in the NDCs of the Nile's riparian countries can be considered as entry points to foster transboundary cooperation in this region. Regional cooperation would help in enhancing the coherence, efficiency, and effectiveness of climate and development actions and would reduce the required capital cost (Dazé et al., 2018). Future research may focus on two main directions (i) of actual development plans of the riparian countries, especially in the water, energy, and food sectors, and (ii) detailed mapping of climate finance sources and destination/priority sectors. These would be key information for investigating mechanisms for advancing joint climate action and benefit sharing in the Nile transboundary basin beyond political borders.

Acknowledgments

The authors would like to thank the Frankfurt School of Finance and Management for the fellowship provided to the first author to attend the course on "Certified Expert in Financing NDCs," during which the main analysis of this research was conducted. We would like to extend our gratitude to the NDC-SDG Connection platform, developed and maintained by the German Development Institute (DIE) and the Stockholm Environment Instiute (SEI) for making their data publicly available through their website.

References

AQUASTAT, 2020. AQUASTAT - FAO's Global Information System on Water and Agriculture. https://www.fao.org/aquastat/statistics/query/index.html?lang=en (Accessed 31 March 2022).

Choudhury, E., Islam, S., 2015. Nature of transboundary water conflicts: issues of complexity and the enabling conditions for negotiated cooperation. Journal of Contemporary Water Research and Education Issue 155, 43–52.

Coffel, E.D., Keith, B., Lesk, C., Horton, R.M., Bower, E., Lee, J., Mankin, J.S., 2019. Future hot and dry years worsen Nile Basin water scarcity despite projected precipitation increases. Earth's Future 7. https://doi.org/10.1029/2019EF001247.

Dazé, A., Terton, A., Maass, M., 2018. Alignment to Advance Climate-Resilient Development, Overview Brief 1 - Introduction to Alignment. International Institute for Sustainable Development (IISD). Available online at: https://napglobalnetwork.org/resource/alignment-to-advance-climate-resilient-development-2/ (accessed on 23.12.2020).

De Stefano, L., Duncan, J., Dinar, S., Stahl, K., Strzepek, K., Wolf, A., 2010. Mapping the resilience of international river basins to future climate change-induced water variability. World Bank Water Sector Board Discussion Paper Series 15.

Dessu, S., 2019. The Battle for the Nile with Egypt over Ethiopia's Grand Renaissance Dam Has Just Begun, Quartz Africa. Available at: www.qz.com/africa/1559821/ethiopias-grand-renaissance-dam-battles-egypt-sudan-onthe-nile/.

DIE & SEI, 2020. NDC-SDG Connection. German Development Institute (DIE) and Stockholm Environment Institute (SEI). https://klimalog.die-gdi.de/ndc-sdg/ (accessed on 20.12.2020).

El Bastawesy, M., Gabr, S., Mohamed, I., 2015. Assessment of hydrological changes in the Nile river due to the construction of renaissance dam in Ethiopia. The Egyptian Journal of Remote Sensing and Space Science 18 (1), 65–75. https://doi.org/10.1016/j.ejrs.2014.11.001.

FAOSTAT, 2020. Crop production data. Food and Agriculture Organization of the United Nations. http://www.fao.org/faostat/en/#data/QC (accessed on 17.12.2021).

Farinosi, F., Giupponi, C., Reynaud, A., Ceccherini, G., Carmona-Moreno, C., Gonzalez-Sanchez, D., Bidoglio, G., 2018. An innovative approach to the assessment of hydro-political risk: a spatially explicit, data driven indicator of hydro-political issues. Global Environmental Change 52, 286–313 (Link).

Hoff, H., 2011. Understanding the Nexus. Background Paper for the Bonn2011 Conference: The Water, Energy and Food Security Nexus. Stockholm Environment Institute, Stockholm. Available online at: https://www.sei.org/publications/understanding-the-nexus/ (accessed on 17.01.2021).

IEA, 2020. Electricity Access Database. International Energy Agency. https://www.iea.org/reports/sdg7-data-and-projections/access-to-electricity.

IPCC, 2007. Climate Change 2007: Impacts, Adaptation and Vulnerability, Contribution of Working Group II to the Fourth Assessment Report of the Intergovernmental Panel on Working Climate Change. Cambridge University Press.

Kasymov, S., 2011. Water resource dispute: conflict and cooperation in drainage basins. International Journal on World Peace 28 (3), 81–110.

Kimenyi, M., Mbaku, J., 2015. Governing the Nile River Basin: The Search for a New Legal Regime. Brookings Institution Press, Washington, D.C. http://www.jstor.org/stable/10.7864/j.ctt130h973.

Mahlknecht, J., González-Bravo, R., Loge, F.J., 2020. Water-energy-food security: a Nexus perspective of the current situation in Latin America and the Caribbean. Energy 194, 116824. https://doi.org/10.1016/j.energy.2019.116824.

McCracken, M., Wolf, A., 2019. Updating the register of international river basins of the world. International Journal of Water Resources Development 35 (5). https://doi.org/10.1080/07900627.2019.1572497.

MDBs, 2019. Joint Report on Multilateral Development Banks' on Climate Finance. Multilateral Development Banks (MDBs). Available online at: https://www.ebrd.com/2019-joint-report-on-mdbs-climate-finance (accessed on 12.11.2020).

Motsert, E., 2005. How Can International Donors Promote Transboundary Water Management? Discussion Paper. Deutsches Institut für Entwicklungspolitik, 8/2005.

Sayed, M.A.-A., 2004. Impacts of Climate Change on the Nile Flows. Ph.D. Thesis. Faculty of Engineering, Ain Shams University, Cairo, Egypt.

Sutcliffe, J.V., Parks, Y.P., 1999. The Hydrology of the Nile. Oxfordshire. The International Association of Hydrological Science (IAHS).

Swain, A., 2011. Challenges for water sharing in the Nile basin: changing geo-politics and changing climate. Hydrological Sciences Journal 56 (4), 687–702. https://doi.org/10.1080/02626667.2011.577037.

Tahiya, A., 2019. Transboundary water conflict resolution mechanisms: substitutes or complements. Water 11, 1337.

Tate, E., Sutcliffe, J., Convay, D., Farquharson, F., 2004. Water balance of Lake Victoria: update to 2000 and climate change modelling to 2100. Hydrological Sciences Journal 49, 563–574. https://doi.org/10.1623/hysj.49.4.563.54422.

UN, 2015. Transforming Our World: The 2030 Agenda for Sustainable Development. United Nations, New York (2015). https://sustainabledevelopment.un.org/post2015/transformingourworld.

UNDESA, 2020. World Population Prospects. https://population.un.org/wpp/Download/Standard/Population/ (Accessed 31 March 2022).

UN Economic Commission for Europe (UNECE), 2009. Guidance on Water and Adaptation to Climate Change. Available online at: www.unece.org/index.php?id=11658/ (accessed on 12.11.2020).

UNEP, 2013. Adaptation to Climate-Change Induced Water Stress in the Nile Basin: A Vulnerability Assessment Report. Division of Early Warning and Assessment (DEWA). United Nations Environment Programme (UNEP), Nairobi, Kenya.

UNFCCC, 2015. Paris Agreement. United Nations Framework Convention on Climate Change (UNFCCC). In: https://unfccc.int/process/conferences/pastconferences/paris-climate-change-conference-november-2015/paris-agreement.

Wheeler, K.G., Jeuland, M., Hall, J.W., Zagona, E., Whittington, D., 2020. Understanding and managing new risks on the nile with the Grand Ethiopian renaissance dam. Nature Communications 11, 5222. https://doi.org/10.1038/s41467-020-19089-x. Available online at: https://doi.org/10.1038/s41467-020-19089-x (accessed on 17.01.2021).

Whittington, D., Wu, X., Sadoff, C., 2005. Water resources management in the Nile basin: the economic value of cooperation. Water Policy 7 (3), 227−252. https://doi.org/10.2166/wp.2005.0015/. Available online at: (accessed on 24.01.2021).

WHO and UNICEF, 2019. Joint Monitoring Programme for Water Supply, Sanitation and Hygiene (JMP). https://washdata.org/.

Wolf, A.T., 2009. A long term view of water and international security. Journal of Contemporary Water Research and Education 142, 67−75. https://doi.org/10.1111/j.1936-704X.2009.00056.x.

World Bank, 2020a. GDP Per Capita (Current US$). https://data.worldbank.org/indicator/NY.GDP.PCAP.CD.

Wold Bank, 2020b. Prevalence of Moderate or Severe Food Insecurity in the Population (%). https://data.worldbank.org/indicator/SN.ITK.MSFI.ZS.

Yoffe, S., Aaron, T.W., Giordano, M., 2003. Conflict and cooperation over international freshwater resources: indicators of basins at risk. Journal of the American Water Resources Association (JAWRA) 39 (5), 1109−1126.

Zhang, W., Pan, X., 2016. Study on the demand of climate finance for developing countries based on submitted INDC. Advances in Climate Change Research 7 (1−2), 99−104. https://doi.org/10.1016/j.accre.2016.05.002.

CHAPTER

22 Energy and environment: sustainable development goals and global policy landscape

Liliana N. Proskuryakova
National Research University Higher School of Economics, Moscow, Russia

Chapter outline

1. Introduction: the process of shaping energy and environment policy landscape

Environmental and sustainability concerns typically appear on the agenda when researchers and policy-makers consider market and behavioral failures. The market failures include the public good nature of energy resources, environmental externalities, high capital costs, deficiencies in marginal cost pricing for electricity, efficiencies in prduction and allocation, and lack and asymmetry of information (Siebert, 2008; Gillingham et al., 2009; Sovacool, 2011). Behavioral failures are associated with barriers and constraints to pro-ecological behavior (Gaspar, 2013).

Energy policy is closely related with economic policy, foreign policy, and national and international security policy. Traditionally, the goals of energy policy were closely linked with assuring energy security, including "security of supply, affordability, and limited impact on the environment" (Kohl, 2004). Thus, environmental considerations have been integrated in energy policy and senergy security, and this interlink has been accentuated in the course of the past decades. In the same period, environmental and climate issues have gained importance and, similar to energy considerations, have

Handbook of Energy and Environmental Security. https://doi.org/10.1016/B978-0-12-824084-7.00006-0
Copyright © 2022 Elsevier Inc. All rights reserved.

also been included in the economic, international relations, and security domains. Additionally, energy and environmental policy discussions are increasingly intertwined with technological change, as energy and environment policy interventions may create significant effects (constraints or incentives) for technological progress (Kohl, 2004; Proskuryakova, 2018).

Environmental policy, like energy policy, is closely connected with a number of other policy domains, including economic, energy, and industrial policy. The interactions between ecosystems and mankind result in environmental degradation, the increasing economic and social demand for natural resources, ecosystem services, and environmental goods, and represent the major challenges for environment security (Zurlini and Müller, 2008). Thus, environmental security involves threats to national and human security posed by environmental change (Barnett, 2020). According to the AC/UNU Millennium Project, environmental security incorporates the following components: "adverse impact of human activities on the environment," "the direct and indirect effects of various forms of environmental change (especially scarcity and degradation), which may be natural or human-generated on national and regional security," and "insecurity individuals and groups (from small communities to humankind) experience due to environmental change such as water scarcity, air pollution, global warming, and so on" (Zurlini and Müller, 2008).

The process of national energy and environment policy-making follows the same cycle as most other policy processes (Fig. 22.1). The main difficulties arise from the need to integrate priorities that may be diverse or even polar. Another complication is the need to take into account the interests and opinions of a wide range of stakeholder groups, where the position of energy stakeholders (especially large energy companies) may prevail.

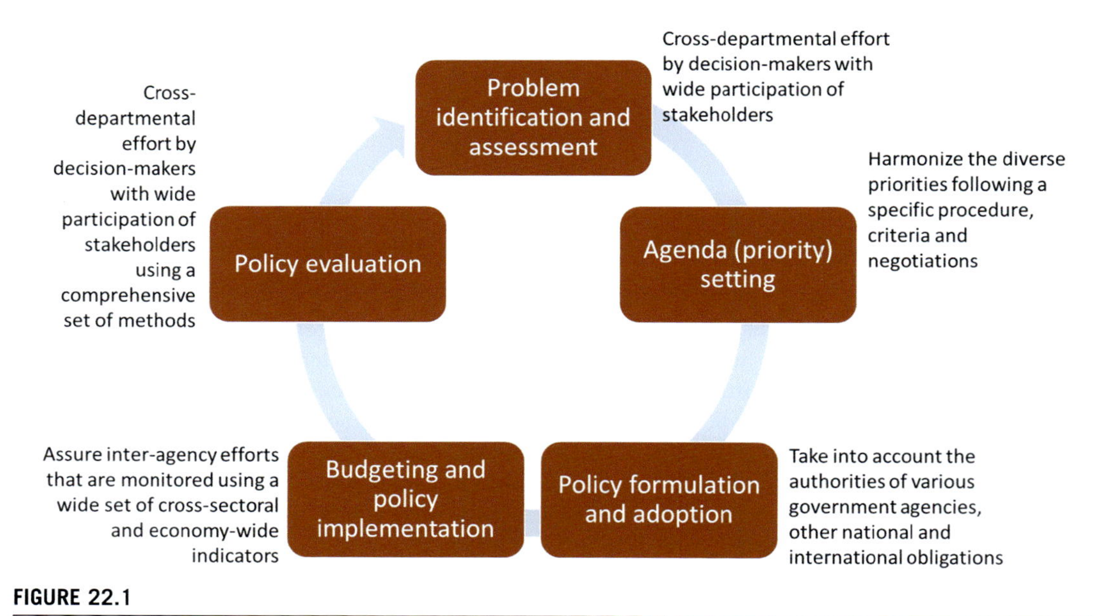

FIGURE 22.1

Energy and environment policy-making cycle.

Source: Author's analysis.

Global policy-making is more complicated. International arena is the area where national interests clash, countries try to move forward their priorities by establishing alliances or exercising pressure, seeking compromise or using other (i.e., economic) issues as a leverage. Energy security has traditionally caused multiple diplomatic and military conflicts among nations. The Organization of the Petroleum Exporting Countries (OPEC) and International Energy Agency (IEA) were consecutively established as a response to the 1970s oil crisis.

Global energy and environmental policies are developed not only by policymakers. Other stakeholders also initiate new policies or express positions on existing laws and regulations. Business offers advice and good practice (such as closed-loop technologies) and addresses grand challenges as part of corporate social responsibility and charity programs (such as Bill and Melinda Gates Foundation). Companies could also react to security risks by establishing international organizations, as it was the case with the World Energy Council (Russ, 2020).

At the global level, green growth was put forward at the G20 Los Cabos Summit in Mexico in 2012 (G20 Information Center, 2012) and provided the basis for a business alliance of corporations (B20). The B20 Group is made up of business leaders from 19 developed and developing countries and the European Union. Until 2019, the B20 Group has addressed issues of energy and environmental security numerous times. In 2020, B20 and IEA issued a joint call on the G20 to accelerate clean energy transitions for a resilient economic recovery, building on the energy demand and related greenhouse gas (GHG) emissions downward trends set by the COVID-19 pandemic. The joint statement calls for accelerated deployment of existing and new low-emissions and emissions-neutral technologies, including hydrogen, energy storage, and carbon capture, storage, and use CCSU. B20 and IEA also underline the need "to secure energy systems and provide access to affordable and uninterrupted flow of clean energy for all" (IEA, 2020).

Global civil society is also an active player in advancing and monitoring sustainable policy agenda. Environmental civil society organizations (CSOs), like Greenpeace and WWF, point out government or corporate actions that violate environmental law. They are also instrumental in raising public awareness on environmental security issues. In some instances, CSOs may put forward policy initiatives aimed at putting energy-related issues on political agenda and ultimately changing energy systems. Given the close relations between environmental issues, climate change, and energy, the same CSOs are often involved in setting both environmental and energy security policy agenda (Biagini and Sagar, 2004; Blanchet, 2015).

This chapter is structured as follows. Section 2 discusses multilateral agreements that set up environmental and energy goals. Section 3 analyzes options for harmonizing environmental and energy security priorities, including the integration of environment and energy policy-making under one administration, close interdepartmental cooperation, and adoption of single policy documents (international agreements, laws, and regulation, etc.). Section 4 looks into the future of international relations around energy and environment security.

2. The main international agreements with environmental and energy goals

The international agreements, negotiations around them, and national strategies provided a significant impetus for advancing renewable and other low-carbon technologies: hydrogen; CCSU; solar and wind; geothermal, wave, and other, including revolutionary ones. This section features several multilateral agreements with environmental and energy goals, with a focus on those that are universal in nature, i.e., signed by all (or nearly all) countries in the world under the auspices of the United Nations, the world's only organization with universal representation. The documents are arranged by subject area.

2.1 Green growth and sustainable development agreements

The Johannesburg Declaration and Plan of Implementation, agreed at the World Summit on Sustainable Development, identify global challenges and specific timetable to address some of them, as well as underline the continuous international oversight on sustainability agreements. Both documents emphasize the human dimension of sustainability and the cross-sectoral security aspects, such as food security and secure housing. With regard to energy security, the focus is on energy efficiency, affordability and accessibility, and the integration of energy issues into socioeconomic programs. The interlink of energy and environmental security is made through improved "*access to reliable, affordable, economically viable, socially acceptable, and environmentally sound energy services and resources, taking into account national specificities and circumstances, through various means, such as enhanced rural electrification and decentralized energy systems, increased use of renewables, cleaner liquid and gaseous fuels, and enhanced energy efficiency, by intensifying regional and international cooperation in support of national efforts, including through capacity-building, financial and technological assistance, and innovative financing mechanisms, including at the micro- and meso-levels, recognizing the specific factors for providing access to the poor*" (UN, 2002).

In 2009, the green economic growth policy was officially adopted by the Organization for Economic Cooperation and Development (OECD) as strategic medium-term (until 2030) and long-term (until 2050) development paths (OECD, 2011). Forty-seven countries, including those that are not members of OECD, adhered to the Declaration. There is no "one-size-fits-all" solution for fostering greener growth that depends on economic development, policy and institutional context, societal issues, environmental pressures, and other factors. Therefore, countries put forward green growth plans and strategies that should take into account national social and economic priorities, and social equity considerations (OECD, 2009, 2021). For example, Vietnam National Green Growth Strategy (Viet Nam National, 2012) "*must be based on science and modern technologies*" and "*contributes to employment, poverty reduction, and improving the material and spiritual life of all people.*" The document is also linked with the national climate change strategy.

"The Future We Want," the outcome document adopted at Rio+20, focuses on green economic growth without negative externalities for the environment (UN, 2012). Similarly to the Johannesburg Declaration and Plan of Implementation, the document emphasizes the human dimension of sustainability (including access to sustainable modern energy services for all) and the cross-sectoral security aspects, such as food security. Energy efficiency measures are called for in "*urban planning, buildings, and transportation, and in the production of goods and services and the design of products.*" Sustainable development is regarded as integration of economic, social, and environmental aspects, and their interlinkages at all levels. The document underlines the use of renewable and other low-emission energy technologies, increasing energy efficiency and more sustainable use of fossil fuels, access to electricity, and sustainable cooking and heating solutions. According to the document, the listed technologies are important for securing development, meeting the climate targets, and promoting research and development in all countries.

One of the key documents is the UN General Assembly Resolution that set the Sustainable Development Goals (SDGs) in September 2015 (UN, 2015). The SDGs for 2015–30 became the new agenda and a plan of action "for people, planet, and prosperity." Energy issues are in the focus of SDG7 "Ensure access to affordable, reliable, sustainable, and modern energy for all," while a number of other SDGs also tackle sustainable generation, distribution and use of energy and natural resources. Environmental concerns are integrated in SDG 15 "Protect, restore, and promote sustainable use of terrestrial ecosystems, sustainably manage forests, combat desertification, and halt and reverse land degradation and halt biodiversity loss," SDGs 6 and 12. The global effort to attain SDGs sets the international policy agenda around sustainable development that addresses energy, climate, and environmental challenges together (Proskuryakova and Loginova, 2021).

Energy and environment security is addressed in SDG7 through the energy access and affordability aspects, in SDG15 by combatting environmental problems caused by climate change and anthropogenic activity, restoration of degraded ecosystems, and addressing threats to human security caused by the loss of biodiversity, resource scarcity, and ecosystem change. There are 12 environment targets related to SDG15 that address a wider array of security issues compared with five energy targets for SDG7. If considered together with eight targets of SDG6 on water and sanitation, it may be noted that there are many more environmental targets that cover a wide array of security aspects. SDG 12 merges environment and energy targets and security elements under the common umbrella of sustainable production and consumption (UN, 2021).

2.2 Environment and climate agreements

There are several conventions that deal with pollution of air, water, and soil (i.e., Convention On Long-Range Transboundary Air Pollution 1979) and a few that limit and control the pollutants—waste generation, disposal, and movement (i.e., Basel Convention on the Control of Transboundary Movements of Hazardous Wastes and Their Disposal 1989). These agreements are aimed at minimizing damage to the environment. The implications for the energy industry most commonly include the shift to clean energy sources, application of pollution prevention and resource efficient technologies, and adoption of circular economy solutions (such as waste-to-energy). Importantly, multilateral agreements underline the transboundary nature of energy and environmental security challenges, and list solutions that need to be applied by all parties.

Agreements and conventions that focus on climate change also state energy and environment-related goals. The Vienna Convention for the Protection of the Ozone Layer (1985) is a framework convention that requires countries to take control actions to protect the ozone layer. The Vienna Convention principles were later laid down in the Montreal Protocol on Substances that Deplete the Ozone Layer (1987). The 1997 Kyoto Protocol operationalized the United Nations Framework Convention on Climate Change (UNFCC): its signatories commited to limit GHG emissions in line with agreed national targets. In 2015 the Kyoto Protocol was replaced by the Paris Agreement, a legally binding international treaty that aims to prevent and mitigate global warming. It has multiple implications for environment and energy security, including advancement of renewable and low-carbon energy technologies, the sustainable use of fossil fuels and large hydropower plants, as well as increased energy efficiency. As stated by Yvo de Boer, UNFCC Executive Secretary, at the Major economies meeting on climate change and energy security in September 2007: "*Overall, mitigation action needs to ensure that the aggregate result of mitigation initiatives is commensurate with the challenge of maintaining climate security*" (UN Climate Change, 2007).

The UN Convention to Combat Desertification in Those Countries Experiencing Serious Drought and/or Desertification, Particularly in Africa (1994) specifies the following actions that have to be included in national action plans "*sustainable management of natural resources, especially the rational management of drainage basins*," "*sustainable management of natural resources in high-altitude areas*," "*rational management and conservation of soil resources and exploitation and efficient use of water resources*." Additionally, convention signatories need to develop and apply emergency plans to address risks related to desertification, such as droughts, by using specially designed instruments like early warning systems, and efficient use of available energy sources, including nontraditional ones. The agreement also underlines the need to enhance the availability of scarce natural resources in affected areas, such as water, through the application of new technologies, including cloud seeding. Information about best available technologies, including the new ones could be made accessible through specific information systems and centers along with terms and conditions of their acquirement.

Summarizing the main features of multilateral environmental agreements, it may be noted that agreements with the largest number of signatories are environmental conventions, since they tackle issues of global public good. Issues like climate change and transboundary pollution cannot be addressed by few countries and require a global approach (Besedeš et al., 2020). Agreements that regulate the use of resources (i.e., fisheries) are commonly concluded with neighboring countries, trade partners, or counties that share common interest. The topics covered by international environmental agreements are very diverse: about one-third of them is related to species, one-third to pollution and energy, and the remaining third are devoted to wide range of other issues (Mitchell et al., 2020).

2.3 Energy agreements

One of the key international energy documents is the International Energy Charter Treaty of 1994 that provides a multilateral framework for cross-border energy cooperation (Energy Charter, 1994). The Treaty aims at promoting energy security by establishing more open and competitive energy markets, with due consideration of sustainable development principles and energy resources sovereignty. Energy security is enhanced through "the protection of foreign investments..., nondiscriminatory conditions for trade in energy materials, products and energy-related equipment based on WTO rules..., the resolution of disputes between participating states..., the promotion of energy efficiency, and attempts to minimize the environmental impact of energy production and use." There are a number of related documents that besides the consolidated Energy Charter Treaty,[1] include The International Energy Charter of May 20, 2015, The European Energy Charter of December 17, 1991, and the Protocol on Energy Efficiency and Related Environmental Aspects of December 17, 1994.

Energy obligations are included in multinational sustainable development and green growth documents—UN resolutions, Council of the European Union, and European Commission documents (Council of the European Union 2017; EUR-Lex, 2013 and other). Unlike environmental issues, energy problems often cause more polarized views, creating obstacles for multinational agreements. Thus, energy agreements and international agreements that incorporate energy targets are typically less binding, have smaller membership, or concluded at regional level. For instance, the International

[1]Includes understandings, declarations and decisions contained in the Final Act of 1994, and changes introduced by the Protocol of Correction of 2 August 1996.

Energy Charter Treaty has 53 signatories, while Vienna Convention for the Protection of the Ozone Layer, Montreal Protocol, UN Framework Convention on Climate Change, UN Convention to Combat Desertification, Convention on Biological Diversity, and Transforming our world: the 2030 Agenda for Sustainable Development (Sustainable Development Goals 2030) have over 190 signatories.

Overall, the universal nature of international energy and environmental agreements creates a level playing field for all countries around the world. Even though universal agreements are more difficult to agree upon and they are typically more general, they lower perceived risks of zero-sum game and provide a much more significant outcome. Universal agreements also offer a solid framework for discussions, information exchange, and cooperation around energy and environmental security issues for all parties involved. If all countries are in, no states would feel discriminated against or attempt to be a free rider.

Whenever it is not possible or feasible to conclude universal agreements, regional or local treaties are made. Undoubtedly, they resolve significant security matters, while at the same may create tensions with other countries. Moreover, close regional energy integration may decrease the demand for cumulative installed capacity and decrease negative geographic and socio-environmental impacts due to resource optimization and greenhouse emissions mitigation (Santos, 2021). However, this conclusion may be region-specific and may not apply to some regions. For example, a study of intraregional trade among the South Asian economies points to a negative interlink between trade increase and renewable energy consumption (Murshed, 2021).

Lessons learned from the past agreements (i.e., the Millennium Development Goals, the Kyoto Protocol) are relevant for the treaties and declarations that came to replace them (SDGs, The Paris Agreement). The criticism of these documents focuses on incentives for participation and compliance, and several issues have to be taken into account in future accords. First, universal global goals need to be tailored for each country and rely on specific policy, environmental, economic, and social context. The value of this approach is to make high-level and general targets relevant and specific for each country; ambitious, but achievable. The challenge of this approach is to make sure that these national goals and targets that nation-states set for themselves are high enough, and countries are not imitating progress toward common goals. Second, they have to be supplemented by national action plans with clear division of responsibility, allocation of financial resources, cross-sectoral and interdepartmental coordination, and cross-country cooperation to address transboundary issues. Third, there should be different approaches to countries depending on their level of development (GDP per capita) and the volume of emissions. Finally, national and regional level goals and targets should be designed to ensure accountability. Not all agreements preview a mechanism to enforce compliance with commitments. Additional difficulty is to gather high-quality data to monitor progress in implementation and compare these data across countries (Philibert, 2004; Diniz, 2007; UNECE, 2015).

3. Harmonizing environmental and energy policies from the security perspective

As noted in Section 1, the energy and environmental policy-making process is similar and could be unified. As stated in Section 2, many environmental and energy targets set in multilateral agreements are closely connected. This section looks into various approaches to harmonizing environmental and energy security policies at institutional and instrumental levels.

There is a set of security factors in energy policy that may contradict those in environmental policy: the need to provide affordable energy for all; budget constraints and the choice of economically feasible energy sources (energy technologies); social and economic acceptability by various stakeholders; the need to assure accessible energy; and good governance (Indriyanto et al., 2011). To give some examples, the dirtiest energy carrier—coal—may be the most affordable and economically feasible energy resource in certain locations. Environment and climate friendly large windmills may face protests from local communities, and certain territories covered by solar panels may be alternatively used for agriculture. Access to energy could imply more high-voltage transmission lines or gas pipelines across an area that may be regarded as problem by local dwellers or have negative implications for nature and wildlife. Social acceptance of certain energy technologies (like nuclear) may also significantly limit policy options.

There are several approaches to harmonizing energy and environmental policies, while assuring security considerations for both domains. Often these issues are grouped under the overarching "sustainability" topic. First is the integration of environment and energy policy-making under one administration (government agency). Second, close interdepartmental cooperation could help government agencies responsible for energy and environment to achieve coherence of both agencies' policies. This could be done through governmental commissions or consultations. Third, one document (international agreement, law, regulation, etc.) could cover both energy and environmental domains, as well as joint security implications. These options are discussed below and illustrated with Fig. 22.2.

The integration of environment and energy policy-making under one administration is an attempt to eliminate some of the inherent contradictions at institutional level. Some international organizations and national authorities organize special units that integrate energy and environment policy-making. A good example here is the US Department of Agriculture that set up the Office of Energy and

Common administration

- One ministry or agency
- Department or unit within a ministry or agency

Inter-departmental cooperation

- Inter-departmental (government) commissions and working groups

Integration of priorities in one document

- Legal regulations
- Common strategy documents
- Action plans and roadmaps

FIGURE 22.2

Approaches to harmonizing energy and environmental policies.

Source: author's own analysis.

Environmental Policy (OEEP). OEEP is a focal point that integrates the Department's energy, environmental, and climate change activities and assures harmonized energy and environmental policy analysis, long-term planning, research priority setting, and response strategies. This single unit accommodates the Office of Environmental Markets (OEM), the Climate Change Program Office (CCPO), and Energy Policy and New Uses (OEPNU). The range of issues covered by OEEP is vast and includes the development of renewable energy and resources; construction of climate resilient and sustainable farms, rural lands, and rural communities; expansion of markets for conservation and environmental goods, and more (US Department of Agriculture, 2021).

Close interdepartmental cooperation is an option when setting up a single energy and environment administration is not deemed possible or feasible. For example, in Russia, the Commission for Strategic Development of the Fuel and Energy Sector and Environmental Security was established to assure interdepartmental cooperation, to coordinate the administrative efforts at federal, regional, and local levels in advancing the energy sector; ensuring industrial, energy, and environmental security; and the rational use or resources. Among the issues discussed at the Committee meetings are the wide application of cutting-edge technologies in coal mining to increase efficiency and safety. The industry should comply with stringent environmental standards along the entire supply chain from mining to coal-handling places, including sea ports, to ensure improvements in quality of life and health, as well as urban environment (President of Russia, 2021).

At the instrumental level, policy-makers tend to integrate energy security concerns in environmental legal and regulatory documents and vice versa. For example, Canada has developed Defense Energy and Environment Strategies (DEES) to integrate energy efficiency and sustainability in the activities of the Department of National Defense (DND) and the Canadian Armed Forces (CAF). DEES is harmonized with a range of other national documents related to sustainable development, green growth, and climate, as well as international agreements such as the 2030 United Nations Agenda for Sustainable Development and the United Nations Declaration on the Rights of Indigenous Peoples. The latest strategy outlines priorities for 2020–2023 and focuses on sustainable use of defense assets, such as land and buildings; green procurement; investments in green technologies and nontechnological innovations; and input in Canada's goal of net-zero GHG emissions by 2050 (Government of Canada, 2017).

In addition to the three approaches to harmonizing energy and environmental policies described above, certain efforts may be applied by international organizations. For instance, the IEA has followed several strategies since 2015. The organization attempted to widen its membership beyond the OECD base, and involve new members, including emerging economies. The IEA also sought to broaden its mandate so that it covers clean energy transition and issues of interest to the new member states, such as energy access. It also attempted to increase its influence at the international arena, in particular, through partnerships with other forums, such as G20 (Downie, 2020).

With the advent of COVID-19 pandemic, international organizations initiate policies to help their member states in preserving energy and environmental security simultaneously. For instance, the UN Economic and Social Commission for Asia and the Pacific (ESCAP) has issued a note summarizing the analytical and research work on energy security and resilience for implementation of the 2030 Agenda. ESCAP approaches energy and environmental security in the context of international cooperation related to integration of new technologies, platforms, and frameworks, and addressing associated resilience challenges. The reviewed research serves as a basis for the development of evidence-based energy security and resilience policy in the region (ESCAP, 2021).

4. Universalism, regionalism, or nation-state protectionism: what is the future of international relations around energy and environment?

With the appearance of new challenges, energy and environmental security experiences more pressure than ever before. Traditionally, security issues have been handled at international and national levels. However, in recent years, there is an increasing influence of subnational and virtual organizations, i.e., cyber terrorism. Not only security issues but also economic and governance issues are increasingly handled at subnational level, thus contributing to greater regionalization.

The changing balance of power among global, middle, and regional states and world regions also have an impact for energy and environmental security. Global powers (world powers and superpowers) are typically characterized by superiority in military, economic, and political power in addition to geographic outreach (O'Loughlin, 2009). They address global security challenges related to price and supply stability at global energy markets, international trade of energy resources and energy technologies, and international energy cooperation. Other global challenges include, but not limited to rising capital needs and misinformation in the energy industry, energy and environmental aspects of rising vehicle use, rising global energy demand and the impact of largest world energy consumers (e.g., China and India), energy poverty, nuclear technologies proliferation, global environment and climate (and related clean energy) targets, resource scarcity, and the deployment of energy infrastructure (Dorian et al., 2006). The examples of superpowers are the United States and USSR during the Cold War period and the British Empire at the end of the 19th century.

Global incidents of market failure make up rationale for global public policy interventions and actions related to energy security. At least four key areas should be subject to global public policy regulation in the energy sphere: "market transparency and planning security; negative spill-over effects of a global scale; free rider problems; and the existence of global market imbalances caused by cartels" (Sovacool, 2011). With regard to environment and climate, global governance should be exercised in relation to any issues of planetary or interregional scale, as all environmental issues may be classified as public good. Having this in mind, it should be noted that there is an inherent collective action problem related to investments in energy infrastructure aimed at increasing energy security due to its public goods character (Sovacool, 2011).

The terms "middle powers" and "regional powers" are somewhat vaguer, and their meanings are debatable. Middle powers (or major powers) are typically referred to states that have somewhat less power in the international arena than superpowers. Although these states lack superpower capabilities, they still exercise significant influence and possess the capacity necessary to shape international relations. The examples of middle powers are Australia, Canada, the United Kingdom, China, and Germany. The middle powers cannot act on their own (as superpowers do) and rely on alliances with other states and "niche diplomacy" to influence global politics in many aspects including energy and environment (Yilmaz, 2019).

Regional powers exercise their influence within a certain geographic region. The regional power capabilities cannot be challenged by any other country in its region making it the regional hegemon. The regional powers shape regional energy and environment security agendas. Regional alliances are instrumental in addressing environmental issues (i.e., a river and its basin) or energy resources (OPEC countries) limited to a certain locality (Yilmaz, 2019). Examples of regional powers are the Mexico, Indonesia, Egypt, Saudi Arabia, and Turkey.

One example of regional alliances with an energy security agenda is the Eurasian Economic Union (EEU). The EEU seeks to create common energy markets for oil and petroleum products, natural gas, and electricity. The close cooperation in the energy area is driven by economic, policy, and development issues. EEU's development institution—the Eurasian Development Bank—provides funds for projects that benefit member states, such as those related to green energy, energy access, and energy security (Bianco et al., 2021). As stated in the EEU 2025 Strategy, "one of integration priorities shall be the improvement of energy saving and energy efficiency, the resolution of existing environmental problems, and the ensuring of sustainable development" (Eurasian Economic Council, 2020).

Within larger countries, policy-makers and researchers also differentiate between various regions that have diverse energy mix, energy supply and demand, and environmental issues. In countries like the United States of America, Canada, China, and Russia, certain regions may be energy poor and other regions may generate the bulk of national energy resources, some regions may accommodate large energy consumers, while others may be sparsely populated with few small economic actors (Proskuryakova, 2017). Researchers employ the concept of "energy regions" with regard to regions within a country to indicate common policy approach based on long-term visions that frame short-term objectives and policy evaluation. These "guiding visions" play an important role in energy transition management for mobilizing social actors and the interagency coordination (Späth and Rohracher, 2010).

Over two decades starting from the year 2000 trends in international relations suggested more regionalism and nation-state protectionism in dealing with energy and environment security. This statement may be illustrated with new OPEC and OPEC + deals, the inability of international community (and the largest emitters) to agree on a new climate change accord, and many countries failing to meet the Paris Agreement goals. The US President Biden announced in 2021 ambitious plans to decrease emissions by 50%–52% below the 2005 level by 2030, thus doubling the commitment of the previous administration. If this happens, it will be done outside the Paris Agreement, as the United States ceased its participation in 2017.

With the advent of the COVID-19 pandemic, most countries faced new challenges related to economic slowdown, overburden of national healthcare systems, and a major decline in international trade (UNCTAD, 2020). Energy systems have also experienced difficulties due to a drop in demand, lower prices, deterrent and nonpayments, and disconnection bans (Proskuryakova et al., 2021). These changes contributed to a more prominent role of nation-states and greater concerns about energy security. Environment pollution and greenhouse gas emissions have reduced due to decreasing economic activity, and this triggered discussions about a more ambitious and consistent green growth strategy (Ranjbari et al., 2021). Green recovery is particularly visible in the financial markets (Naeem et al., 2021), agriculture, tourism, energy, and a few others (Ranjbari et al., 2021).

Even though the recent developments contributed to the advancement of regionalism and nation-state protectionism, the future of energy and environment security is associated with universal agreements. Institutional settings that regulate energy and environmental security should be treated as transnational public goods and require global cooperative efforts. Researchers note that public character of energy security may limit coordination in other areas of global governance, such as climate change and environment pollution (Valdes, 2021). Therefore, energy, climate, and environment security in the global security policy agenda should be treated together, in a coherent and comprehensive manner.

5. Conclusions

Over the past decades, energy and environmental policy have been closely integrated. Moreover, there is an interlink with other policy domains, such as economic policy and foreign and security policy. Although policy agenda in this area is mainly set by government agencies, businesses and civil society are also active players. Environmental and energy issues have been connected not only in policy-making but also in research and public discourse.

The national and international policy-making cycle in energy and environmental security is similar to that in other policy domains. The differences are associated with the necessity to integrate security concerns that in certain cases may be opposite for energy and environment. This integration may be pursued through international agreements, common energy and environment administration, inter-departmental cooperation, and common strategic documents. Additionally, international organizations could also make an input in addressing this challenge.

There are clear benefits in concluding universal energy and environmental agreements. Besides setting the harmonized energy and environmental goals common for most countries around the world, they lower the probability of zero-sum game and free rider risks, limit discrimination, and result in more substantial outcomes. These agreements provide a basis for discussions, information exchange, and cooperation around energy and environmental security issues for all signatories. Regional agreements may also result in security gains for countries involved, including a more optimal use of resources and production capacities among member states, while at the same time may create tensions with the third parties.

The changes in economy, trade, and society that became visible over the last decades contributed to a more prominent role of nation-states and greater concerns about energy security. There has also been a growing trend toward regionalization. At the same time, the future of energy and environment security is associated with universalism and global accords. The public goods nature of both energy and environment, grand challenges that international community faces (climate, pandemic, etc.), and the need to address them in most efficient and timely manner are the arguments in favor of this approach.

Acknowledgment

The chapter is based on the study funded by the Basic Research Program of the National Research University Higher School of Economics.

References

Barnett, J., 2020. Environmental security. In: Kobayashi, A. (Ed.), International Encyclopedia of Human Geography, second ed. Elsevier, pp. 247–251.

Besedeš, T., Johnson, E.P., Tian, X., 2020. Economic determinants of multilateral environmental agreements. International Tax and Public Finance 27 (4), 832–864.

Biagini, B., Sagar, A., 2004. Nongovernmental Organizations (NGOs) and Energy. Encyclopedia of Energy, pp. 301–314.

Bianco, V., Proskuryakova, L., Starodubtseva, A., 2021. Energy inequality in the Eurasian economic union. Renewable and Sustainable Energy Reviews 146, 111155.

Blanchet, T., 2015. Struggle over energy transition in Berlin: how do grassroots initiatives affect local energy policy-making? Energy Policy 78 (C), 246–254.

Council of the European Union, 2017. Implementing the EU Global Strategy - Strengthening Synergies between EU Climate and Energy Diplomacies and Elements for Priorities for 2017. Council of the European Union. https://data.consilium.europa.eu/doc/document/ST-6981-2017-INIT/en/pdf (Accessed 01 July 2021).

Diniz, 2007. Lessons from the Kyoto Protocol. Ambiente Sociedade 10 (1), 27–38.

Dorian, J.P., Franssen, H.T., Simbeck, D.R., 2006. Global challenges in energy. Energy Policy 34 (15), 1984–1991.

Downie, C., 2020. Strategies for survival: the international energy Agency's response to a new world. Energy Policy 141, 111452.

Energy Charter, 1994. The Energy Charter Treaty. In: https://www.energycharter.org/process/energy-charter-treaty-1994/energy-charter-treaty/ (Accessed 1 July 2021).

ESCAP, 2021. Enhancing Energy Security in The Context of the Coronavirus Disease Pandemic for a Greener, More Resilient and Inclusive Energy Future in the Region. Note by the Secretariat, p. 11.

EUR-Lex, 2013. Report from the Commission to the European Parliament, the Council and the European Economic and Social Committee. Implementation of the Communication on Security of Energy Supply and International Cooperation and of the Energy Council Conclusions of November 2011. https://eur-lex.europa.eu/legal-content/EN/ALL/?uri=CELEX%3A52013DC0638 (Accessed 1 July 2021).

Eurasian Economic Council, 2020. Strategic Directions for Developing Eurasian Economic Integration until 2025 Approved by Heads of EAEU States. http://www.eurasiancommission.org/en/nae/news/Pages/11-12-2020-02.aspx (Accessed 30 June 2021).

G20 Information Center, 2012. G20 Los Cabos 2012: G20 Leaders Declaration, G20 Information Centre. http://www.g20.utoronto.ca/2012/2012-0619-loscabos.html (Accessed 24 February 2021).

Gaspar, R., 2013. Understanding the reasons for behavioral failure: A process view of psychosocial barriers and constraints to pro-ecological behavior. Sustainability 5, 2960–2975. https://doi.org/10.3390/su5072960.

Gillingham, K., Newell, R.G., Palmer, K., 2009. Energy efficiency economics and policy. Annual Review of Resource Economics 1 (1), 597–620.

Government of Canada, 2017. Defense Energy and Environment Strategy. https://www.canada.ca/en/department-national-defence/corporate/reports-publications/dees.html (Accessed 1 July 2021).

IEA, 2020. B20 and IEA Call on the G20 to Accelerate Clean Energy Transitions for a Resilient Economic Recovery - News. IEA. https://www.iea.org/news/b20-and-iea-call-on-the-g20-to-accelerate-clean-energy-transitions-for-a-resilient-economic-recovery (Accessed 1 July 2021).

Indriyanto, A.R.S., Fauzi, D.A., Firdaus, A., 2010. The sustainable development dimension of energy security. In: The Routledge Handbook of Energy Security. Taylor & Francis Group.

Kohl, W.L., 2004. National Security and Energy. Encyclopedia of Energy. Elsevier Science, pp. 193–206.

Mitchell, R.B., et al., 2020. What we know (and could know) about international environmental agreements. Global Environmental Politics 20 (1), 103–121.

Murshed, M., 2021. Can regional trade integration facilitate renewable energy transition to ensure energy sustainability in South Asia? Energy Reports 7, 808–821.

Naeem, M.A., et al., 2021. Comparative efficiency of green and conventional bonds pre- and during COVID-19: an asymmetric multifractal detrended fluctuation analysis. Energy Policy 153, 112285.

O'Loughlin, J., 2009. Superpower. International Encyclopedia of Human Geography, pp. 82–86.

OECD, 2009. Declaration on Green Growth. OECD. https://www.oecd.org/env/44077822.pdf (Accessed 1 July 2021).

OECD, 2011. Towards Green Growth'. OECD. https://www.oecd.org/greengrowth/48012345.pdf (Accessed 1 July 2021).

OECD, 2021. Green Growth in Countries and Territories. OECD. https://www.oecd.org/greengrowth/greengrowthincountriesandterritories.htm (Accessed 1 July 2021).
Philibert, C., 2004. Lessons from the Kyoto Protocol: implications for the future. International Review for Environmental Strategies 5 (1), 311−322.
President of Russia, 2021. Presidential Commissions, President of Russia. http://en.kremlin.ru/structure/commissions (Accessed 1 July 2021).
Proskuryakova, L.N., 2017. Russia's Energy in 2030: future trends and technology priorities. Foresight 19 (2), 139−151.
Proskuryakova, L., 2018. Updating energy security and environmental policy: energy security theories revisited. Journal of Environmental Management 223, 203−214.
Proskuryakova, L.N., Loginova, I., 2021. Energy and environment: sustainable development goals and global policy landscape. In: Asif, M. (Ed.), Energy and Environmental Security in Developing Countries. Springer International Publishing (Advanced Sciences and Technologies for Security Applications, Cham, pp. 355−374.
Proskuryakova, L., Kyzyngasheva, E., Starodubtseva, A., 2021. Russian electric power industry under pressure: post-COVID scenarios and policy implications. Smart Energy 3, 100025.
Ranjbari, M., et al., 2021. Three pillars of sustainability in the wake of COVID-19: a systematic review and future research agenda for sustainable development. Journal of Cleaner Production 297, 126660.
Russ, D., 2020. Speaking for the "world power economy": electricity, energo-materialist economics, and the World Energy Council (1924−78). Journal of Global History 15 (2), 311−329.
Santos, T., 2021. Regional energy security goes South: examining energy integration in South America. Energy Research and Social Science 76. Article 102050.
Environmental quality as a public good. In: Siebert, H. (Ed.), 2008. Economics of the Environment: Theory and Policy. Springer, Berlin, Heidelberg, pp. 59−95.
Sovacool, B.K., 2011. The Routledge Handbook of Energy Security, first ed. Taylor & Francis Group https://www.routledge.com/The-Routledge-Handbook-of-Energy-Security/Sovacool/p/book/9780415721639 (Accessed 1 July 2021).
Späth, P., Rohracher, H., 2010. Energy regions": the transformative power of regional discourses on socio-technical futures. Research Policy 39 (4), 449−458.
UN, 2002. Plan of Implementation of the World Summit on Sustainable Development. World Summit on Sustainable Development (WSSD). https://www.un.org/esa/sustdev/documents/WSSD_POI_PD/English/WSSD_PlanImpl.pdf (Accessed 17 June 2021).
UN, 2012. The Future We Want. Outcome Document of the United Nations Conference on Sustainable Development. https://sustainabledevelopment.un.org/content/documents/733FutureWeWant.pdf (Accessed 17 June 2021).
UN, 2015. Resolution Adopted by the General Assembly on 25 September 2015. Transforming Our World: The 2030 Agenda for Sustainable Development', UN. https://www.un.org/ga/search/view_doc.asp?symbol=A/RES/70/1&Lang=E (Accessed 1 July 2021).
UN, 2021. Department of Economic and Social Affairs. Sustainable Development. The 17 Goals. https://sdgs.un.org/goals (Accessed 1 July 2021).
UN Climate Change, 2007. Major Economies Meeting on Climate Change and Energy Security. UNFCCC. https://unfccc.int/news/major-economies-meeting-on-climate-change-and-energy-security (Accessed 17 June 2021).
UNCTAD, 2020. COVID-19 Drives Large International Trade Declines in 2020. UNCTAD. https://unctad.org/news/covid-19-drives-large-international-trade-declines-2020 (Accessed 1 July 2021).
UNECE, 2015. The Millennium Development Goals in Europe and Central Asia Lessons on Monitoring and Implementation of the MDGs for the Post-2015 Development Agenda, Based upon Five Illustrative Case

Studies. http://www.un-rcm-europecentralasia.org/fileadmin/DAM/RCM_Website/Publications/RCM_MDG_full_report_final.pdf (Accessed 08 October 2021).
US Department of Agriculture, 2021. Energy and Environmental Policy. https://www.usda.gov/oce/energy-and-environment (Accessed 1 July 2021).
Valdes, J., 2021. Participation, equity and access in global energy security provision: towards a comprehensive perspective. Energy Research and Social Sciences 78, 102090.
Viet Nam National, 2012. Green Growth Strategy. https://www.giz.de/en/downloads/VietNam-GreenGrowth-Strategy-giz2018.pdf (Accessed 1 July 2021).
Yilmaz, S., 2019. Middle Powers and Regional Powers. Oxford Bibliographies.
Zurlini, G., Müller, F., 2008. Environment security. In: Jørgensen, S.E., Brian, D., Fath, B.D. (Eds.), Encyclopedia of Ecology. Elsevier.

Further reading

Jaffe, A.B., Newell, R.G., Stavins, R.N., 2003. Environmental degradation and institutional responses. In: Handbook of Environmental Economics, pp. 461–516.

CHAPTER

Energy and environment: sustainability and security

23

Tri Ratna Bajracharya[1], Shree Raj Shakya[1,2] and Anzoo Sharma[1,3]

[1]*Center for Energy Studies (CES), Institute of Engineering, Tribhuvan University, Pulchowk, Lalitpur, Nepal;* [2]*Institute for Advanced Sustainability Studies (IASS), Potsdam, Germany;* [3]*Center for Rural Technology (CRT/N), Kathmandu, Nepal*

Chapter outline

1. Introduction to environmental sustainability

John Morelli (2011) on the first issue of the *Journal of Environmental Sustainability* defines environmental sustainability (EnvS) as "*meeting the resource and services needs of current and future generations without compromising the health of the ecosystems that provide them.*" He further elaborates it as "*a condition of balance, resilience, and interconnectedness that allows human society to satisfy its needs while neither exceeding the capacity of its supporting ecosystems to continue to regenerate the services necessary to meet those needs nor by our actions diminishing biological diversity.*" EnvS is about the responsible collaboration with the environment to support human life and meet the needs of development, both social and economic, without depleting or degrading the natural resources so that this generation does not jeopardize the future generations' ability to meet their needs.

Among four kinds of capital: human, human-made, social, and natural, EnvS deals with *natural capital*. Natural capital is the assets provided by the environment like water, woodlands, soil,

Handbook of Energy and Environmental Security. https://doi.org/10.1016/B978-0-12-824084-7.00010-2
Copyright © 2022 Elsevier Inc. All rights reserved.

atmosphere, and petroleum. EnvS necessitates the natural capital be sustainably utilized as a source of economic inputs and as a sink for waste, which implies natural equilibrium be maintained by not harvesting natural resources faster than they can be regenerated, not depleting nonrenewable resources at the rate higher than renewable substitute be developed, and not emitting wastage faster than they can be assimilated by the environment (Diesendorf, 2000; Goodland and Daly, 1996). EnvS ensures ecosystem integrity and carrying capacity of the natural environment such as it remains productively stable and resilient to support human life and development (Brodhag and Taliere, 2006).

Development is desired by every human and society, and in doing so, the environment is so heavily exploited that the limiting factor for development has become the availability of natural capital. For example: in the furniture industry, production is limited by timber in remaining forest rather than sawmills, electricity generation is limited by the availability of fuel/water rather than turbines and so on, and petroleum is limited by the oil in reserve and absorbing capacity of CO_2 rather than the capacity of refineries. When the resources are consumed in an environmentally sustainable manner, the harvest of renewable resources, depletion of nonrenewable resources and pollution absorption can be maintained for the foreseeable future.

2. The linkage between energy and environment

2.1 Environmental issues of fossil fuel

Though the energy sources such as fossil fuel, hydro, and nuclear play an enormous role in structuring today's society and most likely will collapse on its unavailability, a large function of our present and anticipated environmental problems is quite directly related to energy utilization which contributes to the warming of the Earth and consequently climate change. The environmental impacts associated with each activity of exploration (Fig. 23.1), production, refining, transportation (Fig. 23.2), and utilization (Fig. 23.3) are immense. Oil and gas contribute to 54% of the total global energy supply (IEA, 2020c), and in the process, the exploration for and production of petroleum soil, surface, and

FIGURE 23.1

A fire aboard the mobile offshore oil drilling unit Deepwater Horizon, located in the Gulf of Mexico, 2010.

Source: US Coast Guard/EPA.

FIGURE 23.2

Ixtoc oil spill in 1979 that ran wild for nine months and spilled over 140 million gallons of oil into the Bay of Campeche in the Gulf of Mexico.

Source: National Oceanic and Atmospheric Administration (2014).

FIGURE 23.3

Coal-fired power station Neurath in Grevenbroich, North Rhine-Westphalia, Germany.

Photo Credit: Pixabay.

groundwater are polluted extremely. Some potential environmental impacts of the oil and gas industry are listed below (Table 23.1).

In addition, the routine flaring[1] performed at small to medium oil industries causes air pollution.

The oil spill is another major accidental disaster that tends to happen when pipelines break, drilling operations go wrong, or big oil tanker ships (Figs. 23.1 and 23.2).

The addition of heat to the Earth's environment as a result of human activity is completely due to energy utilization. Climate changing is a global threat to the world. The causes for this global threat are many, among them, greenhouse gas (GHG) emission is one of them. The biggest source of human-caused GHG emission is the energy consumption: 73% (World Resources Institute, 2021)

[1]Flaring is the burning of byproduct gases generated during oil extraction.

Table 23.1 Potential environmental impacts of the oil and gas industries.

Impact	Causes
Water contamination	• Effluent rich in inorganic salt, wash water, and cooling water discharges • Seepage from storage and waste tanks • Oil spills
Thermal pollution	• Effluents discharge at a temperature greater than the recipient bodies temperature
Air pollution	• Particulate emission • Sulfur and nitrogen oxides, ammonia, acid mist, and fluoride compounds gas emissions • Routine flaring at small and medium oil industries
Land and surface/ groundwater pollution	• Disposal of solid wastes like sludge and particulate matter from chemical processes
Noise pollution	• Equipment and operations
Overall environmental pollution	• Accidents like large oil spills, leaks, and explosions

Reproduced from Barboza Mariano et al. (n.d.).

(Fig. 23.4); within the energy sector, electricity and heat producers contributed to 15 $GtCO_2e$, transportation 7.7 $GtCO_2e$, and industries, manufacturing, and construction sector 11.7 $GtCO_2e$. CO_2 makes up to 74% of GHG emissions, and fossil fuel combustion is the source of 89% of CO_2 emission. In 2018, the CO_2 emission from the combustion of fuel was 33.5 Gt (IEA, 2020a). Because of global warming, the snow cover in the Northern Hemisphere and Himalayas and floating ice in the Arctic Ocean are reported to decline.

The top 10 emitters are responsible for more than two-third of annual global greenhouse emission, 50% of the global population and around 60% of the world's GDP. The 27 countries of European Union emit 7.8% while China, the biggest emitter, contributes to 26% of global GHG emissions, followed by the United States at 13%, and India at 6.7% (World Resources Institute, 2021) (Fig. 23.5).

2.2 Environmental issues of renewables

All energy sources have some effect on the environment. There is substantially more harm of use of fossil fuel, coal, oil, and natural gas as compared to the renewable energy sources. The impacts can differ at air and water pollution level, impact to public health, wildlife and habitat loss, water usage, land usage, and global warming potential. Nevertheless, renewable sources such as wind, solar, geothermal, biomass, tides, and hydropower also come up with some substantial environmental impacts. The actual form and intensity of impacts differ with the technology employed, the geographic location, and several other factors. As renewable is taking a larger portion of our electricity system,

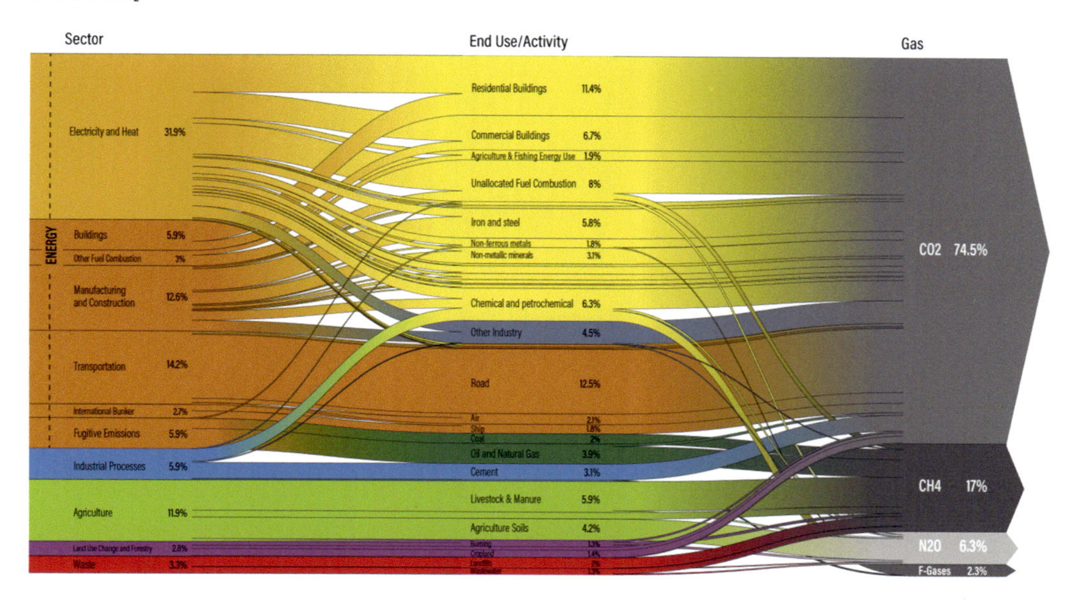

Source: Greenhouse gas emissions on Climate Watch. Available at: https://www.climatewatchdata.org

WORLD RESOURCES INSTITUTE

FIGURE 23.4

World Greenhouse Gas Emission in 2018 (Sector/End use) (48.9 $GtCO_2e$).

Ge and Friedrich (2021).

addressing the potential environment issues associated with the technology production and installation is essential for EnvS (EnergySage, 2021).

There are advantages to renewable energy, and at the same time, several disadvantages are to be taken into consideration. The lifecycle emissions of renewable energy are presented below (Table 23.2).

It is noted that all energy sources have environmental impacts. As the source of nonrenewable energy is limited, the need of turning it into renewable energy is inevitable. As renewable energy also has some negative impacts, more effective and efficient ways to produce, distribute, and utilize energy be necessary to investigate.

3. Environmental security

Since the beginning of the 21st century, environmental security has become an integral part of national security and foreign policy as sustainability and natural resource protection have become the essential elements. The 1994 UN Development Program's Human Development Report was the turning point in

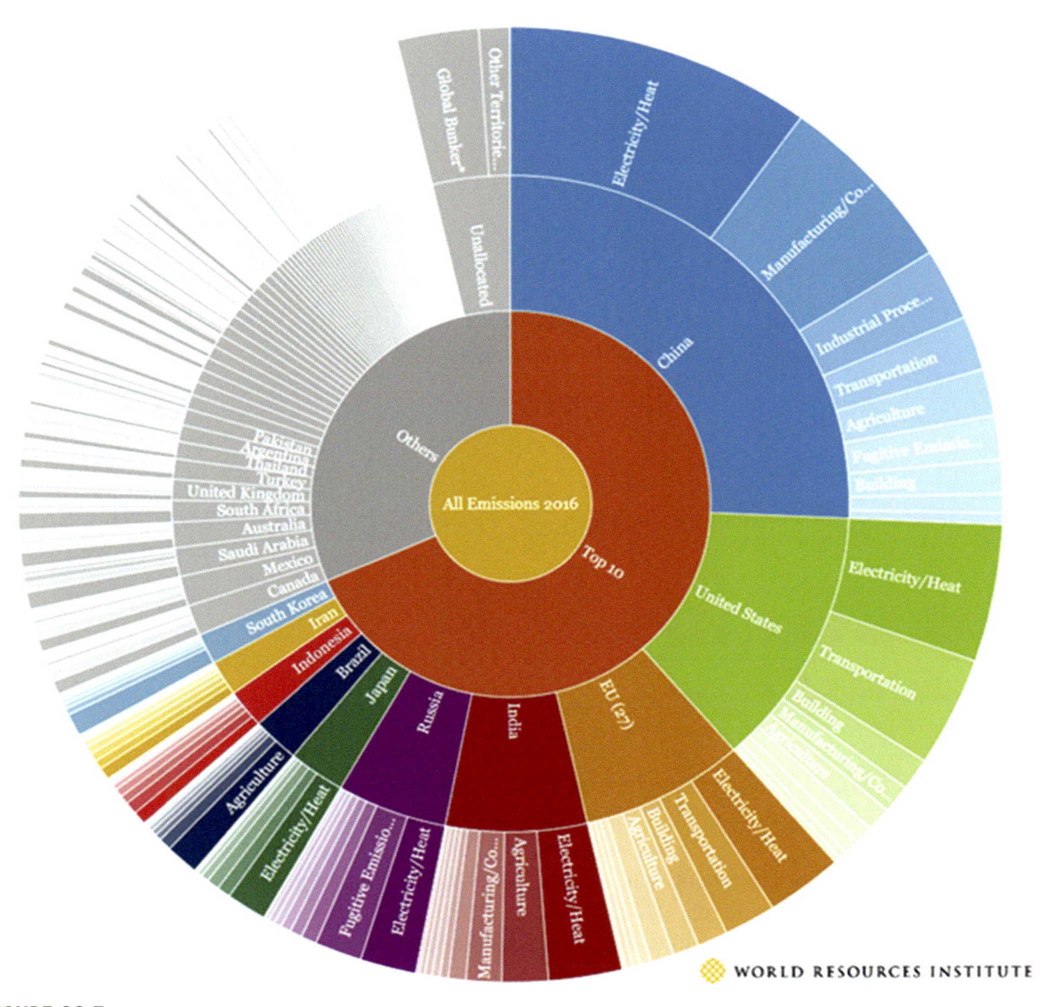

FIGURE 23.5

The top 10 greenhouse gas emitters.

Source: World Research Institute (2021).

recognizing the human condition at the center of national and international security. Toward the early 1990s, the US security interests were increasingly affected by the human security and instability triggered by environmental crises over the scarce resources and was worsened by the failure of governments to address citizen demands for political, social, and economic freedoms, and extremist ideologies (The Concept of Environmental Security –, 2016).

It has always been implicitly recognized that when people are deprived of enough food, water, shelter, or the natural resources essential to survive leads to unstable situations. Over the past two

Table 23.2 Lifecycle emissions from renewable energy power plants.

Technology	Wind	Hydropower	Concentrated solar (solar thermal)	Nuclear	Geothermal	Solar PV
Lifecycle emissions (gCO_2eq/kWh)	11	24[a]	27	12	38	48

[a]Environmental consequences associated with hydropower are being studied, and their actual lifecycle emissions can be expected to be significantly higher than 24 gCO_2eq/kWh.
Source: EnergySage, 2021.

decades, this has been more explicitly understood both in defense strategy and in the environmental community. Consequently, the traditional national security theory and environmental protection and policy development are converging. The convergence has given a new dimension, the environmental security (Yale Insights, 2012).

It is a universal truth that energy security and environmental security come together. The notion of the environmental security is context-specific, but it generally examines the threat posed by environmental events and trends (caused by nature or human process) to individuals, communities, or nations. They entail a wide range of issues such as resource access, equity, economics, land tenure, and property rights. Environmental security is one of the key policy concerns in developing counties where the livelihood of most people is dependent on natural environmental resources such as forests (Bajracharya et al., 2021).

4. Complementarities and conflicts issues of energy security and environmental security

The two complicatedly linked terms, energy and environmental security, are important issues that must be addressed. The national security of any country is associated with the exploitation of its available resources. National security does not entirely rely not on military power projection, but the protection of energy infrastructure at home and outside (Triola, 2008). On the other hand, environmental security is safeguarding the natural environment and vital interests of citizens and society. Specifically, it is also an interest of the state to be alert to the internal and external influences, adverse activities and developmental trends that risk human health, biodiversity, ecosystems, and humankind existence.

The energy security (ES) and environment (ENV) have a direct or indirect relationship with each other for which the functional form is specified in Eqs. (23.1 and 23.2).

$$ES = f(ENV) \tag{23.1}$$

$$ENV = f(ES) \tag{23.2}$$

The Eqs. (23.1 and 23.2) are always the conditional basis for its direct impact or adverse impact of the exploitation of energy resources. If a country's energy supply is based on fossil fuel resources, there might be environmental insecurity. Even by exploiting renewable energy resources, there are some environmental impacts if the resources are not harnessed properly. As such, the complimentary or conflicts issues of energy security and environmental security exist and will be there for a long time and shall probably grow greater in the year ahead. For example, with the rise in economies of China and India, the rate of energy consumption is increasing at remarkable rate. The oil consumption in China increased by almost half a million barrels per day since 2006 and accounts for the 38% of the total global oil consumption. Electricity consumption in India is increasing from 562 billion kilowatt-hours in 2000 to 1593 billion kilowatt-hours in 2019. Overall, the EIA forecasts that worldwide oil consumption will reach 118 million barrels per day in 2030.

Though energy and environmental security seem to oppose each other, measures of climate change mitigation are the most effective means to achieve energy security. Instituting a reliable, pragmatic, and effective framework for collaboration on climate change is the key means to address the energy and environmental security for creating an immediate influence (Mckibbin and Wilcoxen, 2007).

Energy conservation policies can potentially reduce extreme fossil fuel dependence with little to less detrimental consequences on economy and lead toward sustainable environment. However, the crude oil—exporting countries and especially the Organization of the Petroleum Exporting Countries (OPEC) are more prone to experience a difficult energy transition regime than any other nations, as the economies of these countries are highly dependent on crude oil revenue (Mckibbin and Wilcoxen, 2007).

Energy security is always the priority area of any nation as energy is an important driver of economic growth and prosperity. It is not only important domestically but also sometimes depending on the geolocations and the resources available; access to energy sources is a strategic foreign policy concern as well. Energy security is more concerned with domestic or national interest, whereas environmental security is a global concern.

5. Selected case studies on energy, environmental sustainability, and security

Though the preindustrial era is accepted as the reference for global temperature (1.5°C above for livable climate), the changes in climate and environment can be widely sensed by recalling days a decade ago. The Earth's average temperature has risen more than 1°C since the late 19th century, and if the trend continues, the temperature can be expected to rise 3.2°C above preindustrial levels by the end of this century, sea level has risen 4—8 inches, and the precipitation over land has increased by about 1% (UNEP, 2019). For global warming to stay within 1.5°C, the world must achieve net-zero GHG emissions by 2050 and a total of 131 countries have pledged for or are considering some sort of commitment toward the target. However, with the existing policies and recently purposed commitments, Nationally Determined Contributions targeting at Sustainable Development Goal 7 (SDG 7) (ensuring universal access to affordable, reliable, sustainable, and modern energy services by 2030), SDG 3.9 (substantially reduce air pollution), and SDG 13 (taking effective action to combat climate change), the world is still not on the track to achieve the target. Nevertheless, exemplary projects are being carried out across the globe and a few of them are as follows.

5.1 Renewable energy—powered desalination plant: Al Khafji in Saudi Arabia and the Chtouka Ait Baha in Morocco

It is a well-known fact that water accounts for 71% of the Earth's surface, and only 3% of the Earth's water is freshwater of which 2.5% is unavailable: found in glaciers, polar ice caps, atmosphere and soil, and hence only 0.5% is readily available. The Middle East and North Africa region is the most water-scarce zone due to its hyperarid climate and imbalance in water supply and demand. The countries in this region especially Middle East heavily rely on the desalination of seawater, which is both energy and cost-intensive in nature: seawater reverse osmosis technique requires 3-7 kWh/m^3, whereas the distillation process requires 5.2 to 12.5 kWh/m^3 equivalent electrical energy (World Bank, 2019). According to the World Bank report on desalination (2019), until 2018, there were 18,426 desalination plants operating in over 150 counties accounting for daily production of 87 million cubic meters supplying for over 300 million people. The raw material, seawater, of the desalination process is limitless, free of exogenous risks like dependency and climate change risks. Meanwhile, the impact from intake facilities, from hypersaline brine effluent and carbon footprint, as most desalination plants are powered by fossil energies, draws concerns for environmental sustainability and energy security.

Rich countries like the United Arab Emirates, Saudi Arabia, Israel, South Africa, and Australia have been investing in cutting-edge technologies to better the efficiency, reduce the cost, and lessen the environmental footprint of the desalination process. The Middle East region is not just rich in oil and gas, but it is much richer in another power source: the sun. The solar-powered sea water reverse osmosis without storage facilities is observed to be the most cost-effective technology (Kettani and Bandelier, 2020), but the major drawback with the system is the intermittent nature of solar as the utility-scale desalination plants require to be in operation round the clock.

Al Khafji Desalination Project that commissioned in late 2017 is the first large-scale desalination plant in the world. The plant is producing 60,000 m^3 water per day and has an expansion plan of up to 90,000 m^3 per day. It is powered by 20 MVA solar farm during the day and grid electricity during the night. The utilization of abundant solar energy has reduced the operational cost and CO_2 emission.

The Chtouka Ait Baha Project is so far the largest renewable energy desalination project with the production capacity of 275,000 m^3 per day and 43 MWp solar PV system. The capacity is expected to expand to 400,000 m^3 by 2030.

5.2 Hydrogen economy of Iceland

Iceland, a Nordic island country with a population of 366,425 over an area of 103,000 sq km has the share of renewables in final energy consumption is as high as 76.7% in 2019, most of which comes from hydro for electricity (13,459 GWh) and geothermal for both electricity (6019 GWh) and space heating (33,917 TJ) (IEA, 2020b). Oil (530ktoe) contributes to 8.75% of the total energy supply and is predominately consumed in transport (64%, 337 ktoe) and the fishing industry (33%, 175 ktoe). Back in the 1990s, the country relied on imported fossil fuel for almost 30% of its energy supply (Fig. 23.6).

Iceland has the target to have 40% of transport energy needs sourced through renewables by 2030 and aims to be a carbon neutral country by 2040, and the journey began since 1998, when a letter of intent stating, "Iceland wish to pursue a hydrogen economy," was endorsed by the government (Jonsson et al., 2003). The plan is to replace oil-based transportation system with hydrogen or hydrogen-based fuels, produced through water electrolysis, utilizing electricity generated from geothermal and hydropower, and the first hydrogen bus was taken to roads in the early 2000s. Since 1990, the CO_2 emission of the country has remained almost constant: 1.9—2.3 Mt.

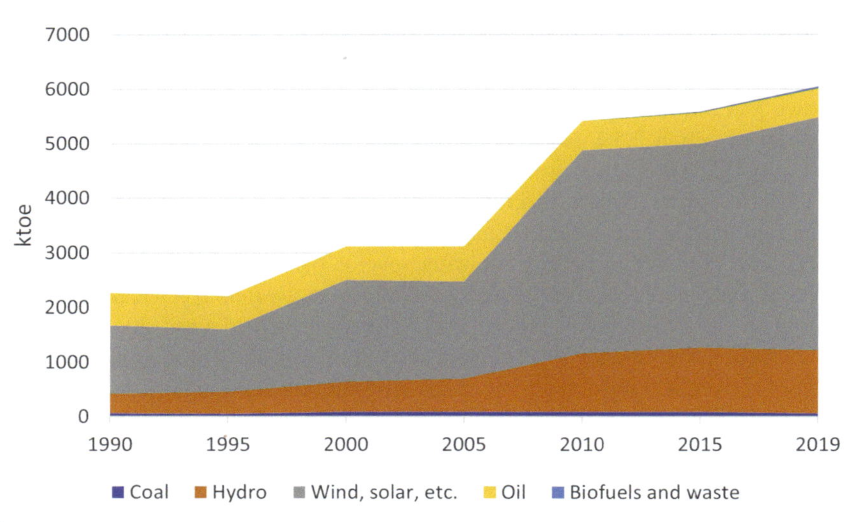

FIGURE 23.6

Total energy supply by source, Iceland 1990–2019.

Source: IEA (2020).

5.3 Three Gorges of China

The Three Gorges Dam, whose official construction began in 1994, is the largest dam and the largest hydropower project (22.5 GW) in the world. The dam is 2335 m long, 185 m high, and has a surface area of 1045 square kilometers (Singapore is 728.3 square kilometers). According to the initial proposal, the primary reason for the dam was preventing flood rather than power generation. The plant is designed to annually generate 88.2 billion kWh electricity. In the year 2020, 34 turbogenerators of the Three Gorges Dam generated 111.8 billion kWh of electricity, the highest in the world so far. The volume of the clean energy generated in 2020 is equivalent to saving 34.39 million tons of standard coal and reducing the emission of 94.02 million tons of CO_2, 22,400 tons of SO_2, and 22,000 tons of nitrogen oxides (CTG, 2021). The largest technological achievements have some major backlashes including issues of environmental impacts like water quality degradation in the higher banks, soil erosion around the river banks, and other is the risks associated with the location of the dam on a seismic fault line, though it is designed to withstand an earthquake up to a magnitude of 7 (Zhang, 2014). A study has found that about 530 million tons of silt are deposited into the reservoir on average annually (Yang et al., 2006). And lack of silt is likely to make the downstream river more vulnerable to flooding.

Environmental sustainability of energy systems represents the "transition of a country's energy system toward avoiding potential environmental harm and adapting to climate change impacts." As the consequences of global warming have already begun and in order to curb the impacts, energy transition at earliest is essential: switching to renewable energy, electrifying the carbon-intensive sectors like transportation and industries, an adaptation of more sustainable agriculture practices, limitation on deforestation and forest degradation, and sustainable production and consumption are a few important steps toward cutting down GHG released to the atmosphere and set the world on a better course and safer future.

6. Summary

Environmental sustainability is about the responsible collaboration with the environment to support human life and meet the needs of development, both social and economic, without depleting or degrading the natural resources so that this generation does not jeopardize the future generations' ability to meet their needs. Although the renewable energy technology cannot be excluded for its impact on environmental, the effects are minimum as compared to that of nonrenewables. As the rising economies are in race to secure their fossil needs, the best approach for energy security could be adopting climate change mitigation strategies. Renewable energy–based desalination plants in Saudi Arabia and Morocco, hydrogen economy of Ireland and hydro-power plants are few examples that create the synergy between environment and energy.

References

Bajracharya, Ratna, T., Darlami, S., 2021. Energy security in the context of Nepal. In: Asif, M. (Ed.), Energy and Environmental Outlook for South Asia. CRC Press, pp. 141–174.

Barboza Mariano, J., Lèbre La Rovere, E. (n.d.). Environmental Impacts of the Oil Industry. Retrieved from: https://www.eolss.net/sample-chapters/c08/e6-185-18.pdf.

Brodhag, C., Taliere, S., 2006. Sustainable development strategies: Tools for policy coherence. Natural Resources Forum 30 (2), 136–145. https://doi.org/10.1111/j.1477-8947.2006.00166.x.

CTG, 2021. Three Gorges Dam Sets World Record for Annual Power Generation in 2020. Retrieved July 18, 2021, from: https://www.ctg.com.cn/en/media/press_release/1084387/index.html.

Diesendorf, M., 2000. Sustainability and Sustainable Development by Mark Diesendorf. Sydney. Retrieved from: http://markdiesendorf.com/wp-content/uploads/2015/09/CorpSust2000.pdf.

EnergySage, 2021. Environmental Impacts of. Nuclear Energy.

Ge, M., Friedrich, J., 2021. World Greenhouse Gas Emissions: 2018. Retrieved from: https://www.wri.org/data/world-greenhouse-gas-emissions-2018?auHash=VOGL49W6tKejUPqq7wQWTERbg9PfurqCmiALyO7WWc8.

Goodland, R., Daly, H., 1996. Environmental sustainability: universal and non-negotiable. Ecological Applications 6 (4), 1002–1017. https://doi.org/10.2307/2269583.

IEA, 2020a. CO_2 Emission from Fuel Combustion. Retrieved June 30, 2021, from: https://www.iea.org/data-and-statistics/data-browser?country=WORLD&fuel=CO2 emissions&indicator=CO2BySector.

IEA, 2020b. Data & Statistics - Iceland. Retrieved July 18, 2021, from: https://www.iea.org/data-and-statistics/data-browser?country=ICELAND&fuel=Electricity and heat&indicator=HeatGenByFuel.

IEA, 2020c. Key World Energy Statistics 2020. Retrieved from: https://www.iea.org/reports/key-world-energy-statistics-2020.

Jonsson, H., Arnason, B., Sigfusson, T.I., 2003. Towards Hydrogen Economy in Iceland. Retrieved July 18, 2021, from: https://iea-etsap.org/workshop/worksh_6_2003/jonsson.pdf.

Kettani, M., Bandelier, P., 2020. Techno-economic assessment of solar energy coupling with large-scale desalination plant: the case of Morocco. Desalination 494. https://doi.org/10.1016/J.DESAL.2020.114627.

Mckibbin, W., Wilcoxen, P., 2007. Energy and Environmental Security. https://www.brookings.edu/wp-content/uploads/2016/07/200702_01energy.pdf.

Morelli, J., 2011. Environmental sustainability: a definition for environmental professionals. Journal of Environmental Sustainability 1 (1). https://doi.org/10.14448/jes.01.0002.

National Oceanic and Atmospheric Administration. (March 15, 2014). Ixtoc I oil spill. This spill was caused by a blownout oil well. The well ran wild for 9 months and spilled over 140 million gallons of oil into the Bay of Campeche, ID: 13949415013677, PublicDomainFiles.com. Retrieved October 3, 2021, from: http://www.publicdomainfiles.com/show_file.php?id=13949415013677.

The Concept of Environmental Security —. The Solutions Journal, 2016.

Triola, L.C. "Energy & National Security: An Exploration of Threats, Solutions, and Alternative Futures," 2008 IEEE Energy 2030 Conference, 2008, pp. 1—47, https://doi.org/10.1109/ENERGY.2008.4781047.

World Bank, 2019. The Role of Desalination in an Increasingly Water-Scarce World. Washington DC. Retrieved from: www.worldbank.org/gwsp.

World Resources Institute, 2021. Climate Watch Historical GHG Emissions. Retrieved June 30, 2021, from: https://www.climatewatchdata.org/ghg-emissions.

Yang, Z., Wang, H., Saito, Y., Milliman, J.D., Xu, K., Qiao, S., Shi, G., 2006. Dam impacts on the Changjiang (Yangtze) River sediment discharge to the sea: the past 55 years and after the three Gorges dam. Water Resources Research 42 (4). https://doi.org/10.1029/2005WR003970.

Zhang, W., 2014. Weighing the Pros and Cons: transformation of Angle of View for three Gorges dam. Natural Resources 05 (16), 1048—1056. https://doi.org/10.4236/NR.2014.516088.

What Is Environmental Security?, 2012. Yale Insights.

CHAPTER

24

Circular economy—A treasure trove of opportunities for enhancing resource efficiency and reducing greenhouse gas emissions

Maksud Bekchanov[1], Mayuri Wijayasundara[2] and Ajith de Alwis[3]

[1]*Research Unit Sustainability and Global Change (FNU), Center for Earth System Research and Sustainability (CEN), University of Hamburg, Hamburg, Germany;* [2]*Faculty of Science Engineering and Built Environment, Deakin University, Melbourne, VIC, Australia;* [3]*Department of Chemical and Process Engineering, University of Moratuwa, Moratuwa, Sri Lanka*

Chapter outline

1. Introduction

Intensifying risks accompanied by global warming and depletion of nonrenewable resource stocks create barriers to achieve sustainable development goals (European Commission, 2018; UNEP, 2011). Increasing use of fossil fuels and consequent raise of greenhouse gas (GHG) emissions are key drivers behind global warming, which is a major threat to economy and environment (IPCC, 2014). Increasing water scarcity due to climate change and depletion of resources stocks such as groundwater and

Handbook of Energy and Environmental Security. https://doi.org/10.1016/B978-0-12-824084-7.00016-3
Copyright © 2022 Elsevier Inc. All rights reserved.

phosphorus deposits influences food production systems and threatens food and income security (Cordell et al., 2009; IPCC, 2019). Pollution of oceans with plastics, eutrophication of inland water bodies, air pollution especially in urban areas occurs due to poor management of waste from factories, farmlands, and residential areas (Lebreton et al., 2017; Kaza et al., 2018; Strokal et al., 2020). Circular economy (CE) concept has recently gained traction to address these societal challenges and reduce the related economic and environmental risks (Donati et al., 2020). Differing from a recently dominant linear economic system that is based on "take-make-waste" approach and triggers undesired environmental externalities, CE is an economic system that is "...restorative and regenerative by design, and aims to keep products, components, and materials at their highest utility and value at all times." (Ellen MacArthur Foundation, 2013). Transformation toward CE goes along with restructuring value chains, changing the ways of producing, and using materials and products as well as recovering them at the end of life (EOL). CE strategies such as upcycling, recycling, recovering, reusing, remanufacturing, and refurbishing enhance the availability of cheaper materials and allow for gaining additional earnings.

CE options greatly reduce waste disposal and environmental damages. However, the influence of CE strategies on climate change mitigation prospects as reflected through GHG emission reduction opportunities has gained little attention. Current climate change mitigation studies also mostly focus on reducing GHG emissions by reforming the energy system, e.g., by decreasing the uses of fossil fuel—based energy, widely adopting renewable energies, and reducing energy losses along the energy supply chain (Material Economics, 2018). These studies have not considered material and energy uses and related GHG emissions related with the production of technologies for renewable energy generation, carbon capture, and storage. Consequently, climate policy studies have not adequately addressed energy demand reduction and climate change mitigation potentials of CE options that target efficient uses and reuses of materials and goods to effectively meet consumer needs. It is important to note that energy saving or conservation occurs not only through improved energy use efficiency but also through the dematerialization of the economy as the same economic output can be produced with the uses of less resources.

Assessments of the potentials of implementing CE options in heavy industries in the European Union (EU) indicated that expected GHG emission reductions due to improved material use efficiency and recovery are as high as climate mitigation potentials from energy-producing sectors (Ellen MacArthur Foundation, 2019). In less industrialized developing countries, improved materials and resource use efficiency and recycling waste in agriculture, textile, and waste management sectors also play an important role in reducing GHG emissions. CE strategies are, therefore, instrumental in supporting the efforts of achieving the Paris Agreement's climate policy targets (Material Economics, 2018). This study, therefore, reviews various CE strategies implementable in different sectors of economy and along the entire supply chain of commodity production. Particularly, we discus different CE options implementable in agriculture, textile, manufacturing, transport, construction, hotel industry, and municipal solid waste (MSW) management sectors. These options serve to reduce GHG emissions through enhancing wider adoption of renewable resource use technologies, decreasing carbon footprints related to consumption of materials and goods, improving material use efficiency, and increasing durability and usefulness of consumed goods.

2. "Circular economy" strategies and GHG emission reduction prospects

Economic system functions within environmental system, and its sustainability depends on the availability of natural resources (energy and material) and capacity of environment to rejuvenate after disposal of waste from production systems. Although recent economic development—which became possible due to enormous availability of fossil fuel–based energy—greatly improved human welfare (consumption, education, leisure, and fulfilled life), its environmental burden has been also extravagant as extraction of the resources and waste generation became beyond the carrying capacity of environmental systems. Linear economy (LE) which is based on "take-make-waste" approach has been criticized due to wasteful uses of natural resources and enormous environmental pollution effects. LE approach thus disregards the finiteness of the Earth's resources to produce economic output continuously and negligence of the welfare of the upcoming generations (Fig. 24.1).

Alternatively, CE which is based on "make-use-return" (cradle-to-cradle) approach considers EOL management and relevant design of products, components, and materials proactively while opening up new repurposing opportunities for constituent resources (or nutrients) and achieving increased resource efficiencies (Braungart et al., 2007). With improved design aspect of products, CE decreases nonvalue adding processing of postconsumption waste, increases resource recovery potential of products, and hence minimizes the demand for the extraction of natural resources and disposal of waste. As CE leads to a dematerialized economy through the use of product-service systems that

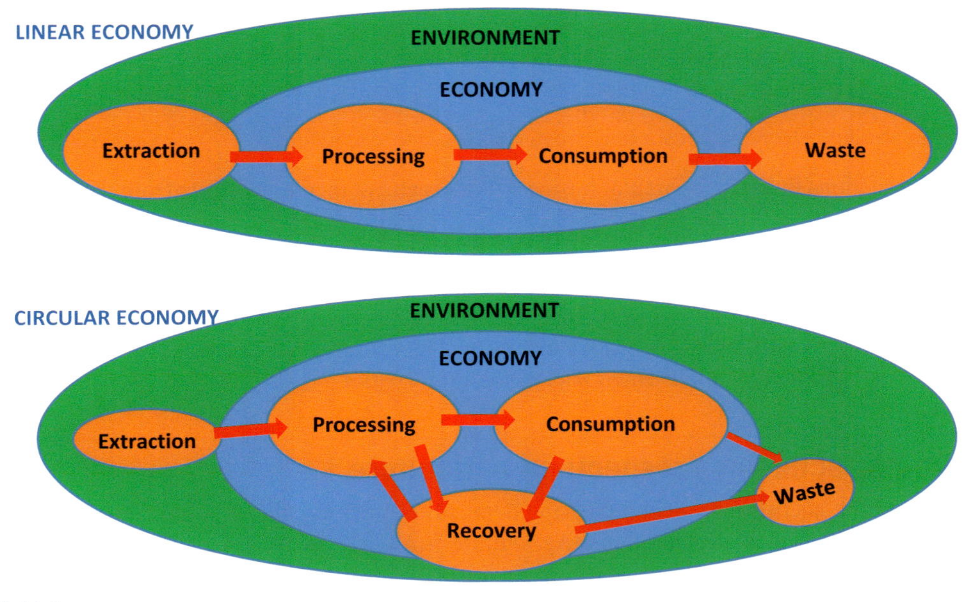

FIGURE 24.1

Linear economy versus circular economy.

Source: Authors' presentation.

provide access on products rather than on ownership, less amount of resources will be in circulation to provide the same economic output, reducing the energy demand of the economic system (Kasulaitis, 2019). Due to this, resource extraction, manufacturing, distribution of products as well as recovery of postconsumption waste will result in less environmental externalities.

"CE" strategies comprise measures beyond resources recycling and reuses and include all options to increase useful services per unit of resources (material) used. Rather than economic value of the commodity or services consumed, "CE" focuses on useful services (utility or cognitive satisfaction (happiness)) from the consumed goods and the ability of goods to address the fundamental human need with consumption. "CE" allows for satisfying needs for useful services with minimal uses of natural resources and minimal pollution (including GHGs). Relationship among GHG emissions (E), material uses (M), production of goods (Q), and useful services (W) can be expressed through this formula (Material Economics, 2018):

$$E = \frac{E}{M}\frac{M}{Q}\frac{Q}{W}W$$

GHG emission reduction, therefore, can be achieved by implementing policies that reduce either of these ratios $\left(\frac{E}{M}\right)$, $\left(\frac{M}{Q}\right)$, and $\left(\frac{Q}{W}\right)$. Three major CE strategies related are materials recovery and reuse (closing supply chains), improved resources use efficiency (eco-efficiency), and circular business models (e.g., product lifetime extension, access over ownership business models). Materials recovery and reuse strategies reduce demand for extraction of fresh resources and decrease pollution per material use $\left(\frac{E}{M}\right)$. In other words, more resources will be available without much increase of GHG emissions under this strategy. Improved resources use productivity can be achieved by reducing the input uses to produce the same amount of output or increasing output with the same uses of inputs. Thus, this strategy decreases resources (materials) use to output ratio $\left(\frac{M}{Q}\right)$, contributing to GHG emission reductions indirectly. Circular business models aim at increased utility from the consumption of goods and thus reduce the ratio of goods per utility $\left(\frac{Q}{W}\right)$. Sharing the goods (e.g., cars, houses, etc.), designing long-durable goods, providing access to goods rather than ownership, and extending the lifetime of goods through improved maintenance are examples for CE model strategies (Kjaer et al., 2018). As the consumption needs are met through consuming less amount of goods, this strategy also indirectly contributes to lowering GHG emissions.

3. Circular economy prospects in agriculture

While agriculture is critical to meet increasing demands of growing global population, environmental externalities of the sector are enormous. Agriculture is a major consumer of fresh water resources; production depends on tremendous uses of chemical fertilizers; agricultural machinery, irrigation operations, extraction of mineral rocks, and food processing and transportation rely on extensive uses of energy resources (Pimentel et al., 2008; Maupin et al., 2010). Agricultural sector is responsible for about 15% of GHG emissions; also, overuse and leakage of chemical fertilizer residues in the system is a major cause of water and soil pollution (Pelletier et al., 2011; Maupin et al., 2010). However, food losses in the system is 30%–50% implying large potential for improving resources use efficiency (Gustavsson et al., 2011). CE opens multiple opportunities to increase efficiency of input uses (water, fertilizer, and energy) and reduces GHG emissions in the agricultural sector.

As the largest consumer of water resources, agriculture is very vulnerable to reduced water availability induced by climate change. Circular water reuses and improved water use efficiency are, therefore, essential for coping with water scarcity issues and enhancing the resilience of food production systems (IWMI, 2007). Cascaded water uses as observed in rice production systems, reuses of agricultural return flows, and reuses of partially treated wastewater water from municipal and industrial systems are effective solutions to enhance water availability in arid regions of the world. Water use efficiency can be improved also through adoption of water conservation technologies such as drip irrigation, improved management of furrows, and mulching (Bekchanov et al., 2010, 2016a). Recently, "virtual water trade" concept gained traction to address water scarcity issues (Bekchanov et al., 2016b; Lenzen et al., 2013). Despite some clear shortcomings, "virtual water trading" may have some implications to improve food distribution by reallocating food from water abundant regions to water scarce areas.

As agricultural activities require enormous energy uses due to the operations of agricultural machinery, wider implementation of renewable energies can greatly reduce GHG emissions. Renewable energies such as biofuel can be used to run the tractors and harvesters, for instance. Nitrogen-fixing crops which are used for biofuel generation additionally improve soil health and decrease nitrogen fertilizer costs. Biogas from livestock waste and crop residues can be used for normalizing temperature inside the greenhouses or livestock-handling facilities in colder seasons. Solar energy can be applied for groundwater pumping and operating drip irrigation systems (Burney et al., 2010; Gao et al., 2013). Adoption of advanced technologies such as laser-guided land leveling improves the field texture consequently allowing for efficient distribution of irrigation water in the field and thus reducing water and energy losses (Abdullaev et al., 2007). Remote sensing technologies including the uses of drones enhance monitoring soil health and vegetation, and help in improving fertilizer use efficiency through targeted fertilization. Wider implementation of digital technologies in water sector also enhances improving water distribution along irrigation canals and supports the management of river basin systems consequently contributing to water and energy loss reductions.

Agricultural system productivity largely depends on the availability of soil nutrients. Soil nutrient depletion or soil contamination is a severe problem that increases food security risks in many parts of the world (Bekchanov and Mirzabaev, 2018). Organic fertilizer such as humus produced through recycling organic household waste, livestock manure, or crop residues is an effective way of improving soil organic content and improving crop yields. Organic fertilizer can create favorable conditions for crop root growth and, therefore, improves the crop nutrition intake consequently increasing crop yields. The effect can be higher when organic fertilizer is used in combination with chemical fertilizer. As organic matter increases soil water retention capacity, it enhances soil humidity over longer time period and helps in coping with drought impacts in hot seasons. Organic fertilizer from waste is also an effective solution to waste management issues, especially in developing countries (Fig. 24.2). Policies such as restricting dumping organic waste into environmental sinks and landfills enhance recycling organic waste. Through the recycling, carbon and nitrogen matter can be returned to soil rather than contributing to GHG emissions.

FIGURE 24.2

Packing the ready compost.

Photo by authors (2017).

Manure from livestock sector and crop residues can be also used in biogas reactors to produce biogas, which is an invaluable source of supplementary energy for rural households or livestock farms (Bekchanov et al., 2019). Decay of the organic waste and its digestion by specific bacteria inside the biogas reactor produces gases rich in methane that can be burned for cooking, heating, and even producing electricity. Although electricity option from biogas is currently characterized with low efficiency due to energy losses during conversion processes, biogas is an effective alternative to wood or crop residues used for cooking. Biogas generation from organic waste is effective especially in tropical areas with hot climate. Bioslurry from the reactor can be also used as a soil conditioner to improve soil nutrition content. Retention of nitrogen after anaerobic digestion makes biogas technology better option than burning dung cake for cooking due to considerable nitrogen losses when burning. Some countries such as Italy provide additional subsidies to the nitrogen recovered through anaerobic digestion.

Recycling organic waste can reduce GHG emissions due to reduced open dumping (Bekchanov and Mirzabaev, 2018). As chemical fertilizer production is associated with enormous amount of fossil fuel−based energy consumption and related GHG gases, increased uses of organic fertilizer decrease chemical fertilizer demand and related GHG emissions. In case of waste recycling to produce biogas, fossil-based energy demand can be decreased due to availability of energy from biogas; this would lead to reduced GHG emissions related with fossil fuel extraction. Similarly, improved water use efficiency and reuses of return flow and wastewater can decrease demand for fresh water diversion and energy required for surface or groundwater pumping, consequently reducing GHG emissions related to fossil fuel energy consumption.

Large amount of indirect GHG emission reductions is also possible in downstream food value chain since amount of postharvest losses (during harvesting, storage, transportation, and consumption) is enormous (HLPE, 2014). Minimizing postharvest losses can reduce demands for farming outputs and inputs (water, energy, and fertilizer) and related GHG emissions (FAO, 2013). In addition to management and logistic measures, reduction of harvesting and storage losses includes protection against pests and torrents. Emerging technologies such as vertical agriculture shorten the distance between producer and consumer consequently decreasing transportation and storage losses. Reducing food losses during consumption can be mostly achieved through improving customer behavior and raising awareness, especially in the developed countries. Consumption of balanced diet is also essential for enhancing health and preventing overconsumption of food with high fat or sugar content. Reduced meat consumption decreases GHG emissions from the land use sector as the carbon footprint of livestock commodities (meat, cheese, and milk) is higher than carbon footprint of plant-based food (Garnett et al., 2017).

Shared uses of tractors, harvesters, and pumps not only make them affordable by farming communities but also reduce GHG emissions related to the production of these mechanisms. Digitalization greatly reduces the transaction costs related with the implementation of shared machinery services (Daum et al., 2021). For example, digital tools help in reducing risks for customers, by showing the location of the nearest machine (tractor, harvester, etc.), rating of the operator (driver), and the price of the service. For the owners of the machinery services, digital tools provide maps of places with high demand and allow for minimizing travel time and transport costs. Digital technologies also reduce enforcement costs as they improve connectedness, enhance social capital (through rating service providers), and reduce information asymmetry (Benkler, 2004). These technologies are thus very relevant in places where enforcements rely on state authorities and governance capacity is weak.

4. Circular economy prospects in textile industry

Large potential exists for implementing CE strategies and reducing GHG emissions and environmental pollution along the whole value chain in the textile sector. As reported, closed-loop recycling rate in the textile sector is less than 1% (Wicker, 2016) and cascaded recycling is around 12% (Fig. 24.3). 73% of the material embedded in clothing end up in landfills or incineration processes. Key materials used in the sector are plastic (63%) and cotton (26%).

Cotton production is associated with heavy uses of water, fertilizer, and fuel for machinery operation. Cotton processing and dying processes also use a lot of water and energy and are key sources of water pollution in Turkey, China, India, Bangladesh, and Indonesia. Globally, the textile

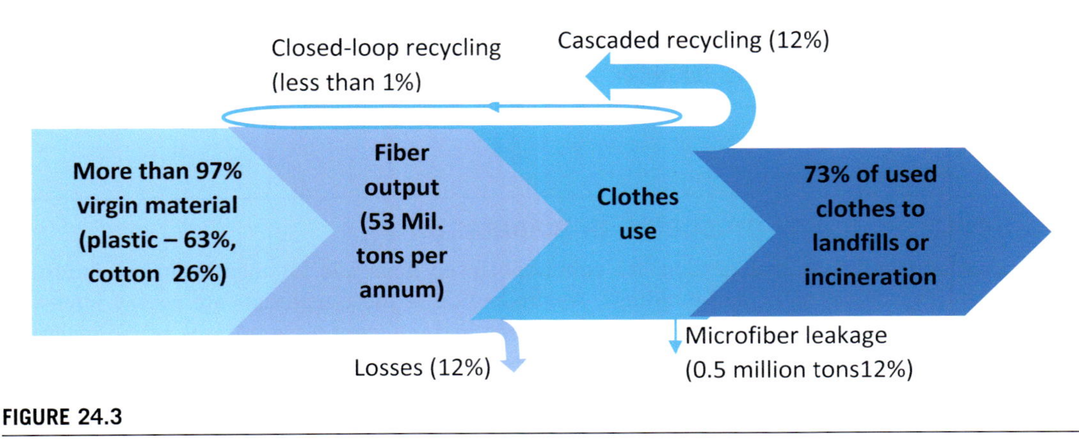

FIGURE 24.3

Global value chain of textile sector.

Adapted from Ellen MacArthur Foundation (2017).

sector was found out as a major contributor to the pollution of ocean with plastics; world ocean annually receives about half a million tons of plastic microfibers through wastewater generated during the washing plastic-based textile (IUCN, 2017). Improving water use efficiency in various phases of production operation processes and increased uses of recyclable virgin materials can greatly reduce unwanted environmental externalities.

Processes along the value chains of the textile industries highly rely on the consumption of nonrenewable materials and fossil fuel–based energy. Annually 342 million barrels of oil are used for producing plastic-based fibers in the sector globally (Nordic Fashion Association, 2017; Muthu, 2014). GHG emissions from the textile production were estimated to be 1.2 billion tons of carbon dioxide equivalent (as of 2015), which is higher than the emissions from international flights and maritime shipping (IEA, 2016). Considerable amount of GHG emissions also occurs during the use of textile products; washing and drying clothes require energy which is accompanied with GHG emissions of 120 million tons of CO_2 equivalent (Pakula and Stamminger, 2009). Increased uses of renewable energy sources and more efficient uses of water resources along textile value chain would greatly reduce GHG emissions from the sector.

As clothing industry is one of the highly polluting industries with a significant environmental footprint, maximizing utilization of clothes through shared use, extension of product life or versatile use is of paramount importance in line with principles of CE (Muthu, 2014). However, underutilization of clothing is high since clothes are not used till the end of their durability, and some garments are disposed after just seven to ten times of wearing (Morgan and Birtwistle, 2009; Barnardo's, 2015). Many people also own more clothes than they actually need (Greenpeace, 2017). There are some business models that improve useful services per textile product. Renting specific clothes (baby clothes, sportswear, and luxury items) that can be used once in a while by an individual for certain occasion can increase utilization of the clothes. People who like changing their outfits frequently may subscribe to rental agencies, which allow for renting certain amount of wearing to their members in each month. The concept of shared economy is gaining a foothold in many areas, and owning assets is becoming no longer an aspiration of the young. Resale of highly durable clothes that come with a warranty and can be easily repaired also improve useful services of such products. Secondhand markets can facilitate the resale of garments used by others before, and customers benefit from much cheaper prices for such items. For customers who want to keep and use their clothes for longer period, appropriate care services (repair, refurbish, and restyle) should be facilitated. Digitalization opens gates for broader implementation of the described business models by facilitating the matches between supply and demand.

5. Circular economy prospects in manufacturing

Steel, aluminum, plastics, and cement are commonly used materials for manufacturing industrial goods such as machines, cars, and buildings. As production of these materials are energy intensive, improving material use efficiency through recirculation and reducing wasteful uses of the materials during the processing has implications for reducing related GHG emissions substantially. As recirculating materials has much less energy demand than extracting and processing virgin materials, it is characterized with lower GHG emissions. In the EU, demand for 75% of steel, 50% of aluminum materials, and 56% of plastics can be met through recirculating what has been already produced

(Material Economics, 2018). Yet, current practices should be largely transformed to upscale CE strategies. New materials should be added to the production cycle just for replacing material losses and compensate lowered quality.

Improving material uses efficiency is achieved through reducing total material inputs per unit of finished products. Minimizing losses of materials during manufacturing processes is one of the options to improve material use efficiency. For instance, half of the aluminum produced does not reach the final product and 15% of construction materials turn into waste; therefore, production systems should be redesigned to eliminate these losses (Material Economics, 2018). Material use efficiency can be also enhanced by reducing material overuse which is defined as extra use relative to strict requirements to meet design standards. For instance, steel overuse in construction sector was estimated as about 100% indicating considerable potential to cut the material overspendings (Material Economics, 2018). In car industry, reducing steel overuse allows for producing lightweight cars that are cheaper and more fuel-efficient.

Digitalization has a great potential to reduce material costs and improve the utility services of goods from manufacturing sector. Particularly, electronic books, music and video products can be offered through web platforms at very cheap prices, and material uses for the production and distribution of these products decrease manifold. Virtualization of books and study materials also greatly reduces the costs of education and trainings. Improved availability and quality of information services through digitalization increase efficiency in various sectors including mass media services, administrative and support services, research and development activities, arts and entertainment, financial and insurance processes, distributive trade and social security.

6. Circular economy prospects in transport sector

Transport sector accounts for substantial share of GHG emissions from overall economic system. Intensified international travels and transportation through air or marine vessels following to economic development and globalization processes trigger carbon footprint. Pollution of air and related cardiovascular illnesses especially in urban areas are often associated with intensified ownership and uses of transportation facilities.

Electrification of the transportation systems and replacing traditional vehicles with the hybrid and electric cars can reduce fossil fuel—based energy consumption in transportation and thus reduce related GHG emissions. Substituting fossil fuel—based energy sources with biofuel also contributes to reducing GHG emissions, yet other externalities such as deforestation and hunger aggravation effects should be controlled. Although biogas from MSW is currently at experimental stage and comes with high costs, this option is demonstrated as effective green fuel for vehicles and public transport (buses). For short distance travels such as work commute and city tour, bikes and electric bikes are promoted as green and healthy option.

Efficiency of fuel uses for transportation can be increased through improved designs of engines, traffic regulation, or better road quality. Effective traffic regulation systems contribute to reducing traffic congestions and energy demand related to waiting. In addition to effective uses of road signs, road design such as road width and availability of bridges or pedestrian passways in road conjunctions impact on traffic motion. In countries with heavy monsoon rains, effective drainage systems are essential to remove storm water rapidly for preventing slow motion and less efficient fuel use in vehicles. Road quality as defined by roughness, texture, and reflection also impact on fuel uses per distance traveled. Improved quality of roads not only improve fuel use efficiency but also extend the lifetime of the vehicles and thus improve useful services per vehicle.

Other ways of improving useful services per vehicle are shared uses of cars and intensified uses of public transport rather than individual ownership and uses of cars. Shared vehicle and bike use and shared ride services are being common in major cities across the world due to improved connection of suppliers and customers to virtual network. Through digital platforms, customers can hire private or shared car to travel to certain location (i.e., Uber), rent vehicles for short-term period, or share individually owned cars for compensation. For this type of mechanisms to work, the increased implementation of instruments to incentivize using shared transport modes is necessary. Priority and exclusive use of road lanes for buses, higher tariffs for city parking are some of those instruments that encourage increased adoption of public and shared transport (Haitao et al., 2019). The use of congestion charge in London for reducing traffic as well as pollution is well known. Curitiba in Brazil is an excellent example of planning public transport alongside developing city design.

With the increased uses of digital technologies and more recently due to the impact of COVID-19 pandemic, studying and working from home and virtual studying and working spaces are also being common. Hybrid form of events and especially higher education is likely to be a permanent feature of the post-COVID normal period. Most international events and conferences are commonly being held in virtual world and without requiring international travels. Consequently, demand for transport systems and services are being decreased. This in turn would largely reduce energy demand and related GHG emissions in transport sector. Increased uses of drones for postal services and restaurant food delivery also affect transportation moods and related energy consumption.

7. Circular economy prospects in the built environment

The built environment sector comprises of the surroundings modified by man and include all buildings and infrastructural objects such as transportation and telecommunication systems, energy and water supply, and waste management systems. This sector accounts for substantial share of natural resources consumption and GHG emissions (ARUP, 2016). CE strategies would greatly reduce environmental impact and carbon footprint of the buildings and other commercial or public infrastructure. Effective design, planning, and construction of the built environment maintain harmony between nature and settlements and harness human health and well-being. Buildings would generate food and energy instead of consuming them. Closed water, energy, nutrient and material loops would improve resources use efficiency. Shared and flexible living and working spaces would increase useful services per square meter of the buildings.

Built environment sector can reuse waste material and also generate waste that can be recycled and reused. Kitchen and bathroom furniture, heating system components, doors, windows, and bricks in the buildings under demolishment can be removed, refurbished, and reused for equipping existing or newly constructed buildings, for example. Waste tires can be used for building walls in rural areas and plastic waste for constructing roads. Implementing circular water and energy systems also greatly enhance resource efficiency in the buildings. Organic and solid waste (paper, wood, and garment) can be used for generating energy, for instance. Wood waste generated from discarded furniture and faulty wooden pallets along with other organic waste can be converted into biochar by anaerobic incineration methods (pyrolysis), and the resulting biochar residue is used in farms to improve soil functions by sequestering carbon for longer periods (Verheijen et al., 2010). Closed circular water system largely reduces demand for fresh water from residential areas. Some of the residential areas in India have already implemented the use of partially treated gray water for washing out human waste from toilets or for gardening purposes (Godfrey et al., 2009; Ravishankar et al., 2018). Anaerobic digestion technologies can be also used to produce biogas from household sewage, kitchen waste, and livestock waste.

Production of cement (key ingredient for concrete) for construction sector alone contributes to 8% of global GHG emissions. Concrete manufacturing also uses natural aggregates extracted from rock formations which form over millions of years. Thus, recycling the concrete aggregates from demolished buildings would essentially conserve quarry resources and supplement the demand for aggregates as road base fillers and as constituents in concrete manufacturing. Recycling construction materials demands lower energy and generates lower GHG emissions as compared to manufacturing new ones (Wijayasundara et al., 2017a, 2017b). The construction sector also demands other key materials such as sand, steel, limestone, etc. Manufactured sand is a proven substitute for river sand, but it uses virgin granite deposits as raw material. Other viable replacements of conventional sand are copper and ferrous slags, quarry dust, foundry sand, and processed waste plastics (Somvanshi, 2020).

Using alternative materials with low carbon footprint such as engineered wood product (e.g., cross laminated timber) is currently gaining traction to replace concrete and steel. Wood products or wood residues can be also used for producing insulation material to equip external walls for reducing energy demand for heating homes in cold and cooling in hot period. Another key material used by the construction industry is gypsum plaster, which is used mainly for interior decoration and as a protective coat for indoor surfaces. The key environmental issue with the use of gypsum plaster is the inappropriate disposal of the material. This can be addressed by appropriate recycling methods as gypsum plaster is highly recyclable (Geraldo et al., 2017; Camarini et al., 2016). The recovery of construction materials from existing buildings and infrastructure units is best possible by using selective demolition techniques. However, these methods are energy intensive, and a case-by-case analysis is essential to draft the appropriate material recovery plan in the build environment (Pantini and Rigamonti, 2020).

Autonomous energy and water generation and use systems allow for saving plenty of resources in residential areas. With the decreased investment costs of solar power, house roofs are being equipped with solar panels that can generate sufficient energy for household activities (e.g., lighting, TV, heating or cooling, and kitchen uses). With the advancement of the technologies, even windows that can generate energy under sunlight are being implemented in multistorage modern buildings. Building-integrated photovoltaics are an example of such a technology, wherein PV modules are integrated with flexible polymers, ceramic roof tiles, or glass depending on the application. PV modules are also embedded with facades attached to the exterior of existing buildings to generate power and reduce building carbon footprint (Shukla et al., 2017). In areas with sufficient humidity and rainfall, solar roofs also can be used for harvesting rainwater that can be used for various household activities (e.g., bathing and washing). The mechanical, electrical, and plumbing (MEP) systems in buildings, however, use PVC or metal-based products. The philosophy of recycling, reusing, reducing, and rethinking of material and energy use needs to be implemented during the construction, operational, and deconstruction phases of the buildings to minimize and recover the material use in these MEP systems.

To maintain healthy lifestyles, greening the buildings through applying green facade and roofs are also introduced. In addition to decorating the buildings and providing valuable ambience, greening the buildings also can supply additional food (fresh fruits and vegetables) to the households. Taisugar's circular village in Taiwan and Kamikatsu in Japan are examples of developing a circular built environment. The residential complex consists of 400 apartments and uses solar panels, roof farming, aquaponics, urban farming, space sharing, green transport, and conservation of local ecology to become circular. The developers also address zero-waste, zero-carbon, and zero-accidents targets through enhancing self-sufficiency in food resource and energy systems (Tserng et al., 2021, Taisugar, 2018). Other examples of CE-based build environments are Park20|20, Venlo City Hall, ABN AMRO CIRCL based in the Netherlands, and Nangang public housing project by the Taipei government. Vertical or closed buildings are also used for large-scale food production. One example is from the Netherlands, which is the world's second largest exporter of agricultural products and where indoor farming and vertical agricultural methods are widely practiced. Similarly, large-scale indoor farms are being developed in Japan, the United States, Abu Dhabi, and several other places (Kalantari et al., 2018).

Sharing spaces and allowing flexible working hours allow companies to effectively use available spaces and reduce renting and utility payment costs. With the growing population and increasing demand for houses, co-living option is gaining traction that allows for effectively using living spaces and reduce the renting costs. In accordance with co-living concept, while private bedrooms are available individually, dining and bath rooms are shared among the community members. Efficiency of space utilization is also increased when house owners also consider renting extra rooms in their house for short- or long-term period.

Like in all other economic sectors, digitalization is a game changer to improve material use efficiency in the built environment. Web platforms such as Airbnb are effective tools to match suppliers and customers of shared flat services. Energy and water consumption of shared spaces are also much lower than those of individually owned spaces. Introduction of 3D printers is offering precise designs and construction processes consequently greatly reducing construction material waste. Increased uses of virtual spaces for shopping, watching movies, and subscribing to sport and leisure services are enriching consumption opportunities but also reducing demand for physical space.

8. Circular economy prospects in hotel industry

Hotels are using environmentally sustainable technologies to enhance their brand and expand their customers. As heating water, air conditioning, washing clothes, and cooking requires enormous amount of water and energy in hotels, adoption of circular water use systems and renewable energy technologies greatly improves resources use efficiency. Due to reduced amounts of water and energy uses, related GHG emissions also decline. Hotels also can contribute to sustainable development through establishing relationships with upstream suppliers of goods that are remanufactured and reusable. Similarly, relationships with downstream suppliers that recycle discarded goods from the hotel enhance sustainable development (Manniche et al., 2017).

Hotel facilities are centralized points for food waste generation as much as restaurants that provides opportunity for implementing waste management practices for segregated collection and for installing efficient units for recovery and repurposing. CE strategies can influence not only from the perspective of effective waste management but also from the perspective of minimizing food waste generation. Some examples are monitoring food consumption pattern using tracking of served portions, better planning and standardization of preparation techniques, and conservation of resources through choosing alternative food choices that are less material and energy intensive (e.g., locally grown food to encourage local circulation of nutrients or increased use of food items with a lower environmental footprint (Ziegler et al., 2021).

Circular water uses that consider cascaded uses of water in multiple production processes have a great potential to improve water use efficiency in a hotel. Water from swimming pool and water heated for washing clothes can be further used in the heating systems consequently reducing energy requirements. Return water from kitchen and laundry can be minimally treated and further used for flushing toilets. Return waters from various production operations can be effectively used for irrigating plants in courtyards and surroundings of a hotel. Compost produced through recycling organic waste and food waste from hotel can be also used for improving soil quality and plant growth in hotel yards.

Various renewable energy technologies such as solar panels, biogas digesters, and bioenergy plants can also largely replace the consumption of fossil fuel—based electricity and heating energy. Food waste from hotel restaurants and wastewater from rooms can be used to generate biogas, which can be further used for producing heating energy. Energy from biogas generation and wood incineration can be used for heating water for laundry and heating rooms (Fig. 24.4). Solar panels are not only source of electricity for lights in a hotel yard but also means for effective decoration. In tropical countries, solar energy can be directly used for heating water and thus reduce demand for grid-based electricity.

Hotels are also very innovative in influencing on customer behavior to improve resources use efficiency and overall sustainability of the consumption. For instance, asking customers agreement in "washing towels and bed cover less frequently" can contribute to reduced consumption of water and energy. An offer of shared rooms is another option that improve resources use productivity but also cost effective from customer perspective. Implementation of sensor technology in washing and drying hands and for automatically switching off lights when there is nobody in the room also improves water and energy productivity. Intelligent building climate control system connected to the booking system regulates heating, cooling, and air ventilation consequently minimizing the energy and water usage when a hotel room is not in use (Manniche et al., 2017).

Tourism services trigger long-distance international travels and are, therefore, associated with enormous GHG emissions (Manniche et al., 2017). Digital technologies such as virtual 3D and

FIGURE 24.4

Wood incineration for heating water in a hotel.

Photo by authors (2017).

multisensor technologies that enable people to attend events and experience certain sightseeing locations without actual travel opens new opportunities to reduce travel costs and related environmental impact in the tourism industry. COVID-19 lockdowns sparked many such developments from museums, wildlife parks, and libraries. However, replacing physical presence with virtual presence in such way transforms the tourism product totally into a service product such as leisure, entertainment, or education. The application Internet of Things (IoT) and cyber-physical system platforms to optimize space utilization is gaining traction. The technology uses motion sensors based on infrared or ultrasounds to detect occupancy and shares these data to a cloud-based repository. IoT-based sensors can calculate number of users, movement patterns, and duration of occupancy with the highest accuracy. These sensors also monitor the room temperature, lighting conditions, noise and air quality levels, and thereby enable efficient energy and space use (Wexlar, 2019). Digitalization also helps for improving utilization services of home space uses. Airbnb is an example for arranging matches between hosts and travelers, where a host can share a space for a stay while travelers can book affordable accommodation.

9. Circular economy prospects in MSW management

Increased volume of MSW is a global challenge and serious environmental threat especially in vast areas of the developing world (Kaza et al., 2018). Total generation of MSW is 2.01 billion tons per annum globally of which 33% at least is not managed in environmentally safe way (as of 2016). Global waste generation is expected to increase further reaching 3.4 billion tones by 2050. Despite developed countries account for 16% of global population, these countries generate about 34% of global MSW.

Open waste dumping sites and unsanitary landfills are sources of air, water, and soil pollution. GHG emissions from MSW sector are expected to increase from current levels of 1.6 billion tons (as of 2016) to 2.6 billion tons by 2050 (Kaza et al., 2018). Emissions from MSW account for about 5% of global GHG emissions. Since waste is often dumped on the banks of rivers and wetlands, it heavily pollutes the water bodies and makes them less convenient for habitation. Accumulation of waste in the dumping site over time also reduces flood retention areas or block the ways of water consequently increasing flooding risks. Undesirable odors and insects often disturb the communities residing in the close vicinity of the dumbing sites. Aggravation of waste mountains due to limitedness of area for waste dumping in large cities increases the damage risks to the surrounding houses and families. For example, as recently observed in the Meethotamulla dumping site in Colombo Municipality of Sri Lanka, landsliding in the garbage mountain destroyed the neighboring houses and took the lives of several people (Jayaweera et al., 2019).

Recycling MSW can greatly reduce the pressure of waste aggravation and additionally allow for recovering some useful materials for further use and reducing environmental pollution. Especially, organic waste such as food and plant residues can be composted to produce organic fertilizer. Waste materials such as metal, glasses, plastic bottles, and paper are valuable resources that can be returned to production process with minimal treatment. Electronic waste can be recycled and delivered as a cheap alternative material to the industries. Used tires of cars and trucks are used in road construction. Wood, wooden furniture, and clothes can be remanufactured or reused. Alternatively, they can be also burned to produce heat energy. Yet, burning can be applied only as the last choice since the reuses of wood may have much greater value than energy generated through burning.

Composting MSW is a win-win option in countries with rapid urbanization rates and high dependence on fertilizer imports to sustain agriculture (Bekchanov and Mirzabaev, 2018). However, metal and salt content of compost made of MSW should be minimized to acceptable levels before applying in agriculture (Hargreaves et al., 2008). Separation of organic and nonorganic solid waste before or immediately after waste collection can reduce the metal content of waste. Voluminous waste can be also used for soil surface reclamation measures after the full exploitation of mining sites, or for establishing green parks, or maintaining forests.

Plastic waste is a major threat to environmental security and wildlife when uncollected or openly dumped as natural decay of plastics takes quite long time (Kaza et al., 2018). Plastic litter often leaks to water bodies consequently blocking water bodies and causing flooding; wild animals feeding in open waste landfills may confusingly consume them and lose life; plastic waste thrown away to nature can turn to trap to some animals causing various injuries; plastics are key pollutants in ocean threatening the marine ecosystems. Yet, when properly collected, plastics are easy to recycle, and recovered plastic can be used as material to make household items such as baskets, construction blocks, and roads (Fig. 24.5). Designing new materials which are easily degradable such as bioplastics to replace plastics can also greatly reduce environmental burden of excessive uses of plastics.

The incineration is a controlled process in which combustible waste is burned and converted into heating energy. The collected solid waste should be initially separated into organic and nondegradable content at the gate of the plant to maximize the energy production. The residual ashes can be safely dumped into landfills. This option greatly reduce demand for lands required for dumping waste. Yet, the costs of constructing incineration plant can be high. Especially, in areas where the wet content of MSW is high, additional energy requirements may further increase the costs. GHG emissions are indirectly reduced as heating energy through incineration can be an effective alternative to fossil fuel—based energy. As mentioned above, incineration is only the last choice for wood, cloth, and paper waste since these materials can be recycled to generate assets with higher value.

FIGURE 24.5

Sri Lanka's first plastic road.

Photo by authors (2018).

10. Conclusions

This study presented a review of CE strategies of improving material use efficiency, reducing carbon footprint, implementing circular resource uses, and improving utility per unit of consumption goods in various economic sectors including agriculture, textile industry, manufacturing, transport, construction, hotel industry, and MSW. These strategies have substantial potential to decrease resource depletion and environmental pollution effects through transforming the current "linear" production systems into more sustainable and welfare-enhancing economic systems. The CE strategies decrease GHG emissions not only through broader implementation of renewable energy uses and improved energy efficiencies but also through increased material use efficiency in various sectors. Differing from conventional economic approach of maximizing value added or GDP, a new economic development approach based on the CE strategies aims at achieving the highest utility (welfare or satisfaction) from the consumed services using minimal amount of resources (materials and energy). Thus, resources recovery and reuse, integrated management of systems (e.g., water-energy-food nexus), sharing consumption goods (houses, cars, clothes, etc.) gain traction to achieve social and environmental security goals. Digital technologies go hand in hand with this transformation creating favorable enabling environment for the sustainable CE strategies.

References

Abdullaev, I., Ul Hassan, M., Jumaboev, K., 2007. Water saving and economic impacts of land leveling: the case study of cotton production in Tajikistan. Irrigation and Drainage Systems 21, 251–263.

ARUP, 2016. The Circular Economy in the Built Environment. London, the UK. Available online at: www.arup.com/perspectives/publications/research/section/circular-economy-in-the-built-environment.

Barnardo's, 2015. Once Worn Thrice Shy – British Women's Wardrobe Habits Exposed!, 11 June 2015 Barnardo's. Available online at: http://www.barnardos.org.uk/news/press_releases.htm?ref=105244.

Bekchanov, M., Mirzabaev, A., 2018. Circular economy of composting in Sri Lanka: opportunities and challenges for reducing waste related pollution and improving soil health. Journal of Cleaner Production 202, 1107–1119.

Bekchanov, M., Lamers, J.P.A., Martius, C., 2010. Pros and cons of adopting water-wise approaches in the lower reaches of the Amu Darya: a socio-economic view. Water 2, 200–216.

Bekchanov, M., Ringler, C., Bhaduri, A., Jeuland, M., 2016a. Optimizing irrigation efficiency improvements in the aral sea basin. Water Resources and Economics 13, 30–45.

Bekchanov, M., Lamers, J.P.A., Bhaduri, A., Lenzen, M., Tischbein, B., 2016b. Input-output model-based water footprint indicators to support IWRM in the irrigated drylands of Uzbekistan, Central Asia. In: Borchardt, D., Bogardi, J., Ibisch, R. (Eds.), Integrated Water Resources Management: Concept, Research, and Implementation. Springer, Berlin and Heidelberg, Germany, pp. 147–168.

Bekchanov, M., Mondal, AH Md, de Alwis, A., Mirzabaev, A., 2019. Why adoption is slow despite promising potential of biogas technology for improving energy security and mitigating climate change in Sri Lanka? Renewable and Sustainable Energy Review 105, 378–390.

Benkler, Y., 2004. Sharing nicely: on shareable goods and the emergence of sharing as a modality of economic production. The Yale Law Journal 114 (2).

Braungart, M., McDonough, W., Bollinger, A., 2007. Cradle-to-cradle design, creating healthy emissions: a strategy for eco-effective product and system design. Journal of Cleaner Production 15, 1337–1348.

Burney, J., Woltering, L., Burke, M., Naylor, R., Pasternak, D., 2010. Solar-powered drip irrigation enhances food security in the Sudano-Sahel. Proceedings of the National Academy of Sciences of the United States of America 107 (5), 1848–1853.

Camarini, G., Dos Santos Lima, K.D., Pinheiro, S.M., 2016. Investigation on gypsum plaster waste recycling: an eco-friendly material. Green Materials 3, 104–112.

Cordell, D., Drangert, J.-O., White, S., 2009. The story of phosphorus: global food security and food for though. Global Environmental Change 19 (2), 292–305.

Daum, T., Villalba, R., Anidi, O., Mayienga, S.M., Gupta, S., Birner, R., 2021. Uber for tractors? Opportunities and challenges of digital tools for tractor hire in India and Nigeria. World Development 144, 105480.

Donati, F., Aguilar-Hernandez, G.A., Sigüenza-Sánchez, C.P., de Koning, A., Rodrigues, J.F.D., Tukker, A., 2020. Modeling the circular economy in environmentally extended input-output tables: methods, software and case study. Resources, Conservation and Recycling 152, 104508.

Ellen MacArthur Foundation, 2013. Towards the Circular Economy. Economic and Business Rationales for an Accelerated Transition, Cowes, UK.

Ellen MacArthur Foundation, 2017. A New Textiles Economy: Redesigning Fashion's Future. Available online at: www.ellenmacarthurfoundation.org/publications.

Ellen MacArthur Foundation, 2019. Completing the Picture: How the Circular Economy tackles climate change. Available online at: www.ellenmacarthurfoundation.org/publications.

European Commission, 2018, 2018. Report on Critical Raw Materials and the Circular Economy. Commission Staff Working Document. SWD, 36.

FAO, 2013. Food Wastage Footprint: Impacts on Natural Resources. FAO, Rome, Italy.

Gao, X., Liu, J., Zhang, J., Yan, J., Bao, S., Xu, S., Qin, T., 2013. Feasibility evaluation of solar photovoltaic pumping irrigation system based on analysis of dynamic variation of groundwater table. Applied Energy 105, 182–193. https://doi.org/10.1016/j.apenergy.2012.11.074.

Garnett, T., Godde, C., Muller, A., Röös, E., Smith, P., de Boer, I.J.M., Ermgassen, E., Herrero, M., van Middelaar, C., Schader, C., van Zanten, H., 2017. Grazed and Confused? Ruminating on Cattle, Grazing Systems, Methane, Nitrous Oxide, the Soil Carbon Sequestration Question - and what it All Means for Greenhouse Gas Emissions: Environmental Change Institute. University of Oxford.

Geraldo, R.H., Pinheiro, S.M., Silva, J.S., Andrade, H.M., Dweck, J., Gonçalves, J.P., Camarini, G., 2017. Gypsum plaster waste recycling: a potential environmental and industrial solution. Journal of Cleaner Production 164, 288–300.

Godfrey, S., Labhasetwar, P., Wate, S., 2009. Greywater reuse in residential schools in Madhya Pradesh, India—a case study of cost—benefit analysis. Resources, Conservation and Recycling 53, 287–293.

Greenpeace, 2017. After the Binge, the Hangover: Insights into the Minds of Clothing Consumers. Available online at: www.greenpeace.de/sites/www.greenpeace.de/files/publications/2017-05-08-greenpeace-konsum-umfrage-mode.pdf.

Gustavsson, J., Cederberg, C., Sonesson, U., Van Otterdijk, R., Meybeck, A., 2011. Global Food Losses and Food Waste. Food and Agriculture Organization of the United Nations, Rome.

Haitao, H., Yang, K., Liang, H., Menendez, M., Guler, S.I., 2019. Providing public transport priority in the perimeter of urban networks: a bimodal strategy. Transportation Research C: Emerging Technologies 107, 171–192.

Hargreaves, J.C., Adl, M.S., Warman, P.R., 2008. A review of the use of composted municipal solid waste in agriculture. Agriculture, Ecosystems and Environment 123, 1–14.

HLPE, 2014. Food Losses and Waste in the Context of Sustainable Food Systems. A Report by the High Level Panel of Experts on Food Security and Nutrition of the Committee on World Food Security. FAO, Rome. http://www.fao.org/3/a-i3901e.pdf.

Intergovernmental Panel on Climate Change (IPCC), 2014. Climate Change 2014: Mitigation of Climate Change. Working Group III Contribution to the Fifth Assessment Report of the Intergovernmental Panel on Climate Change. Cambridge University Press, Cambridge, United Kingdom and New York, NY, USA.

Intergovernmental Panel on Climate Change (IPCC), 2019. Special Report on Climate Change and Land: Summary for Policymakers. Geneva.

International Energy Agency (IEA), 2016. Energy, Climate Change & Environment: 2016 Insights.

International Union for Conservation of Nature (IUCN), 2017. Primary Microplastics in the Oceans. A global evaluation of sources.

International Water Management Institute (IWMI), 2007. Comprehensive Assessment of Water Management in Agriculture. Water for Food, Water for Life. A Comprehensive Assessment of Water Management in Agriculture, London, Earthscan, and Colombo, IWMI.

Jayaweera, M., Gunawardana, B., Gunawardana, M., Karunawardena, A., Dias, V., Premasiri, S., Dissanayake, J., Manatunge, J., Wijeratne, N., Karunarathne, D., Thilakasiri, S., 2019. Management of municipal solid waste open dumps immediately after the collapse: an integrated approach from Meethotamulla open dump, Sri Lanka. Waste Management 95, 227–240.

Kalantari, F., Tahir, O.M., Joni, R.A., Fatemi, E., 2018. Opportunities and challenges in sustainability of vertical farming: a review. Journal of Landscape Ecology 11, 35–60.

Kasulaitis, B.V., Babbitt, C.W., Krock, A.K., 2019. Dematerialization and the circular economy: comparing strategies to reduce material impacts of the consumer electronic product ecosystem. Journal of Industrial Ecology 23, 119–132.

Kaza, S., Yao, L.C., Bhada-Tata, P., Van Woerden, F., 2018. What a Waste 2.0: A Global Snapshot of Solid Waste Management to 2050. World Bank, Washington, DC. https://openknowledge.worldbank.org/handle/10986/30317.

Kjaer, L.L., Pigosso, D.C.A., Niero, M., Bech, N.M., McAloone, T.C., 2018. Product/service-systems for a circular economy: the route to decoupling economic growth from resource consumption? Journal of Industrial Ecology 23 (1), 22–35.

Lebreton, L., van der Zwet, J., Damsteeg, J.W., Slat, B., Andrady, A., Reisser, J., 2017. River plastic emissions to the world's oceans. Nature Communications 8, 15611.

Lenzen, M., Moran, D., Bhaduri, A., Kanemoto, K., Bekchanov, M., Geschke, A., Foran, B., 2013. International trade of scarce water. Ecological Economics 94, 78–85.

Manniche, J., Larsen, K.T., Broegaard, R.B., Holland, E., 2017. Destination: A Circular Tourism Economy A Handbook for Transitioning toward a Circular Economy within the Tourism and Hospitality Sectors in the South Baltic Region. Centre for Regional & Tourism Research (CRT), Nexoe, Denmark.

Material Economics, 2018. The Circular Economy - a Powerful Force for Climate Mitigation. Transformative Innovation for Prosperous and Low-Carbon Industry, Stockholm, Sweden.

Maupin, M.A., Kenny, J.F., Hutson, S.S., Lovelace, J.K., Barber, N.L., Linsey, K.S., 2010. Estimated use of water in the United States in 2010: U.S. Geological Survey Circular 1405, 56. https://doi.org/10.3133/cir1405.

Morgan, L.R., Birtwistle, G., 2009. An Investigation of Young Fashion Consumers' Disposal Habits.

Muthu, S., 2014. Roadmap to Sustainable Textiles and Clothing: Eco-Friendly Raw Materials, Technologies and Processing Methods. Springer.

Nordic Fashion Association (NFA), 2017. Polyester and Synthetics; Statista, Global Oil Production from 1998 to 2015. https://www.statista.com/statistics/265203/global-oil-production-since-in-barrels-per-day/.

Pakula, C., Stamminger, R., 2009. Electricity and water consumption for laundry washing by washing machine worldwide. Energy Efficiency 3, 365–382.

Pantini, S., Rigamonti, L., 2020. Is selective demolition always a sustainable choice? Waste Management 103, 169–176.

Pelletier, N., Audsley, E., Brodt, S., Garnett, T., Henriksson, P., Kendall, A., Kramer, K.J., Murphy, D., Nemecek, T., Troell, M., 2011. Energy intensity of agriculture and food systems. Annual Review of Environmental Resources 36, 223–246.

Pimentel, D., Williamson, S., Alexander, C.E., Gonzalez-Pagan, O., Kontak, C., Mulkey, S.E., 2008. Reducing energy inputs in the US food system. Human Ecology 36, 459–471.

Ravishankar, C., Nautiyal, S., Seshaiah, M., 2018. Social acceptance for reclaimed water use: a case study in Bengaluru. Recycling 3, 4.

Shukla, A.K., Sudhakar, K., Baredar, P., 2017. Recent advancement in BIPV product technologies: a review. Energy and Buildings 140, 188–195.

Somvanshi, A., 2020. Concrete Without Sand? [Online]. Available: https://www.downtoearth.org.in/blog/concrete-without-sand-41849.

Strokal, M., Kahil, T., Wada, Y., Albiac, J., Bai, Z., Ermolieva, T., Langan, S., Ma, L., Oenema, O., Wagner, F., Zhu, X., Kroeze, C., 2020. Cost-effective management of coastal eutrophication: a case study for the Yangtze river basin. Resources, Conservation and Recycling 154, 104635.

Taisugar, 2018. TaiSugar's Circular Village [Online]. Available: https://www.taisugarcircularvillage.com/.

Tserng, H., Chou, C.-M., Chang, Y.-T., 2021. The key strategies to implement circular economy in building projects—a case study of taiwan. Sustainability 13, 754.

UNEP, 2011. Decoupling Natural Resource Use and Environmental Impacts from Economic Growth. International Resource Panel.

Verheijen, F., Jeffery, S., Bastos, A., Van Der Velde, M., Diafas, I., 2010. Biochar Application to Soils. A Critical Scientific Review of Effects on Soil Properties, Processes, and Functions. Institute for Environment and Sustainability (Joint Research Centre).

Wexlar, K., 2019. How IoT Will Transform Office Space Utilization.

Wicker, A., 2016. Fast Fashion Is Creating an Environmental Crisis, Newsweek (2016).

Wijayasundara, M., Crawford, R.H., Mendis, P., 2017a. Comparative assessment of embodied energy of recycled aggregate concrete. Journal of Cleaner Production 152, 406–419.

Wijayasundara, M., Mendis, P., Ngo, T., 2017b. Comparative assessment of the benefits associated with the absorption of CO_2 with the use of RCA in structural concrete. Journal of Cleaner Production 100, 285–295.

Ziegler, F., Nilsson, K., Levermann, N., Dorph, M., Lyberth, B., Jessen, A.A., Desportes, G., 2021. Local seal or imported meat? Sustainability evaluation of food choices in Greenland, based on life cycle assessment. Foods 10 (6), 1194.

CHAPTER

Business climate for energy regaining and environmentally sustainable waste-to-resource technologies

25

Maksud Bekchanov[1,2] and Daphne Gondhalekar[3]

[1]*Research Unit Sustainability and Global Change (FNU), Center for Earth System Research and Sustainability (CEN), University of Hamburg, Hamburg, Germany;* [2]*Center for Development Research (ZEF), University of Bonn, Bonn, Germany;* [3]*Chair of Urban Water Systems Engineering, Technische Universität München, Munich, Germany*

Chapter outline

Handbook of Energy and Environmental Security. https://doi.org/10.1016/B978-0-12-824084-7.00008-4
Copyright © 2022 Elsevier Inc. All rights reserved.

1. Introduction

Poor sanitation systems, lack of hygiene, and improper management of waste and wastewater are key causes of heavy environmental pollution and health degradation in developing countries (Gondhalekar et al., 2013; Wintgens et al., 2016; Singh et al., 2017). Discharge of hazardous waste into fresh water bodies, open fields, and sites neighboring residential areas due to lack of proper sanitation systems and wastewater treatment leads to heavy pollution of groundwater aquifers, surface water bodies, and soil. Air and water pollution in turn contribute to increasing incidents of diarrhea, malaria, childhood stunting, child mortality, and premature death. Unpleasant view and poor living environments in residential areas because of environmental pollution reduce potential incomes from tourism. Economic damage costs of poor sanitation were estimated to be at the level of US$ 222.9 billion in developing countries (as of 2015, Oxford Economics, 2016). India solely accounts for almost half of these costs (e.g., US$ 106.7 billion or 5.2% of GDP) due to lack of improved sanitation value chain in many parts of the country.

Changing the approach toward waste streams and minimizing exposure to waste in each stage of the sanitation value chain (containment, collection, transportation, ztreatment, disposal, or end-use) are essential for decreasing the health damage and environmental pollution costs of mismanagement in the sanitation sector (Singh et al., 2017). Each US$ 1 investments in improving sanitation value chain would yield economic returns of US$ 3 to 34 (Oxford Economics, 2016). When handled properly, septage and wastewater can be also a source for recovering energy and nutrients as postulated by the concept of circular economy (Geissdoerfer et al., 2017; Bekchanov, 2017). Growing scarcity of water, energy, and nutrients increases the importance of treating waste as a resource and implementing Resources Recovery and Reuse (RRR) strategies (use of biodegradable waste to produce biogas, compost, etc.). Treating waste as a resource to recover water, energy, and nutrients can also contribute to achieving several Sustainable Development Goals (SDGs) related with improving water and sanitation access (SDG 6), enhancing food and energy security (SDGs 2 and 7), alleviating poverty (SDG 1), mitigating to climate change and ensuring environmental safety (SDG 13), and maintaining healthy livelihoods (SDG 3). In this context, upgrading waste-to-resource value chains and adoption of RRR technologies are gaining importance across the world due to their environmental and economic benefits.

While various options of septage and wastewater treatment and recycling are available, market conditions and governmental institutions and policies for their implementation are not always in place. Given their high investment costs and lack of social awareness about the extent of adverse health and environment effects of waste dumping and untreated wastewater releases, the implementation of RRR technologies is still at rudimentary stages in many developing countries.

Suitability of RRR technologies for any particular location requires a thorough assessment by considering technical, economic, and institutional aspects. Related analyses of business climate, value chains, production relationships, and main stakeholders may reveal the important institutional and infrastructural barriers and opportunities for successful adoption, performance, and expansion of RRR technologies. Given that waste and wastewater recycling processes are strongly related with several sectors such as water, waste and wastewater treatment, energy and fertilizer production, and agriculture, essential interlinkages, synergies, and trade-offs across these sectors should be also taken into account. This paper presents a methodological framework for business climate assessment that considers effluents, energy, and nutrients recovery from waste and wastewater streams by incorporating such intersectoral linkages. The framework is applied to analyze the business climate for RRR options in the case of India where poor sanitation and mismanagement in the waste sector are challenging issues. The country is also a home for 17.5% of the world's population and accounts for almost half of the global damage costs related to poor sanitation. Given its population size and the scope of the environmental and economic issues related to waste/wastewater sector, the country thus can be a representative of the developing world. This study particularly describes the current status of wastewater and waste treatment systems, investigates the economic and environmental benefits of waste-to-resource value chains, and examines the enabling environment (financial feasibility, demand, regulatory framework, and institutions) for their wider adoption.

Expert opinions from various technical reports, research articles, and statistical bulletins of state agencies have been surveyed and synthesized to prepare the sections on the current status of the sanitation sector (Water Aid, 2016; Ministry of Statistics, 2016; Rohilla et al., 2017), Municipal Solid Waste (MSW, Kumar et al., 2017; Ahluwalia and Patel, 2018), and wastewater treatment (Starkl et al., 2013; Ministry of Statistics, 2016). Similar reports and papers provided information on composting (MCF, 2017; Dilkara et al., 2016), biogas (Rao et al., 2010; Kaniyamparambil, 2011; MNRE, 2017; Mittal et al., 2018), and waste to energy (WTE) plants in India (Kalyani and Pandey, 2014; Ahluwalia and Patel, 2018). Additional publications have been examined to get expert viewpoints on socio-institutional aspects of the system such as demand for fertilizer (MAFW, 2017; Dilkara et al., 2016), institutions and policies (Dilkara et al., 2016; Rohilla et al., 2017; Mittal et al., 2018), financing options (Chatri et al., 2012; Rohilla et al., 2017), and governance and business climate (World Bank 2017a, 2018).

2. Review of business climate evaluation approaches

Business environment is an environment where business operates, shaped by public policies and organizations, backward and forward linkages with other economic agents and market participants, in addition to technical, economic, and financial feasibility of the businesses themselves. Business and investment climate studies have been commonly used to evaluate attractiveness of doing business in a particular country through construction of indicators measuring specific aspects of business environment (e.g., number of days it takes to set up a business). These indicators can also help policy makers to determine the flaws in organizational, legal, and regulatory frameworks and market conditions and direct the resources to fix the problems.

General interest in developing a set of indicators which allows for diagnosis of business environment in a particular country appeared in the early 1970s. Since then, several international research organizations have offered different approaches to evaluate the business climate. Ease of Doing Business (EDB) index developed by the World Bank, Global Competitiveness Index (GCI) by the World Economic Forum (WEF), Entrepreneurship Measurement Framework (EMF) by the Organization for Economic Co-operation and Development (OECD), Business Competitiveness Index (BCI) developed by Porter (2004), and Investment Compass Indicator (ICI) by the United Nations Conference on Trade and Development (UNCTAD) are among them. The definitions of business environment by various organizations differ because of their selective emphasis on some particular dimensions of a business climate. For instance, the EDB index puts more emphasis on employment, economic growth, and high return investment opportunities (World Bank, 2018), whereas macroeconomic, institutional, legal, and infrastructural frameworks are key aspects in calculating ICI (Stern, 2002; Christy et al., 2009).

EDB index by the World Bank is measured as an average value of several indexes such as easiness of starting business, obtaining construction permits, registering property, access to electricity, access to credit, protection of small entrepreneurs, procedures of paying taxes, opportunities for external trade, enforcement of agreements, and easiness of closing business (World Bank 2015, 2018). These aspects are assessed in terms of time required, costs, and volume of the procedure. The data are collected from statistical reports, information on laws and regulations, and through surveys of officially registered enterprises on their perceptions of barriers to businesses. The information on regulations is obtained through primary surveys of legal officers. The data behind Doing Business indicator are obtained by surveying formal enterprises in the largest cities. Thus, this reflects only a limited geographic coverage within countries and does not consider the informal sector. Since most calculations are based on ranking the countries, its practical value is low when there is only a limited number of countries with similar legal and cultural conditions. Macroeconomic and institutional frameworks are neglected in calculating this index.

GCI by WEF considers both microeconomic and macroeconomic factors of business climate (WEF, 2015). According to this indicator, competitiveness of the economy which determines its prosperity depends on macroeconomic, institutional, and infrastructural conditions, quality of education, openness of markets, access to technology, investments in innovations, and protection of intellectual property rights. This indicator also considers the stages of development through attributing higher weights to subcomponents which are more relevant in particular stage. For instance, for resource-driven economies, GCI is calculated by considering higher weights for well-functioning markets, developed infrastructure, and stable macroeconomic conditions. In innovative-driven economies, however, competitiveness is assessed through prioritizing quality of research, collaboration between universities and industry, and protection of intellectual property. Given the differences in considered subcomponents and weighting factors, country ranks based on GCI and EDB indicators differ from each other (Christy et al., 2009). Despite several advantages of GCI over EDB indicator, it does not consider industry- or sector-specific assessment of the business climate.

BCI by Porter (2004) focuses on business climate at local level and addresses a rigorous assessment of competitiveness at firm and national levels. Particularly, productivity of the firm is assumed to be largely dependent on factors such as sophistication of production processes, staff training, incentive compensation, level of customer orientation, capacity of innovation, value chain development, and branding. Necessary yet insufficient external business climate conditions as considered in this index are access to inputs, human capital, infrastructure, technologies availability, demand conditions, and context for firm strategy and domestic rivalry. Despite advantages of this approach due to a detailed grass-root level assessment, requirements for intensive data collection impedes its application. Lack of reliable data, especially in developing countries, increases the challenge.

EMF by OECD is other measure to assess the investment attractiveness of an economy (OECD, 2008). Differing from the remaining indicators of measuring business climate that emphasize on the development of entrepreneurship, EMF also considers the impact of entrepreneurship on the overall economy. Main subcomponents of EMF indicator are transparency of laws, intellectual property protection, investment promotion, reduced trade and transaction costs, fair competition, effective tax policy, capacity development, affordability of infrastructural services, and quality of public governance. EMF also does not aim to rank the countries based on averaged indexes. Instead this indicator informs investors and policy makers about important areas which need attention for improving the efficiency of policies and supporting private sector growth (Christy et al., 2009).

ICI by UNCTAD focuses on macroeconomic and infrastructural conditions in assessing business climate. Main subcomponents considered in ICI are human capital (literacy rate and number of science and engineering students), availability of natural resources, market size (GDP and population), infrastructure (information communication technology [ICT] and transportation), wage and administration costs, macroeconomic performance (inflation, growth rates, and unemployment), governance quality, tax, and subsidy incentives, and regulatory framework (property rights, settling disputes, regulations of foreign exchange, labor markets, etc.). While addressing very relevant and essential determinants of business climate, ICI fails to take into account more detailed value chain components.

Business climate indexes discussed above reflect the general economic conditions at national level and often miss sector-specific characteristics of business environments. Despite some subcomponents of the business climate indicators may have important implications also for individual sectors, main limitation of these indicators is that they largely ignore sector- or location-specific factors of business climate. Thus sector-specific implications for institutional, legal, and infrastructural improvements based on these indexes are limited. Considering these limitations, revised set of subcomponents for assessing sector-specific business climate in agro-processing industry has been suggested (Christy et al., 2009; FAO, 2013, World Bank, 2017a,b). This approach addressed relationships and issues along the agricultural value chain and thus looked at socioeconomic (costs and importing options) and institutional (registration, documents required, etc.) issues around supply of inputs (seed, fertilizer, machinery, land, etc.) and markets for outputs (e.g., easiness of exporting procedures, transportation, logistics, etc).

Similar approaches to assess socioeconomic and institutional environment around wastewater management system and business models have been recently developed. A recent report by Swiss Federal Institute of Aquatic Science and Technology (EAWAG) discussed RRR options and enabling environment for implementing these technologies directly focusing on fecal sludge and wastewater management sector (Strande et al., 2014). While providing a detailed presentation of the sanitation and wastewater management systems and potential problems along these value chains, the report also highlighted key enabling environment factors such as institutions, government support, skill capacity, financial arrangements, legal-regulatory frameworks, and sociocultural perceptions. According to the report, in addition to technical parameters of the RRR technologies and their relevance to local physical conditions, these socioeconomic and institutional factors also play a pivotal role for the successful adoption of RRR options and business models.

3. Conceptual-methodological framework

The methodological framework that depicts main factors influencing the course of scaling up RRR options and is presented below builds on the previous approaches to business climate assessment developed in World Bank (2015, 2018), FAO (2013), and EAWAG (Strande et al., 2014). The interlinkages between waste management and agricultural value chains and their relationship to general socioeconomic framework and environmental system are presented in Fig. 25.1. According to the framework, waste management and agricultural value chains are part of a broader economic system and are influenced by various socioeconomic and institutional factors. The value chains and the socioeconomic system function within the physical environment play a pivotal role for the sustainability of the production system and determine the relevance of the production and recycling technologies to a particular location.

The microenvironment depicts the specifics of sanitation and agricultural value chains and their interlinkages. If composting and biogas production from organic waste or safe disposal of waste are underdeveloped, large portion of the generated waste ends up in waterways, lakes, and neighboring

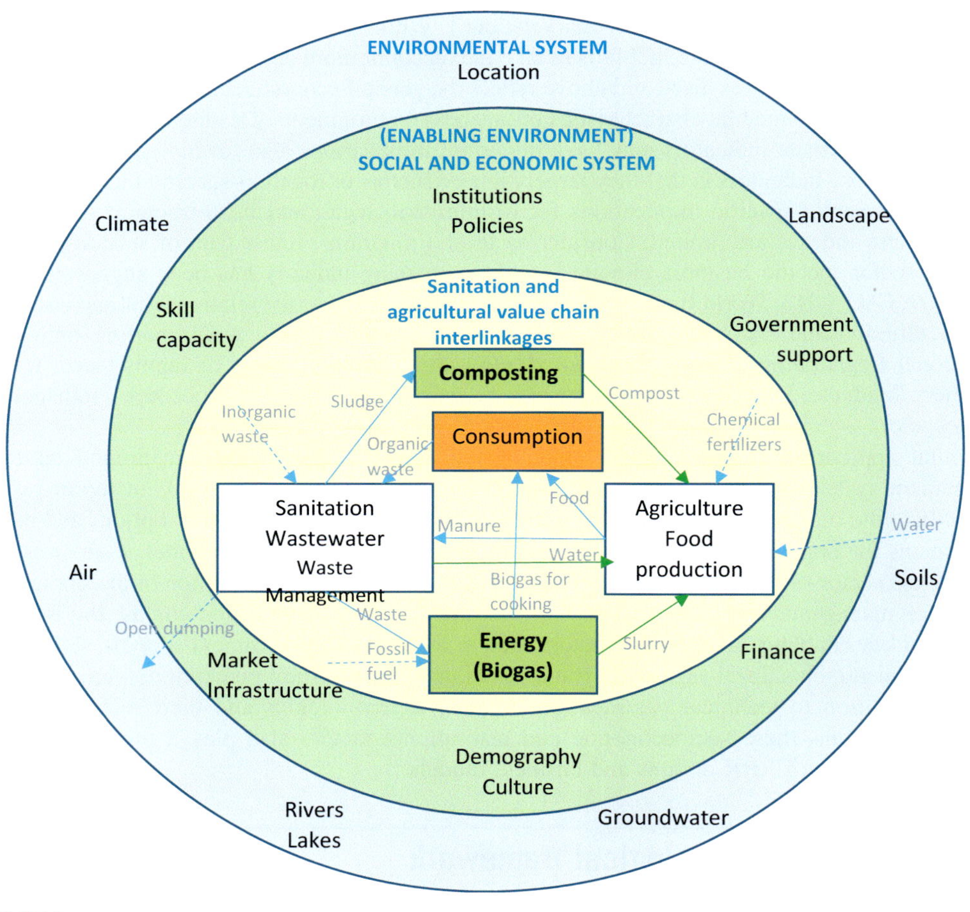

FIGURE 25.1

Waste management and agricultural production value chains as part of a broader socioeconomic, institutional, and environmental system.

Source: Authors' presentation.

areas, polluting water sources, causing toxic gas emissions, and damaging environmental health. RRR options such as compost and biogas production from organic waste greatly help in reducing the amount of waste to be openly dumped and supply additional energy or fertilizer for production processes and household consumption.

Understanding the interlinkages between sanitation and agricultural value chains is essential to assess the amount and value added of input and output flows along the chains concurrently determining any issues related with input supply, high costs, low productivity, demand, and competition for the outputs. For instance, aiming at introducing composting technologies may not yield the expected results due to lack of proper sanitation and waste collection facilities. Or, in water-abundant conditions characterized with heavy rainfall, a reuse of treated wastewater is less likely to be an economically viable option. Likewise, if the costs of chemical fertilizers are much cheaper than producing and applying compost in crop production, promoting demand for compost cannot be easy. Similarly, biogas plants may not be preferred by households and businesses if cheap electricity or fossil gas is available as an energy source.

In addition to production system characteristics, the consideration of geographic conditions is essential for assessing the relevance of any RRR option for particular location. Groundwater levels influence on the choice of sanitation facility, for instance. Fertilizer application requirements may vary depending on soil type, and thus it may determine overall compost demand in particular area. Cold temperature may reduce the technical and economic feasibility of biogas plants. Since compost plants require large area and may cause unpleasant odor in the neighborhood, this option can be less recommendable in densely populated areas.

While multiple environmental and economic benefits of RRR options are unquestionable when technical standards are followed in their installation and operation, and physical-geographical factors are considered, the main issue is related with the socioeconomic and institutional factors that influence on the course of scaling up RRR projects. Lack of proper laws, policies, and institutions to incentivize the implementation of improved sanitation and RRR options may lead the continuation of unsustainable waste management practices and environmental degradation. Even if proper laws for safeguarding environment exist enforcement of these laws and monitoring can be limited. High corruption levels and lack of transparency are key barriers for maintaining the rule of law. In case the laws are supportive to enhance sustainable production and adoption of RRR options, financial feasibility of such changes may depend on income levels of people, access to credit, and subsidization rates. Infrastructural conditions such as access to energy, water, roads, education services, and ICTs are also important for the feasibility of RRR options. Availability of affordable technologies (mechanized compost application) and skilled workers are required for the continuation of the RRR processes. Sociocultural factors play an important role for wider implementation of RRR options. For instance, the use of fecal sludge for biogas generation and biogas use for cooking can be unacceptable in some societies because of social stigma attached to excreta. Moreover, the roles and responsibilities of various stakeholders involved in waste and wastewater management either issuing and enforcing laws or coordinating and monitoring the process should be clearly defined for effective functioning of the institutions.

4. A study area

India with the territory of 3,287,263 km^2 is the seventh largest country in the world (MEF, 2009). The landscape of the country is kaleidoscopic varying from Himalayan highlands in the North to tropical rainforest in the South. The mainland of India is divided into four regions such as the Great Mountain Zone, the Indo-Gangetic Plains, the Desert Region, and the Southern Peninsula.

India has common borders with Afghanistan and Pakistan in the North-West (Fig. 25.2). The country also borders on China, Bhutan, and Nepal in the North, and Myanmar and Bangladesh in the

FIGURE 25.2

Location of India.

Source: https://commons.wikimedia.org/wiki/File:India,_administrative_divisions_-_de_-_colored.svg.

East. In the South, a narrow channel of sea separates India from Sri Lanka. The length of the border on the mainland is about 15,200 km, whereas the total length of the coastline, including the mainland and islands (Lakshadweep, the Andaman and Nicobar) is 7517 km. Population of India is over 1.3 billion people (as of 2016; WB, 2017). Two thirds of the population in the country resides in rural areas.

Average GDP per capita in India was estimated as US$ 1364 (as of 2014). However, the states across the country have different levels of economic development. The wealthiest population with income levels of higher than US$ 2500 is from the smallest states such as Goa and Sikkim. In terms of GDP per capita, Gujarat, Haryana, Maharashtra, and Tamil Nadu also perform well having income levels between US$ 2000 and 2400. The states located in the Ganga Basin such as Uttar Pradesh, Bihar, Jharkhand, and Madhya Pradesh are the poorest in terms of per capita income earning less than US$ 1000 per capita.

Income levels largely vary not only across the regions within India but also across the social groups. In other words, income inequality and poverty rates are quite high in most of the states across the country. In some states such as Chhattisgarh, Jharkhand, and Manipur, incomes of almost 40% of the population are below the poverty line (Fig. 25.3). Poverty rates are greater in rural areas. Even in the states with the lowest poverty rates such as Andhra Pradesh, Goa, Kerala, Punjab, and Sikkim, incomes of almost 10% population are below the poverty line. In India, people living under the conditions of below the poverty line are 21% on average. Additional resources recovered from waste can considerably enhance soil health, energy access, and household incomes for the poor in the country.

Agriculture is the backbone of rural economies, since this sector provides 45% of total employment despite its GDP share of 18% (as of 2015; World Bank, 2017a). The role of the agriculture sector is

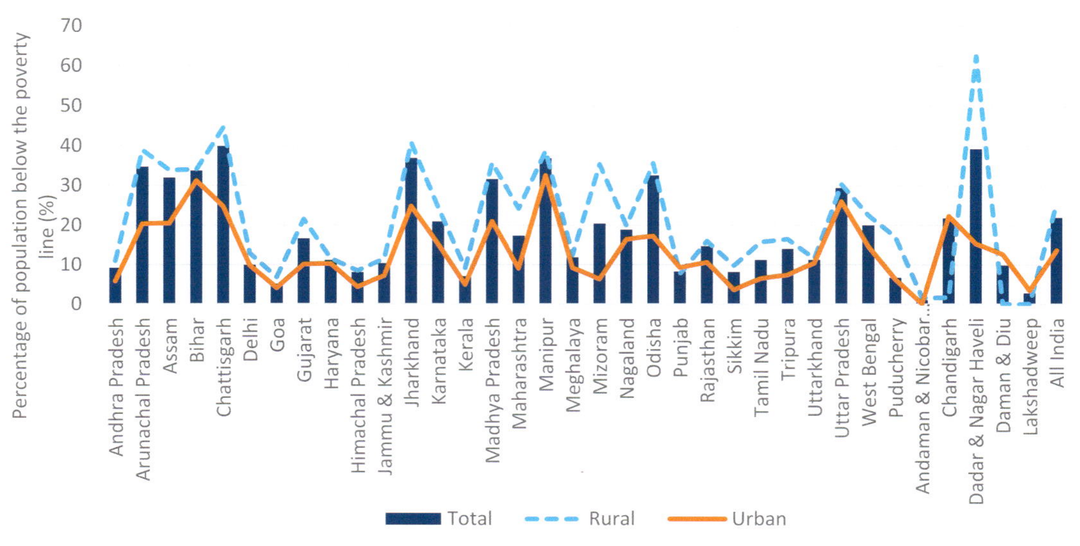

FIGURE 25.3

Poverty rates across India (2011–2012).

Source: Authors' presentation based on Ministry of Statistics (2016).

essential for livelihoods in rural areas across India, since its share in total employment is higher than 35% in all states of the country (Fig. 25.4). Especially in states such as Bihar, Madhya Pradesh, Maharashtra, and Gujarat, the agricultural sector provides more than 80% of employment opportunities in rural areas (Venkatesh et al., 2015).

Madhya Pradesh together with Utter Pradesh owns also the largest areas of croplands in the country having 23.6 and 22.8 million ha cropping lands, respectively. Maharashtra and Rajasthan owns more than 18 million ha lands each. Production of cereals especially rice and wheat is main cropping activity in many states of India. Wheat is mainly cultivated in the Northern states. Cotton is mainly grown in Gujarat, Telangana, and Maharashtra (Ministry of Statistics, 2016).

5. Environmental issues in wastewater and waste management sectors

5.1 Sanitation and wastewater management

Access to water and sanitation is a basic need and essential right. Achieving SDGs related to poverty reduction and gender equality is closely linked with improved access to water and sanitation. In India, overall 43.5% of households have an access to tap water delivery, but 11.6% of them receive untreated water through the tap water supply system (Table 25.1). The rest of the households use wells or hand pumps for water. The percentage of households with access to tap water is much lower in rural areas compared with that of urban areas.

Wastewater generation in metropolitan cities of India is about 9275 million liters per day (or 9.3 million m^3 per day) according to official statistics (Ministry of Statistics, 2016). Some reports indicated 38.3 million m^3 wastewater generation per day in the biggest (metropolitan and large size) cities

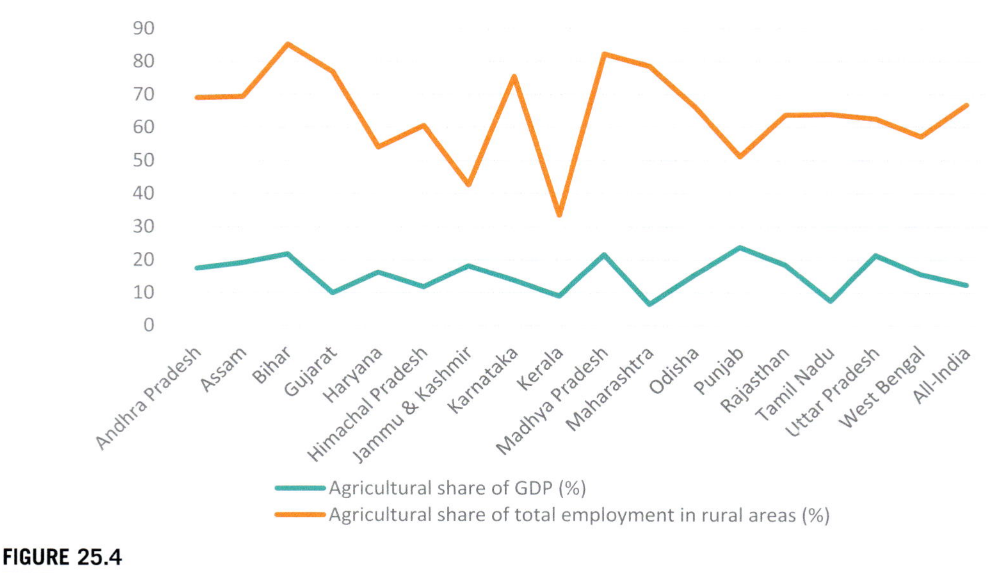

FIGURE 25.4

Agricultural share of GDP and total rural employment (2010).

Authors' presentation based on Venkatesh et al. (2015).

of India (Hingorani, 2011). However, 53% of households lack toilets in their premises (Table 25.1). In rural areas, 69.3% of the households lack toilets within their premises, whereas this applies to about 18.6% households in urban areas.

Open defecation is a major problem due to lack of acces to proper toilets by many poor households across India (Fig. 25.5). More than half of the population were reported to practice open defecation before 2005 (Tarraf et al., 2016). With government support, open defecation was considerably reduced

Table 25.1 Access to water supply and sanitation (2011).

	Number of households (million)	Households with tap water supply (%)			Availability of toilet within the premises (%)	
		Total	Treated	Untreated	With any type of toilet	Without any toilet
Total	246.7	43.5	32	11.6	46.9	53.1
Rural	167.9	30.8	17.9	13	30.7	69.3
Urban	78.9	70.6	62	8.6	81.4	18.6

Source: Ministry of Statistics (2016).

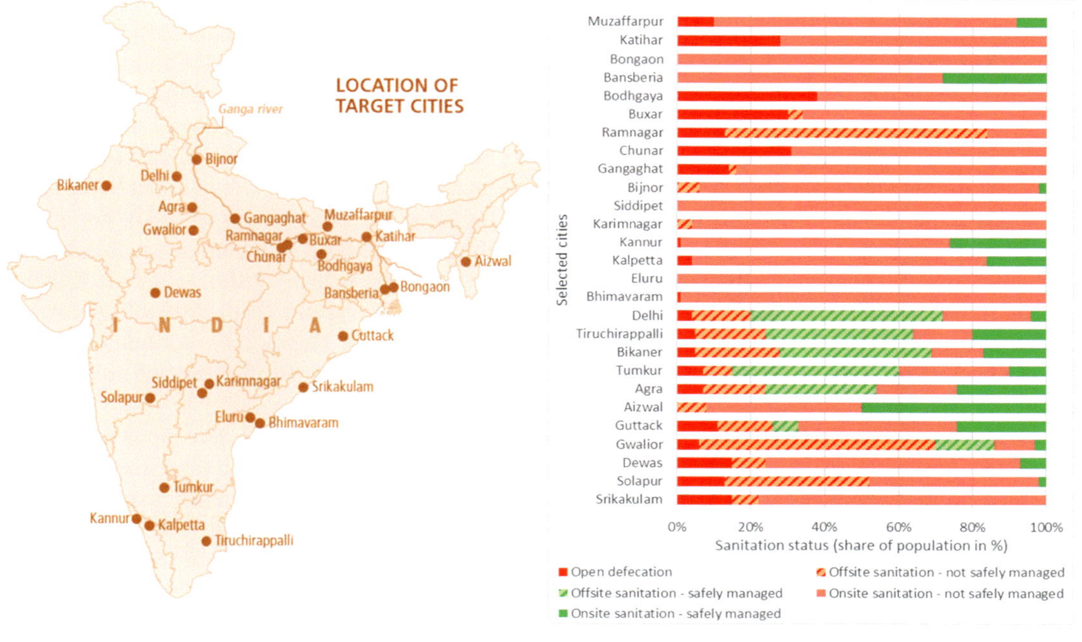

FIGURE 25.5

Open defecation, fecal sludge, and wastewater generation and disposal.

Adapted from Rohilla et al. (2017).

in a short period, and about 9% of the population practices open defecation across the country at present (Balachandran, 2015). However, in some cities such as Chunar, Buxar, and Bodh Gaya located in the northeastern part of the country, open defecation rates are still more than 30%. Offsite sanitation facilities (sewerage network) are lacking in most of the cities across the country. In Solapur and Ramnagar, despite the availability of the off-site facilities, wastewater was not properly managed. Only in few cities such as Agra, Tumkur, Bikaner, and Tiruchirappalli, large portions of the generated fecal sludge and wastewater are properly managed.

Due to the lack of adequate wastewater management, sometimes wastewater stagnates in small ponds and percolates into groundwater aquifers, polluting drinking and irrigation water sources (Hingorani, 2011). Lack of access to improved sanitation is a key cause for environmental pollution and increased health risks (WSP, 2011; Singh et al., 2017). Currently, 70% of surface water is polluted by different types of organic and inorganic waste (Water Aid, 2016). And 33% of the total 45,000 km of rivers are contaminated by fecal coliforms at the level of more than 500 MPN per 100 mL, which is five times higher than the maximum allowable level (fecal coliforms should be less than 104 MPN per 100 mL in unpolluted water; Water Aid, 2016). Damage costs due to poor sanitation are US$ 106.7 billion (or 5.2% of GDP). The costs related with premature morbidity and child mortality were estimated to be at the level of 54% of total health damage costs (WSP, 2011). Diarrhea and related diseases took the lives of 400,000 people annually across India.

5.2 Solid waste generation

Problems of solid waste management is also growing along with increasing population and urbanization (Kumar et al., 2017). Between 2001 and 2013, for instance, the total population in India grew from 1028 million to 1252 million (World Bank, 2017a). Except demographic change, MSW generated also depends on living conditions, eating habits, capacity and patterns of production processes, and season. Per capita MSW generated in a day varies between 0.17 kg per capita in small towns and 0.62 kg per capita in cities. Over 133,000 tons of MSW is generated daily, of which about 91,000 tons is collected and about 16,000 tons is treated (Kumar et al., 2017). Total amount of MSW per annum was estimated to be between 50 and 70 million tons of which 55% to 60% is organic waste (as of 2014/2015; Ahluwalia and Patel, 2018). The six largest metropolitan cities—Delhi, Mumbai, Kolkata, Chennai, Bangalore, and Hyderabad—are main producers of MSW, varying from 4000 tons per day in Hyderabad to 9260 tons per day in Delhi (Ahluwalia and Patel, 2018). These cities account for 21% of MSW generated in urban areas across India despite the fact that their share of the total urban population is 16%.

More than 90% of MSW is dumped improperly (Kumar et al., 2017). Land requirements for disposing MSW are growing rapidly. Open dumping of MSW has hazardous environmental and health effects. Methane generated from the decomposition of biodegradable waste causes explosions and contributes to global warming. Unpleasant odor from dumping sites, especially in summer, and leachate causes heavy air and water pollutions. Water collected in discarded tires or in shallow pools (on improperly managed land areas) creates a condition for mosquito breeding. This in turn opens opportunities for increased incidents of malaria, dengue, and fever. Uncontrolled burning of waste emits pollutants into the atmosphere, subsequently increasing the incidents of infections in breathing organs, asthma, allergies, and reduced immunity.

6. Technology options and current capacities for resources recovery from wastewater and municipal solid waste

In this chapter, different options for treating wastewater and recovering energy and nutrients from wastewater or solid waste are discussed to present the current capacity of RRR facilities across India. Considering commonly used RRR options, they are divided into four broad topic areas as follows: (1) wastewater treatment and reuses; (2) composting organic waste (including dried fecal sludge); (3) biogas from organic waste and wastewater; and (4) other energy generation options using waste. Wastewater treatment capacity and quality signals about the sources and scope of environmental pollution.

6.1 Wastewater treatment and reuses

Treating and reusing wastewater is important under water scarcity conditions and when the cost of delivering water with acceptable quality is quite expensive (WSP/IWMI, 2016). According to government statistics, about 80% of all wastewater generated in metropolitan cities is collected in India (Ministry of Statistics, 2016). However, wastewater collection rates can be as low as 60% in some cities such as Bhopal and Kochi (Table 25.2).

Primary and secondary wastewater treatments are available in most cities though wastewater treatment plants do not exist in some cities such as Kolkata, Kochi, and Madurai. Wastewater treatment capacity is only 4% in Mumbai—the biggest city in terms of population. Nationwide, less than 20% of total wastewater generated is properly treated before disposal because of low rates of access to sewerage system, leakages in conveyance, insufficient capacity, and low loading rate of treatment plants (Hingorani, 2011; Balachandran, 2015).

Waste stabilization ponds are a common way of wastewater treatment and cover over 70% of treatment capacity in towns with more than 100,000 inhabitants (Starkl et al., 2013). In some cities, constructed wetlands maintain secondary or tertiary treatment after anaerobic baffled reactors or anaerobic filters. Biomass (cattail, reed, etc.) produced in the wetlands can be harvested for further uses such as covering roofs or producing particle board (Kumar, 2016). Duckweed ponds are also used for wastewater treatment in some states such as Punjab (Starkl et al., 2013). Advanced wastewater treatment systems have been rarely implemented due to high financial costs.

Reuse of wastewater can considerably reduce the costs of industrial enterprises in case water tariffs for industrial use is quite high. Since cooling thermal power plants or heating the rooms do not require high-quality water, wastewater with little treatment can be supplied to these systems and allow for economizing substantial amount of money which should be spent for freshwater supply otherwise. There are some companies in India such as Chennai Petroleum Corporation Limited and Koradi Thermal Power Station, which uses treated wastewater to meet their water demands since it is cheaper to do so (WSP/IWMI, 2016). Reuses of treated wastewater for irrigating trees in recreational parks during dry seasons have been reported in the case of Nagpur city (Garfí, 2015).

Irrigated agriculture in India largely relies on groundwater pumping, which is getting risky due to the depletion of the groundwater sources. Since agriculture does not require high-quality water, treated wastewater can be used for irrigation after removing metal substances and pathogenic microorganisms. Wastewater contains essential nutrients (nitrogen, phosphate, and potassium), which can be recovered

Table 25.2 Wastewater generation, collection, and treatment in metropolitan cities of India.

Major cities	Population (million)	Wastewater (million liter per day)	Wastewater collection rate (%)	Wastewater treatment capacity (%)	Mode of disposal
Ahmedabad	3.3	556	80	77	Sabarmati river
Bangalore	4.1	400	75	73	V. Valley, Ksc Valley
Bhopal	1.1	189.3	50	46	Agriculture
Mumbai	12.6	2456	90	4	Sea
Kolkata	11.0	1432.2	75.1	0	Hughly river/ fish farm
Coimbatore	1.1	60	75	0	Noyyal river, irrigation
Delhi	8.4	1270	80	77	Agriculture, Yamuna river
Hyderabad	4.3	373.3	80.1	31	River, irrigation
Indore	1.1	145	80	10	Khan river, irrigation
Jaipur	1.5	220	75	12	Agriculture
Kanpur	2.0	200	75	21	Ganga, sewage farm
Kochi	1.1	75	60	0	Cochin back waters
Lucknow	1.7	106	75.5	0	Gomati river
Ludhiana	1.0	94.4	49.8	0	Agriculture
Chennai	5.4	276	93.1	93	Agriculture, sea
Madurai	1.1	48	70	0	Agriculture
Nagpur	1.7	204.8	79.6	22	Agriculture
Patna	1.1	219	74.9	48	River, fisheries
Pune	2.5	432	85	39	River
Surat	1.5	140	80	50	Garden/creek
Vadodara	1.1	140	75	58	River, agriculture
Varanasi	1.0	170	74.7	59	Ganga, agriculture
Vishakhapatnam	1.1	68	80.9	0	–
Total	**71.0**	**9275.0**	**80.6**	**31.5**	

Source: Ministry of Statistics (2016).

and enhance crop growth when wastewater is applied for irrigation. Currently, about 40,000 ha of irrigated lands use untreated wastewater (World Bank, 2010). Wastewater from Class I and II cities is about 14 km^3 which can be sufficient to irrigate 1 to 3 million ha of land (WSP/IWMI, 2016).

6.2 Composting

Composting is another option to generate wealth from organic waste. In addition to being a valuable fertilizer to enhance soil productivity, it reduces the amount of waste to be dumped into landfills, consequently decreasing the generation of harmful greenhouse gases such as methane into the atmosphere. Application of chemical fertilizers solely lead to fast runoff of more than half of nutrients applied into groundwater aquifers, canals, and the drainage system, subsequently contaminating water sources (Ahluwalia and Patel, 2018). Combined use of chemical and organic fertilizers reduces the wastage of nutrient application since organic matter absorbs nutrients and releases them slowly over time.

Composting at small scale has been historically widespread across India (Zurbugg et al., 2004). However, composting urban solid waste at large scales, which was promoted in the 1970s, was found out to be uneconomical (Dulac, 2001). High operation and transportation costs and low quality of compost because of the use of mixed waste did not allow gaining profit for most of the composting firms at that time. Since the 1990s, NGOs and international funding agencies supported community-level composting which was manually operated (Zurbugg et al., 2004). Reduced reliance on poor waste collection services, lower capital and transportation costs, and new employment opportunities for poor people were main advantages of the decentralized composting at community level. Recently, the Policy on Promotion of City Compost was adopted (on 10-20-2016) to scale up and marketing the compost (Dilkara et al., 2016). The Policy allows substantial subsidies for compost production by composting plants and marketing by fertilizer companies.

At present, composting municipal organic waste is implemented across many states of India (Table 25.3). As reported, overall, the compost production capacity of these composting plants is about 1 million tons though annual production of compost is only 0.15 million tons (MCF, 2017). The largest composting facilities are located in Delhi, Gujarat, and Karnataka. Given the environmental and economic benefits of composting, the government has plans for further extending the composting to recycle more organic waste. Large capacities of composting are expected to be established especially in Karnataka and Maharashtra (MCF, 2017, Fig. 25.6).

Opening and closure rates of compost plants to recycle organic municipal waste are considerably high, and the first 10 composting plants established in India are not functioning anymore (Dilkara et al., 2016). Lack of separating waste at the source point and low quality of the produced compost were key reasons for the failure of the plants. Mixed solid waste received by the plants increased the transportation and segregation costs. Produced compost mass was only 6% to 7% of the mixed waste received by the plant. Large amount of inorganic waste separated through initial screening was dumped or sent to landfills at extra cost. Quality of compost produced from organic municipal waste was also low due to contamination with hazardous metal, glasses, and chemicals.

Some plants such as Bruhat Bengaluru Mahanagara Palike (since 1975), Mysore City Corporation (since 2008), Mangalore City Corporation (since 2013), Belagavi City Corporation (since 2007, in Hyderabad), and Shimoga City Corporation (since 2008, in Hyderabad) have been able to cope with the technological, financial, and logistical difficulties (Dilkara et al., 2016). Some municipalities in Suryapet, Warangal, Nagpal, Lucknow, Ahmedabad, Chennai, and Mumbai in partnership with NGOs

Table 25.3 Composting across India (2016).

States	Number of plants	Annual compost Production capacity (ton)
The Andaman and Nicobar Islands	1	90
Chhattisgarh	1	1350
Daman and Diu	1	4050
Delhi	3	350,500
Goa	1	1125
Gujarat	11	337,225
Karnataka	5	79,200
Kerala	1	4950
Madhya Pradesh	1	12,385
Maharashtra	3	59,400
Rajasthan	1	14,850
Tamil Nadu	8	62,569
Telangana	2	2520
Uttar Pradesh	4	20,295
West Bengal	2	49,500
Total	**45**	**1,000,009**

Source: MCF (2017).

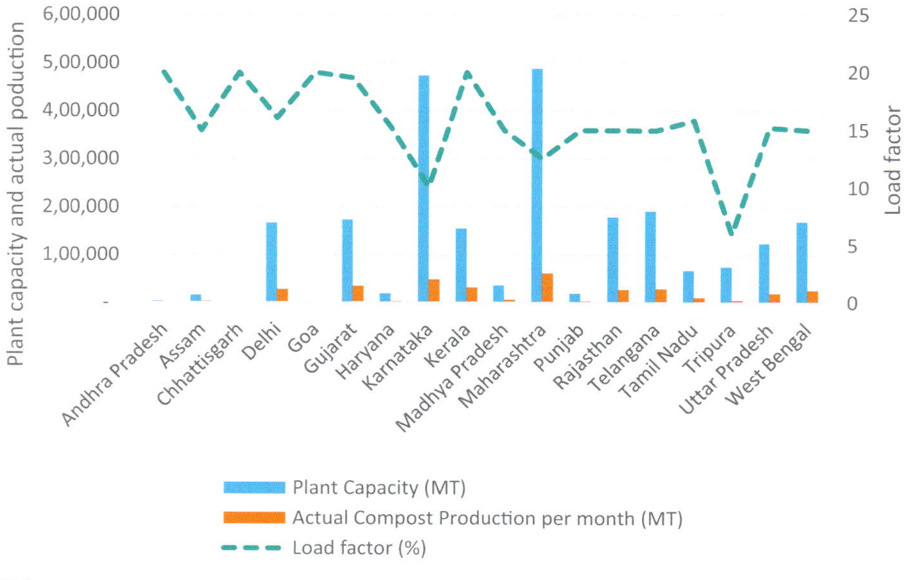

FIGURE 25.6

Planned capacities and loading rates of new composting plants across India.

Source: Authors' presentation based on MCF (2017).

or private companies established door to door services of collecting waste to encourage separation of waste and thus decrease the logistical costs of composting degradable material. The compost produced by these plants was distributed through the network of various fertilizer supply companies.

6.3 Biogas

Energy deficit at the level of 11,436 MW or 12.6% of peak demand increases the importance of the alternative energy production sources in India (Rao et al., 2010). Biogas generation has a substantial energy supply potential in India due to availability of vast amount of agricultural waste, livestock manure, sewage sludge, and wastewater. As estimated earlier, overall production potential of biogas from waste (MSW, crop residues, animal manure, poultry litter, industrial waste from distilleries and dairy plants, excluding wastewater sludge) is about 40,734 million m^3 per annum (Rao et al., 2010). Potential capacity of generating power from this biogas was estimated to be 25,700 MW.

In addition to providing a clean energy for cooking, a digested bioslurry from biogas plants can be used to supplement the chemical fertilizers in crop production. Additionally, biogas generation plants help to improve sanitation when septage tanks are linked with the biogas digester. Hot temperature throughout all seasons makes biogas generation technology favorable and less expensive in tropical countries.

India is the second largest country after China in terms of the number of installed biogas plants. As of 2014, about 4,750,000 biogas plants were installed in the country. Within a single year between 2014 and 2015, about 111,000 biogas plants were planned and 47,490 were constructed. In addition to household biogas plants, several biogas plants were built for supplying off-grid power (Mittal et al., 2018). Madhya Pradesh and Andra Pradesh host the largest number of biogas plants (Fig. 25.7).

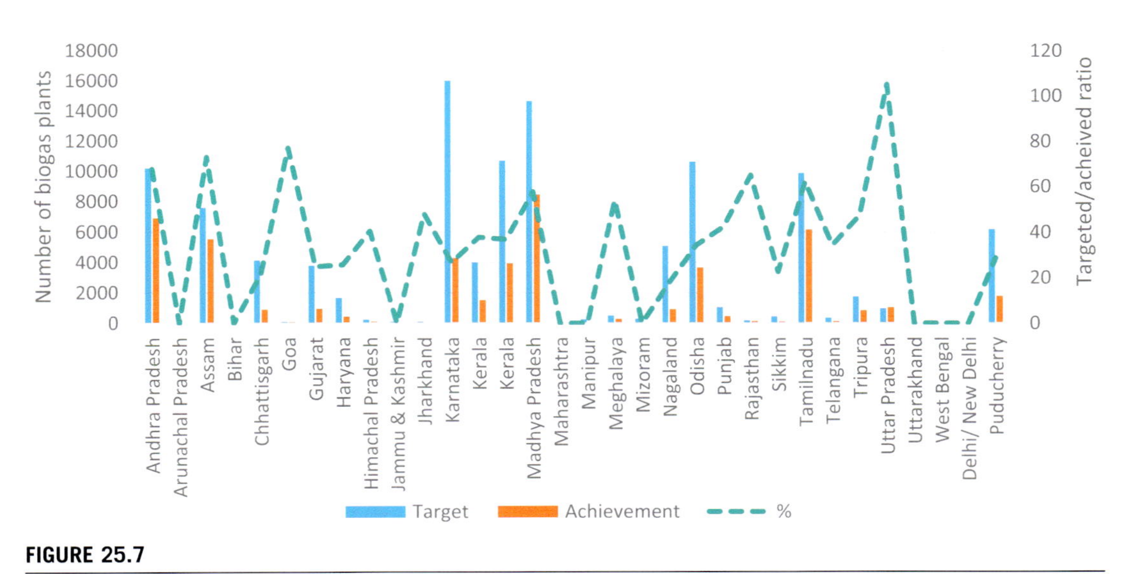

FIGURE 25.7

Biogas plant constructions in 2014/2015.

Source: Authors' presentation based on MNRE (2017).

However, biogas recovery from waste in municipal areas is very low at present because of high capital costs and low financial feasibility compared to alternative waste recycling options (Mittal et al., 2018). Currently, only 56 biogas-based power generation plants are operational across India, and the majority of them are located in Maharashtra, Kerala, and Karnataka. A biogas plant in Pune (Maharashtra), for instance, allows for recycling 150 tons of MSW daily and generates 14,000 m^3 biogas (Kalyani and Pandey, 2014). In Lucknow (Uttar Pradesh), biogas from waste was used to generate 6 MW power but failed due to lack of waste separation at the source. Facilities for capturing landfill gas also exist as exemplified in the case of Gorai dumping site in Mumbai (Maharashtra) in 2008 (Kalyani and Pandey, 2014).

Despite substantial subsidies by the government, upscaling biogas generation technologies are impeded because of various financial, technical, market, sociocultural, and institutional barriers (Mittal et al., 2018). High capital costs of the construction and low financial capability of people were found to be the main barrier for constructing biogas plant in rural areas (Rao and Ravindranath, 2002; Bansal et al., 2013). Although government subsidies are available to support biogas plant constructions by the targeted group of people (the poor), delays in the transfer of the subsidies increase the overhead costs (Chandra et al., 2006). Limited access to credit also imped the installation of biogas plants for household cooking (Ravindranath and Balachandra, 2009).

Biogas generated is in direct competition with electricity from grid and liquid petroleum gas (LPG). Subsidized electricity or abundant supply of LPG were found to be key reasons for lack of interest in utilization of biogas plants. Some people do not like the idea of using night soil or human waste for biogas generation. Women are mostly involved in cooking and prefer cleaner technologies of cooking, but they have less power in decision making and have limited influence on the decision of investing in biogas plants (Mittal et al., 2018).

From technical perspective, inadequate supply of dung or organic waste often led the failure or underperformance of the biogas plants (Mittal et al., 2018). In dry areas, unavailability of water reduces the feasibility of biogas plants. In cold regions or seasons, temperature is a problem since hydraulic retention time is slow and methanogenesis is inhibitive during cold periods (Kalia and Kanwar, 1998). In addition, lack of trained personnel to fix technical defects of biogas plants also decreases the attractiveness of investing in biogas plants in rural sites (Kaniyamparambil, 2011).

6.4 Waste to energy plants

Several WTE plants are in operation to produce power through waste incineration in Delhi, Hyderabad, Chennai, Jabalpur, and Shimla (Fig. 25.8). WTE plants, for instance in Hyderabad, allow for recycling up to 2400 tons of waste per day and run the power generation plants with the capacity of 20 MW. Total capacity power generation of these WTE plants observed across India amounts to 99 MW and is quite low when compared to national power generation capacity of 330 GW (Ahluwalia and Patel, 2018).

Waste can be also used to produce a refuse-derived fuel (RDF), which can be further used by households for heating or by industries, for instance, for combusting in cement kilns (Kalyani and Pandey, 2014). A few RDF plants have been installed in Hyderabad in Telangana, Vijayawada in Andhra Pradesh, Chandigarh in Punjab, Kalaburagi in Karnataka, and Jaipur in Rajasthan. The plant in Hyderabad was commissioned in 1999 and had a capacity of recycling 1000 tons of waste daily, allowing for producing 210 tons of fluff and pellets in addition to 6.6 MW power. To generate 6 MW

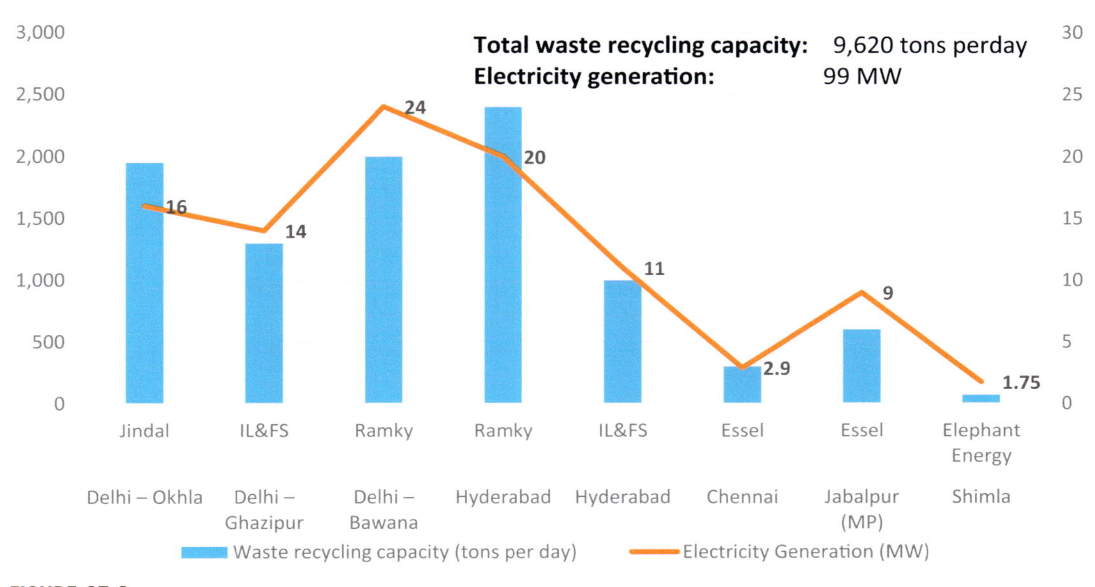

FIGURE 25.8

Capacity of waste to energy plants.

Source: Authors' presentation based on Ahluwalia and Patel (2018).

power in the Vijayawada plant, 500 tons of waste were recycled. These plants are not in use anymore because of financial and logistical issues. Pellets and other solid fuel is produced from 500 tons of waste daily in Chandigarh plant. The RDF project in Karnataka allowed for compacting 50 tons of MSW into 5 tons of fuel pellets. The plant in Jaipur handles 500 tons of waste daily to supply RDF for combusting in cement kilns (Kalyani and Pandey, 2014).

7. Potential niches for RRR options to enhance food, energy, and water security

7.1 Demand for fertilizer

Differing from most of its neighbors, India produces most of chemical fertilizers delivered to domestic market (Table 25.4). Exports of fertilizers are quite low in comparison with total amount of fertilizer produced. Therefore, most of the domestically produced fertilizers are supplied to the market to meet domestic demand. Fertilizer uses are increasing over time, and additional demand for fertilizers are mostly met through increased imports of fertilizers. As of 2014, more than 25% of the consumption of nitrogen fertilizers and about 30% of phosphorus fertilizers are met through the imports (FAO, 2018). Thus, recovering nutrients can help saving substantial amounts of expenditures for fertilizer imports.

Although India produces most of the domestically consumed fertilizers, access to fertilizer and thus application of fertilizer in fields largely vary across the states (Fig. 25.9). Northern states except Himachal Pradesh, Chandigarh, Himachal Pradesh, and Jammu and Kashmir apply large amounts of

Table 25.4 Domestic production, exports and imports of chemical fertilizers (tons).

	Nitrogen			Phosphorus		
Year	**Production quantity in nutrients**	**Exports quantity in nutrients**	**Imports quantity in nutrients**	**Production quantity in nutrients**	**Exports quantity in nutrients**	**Imports quantity in nutrients**
2005	11,218,193	10,127	1,389,864	4,092,561	10,869	1,144,742
2010	12,087,720	17,396	4,547,810	4,303,880	5563	3,698,990
2014	12,329,482	19,136	4,809,303	4,097,197	20,553	1,886,750

Source: FAO (2018).

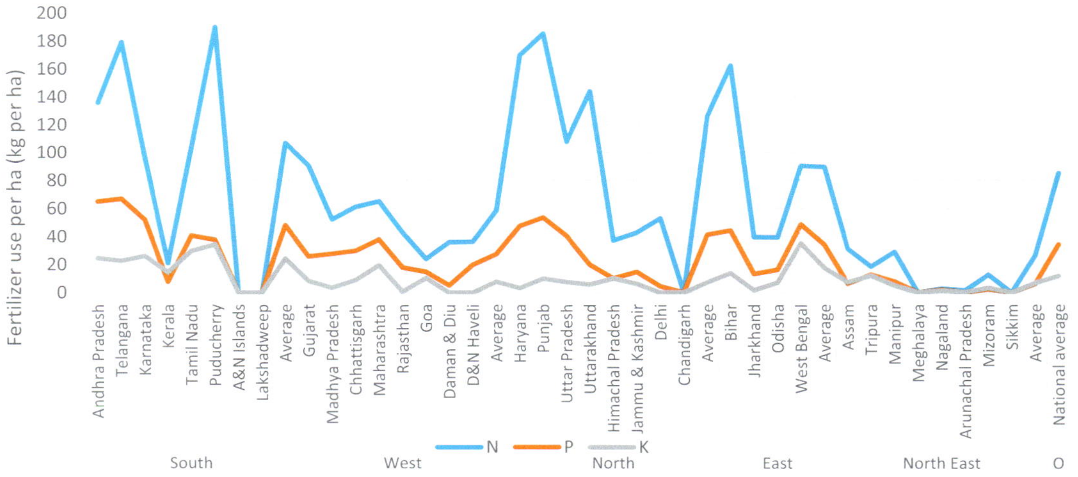

FIGURE 25.9

Fertilizer application per ha across the states of India.

Source: Authors' presentation based on MAFW (2017).

fertilizers per ha of croplands. The lowest application of fertilizers was found in states in the north-eastern part of the country. Medium levels of fertilizer applications were found for the western states of the country. Inadequate levels of fertilizer application reduce soil productivity and crop yields, consequently exacerbating food insecurity and poverty. Thus, compost can be a substitute fertilizer to enhance yields and improve food security in the regions with low agricultural productivity.

Despite the acknowledged benefits of compost in both organic and conventional farming, compost uses are not widespread among farmers in India (Dilkara et al., 2016). Crop residues are usually burned to prevent the transmission of diseases without thinking on possible damage to soil microflora which enhance soil fertility. Although the uses of aged farmyard manure have been known since ancient times, farmers have limited knowledge on improved ways of compost preparation and application. Yet, the situation is changing at present due to government support programs for promoting waste reduction and organic fertilizer use practices.

7.2 Energy demand and water availability

The majority of households in India use wood (49%), LPG (25%), and dung cake (11%) as fuel for cooking (MEF, 2009). Although most households (67.3%) in urban areas use LPG, in rural areas, 67% of the households use wood for cooking, 15% use LPG, and 9.6% use dung cake. Unfortunately, uses of wood, crop stems and residues, and dung cakes for cooking pollute indoor environments exposing the people inside the house to smoke and consequently causing high incidents of eye ailment and respiratory system diseases (Fig. 25.10). Using electricity or gas for cooking can be safer options. Thus, using biogas for cooking would reduce health risks related to indoor air pollution and largely substitute wood, consequently reducing a need for deforestation.

Surface water sources are estimated as 690 km^3 and groundwater sources as 431 km^3 in India (MEF 2009). The Ganges and Brahmaputra are the main rivers and sources of water in the country. Per capita water availability is about 1500 m^3. Despite the vast amount of water supply, groundwater depletion and pollution of rivers are key issues that are increasing water risks in India. At present, about 750 thousand ha lands are classified as drought-prone zones (Ministry of Statistics, 2016). Treated wastewater reuse can be an option in water scarce areas of India to augment water supply. Water treatment before disposal is also important not to pollute freshwater bodies and thus allow more water with acceptable quality to downstream users and environmental systems.

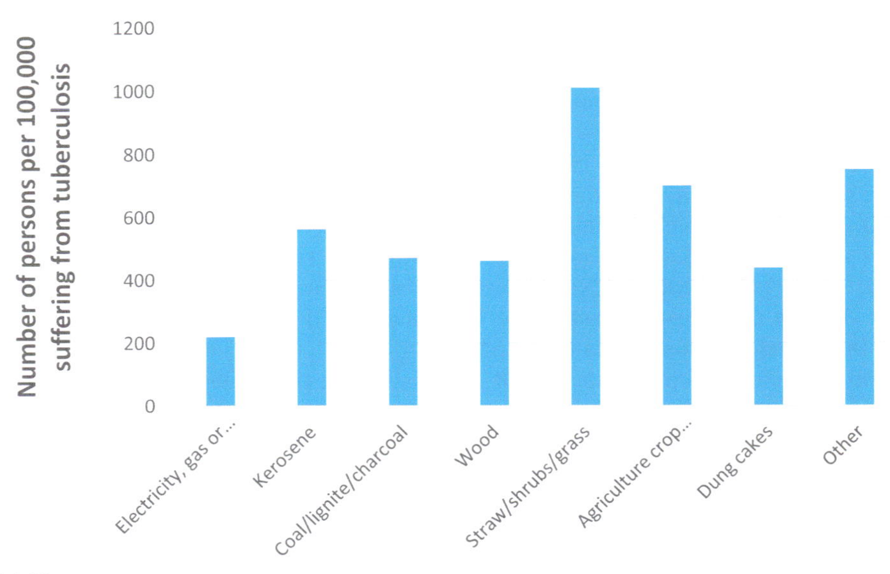

FIGURE 25.10

Energy use for cooking and tuberculosis incidents.

Source: Authors' presentation based on MEF (2009).

8. Enabling environment for RRR options

8.1 Affordability of technological change and access to credit

High poverty rate is one of the main barriers for the adoption of improved sanitation and RRR technologies in the country. Subsidy and credit schemes are therefore essential to raise the required upfront investment costs. Access to bank loans seems favorable for businesses due to sufficiently low inflation and interest rates in India. World Bank (2018) estimated easiness of getting credit index for the country is 75 (within 0 to 100 range). Annual inflation rate is about 5% and interest rate is 9.7% (as of 2016; World Bank 2017a).

8.2 Business opportunities and financing arrangements along the sanitation chain

Along the sanitation value chain, several options of value creation exist (Rohilla et al., 2017). First, demand for improved sanitation facilities creates considerable potential for revenue generation to the producers and constructors of the sanitation facilities (pit latrines, septic tanks, water flush toilets, etc.). Second, services of emptying and transportation of fecal sludge in areas not connected to the sewerage system but with adequate enforcement of environmental protection laws have considerable benefit generation potential through waste collection fees and discharging sludge into crop fields. Third, fecal sludge treatment facilities and landfill sites for disposing sludge can augment benefit through payments to waste disposal and government subsidies. Fourth, facilities to recover nutrients from sludge such as composting plants may recycle waste and sell the produced compost to fertilizer supply companies or farmers. Fifth, energy commodities such as biogas, biochar, RDF, or power can be produced through recycling waste and sold to households or industrial enterprises. Indeed, any combination of these options also can be considered as a business model by private enterprises, NGOs, and state companies to generate revenues.

Given the high capital costs and low prices in RRR sector, governmental financial support through adequate subsidies and tax holidays is important for a successful performance. In India, for supporting municipalities to improve their waste management system, the 13th Finance Commission has endorsed the portion of gross *tax* revenue allocable to Urban Local Bodies (ULBs; Chatri and Aziz, 2012). State governments and funding agencies also considered budgets to support ULBs for the same purpose. Additionally, since 2005, Jawaharlal Nehru National Urban Renewal Mission is distributing government grants through the Urban Infrastructure Governance and Urban Infrastructure Development Scheme for Small and Medium Towns (UIDSSMT).

Long-term agreements on purchase of RRR commodities at guaranteed price may play important role for durance and long-term sustainability of RRR business models (Mittal et al., 2018). For preventing illegal disposal of waste into sites not designed for dumping and thus safeguarding the environment, fines or alternative punishment measures can be also considered for such misconduct while enforcing environmental protection rules. Public–private partnerships (PPPs) are also essential for efficient functioning of RRR facilities. PPPs have been increasing in sanitation sector and RRR sphere, but there are still many difficulties for wider cooperation because of inadequate institutions, inefficient governance, and weak market linkages (Chatri and Aziz, 2012).

8.3 Regulation and policies

For effective management of wastewater, the legislative framework is essential for the proper functioning of the stakeholders involved in sanitation supply chain. Protection of water and environment at state and city levels is adequately addressed in the legislation in India (Rohilla et al., 2017). Maintaining sanitation and public health is assigned as the responsibility of the municipalities, according to the Constitution. The Water (Prevention and Control of Pollution) Act (1974) and Environmental Protection Act (1986) are the main legal documents, which provide provisions for prevention and control of water pollution though septage management was not mentioned in these documents.

National Urban Sanitation Policy (2008), Advisory Note on Septage Management (2013), and National Urban Fecal Sludge Management Policy (2017) are the main documents which prioritize sanitation strategies state-wide and emphasize the need for on-site sanitation facilities in areas with no sewerage system. Particularly, the National Policy on Fecal Sludge and Septage Management aims at developing state-level frameworks and plans for tackling septage management, capacity building and trainings on fecal sludge and septage management, supporting the system of data collection for registered on-site sanitation systems, promoting PPPs in the sector, maintaining integrated citywide sanitation, and safe disposal of wastewater.

Several government programs such as Swachh Bharat Mission (2014) and Pradhan Mantri Awas Yojna (2015) aim at reducing open defecation through providing toilets or fecal sludge containment facilities. The Bureau of Indian Standards prepares standards for building a septic tank. Regulations on sanitation at household level are related with the construction of the sanitation systems and disposal of the waste. The National Building Code of India (1983) (Part IX Plumbing Services, Drainage and Sanitation) and Code of Practice for the Design of Septic Tanks are main documents to manage wastewater at household level. However, the programs on sanitation do not adequately address safe disposal of the collected sludge or proper treatment of wastewater.

No license or permits are required for emptying services, and vehicles license is not required when tractors are used for collecting the septage. The Employment of Manual Scavengers and Construction of Dry Latrines Act (1993) prohibits manual emptying of septic tanks though it is still widely practiced in poor areas. Because of the absence of proper disposal sites, emptied fecal sludge is often discharged into freshwater bodies neglecting possible damage to health and environment. Thus, operators engaged in emptying often face harassment by police or environmental protection authorities for illegal dumping of septage though properly designed dumping sites do not exist (Rohilla et al., 2017).

In terms of solid waste, the Solid Waste Management Rules (2016), which is an improved version of the earlier document adopted in 2000, provide adequate regulatory support for MSW management in India. Composting organic municipal waste was seen as an essential way of returning organic carbon and nutrients embedded in the waste back to the soil and enhance sustainable farming and livelihoods, according to the Inter-ministerial Task Force on Integrated Plant Nutrient Management Using City Compost (Dilkara et al., 2016). The Task Force also emphasized a proper quality and affordability of the produced compost to be marketable and usable for farming. To address quality issues and enforce the quality standards for organic fertilizer in addition to tackling logistical problems in the chain, the National Project on Organic Farming has been established and provides subsidies to capital investments in commercial production of organic fertilizers. The main objectives of the project are to promote organic farming to increase agricultural productivity, reduce dependence on chemical fertilizer uses, and prevent environmental pollution and degradation. The Indian Council for Agricultural Research developed technical programs and resources for promoting organic fertilizer production from organic municipal waste and its further uses to enhance soil health.

In terms of biogas production, government has been supporting policies on a wider implementation of biogas technologies since the 1970s (Mittal et al., 2018). The global oil crisis at that time showed the importance of readiness to shocks in energy markets to prevent dramatic fall in energy access by poor households in remote rural areas. The National Biogas Development Plan (NBDP) was launched in 1981 to alleviate energy crisis in rural areas, and capital subsidies were provided to households owning two to three cows. In 1995, biogas development strategies were expanded to urban areas for reducing pollution from waste and recovering energy from MSW. The NBDP was renamed as National Biogas and Manure Management Programme (NBMMP), and off-grid power generation from biogas projects was promoted. Subsidies as well as short courses for raising awareness were provided within the program. The Ministry of New and Renewable Energy (MNRE) is a main organization which implements the NBMMP across India. MNRE has also always been supportive to WTE programs through providing financial support or organizing trainings. Programs and policies of promoting generation of power from municipal and industrial waste have been also adopted. Although government is in general supportive toward a wider implementation of biogas facilities, there are no legally supported standards of converting biogas into compressed natural gas for vehicles or off-grid power into electricity grid (Mittal et al., 2018).

Procedures related with registering new fertilizer and related costs play important role for the feasibility of composting plants. As shown in Table 25.5, efficiency of regulatory framework for registering new fertilizer seems acceptable. Especially, cost of registering business is not a big deal. However, procedures related with testing new fertilizer and giving permission to produce and sell may take more than two years. Department of Agriculture is responsible for approving the sales of organic fertilizers. Quality control and standards for new fertilizer are unfortunately underdeveloped. Low quality of compost is one of the main reasons for lack of markets for the product in addition to financial difficulties to fund capital investments.

Table 25.5 Easiness of registering new fertilizer (2016).

Indicator	Value
Fertilizer registration index (0–7)	5.0
Quality control of fertilizer index (0–7)	3.5
Time to register a new fertilizer product (days)	804
Cost to register a new fertilizer product (% income per capita)	17.1

Source: Based on Doing Business in Agriculture (World Bank, 2017b)

8.4 Institutions and stakeholders

The main stakeholders involved in wastewater and organic municipal waste management are the Ministry of Urban Development, Ministry of Environment, Forest and Climate Change, Ministry of Social Justice and Empowerment, Central and State Pollution Control Boards, State Governments, ULBs, NGOs, and households (Rohilla et al., 2017).

The Ministry of Urban Development is responsible for formulating fecal sludge management strategies and designing the implementation plans at state and city levels. The Ministry provides technical support to states, conducts trainings for state officials, provides funding, establishes nation-wide awareness programs, promotes PPPs in wastewater and organic municipal waste management system, and organizes nation-wide monitoring of the progress in sanitation system development.

The Ministry of Environment, Forest and Climate Change is responsible for the enforcement of the rules and regulations related with environmental protection and capacity building to control pollution. The ministry ensures the compliance of the environmental laws and regulation in the process of collecting, transporting, disposing, and recycling the fecal and sewage sludge.

The Ministry of Justice and Empowerment helps states in eliminating manual scavenging and manual empting septic tanks. The Ministry monitors the progress in rehabilitating manual scavengers and workers involved in manual emptying. This Ministry also organizes nation-wide campaigns to raise public awareness on the risks related with manual scavenging.

State administrations are responsible to support ULBs technically and financially, coordinate the cooperation among ULBs, and ensure financial stability of the municipalities in improving sanitation services. The state administrations develop plans and strategies of sanitation system improvement at state level, organize capacity building and training of officials engaged in sanitation sector, support research in sanitation system, and organize public awareness campaigns at state level. ULBs are responsible to support and implement sanitation measures at municipal level. Urban municipalities are also in charge of establishing partnerships with NGOs and private companies for conducting waste and wastewater collection services, operating landfills or incineration plants, and organizing composting activities.

8.5 Governance quality and business climate

Quality of governance and institutional framework matters for the performance of business enterprises. Openness and transparency are maintained at moderate level in India as reported by the World Bank (Table 25.6). However, high corruption rates are main issues which impede any policy or technological reforms.

Note: "Cost and Time to start a business" and "Cost to close business" indices are expressed as the "Distance to the frontier" (DTF), i.e., a low score represents a more adverse scenario in terms of starting and closing businesses.

9. Conclusions

Poor sanitation and improper disposal of wastewater and organic waste are main reasons behind heavy environmental pollution and increased health risks across the developing countries such as India. Introduction of RRR technologies is a promising option to prevent open dumping and thus stops

Table 25.6 Governance indicators (2016).

Indicator	Value
Governance	
Corruption index	−0.30
Transparency index (voice and accountability score)	0.41
Governance quality (Government effectiveness)	0.10
Investment climate	
Cost and time to start business	75.4
Cost to close business	40.8
Enabling environment [ease of doing business rank among 190 countries (1 is best)]	100

Source: Based on World Development Indicators (WDI; World Bank, 2017a) and Worldwide Governance Indicators (WGI, 2020).

environmental degradation and prevents health hazards but may require substantial changes along the entire sanitation and waste value chains. This study described a methodological framework to analyze business climate for the adoption of RRR options, implemented the framework to the case of India, and identified main barriers for the technological transformations and ways of coping with them.

Given the high rates of open defecation even in municipal areas, not only adequate establishment of proper sanitation systems such as improved toilets with pit latrines or septage tank but also educative programs to raise the awareness of people about the harmful health and pollution consequences of such behavior are of paramount importance. In addition to the costs of such technologies, their environmental effects and technical feasibility through the consideration of local conditions such as shallowness of groundwater table should be taken into account. Timely and affordable services for the collection of the septage and safe disposal or treatment of this waste are required to prevent overfilled and dysfunctional septage tanks and maintain safe and clean living environment. Availability of septage treatment plants in close location and affordability of emptying waste into these sites are important for preventing illegal dumping of septage into sewerage network or adjacent water bodies.

Separation of solid waste at the source also largely reduces the costs of transportation, screening, and system maintenance. Recycling organic waste into nutrients or energy could provide additional value added. The amount of waste to be converted into nutrients or energy should be determined based on the availability, demand, and prices of the alternative commodities such as chemical fertilizers, electricity, and LPG. Considering interconnectedness of all processes along the sanitation and waste management value chains, a proper planning and integrated management in the system is essential for a proper functioning and successful performance of sanitation services and RRR enterprises.

Availability of lands and easiness of taking permission to use this land for constructing RRR facilities are important factors for modernization in the sanitation and waste management sectors. Improved institutional framework and governance by formulating proper policies for reducing illegal open dumping and fining such behavior as well as restricting waste disposal into landfills may create incentives for wider implementation of the RRR technologies. Centralization or decentralization of wastewater treatments depend on local conditions and financial affordability. Because of low-income levels in India, cheaper sanitation facilities and septage and wastewater treatment options are recommendable. Participation of the private sector in constructing sanitation systems or recycling facilities, collecting and treating the waste should be supported. Given the unaccounted environmental and health benefits (positive externalities) but high capital intensity of improved sanitation and RRR business, adequate subsidies from public funds should be allocated for supporting RRR activities. To improve the marketability of the recovered materials such as compost or compressed natural gas, the quality of such products should be ensured through establishing a trustful certification system. Monitoring the proper functioning of the private companies in the sanitation value chain should be conducted to prevent the pollution due to unsafe disposal or treatment of wastewater and sludge. Enforcement of laws on environmental security and substantial fines for misconduct are important for achieving environmental sustainability goals through improved sanitation and waste treatment. Educational programs and awareness raising campaigns are important for improving knowledge of people and government officials on environmental well-being and increasing social acceptability of improved sanitation facilities and RRR products.

Acknowledgment

This study was funded by Federal Ministry for Economic Cooperation and Development of Germany (BMZ) through a joint research project (PN10200176) of International Water Management Institute (IWMI) and Center for development Research (ZEF) titled "Research and capacity building for inter-sectorial private sector involvement for soil rehabilitation."

References

Ahluwalia, I.J., Patel, U., 2018. Solid Waste Management in India: An Assessment of Resource Recovery and Environmental Impact. ICRIER, New Delhi, India.

Balachandran, B.R., 2015. Citywide sanitation planning: insights from India and Bangladesh. In: Gutterer, B., Reuter, S. (Eds.), Key Elements for a New Urban Agenda Integrated Management of Urban Waters and Sanitation. BORDA, pp. 30–35.

Bansal, M., Saini, R.P., Khatod, D.K., 2013. Development of cooking sector in rural areas in India – a review. Renewable and Sustainable Energy Reviews 17, 44–53.

Bekchanov, M., 2017. Potentials of Waste and Wastewater Resources Recovery and Re-use (RRR) Options for Improving Water, Energy and Nutrition Security. ZEF Working Paper No 157. Center for Development Research (ZEF), Bonn University, Bonn, Germany.

Chandra, R., Vijay, V.K., Subbarao, P.M.V., 2006. A study on biogas generation from non-edible oil seed cakes: potential and prospects in India. In: Proceedings of the 2nd Joint International Conference on Sustainable Energy and Environment. Bangkok, Thailand.

Chatri, A.K., Aziz, A., 2012. Public Private Partnerships in Solid Waste Management Potential and Strategies. Athena Infonomics.

Christy, R., Mabaya, E., Wilson, N., Mutambatsere, E., Mhlanga, N., 2009. In: Da Silva, C., Baker, D., Shepherd, A.W., Jenane, C., Miranda da Cruz, S. (Eds.), Enabling Environments for Competitive Agro-Industries. Agro-industries for development, pp. 136−185.

Dilkara, S., Pamphilon, B., Yousuf, T.B., Islam Shah, M.M., Singh, A., Venkataramaiah, M., Russell, K., 2016. An Exploration of Opportunities to Utilize Urban Organic Waste for the Livelihood Improvement of Rural and Urban Communities in Bangladesh and India. ACIAR, Canberra, Australia.

Dulac, N., 2001. The organic waste flow in integrated sustainable waste management. In: Scheinberg, A. (Ed.), Tools for Decision-Makers — Experiences from the Urban Waste Expertise Programme (1995-2001). WASTE, Nieuwehaven.

FAO- Food, Agricultural Organization, 2013. Enabling Environments for Agribusiness and Agro-Industries Development − Regional and country perspectives. Rome.

Food and Agricultural Organization (FAO), 2018. FAOSTAT − Online Database. Available online at: www.fao.org/faostat/en/#data (Accessed on 12 June 2017).

Garfí, M., 2015. Wastewater Treatment and Reuse in Dayanand Park, Nagpur, Maharashtra, India (NaWaTech) - Case Study of Sustainable Sanitation Projects. Sustainable Sanitation Alliance (SuSanA).

Geissdoerfer, M., Savaget, P., Bocken, N.M.P., Hultink, E.J., 2017. The circular economy − a new sustainability paradigm? Journal of Cleaner Production 143, 757−768.

Gondhalekar, D., Nussbaum, S., Akhtar, A., Kebschull, J., Keilmann, P., Dawa, S., Namgyal, P., Tsultim, L., Phuntsog, T., Dorje, S., Namgail, P., Mutup, T., 2013. Water-related health risks in rapidly developing towns: the potential of integrated GIS-based urban planning. Water International 38 (7), 902−920.

Hingorani, P., 2011. The economics of municipal sewage water by recycling and reuse in India. In: India Infrastructure Report. Oxford University Press. www.idfc.com/pdf/report2011/Chp-21.

Kalia, A.K., Kanwar, S.S., 1998. Long term evaluation of a fixed dome janata biogas plant in hilly conditions. Bioresource Technology 65, 61−63.

Kalyani, K.A., Pandey, K.K., 2014. Waste to energy status in India: a short review. Renew Sustain Energy Reviews 3, 113−120.

Kaniyamparambil, J.S., 2011. A Look at India's Biogas Energy Development Program − after Three Decades, Is it Useful (Doing what it Should) and Should it Be Continued? School of Engineering Practice McMaster University.

Kumar, R., 2016. Eco-friendly wastewater treatment for reuse in agriculture (India). In: Hettiarachchi, H., Ardakanian, R. (Eds.), Safe Use of Wastewater in Agriculture: Good Practice Examples. UNU-FLORES, Dresden, Germany, pp. 139−156.

Kumar, S., Smith, S.R., Fowler, G., Velis, C., Kumar, S.J., Arya, S., Kumar, R., Cheeseman, C., 2017. Challenges and Opportunities Associated with Waste Management in India, vol. 4. Royal Society Open Science, p. 160764.

Ministry of Agriculture and Farmers Welfare (MAFW), 2017. Agricultural Statistics at a Glance 2016. MAFW, New Delhi.

Ministry of Chemicals and Fertilizers (MCF), 2017. Implementation of Policy on Promotion of City Compost. Thirty-fourth report, New Delhi, India.

Ministry of Environment & Forests (MEF), 2009. State of Environment Report India-2009. New Delhi, India.

Ministry of New and Renewable Energy (MNRE), 2017. The State/UT-wise Targets for Setting up of the Family Type Biogas Plants in the Country under National Biogas and Manure Management Programme (NBMMP) during the Last Three Years (2012-13, 2013-14 and 2014-15) and Achievements of Current Year 2015-16 (Up to 31.01.2016). Online Document. Available at: http://mnre.gov.in/schemes/decentralized-systems/schems-2/ (Accessed on 5 February 2018).

Ministry of Statistics, 2016. Compendium of Environment Statistics - India 2016. New Delhi, India.

Mittal, S., Ahlgren, E.O., Shukla, P.R., 2018. Barriers to biogas dissemination in India: a review. Energy Policy 112, 361–370.
OECD-Organization for economic co-operation and development, 2008. A Framework for Addressing and Measuring Entrepreneurship. OECD Statistics Working Paper.
Oxford Economics, 2016. The True Cost of Poor Sanitation.
Porter, M.E., 2004. Building the microeconomic foundations of prosperity: findings from the Business Competitiveness Index. In: The Global Competitiveness Report, 2004–2005. World Economic Forum, New York, pp. 29–56.
Rao, K.U., Ravindranath, N.H., 2002. Policies to overcome barriers to the spread of bioenergy technologies in India. Energy Sustainable Development VI (3).
Rao, P.V., Baral, S.S., Dey, R., Mutnuri, S., 2010. Biogas generation potential by anaerobic digestion for sustainable energy development in India. Renewable and Sustainable Energy Reviews 14, 2086–2094.
Ravindranath, N.H., Balachandra, P., 2009. Sustainable bioenergy for India: technical, economic and policy analysis. Energy 34, 1003–1013.
Rohilla, S.K., Luthra, B., Bhatnagar, A., Matto, M., Bhonde, U., 2017. Septage Management: A Practitioner's Guide. Centre for Science and Environment, New Delhi.
Singh, S., Mohan, R.R., Rathi, S., Raju, N.J., 2017. Technology options for faecal sludge management in developing countries: benefits and revenue from reuse. Environmental Technology & Innovation 7, 203–218.
Starkl, M., Amerasinghe, P., Essl, L., Jampani, M., Kumar, D., Asolekar, S.R., 2013. Potential of natural treatment technologies for wastewater management in India. Journal of Water, Sanitation & Hygiene for Development 3, 500–511.
Stern, N., 2002. A Strategy for Development. The World Bank, Washington, DC.
Strande, L., Ronteltap, M., Brdjanovic, D. (Eds.), 2014. Faecal Sludge Management (FSM) Book - Systems Approach for Implementation and Operation. IWA Publishing, UK.
Tarraf, A., 2016. Social and Behaviour Change. Communication Insights and Strategy Case Study: Open Defecation in India. WPP Government & Public Sector Practice. Available online at: https://www.wpp.com/govtpractice/~/media/wppgov/insights/open%20defecation/wpp_open_defecation_in_rural_india.pdf (Accessed on 24 May 2018).
Venkatesh, P., Nithyashree, M., Sangeetha, V., Pal, S., 2015. Trends in agriculture, non-farm sector and rural employment in India: an insight from state level analysis. Indian Journal of Agricultural Sciences 85 (5), 671–677.
Water Aid, 2016. An Assessement of Faecal Sludge Management Policies and Programmes at the National and Select States Level. Dehli, India.
Water and Sanitation Program (WSP), World Bank, 2011. Economic Impacts of Inadequate Sanitation in India (Falgship Report). World Bank, Washington DC, the USA.
Water and Sanitation Program (WSP), World Bank; International Water Management Institute (IWMI), 2016. Recycling and Reuse of Treated Wastewater in Urban India: A Proposed Advisory and Guidance Document (Resource Recovery and Reuse Series 8). International Water Management Institute (IWMI), Colombo, Sri Lanka, p. 57p. https://doi.org/10.5337/2016.203.
Wintgens, T., Nättorp, A., Elango, L., Asolekar, S.R., 2016. Natural Water Treatment Systems for Safe and Sustainable Water Supply in the Indian Context: Saph Pani. IWA publishing, London, UK.
World Bank, 2010. Improving Wastewater Use in Agriculture: An Emerging Priority. World Bank, Washington DC, the USA.
World Bank, 2015. Doing Business 2015. Going beyond Efficiency. World Bank Group, Washington DC.
World Bank, 2017a. World Development Indicators. Online Database. Available online at: databank.worldbank.org (Accessed on 30 January 2018).

World Bank, 2017b. Enabling the Business of Agriculture 2017. Available online at: http://eba.worldbank.org/~/media/WBG/AgriBusiness/Documents/Reports/2017/EBA-Full-Report.pdf?la=en (Accessed on 2 March 2018).
World Bank, 2018. Doing Business Report 2018: Reforming to Create Jobs. Available online at: http://www.doingbusiness.org/~/media/WBG/DoingBusiness/Documents/Annual-Reports/English/DB2018-Full-Report.pdf (Accessed on 25 January 2018).
World Economic Forum (WEF), 2015. The Global Competitiveness Report. 2015. Geneva World Bank, 2004. World Development Report 2005. A better investment climate for everyone, Washington, DC.
Worldwide Governance Indicators, 2020. http://info.worldbank.org/governance/wgi/Home/Reports (Accessed 31 March 2022).
Zurbrügg, C., Drescher, S., Patel, A., Sharatchandra, H.C., 2004. Decentralised composting of urban waste–an overview of community and private initiatives in Indian cities. Waste Management 24 (7), 655–662.

CHAPTER

Climate change and sustainable energy systems 26

Tri Ratna Bajracharya[1], Shree Raj Shakya[1,2] and Anzoo Sharma[1,3]

[1]*Centre for Energy Studies (CES), Institute of Engineering, Tribhuvan University, Pulchowk, Lalitpur, Nepal;* [2]*Institute for Advanced Sustainability Studies (IASS), Potsdam, Germany;* [3]*Center for Rural Technology (CRT/N), Kathmandu, Nepal*

Chapter outline

1. Global climate change and its impacts

Climate change consists of the global warming effect due to human activities–induced (anthropogenic) emissions of greenhouse gases (GHGs) and those resulting from large-scale changes in weather patterns (Stocker et al., 2013). The trend of global average temperature is rising as revealed by the annual average global surface temperature between 1850 and 2020 (Fig. 26.1). Human influence on the climate system is clear, and recent anthropogenic emissions of GHGs are the highest in history (EEA, 2017). The global anthropogenic CO_2 emissions resulted from forestry and other land use (FOLU) and those from the burning of fossil fuels, cement production, and flaring are shown in Fig. 26.2. The occurrence of climate changes has resulted in widespread impacts on human and natural systems. The ongoing warming process of the climate system since the 1950s is unequivocal, and many of the observed changes are seen as unprecedented over decades to millennia. There is an observation of warming of the atmosphere and ocean, diminishing of the snow and ice sources, and rising of the sea level across the globe (IPCC, 2014b).

The impacts attributed to the climate change can be categorized into physical systems, biological systems, and human and managed systems. The scale of impacts is different place by place as shown in Table 26.1.

Assessment of vulnerability to climate change has been gaining focus in recent days due to the unavoidable need of a society's response to climate change consisting of climate change adaptation, developing resilience, assessments of climate risk, or addressing climate justice concerns. Climate change vulnerability has been defined in the third IPCC report as "the degree to which a system is

Handbook of Energy and Environmental Security. https://doi.org/10.1016/B978-0-12-824084-7.24001-0
Copyright © 2022 Elsevier Inc. All rights reserved.

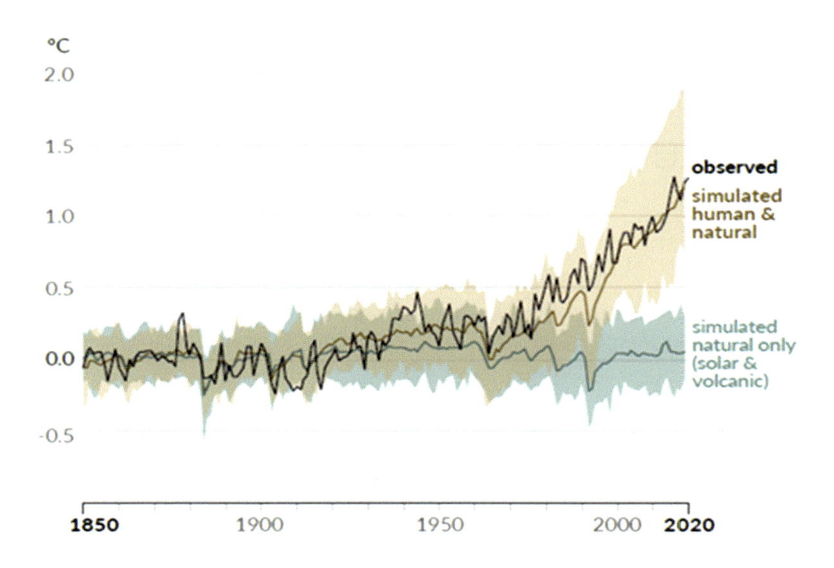

FIGURE 26.1

Changes in annual average global surface temperature between 1850 and 2020.

Source: Figure SPM.1, Panel (b) IPCC (2021).

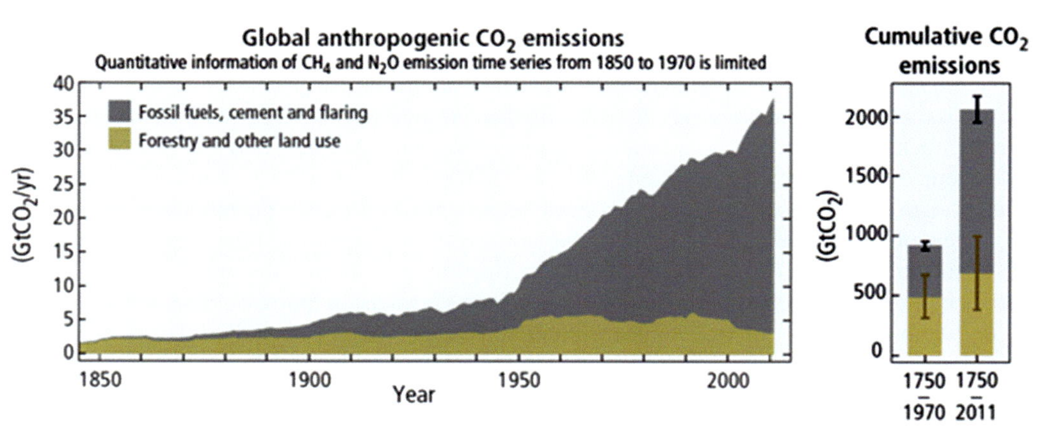

FIGURE 26.2

Global anthropogenic CO_2 emissions.

Source: Figure SPM.1, Panel (d) IPCC (2014b).

Table 26.1 Impacts attributed to climate change in different regions of the world.

	North America	Central and South America	Europe	Africa	Asia	Australasia	Small Islands
Physical systems							
Glaciers, snow, ice, and/or permafrost	M	M	M	M	M	M	
Rivers, lakes, floods, and/or drought	Mm	M	m	M	Mm	Mm	m
Coastal erosion and/or sea level effects	M				M		m
Biological systems							
Terrestrial ecosystems	Mm	m	M	M	M	M	Mm
Wildfires	Mm	m	M	M			
Marine ecosystems	M	Mm	M	M	M	M	Mm
Human and managed systems							
Food production		M	m	Mm	m	Mm	
Livelihoods, health, and/or economics	M	M	M	Mm	Mm	M	m

Note: M, *major contribution of climate change;* m, *minor contribution of climate change; and* Mm, *major to minor contribution of climate change.*
Adopted from IPCC (2014b).

susceptible to, and unable to cope with, adverse effects of climate change, including climate variability and extremes" (IPCC, 2007).

Vulnerability can be categorized into two major types, socioeconomic vulnerability and geographic vulnerability. Most of the literature uses climate vulnerability to describe impacts on the communities, economic systems, or geographies (EEA, 2017). In order to monitor and track the extent of climate change vulnerability effects on human society, various indexes have been developed (de Sherbinin et al., 2019). The climate risk index of the top 10 most affected countries in the last decades is given in Table 26.2 (Eckstein et al., 2021).

Table 26.2 The Climate Risk Index (CRI) for top 10 most affected countries from 2000 to 2019 (annual averages).

CRI 2000–2019 (1999–2018)	Country	CRI Score	Fatalities	Fatalities per 100,000 inhabitants	Losses in million US$ PPP	Losses per unit GDP in %	Number of events (2000–2019
1 (1)	Puerto Rico	7.17	149.85	4.12	4149.98	3.66	24
2 (2)	Myanmar	10.00	7056.45	14.35	1512.11	0.80	57
3 (3)	Haiti	13.67	274.05	2.78	392.54	2.30	80
4 (4)	Philippines	18.17	859.35	0.93	3179.12	0.54	317
5 (14)	Mozambique	25.83	125.40	0.52	303.03	1.33	57
6 (20)	The Bahamas	27.27	5.35	1.56	426.88	3.81	13
7 (7)	Bangladesh	28.33	572.50	0.38	1860.04	0.41	185
8 (5)	Pakistan	29.00	502.45	0.30	3771.91	0.52	173
9 (8)	Thailand	29.83	137.75	0.21	7719.15	0.82	146
10 (9)	Nepal	31.33	217.15	0.82	233.06	0.39	191

Reproduced from Eckstein et al. (2021).

2. Sustainable energy systems for mitigation of climate change

The energy sector is the major source of GHG emissions today with a three-quarter share in the annual emissions and holds the key to averting the worst impacts of climate change (IEA, 2021). According to the International Energy Agency (IEA), global energy–related CO_2 emissions in 2019 were 33 billion metric tons of which 33% contribution were from the advanced economies and the remaining from the rest of the world (IEA, 2020). During the 21st Conference of Parties (COP 21) of the United Nations Framework Convention on Climate Change (UNFCCC), the Paris Agreement had set a historic turning point in global efforts to address climate change by agreeing on specific long-term goals to hold temperature increase to well below 2°C and to pursue efforts to limit the increase to 1.5°C, including its accompanying goals to reach global peaking of GHG emissions as soon as possible and a balance between emission and removals in the second half of the century. As of September 2021, 191 parties (190 countries and the European Union [EU]) representing 97% of GHG emissions have ratified the Paris Agreement.[1] This signifies the very broad support and commitment of countries toward the multilateral process and its importance to resolving the challenges addressed by the Paris Agreement.

Efforts to switch to renewable and environment-friendly sources have flattened the emissions particularly in developed countries. In 2019, developed countries observed a decline in their CO_2 emission by 3.2% from the 2018 level, whereas there was still a 2% growth in emission from the rest of the world during the same period.[2] Power generation, industrial processes, and agriculture, forestry,

[1]https://unfccc.int/process/the-paris-agreement/status-of-ratification.
[2]https://www.iea.org/articles/global-co2-emissions-in-2019.

and other land use are major contributors to GHG emissions, and the recent reduction in GHG emissions was observed as a result of fuel switching to renewable sources in the power generation sector where coal is the primary source of fuel. Global GHG emissions were 49.8 billion metric tons in 2015[3] and are expected to reach 56 billion tons which are twice the targeted emissions in 2030. To limit global warming to 1.5°C, reducing emissions by 7.6% annually is required (UNEP, 2019).

The long-term transition toward net-zero emission by 2050 would require a reduction in energy service demands, decarbonization of electricity and fossil fuels, electrification of end-use services, reduction in agricultural emissions, and offsetting the remaining emission by carbon dioxide removal processes such as carbon storage in land through forests or by carbon capture and storage (CCS) in geological reservoirs (IPCC, 2018). The need for negative emission technologies or carbon sink cannot be overlooked to achieve a net-zero emission target. Demand-side measures such as lowering energy service demands and reducing the consumption of GHG intensive goods which comes from behavioral and lifestyle changes can further help to reach net-zero emission target. The new emerging concept of carbon dioxide capture, utilization, and storage (CCUS) has been seen as the potential option in future as it involves utilization of carbon dioxide in the production of valuable outputs. For example, CO_2 captured can be used as feedstock to produce urea, methanol, formaldehyde, etc., through various chemical processes. Afforestation and reforestation can also offset GHG emissions by carbon sequestration in forests.

Energy efficiency improvement and fuel switching from fossil fuels to electricity in transport, residential, commercial, and agriculture sectors offer deep decarbonization to achieve net-zero target. Transport sector is mainly based on fossil fuels to meet its energy need. Consumer behavioral changes toward avoiding unnecessary long distance travel, preferring modal shift to mass transport, using improved vehicle and engine technologies, low-carbon energy sources, and investment and development of necessary infrastructure combined offer high mitigation potential in the transport sector (Sims et al., 2014). Electric vehicles are on the horizon and gaining traction as the promising technology of the future. In addition to battery electric vehicles (BEVs), fuel cell electric vehicle that runs on hydrogen as fuel can have an important role in reducing emissions from the transport sector. Hydrogen fuel if produced from low-carbon or renewable sources can be deployed for the decarbonization of transport and industry sectors. Biofuel is another low-carbon fuel option in the transport sector. In freight transport, systemic improvement in combination with efficiency improvement in vehicle efficiency can be a potential option. Systemic improvement includes efficient routing, supply chain, and logistics (IEA, 2017). The advancement in biofuel development, fuel-cell vehicles, and electric trucks extends higher mitigation potential in freight transport. The aviation and maritime transport are more challenging to decarbonize, and hydrogen is also seen as the potential fuel substitute for decarbonization.

Energy-efficient building design, improved energy efficiency of technology coupled with electrification of end-use services (such as cooking, space heating, etc.) in residential and commercial sector offer significant emission reduction potential. Heat pump for space heating and light-emitting diode for lighting are some of the most efficient technologies in buildings. Industry sector is the largest final energy consuming as well as GHG-emitting end-use sector. Material industries such as cement, steel, nonferrous metal, paper and pulp, etc., are both energy and emission-intensive industries. The industries use combustible fossil fuels and biomass in bulk for thermal application (such as process heat

[3]https://ourworldindata.org/greenhouse-gas-emissions.

and steam generation), while electricity is used mainly for motive power. In general, the emission reduction measures in industry can be categorized into following five strategies: reducing demand, improving energy efficiency, increasing electrification of energy demand, switch to low-carbon nonelectric fuels, deploying innovative processes, and application of CCS (IPCC, 2018). Material efficiency improvement in the design of buildings and infrastructures reduces the demand of industrial products. Full biomass firing in boilers, electrification of heating application using electric boilers and industrial heat pumps are some of the technical solutions to decarbonize the industry sector.

As of September 2021, 32 countries and the EU have submitted a long-term strategy to attain a net-zero emission target to UNFCCC, and more than 100 countries are planning to do so. Bhutan and Suriname are the only two carbon net-negative countries removing more carbon than they emit. Singapore and Australia have planned to achieve net-zero GHG emission in the second half of the century. The EU and remaining other countries aim to achieve net-zero by 2050.[4] The development of quantitative scenarios that assess various pathways to achieve net-zero emissions is a fundamental to understand the road map to achieving the target and enables discussion with stakeholders regarding the implications of the target in terms of trade-off and opportunities that lie ahead. Transformation of the energy sector is crucial to achieve this target in any country. This would require energy efficiency improvement and adoption of low-carbon technology. In addition, mandatory energy labeling scheme and minimum energy performance standards are equally important to improve energy efficiency. Singapore aims to have 80% green building by 2030 and zero private vehicle growth to reach the target.

The circular economy approach in water and waste management is important to reduce waste and increase recycling. The adoption of CCUS technologies that are still not matured would have a significant role in the future (NCCS, 2020). Bhutan's Intended Nationally Determined Contribution suggests the adoption of climate-smart agriculture and livestock farming as the potential measures to reduce GHG emissions from the agriculture sector. Besides, sustainable forest management and conservation of biodiversity are deemed to be options to increase sequestration potential from the land-use sector. In the United Kingdom, resource and energy efficiency, CCS technology, reducing methane venting or leakage, and fuel switching from fossil fuels to hydrogen, electricity, and bioenergy are considered to be the long-term strategy to become carbon neutral. The use of heat pumps, biomass-based boilers, and hydrogen boilers are considered as the potential mitigation measures in the industry as well as buildings. In transport, phasing out diesel trains and replacing internal combustion engine vehicles by BEV and plug-in hybrid electric vehicles (PHEVs), improving logistic efficiency along with investment in cycling and walking infrastructure are considered as potential options (CCC, 2019). Similarly, in other countries, the low-carbon strategy includes energy efficiency improvement, fuel switching, use of advanced low-emission technologies (existing and under development phase), afforestation/reforestation, and CCUS as potential options.

Long-term net-zero target is a challenge to every country as they need to transform their economies and is also an opportunity toward sustainable economic growth that will avoid possible climate impacts. In addition to setting net-zero targets and identifying viable options, a broad range of policies and measures are also needed such as framework legislation and strategies (e.g., climate laws and long-

[4]https://unfccc.int/process/the-paris-agreement/long-term-strategies.

term strategies), economic instruments (e.g., carbon taxes, subsidy reform, trade policy, and tax incentives), regulatory instruments (e.g., emissions, technology, and product standards), and other approaches, such as information policies, procurement policies, voluntary agreements, and valuation and accountability mechanisms, can play important roles in the broader climate policy package (Levin et al., 2020). The key mitigation measures identified by the IEA and their target to attain global net-zero emission by 2050 is given in Table 26.3.

3. Sustainable energy systems for adaptation to climate change

Adaptation to climate change is the process of adjusting to the current or future expected climate change and its effects (IPCC, 2014a). For people, adaptation measures focus on minimizing or avoiding climate change related harm and exploiting related opportunities. Whereas for natural systems, proactive humans' involvement is expected to help adjustment to unavoidable changed condition. Adaptation measures can be either incremental nature (which preserve the essence and integrity of a system) or transformational nature (which can change the fundamental characteristics of a system in response to climate change and its impacts) (Noble, 2014).

The nature of adaptation measures may vary at different places depending on the sensitivity and vulnerability of humans and natural systems to climate change (Sarkodie and Strezov, 2019). Implementation of effective adaptation measures is especially important in developing countries due to their higher vulnerability to climate change and is quite challenging due to their limited ability for social and economic transformation (UN, 2020). The economic costs of adaptation to climate change are quite high requiring annual investment of billions of dollars for several decades (Margulis, 2010). Besides there may be an indirect cost arising from a diversion of resources from productive toward adaptive capital (adaptive investment effect) (Mohaddes and Williams, 2020).

Sustainable supply of energy plays an important role while implementing measures to adapt to climate change. The energy can be used as the basic commodity, distributed energy generation source, or energy infrastructure development for different adaption plans and programs. Different category of adaptation measures and the role of energy is listed in Table 26.4.

Khanal et al. (2020) highlighted the role of clean and renewable energy to address the adverse impacts of climate change and they include: reducing or controlling GHG emissions and local environmental pollution; improving resilience of the energy system; addressing the local needs of women, marginalized and vulnerable groups; improving the access to the range of households' level service (lighting, heating, and clean air) and productive needs (education, health, agriculture, communication, etc.) during occurrence of climate change. These attributes are crucial to enhance the capacity of an individual or community to improve their living environment, income generation, and information access, which are the integral components of adaptive capacity. Besides, it also helps in the sustainable management of air quality, water and land resources. These collective interactions help in improving the adaptive capacity of the community and help in the reduction of poverty (Khanal et al., 2020). Adaptive capacity also reflects the resilience of communities to variability and change (including but not limited to climate change) and represents the capability to transform various assets consisting of the environmental, social, and financial dimensions into human well-being. Fig. 26.3 shows the energy, human capability, resources and system interactions, and climate change adaption.

Table 26.3 Key mitigation measures to attain global net zero emission by 2050 (IEA, 2021).

	Building	Transport	Industry	Electricity and heat	Other
2025	• Restrict new sales of fossil fuel boilers				
2030	• Universal energy access • Zero-carbon ready for all new buildings	• Share of electric vehicle to reach 60% of global car sales	• Most new clean technologies demonstrated at scale in heavy industry	• 1020 GW of annual solar and wind installation • Advanced economies phasing out all unabated coal plant • 850 GW (150 Mt) low-carbon hydrogen electrolysers	
2035	• Best in class standard for most appliances and cooling systems sold	• 50% of heavy truck sales are electric • Restrict new ICE car sales	• Best in class standard for all industrial electric motor sales	• Advanced economies attaining overall net-zero emissions electricity	• Capture of 4 Gt CO_2 emissions
2040	• Zero-carbon ready level retrofit for 50% of existing buildings	• Aviation sector using low-emission fuel with 50% share	• 90% of existing capacity in heavy industries reaches end of investment cycle	• Globally net-zero emission electricity • Phase-out of all unabated coal and oil power plants being phased out	
2045	• 50% of heating demand met by heat pumps			• 3000 GW (435 Mt) low-carbon hydrogen electrolysers	
2050	• Zero-carbon buildings reaching more than 85% share		• Attaining low-emissions in heavy industrial production with more than 90% share	• Almost 70% of electricity generation globally from solar PV and wind	• Capture of 7.6 Gt CO_2 emissions

Table 26.4 Role of energy in implementation of adaptation measures (IPCC, 2014a).

Adaptation category	Example	Role of energy
Human development	Improvement in access to energy, health facilities, nutrition, education, safe housing and settlement structures, and social support structures; decrease in gender inequality and marginalization.	Energy as basic commodity
Poverty alleviation	Enhancement in access/control of local resources; disaster risk reduction; land tenure; insurance schemes; social safety nets and social protection.	Energy as facilitating input
Livelihood security	Improved infrastructure; Asset, income, and livelihood diversification; access to technology and decision-making opportunities; changed cropping, livestock and aquaculture practices; reliance on social networks; enhanced decision-making power.	Energy as facilitating input
Disaster risk management	Access to benefits from early warning systems; hazard and vulnerability mapping; diversifying water resources; improved drainage; flood and cyclone shelters; building codes and practices; storm and wastewater management; transport and road infrastructure improvements.	Energy as facilitating input, distributed generation
Ecosystem management	Watershed and reservoir management; reduction of other stressors on ecosystems and of habitat fragmentation; maintaining wetlands and urban green spaces; maintenance of genetic diversity; coastal afforestation; community-based natural resource management; manipulation of disturbance regimes.	Energy as facilitating input
Spatial or land-use planning	Access to adequate housing, infrastructure, and services; urban planning and upgrading programs; managing development in flood prone and other high-risk areas; easements; land zoning laws; protected areas.	Energy as facilitating input, energy infrastructure
Structural/physical	***Engineered and built-environment options:*** Water storage; flood control embarkments; improved drainage; sea walls and coastal protection structures; flood and cyclone shelters; storm and wastewater management; building codes and practices; floating houses; transport and road infrastructure improvements; Power plant and electricity grid adjustments. ***Technological options:*** New crop and animal varieties; indigenous, traditional and local knowledge, technologies and methods; efficient irrigation; water-saving technologies; desalinization; conservation agriculture; food storage and preservation facilities; hazard and vulnerability mapping and monitoring; early warning systems; building insulation; mechanical and passive cooling; technology development, transfer, and diffusion.	Energy as facilitating input, energy infrastructure

Continued

Table 26.4 Role of energy in implementation of adaptation measures (IPCC, 2014a).—cont'd

Adaptation category	Example	Role of energy
	Ecosystem-based options: Ecological restoration; soil conservation; Afforestation and reforestation; mangrove conservation and replanting; green infrastructure (e.g., shade trees, green roofs); controlling overfishing; fisheries co-management; assisted species migration and dispersal; ecological corridors; seed banks, gene banks, and other ex situ conservation; community-based natural resource management. ***Services:*** Social safety nets and social protection; food banks and distribution of food surplus; municipal services including water and sanitation; vaccination programs; essential public health services; enhanced emergency medical services.	
Institutional	***Economic options:*** Financial incentives; insurance; catastrophe bonds; payments for ecosystem services; pricing water to encourage universal provision and careful use; microfinance; disaster contingency funds; cash transfers; public—private partnerships. ***Laws and regulations:*** Land zoning laws; building standards and practices; easements; water regulations and agreements; laws to support disaster risk reduction; laws to encourage insurance purchasing; Defined property rights and land tenure security; protected areas; fishing quotas; patent pools and technology transfer. ***National and government policies and programs:*** National and regional adaptation plans including mainstreaming; subnational and local adaptation plans; economic diversification; urban upgrading programs; municipal water management programs; disaster planning and preparedness; integrated water resource management; integrated coastal zone management; ecosystem-based management; community-based adaptation.	Energy as facilitating input
Social	***Educational options:*** Awareness raising and integrating into education; gender equity in education; extension services; sharing indigenous, traditional, and local knowledge; participatory action research and social learning; knowledge-sharing and learning platforms. ***Informational options:*** Hazard and vulnerability mapping; early warning and response systems; systematic monitoring and remote sensing; climate services; use of indigenous climate observations; participatory scenario development; integrated assessments. ***Behavioral options:*** Household preparation and	Energy as facilitating input

Table 26.4 Role of energy in implementation of adaptation measures (IPCC, 2014a).—cont'd

Adaptation category	Example	Role of energy
	evacuation planning; migration; soil and water conservation; storm drain clearance; livelihood diversification; changed cropping, livestock, and aquaculture practices; reliance on social networks.	
Spheres of change	***Practical:*** Social and technical innovations, behavioral shifts, or institutional and managerial changes that produce substantial shifts in outcomes. ***Political:*** Political, social, cultural, and ecological decisions and actions consistent with reducing vulnerability and risk and supporting adaptation, mitigation, and sustainable development. ***Personal:*** Individual and collective assumptions, beliefs, values, and worldviews influencing climate-change responses.	Energy as facilitating input

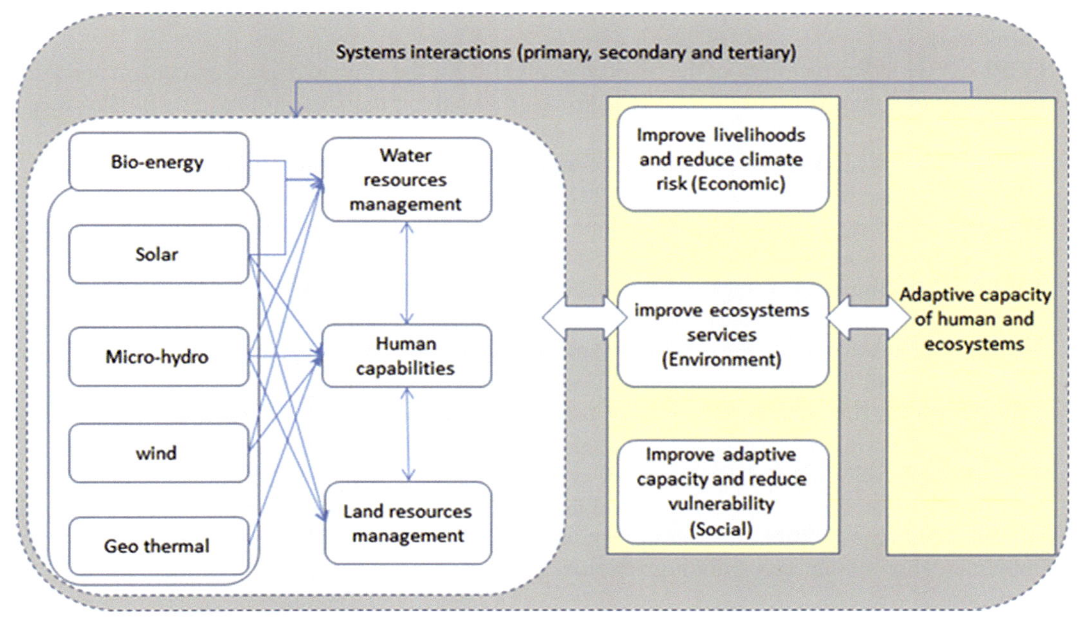

FIGURE 26.3

Energy, human capability, resources and system interactions, and climate change adaption.

Khanal et al. (2020).

4. Global and national initiatives

Climate change discussion started with the establishment of the Intergovernmental Panel on Climate Change (IPCC)[5] in 1988. The United Nations General Assembly at its 45th session decided to start the process of formulation of the framework convention on climate change. As a result, the text of the UNFCCC was prepared which was adopted in May 1992 and entered into force in 1994. The adoption of the UNFCCC embarks on a new history in an effort of the global community to address challenges of climate change.

After the UNFCCC enters into force, 25 COPs to the UNFCCC have been organized and several decisions were made. The significant and most prominent COPs were:

- COP3, 1997, where the Kyoto Protocol was adopted. This provided mandatory target to reduce GHG emission for developed countries.
- COP13, 2007, with Bali Road Map, as an outcome became another milestone that led the negotiation toward a shared vision, mitigation, adaptation, technology transfer, and financing. In this COP, concept of implementation of Reducing Emissions from Deforestation and Forest Degradation was agreed.
- The COP15, 2009, one of the highly talked COP but failed to come up with substantial outcome resulted in the Copenhagen Accord that sets out the climate finance target of USD100 billion per year by 2020.
- COP16, 2010 at Cancun, Mexico, sets out historic milestones with a decision of establishing Green Climate Fund (GCF), the Technology Mechanism (TM), and Cancun Adaptation Framework.
- COP17, 2011 at Durban sets out the timeframe of 2015 for the climate change agreement and thus started the negotiation under the Ad Hoc Working Group on the Durban Platform for Enhance Action (ADP).
- COP 21 at Paris marked the historic adoption of the Paris Agreement. The historic Paris Agreement, after it entered into force in 2016, shifted the climate change discourse to the new level thus speeding up the transition toward low-carbon development across the globe.
- COP 22 in 2016 in Marrakech, Morocco, sets a two-year time frame to agree to rules and procedures for the Paris Agreement, endorsed the decision of the GCF to provide USD 3 million per country to support adaptation plan and adaptation planning process, and provided guidance focusing on Nationally Determined Contributions (NDCs), a transparency framework, technology development and transfer, adaptation, and market and nonmarket approaches.
- COP 23 held in Bonn, Germany, in 2017 focuses the role of local communities and indigenous peoples in achieving the targets and goals set out in the Paris Agreement and the 2030 Agenda for Sustainable Development, mainstreaming gender-responsive climate policies. It also highlights the need for a technical assessment on mitigation and adaptation to improve its effectiveness and enhancing climate technology development and transfer through TM, emphasizes the implementation of capacity-building activities particularly in countries with economies in transition.

[5]The Intergovernmental Panel on Climate Change (IPCC) is the international body for assessing the science related to climate change (http://www.ipcc.ch/news_and_events/docs/factsheets/FS_what_ipcc.pdf).

- COP 24 in 2018 in Katowice, Poland, sets out the guidelines for implementing the 2015 Paris Climate Change Agreement to limit the global warming of 1.5°C as per the special report of the IPCC.
- COP 25 held in Madrid, Spain, in 2019 encourages countries to set more ambitious NDCs target to limit the global average temperature increase to well below 2°C goal.

The 26th UN Climate Change Conference of the Parties (COP 26), to be hosted by the United Kingdom, is going to be organized in Glasgow on November 1, 2021–November 12, 2021. The COP26 is focused on four main thematic components (Fig. 26.1) with the agendas[6] to:

I. Secure global net-zero by mid-century and keep 1.5° within reach
II. Adapt to protect communities and natural habitats
III. Mobilize finance for mitigation and adaptation
IV. Work together to deliver

There is a gap between rhetoric and reality on emissions, and due to various reasons, the world is not on track to limit global warming to 1.5°C. Scientific analyses have shown that there are technically feasible, cost-effective, and socially acceptable pathways to reach net-zero by 2050, though they may look extremely challenging and may be requiring efforts from all the stakeholders—governments, businesses, and citizens (IEA, 2021). As of September 2021, 192 parties have submitted first NDC document and 13 parties have submitted second NDC to UNFCCC, thus indicating their commitment to attain net-zero emissions by 2050. COP26 is expected to be an opportunity for member countries—both from the developed and the developing nations—to move from the rhetorical commitments to real urgent action.

5. Conclusion

Climate change is the global concern for the sustainability of human society and natural systems. The recent human-made, anthropogenic, emissions in the course of the economic development are seen as the major culprit for the increasing global GHGs inventory resulting in negative impacts on the human and natural systems. The impacts of climate change are experienced all over the world with various degrees of disturbance in the physical, biological, and human managed systems. Developing countries are at greater risk of climate change impacts due to their limited capacity to climate change adaptation, developing resilience, and addressing climate justice concerns. The Paris Agreement under the UNFCCC had established a historic target of the long-term transition toward net-zero emission by 2050 to limit the increase in global temperature rise to 1.5°C by end of this century. Energy supply and demand sectors are the major contributors of anthropogenic emissions, and decarbonization of these sectors can play an instrumental role in achieving this target. Sustainable supply of energy plays an important role while implementing measures to adapt to climate change. A clean, renewable, and sustainable energy system can help to address the negative impacts of climate change such as mitigation of GHG emissions as well as environmental pollution; distributed and smart energy generation; improve the resilience of communities; and enhance the capability to transform environmental, social, and financial assets into human well-being and preserve the natural ecosystem.

[6]UN Climate Change Conference, 2021. https://ukcop26.org/cop26-goals/. Accessed 15 June 2021.

References

CCC, 2019. Net Zero - Technical Report. Climate Change Committee. https://www.theccc.org.uk/publication/net-zero-technical-report/.

de Sherbinin, A., Bukvic, A., Rohat, G., Gall, M., McCusker, B., Preston, B., Apotsos, A., Fish, C., Kienberger, S., Muhonda, P., Wilhelmi, O., Macharia, D., Shubert, W., Sliuzas, R., Tomaszewski, B., Zhang, S., 2019. Climate vulnerability mapping: a systematic review and future prospects. In: Wiley Interdisciplinary Reviews: Climate Change, vol. 10(5). Wiley-Blackwell, p. e600. https://doi.org/10.1002/wcc.600.

Eckstein, D., Künzel, V., Schäfer, L., 2021. Global Climate Risk Index 2021. www.germanwatch.org.

EEA, 2017. Climate Change, Impacts and Vulnerability in Europe 2016 — European Environment Agency. https://www.eea.europa.eu//publications/climate-change-impacts-and-vulnerability-2016.

IEA, 2017. Catalyzing Energy Technology Transformations. EPIC. https://epic-staging.uchicago.edu/news/catalyzing-energy-technology-transformations/.

IEA, 2020. Global CO_2 Emissions in 2019 – Analysis. IEA. https://www.iea.org/articles/global-co2-emissions-in-2019.

IEA, 2021. Net Zero by 2050 - A Roadmap for the Global Energy Sector. www.iea.org/t&c/.

IPCC, 2007. In: Pachauri, R.K., Reisinger, A. (Eds.), Climate Change 2007 Synthesis Report. Contribution of Working Groups I, II and III to the Fourth Assessment Report of the Intergovernmental Panel on Climate Change [Core Writing Team], vol. 104. https://www.ipcc.ch/site/assets/uploads/2018/02/ar4_syr_full_report.pdf.

IPCC, 2014a. Chapter Climate Change 2014 Synthesis Report Summary for Policymakers Summary for Policymakers.

IPCC, 2014b. Climate Change 2014: Synthesis Report. Contribution of Working Groups I, II and III to the Fifth Assessment Report of the Intergovernmental Panel on Climate Change. Gian-Kasper Plattner. http://www.ipcc.ch.

IPCC, 2018. Global Warming of 1.5 oC — an IPCC Special Report on the impacts of global warming of 1.5°C above pre-industrial levels and related global greenhouse gas emission pathways. In: The Context of Strengthening the Global Response to the Threat of Climate Change. https://www.ipcc.ch/sr15/.

IPCC, 2021. Summary for policymakers. In: Masson-Delmotte, V., Zhai, P., Pirani, A., Connors, S.L., Péan, C., Berger, S., Caud, N., Chen, Y., Goldfarb, L., Gomis, M.I., Huang, M., Leitzell, K., Lonnoy, E., Matthews, J.B.R., Maycock, T.K., Waterfield, T., Yelekçi, O., Yu, R., Zhou, B. (Eds.), Climate Change 2021: The Physical Science Basis. Contribution of Working Group I to the Sixth Assessment Report of the Intergovernmental Panel on Climate Change. Cambridge University Press (in press).

Khanal, R.C., Shakya, S.R., Bajracharya, T.R., 2020. Contribution of renewable energy technologies (RETs) in climate resilient approach and SDG 7. Journal of the Institute of Engineering 15 (3), 393–401. https://doi.org/10.3126/jie.v15i3.32230.

Levin, K., Rich, D., Ross, K., Fransen, T., Elliott, C., 2020. Designing and Communicating Net-Zero Targets. www.wri.org/design-net-zero.

Margulis, S.N.U., 2010. The Costs to Developing Countries of Adapting to Climate Change : New Methods and Estimates - the Global Report of the Economics of Adaptation to Climate Change Study. https://documents.worldbank.org/en/publication/documents-reports/documentdetail/667701468177537886/the-costs-to-developing-countries-of-adapting-to-climate-change-new-methods-and-estimates-the-global-report-of-the-economics-of-adaptation-to-climate-change.

Mohaddes, K., Williams, R.J., 2020. The adaptive investment effect: evidence from Chinese provinces. Economics Letters 193, 109332. https://doi.org/10.1016/j.econlet.2020.109332.

NCCS, 2020. Charting Singapore's Low-Carbon and Resilient Future. Singapore: National Climate Change Secretariat (NCCS). https://unfccc.int/sites/default/files/resource/SingaporeLongtermlowemissionsdevelopmentstrategy.pdf.

Noble, I.R., Huq, S., Anokhin, Y.A., Carmin, J., Goudou, D., Lansigan, F.P., Osman-Elasha, B., Villamizar, A., 2014. Adaptation needs and options. In: Field, C.B., Barros, V.R., Dokken, D.J., Mach, K.J., Mastrandrea, M.D., Bilir, T.E., Chatterjee, M., Ebi, K.L., Estrada, Y.O., Genova, R.C., Girma, B., Kissel, E.S., Levy, A.N., MacCracken, S., Mastrandrea, P.R., White, L.L. (Eds.), Climate Change 2014: Impacts, Adaptation, and Vulnerability. Part A: Global and Sectoral Aspects. Contribution of Working Group II to the Fifth Assessment Report of the Intergovernmental Panel on Climate Change. Cambridge University Press, Cambridge, United Kingdom and New York, NY, USA, pp. 833–868.

Sarkodie, S.A., Strezov, V., 2019. Economic, social and governance adaptation readiness for mitigation of climate change vulnerability: evidence from 192 countries. Science of the Total Environment 656, 150–164. https://doi.org/10.1016/j.scitotenv.2018.11.349.

Transport. In: Sims, R., Schaeffer, R., Creutzig, F., Cruz-Núñez, X., D'Agosto, M., Dimitriu, D., Tiwari, G. (Eds.), 2014. Climate Change 2014: Mitigation of Climate Change. Contribution of Working Group III to the Fifth Assessment Report of the Intergovernmental Panel on Climate Change. Cambridge University Press. https://www.researchgate.net/publication/274897242_Transport_In_Climate_Change_2014_Mitigation_of_Climate_Change_Contribution_of_Working_Group_III_to_the_Fifth_Assessment_Report_of_the_Intergovernmental_Panel_on_Climate_Change.

Stocker, T., Qin, D., Plattner, G., Tignor, M., Allen, S., Boschung, J., Nauels, A., Xia, Y., Bex, V., Midgley, P., Alexander, L.V., Allen Switzerland, S.K., Zealand, N., Bindoff, N.L., Allen, M.R., Boucher, O., Chambers, D., Hesselbjerg Christensen, J., 2013. Summary for policymakers. In: Climate Change 2013: The Physical Science Basis. Contribution of Working Group I to the Fifth Assessment Report of the Intergovernmental Panel on Climate Change. Monika Rhein. Gian-Kasper Plattner.

UN, 2020. The Health Effects of Global Warming: Developing Countries Are the Most Vulnerable. United Nations. https://www.un.org/en/chronicle/article/health-effects-global-warming-developing-countries-are-most-vulnerable.

UNEP, 2019. Emissions Gap Report 2019 | UNEP - UN Environment Programme. https://www.unenvironment.org/resources/emissions-gap-report-2019.

Index

Note: 'Page numbers followed by 'f' indicate figures those followed by 't' indicate tables and 'b' indicate boxes.'

A

B

D

E

F

S

T

U

V

W